Fundamentals of Salt Water Desalination

Fundamentals of Salt Water Desalination

Hisham T. El-Dessouky
Hisham M. Ettouney

Department of Chemical Engineering
College of Engineering and Petroleum
Kuwait University

2002
ELSEVIER
AMSTERDAM – LONDON – NEW YORK – OXFORD – PARIS – SHANNON – TOKYO

ELSEVIER SCIENCE B.V.
Sara Burgerhartstraat 25
P.O. Box 211, 1000 AE Amsterdam, The Netherlands

First edition 2002

Library of Congress Cataloging in Publication Data
A catalog record from the Library of Congress has been applied for.

ISBN: 0-444-50810-4

♾ The paper used in this publication meets the requirements of ANSI/NISO Z39.48-1992 (Permanence of Paper).
Printed in The Netherlands.

Acknowledgment

Professor El-Dessouky dedicates this book to the memory of his parents. Also, he dedicates the book to his family, Fatma, Sarah, Sameh, and Sanyia.

Professor Ettouney dedicates this book to his parents, Mohamed and Fatma, his brothers Sayed, Mohamed, and Osama, his family, Kareema, Osama, and May, and his friends Reem, Kareem, and Amira.

Preface

Industrial desalination of sea and brackish water is becoming an essential part in providing sustainable sources of fresh water for a larger number of communities around the world. Desalination is a main source of fresh water in the Gulf countries, a number of the Caribbean and Mediterranean Islands, and several municipalities in a large number of countries. As the industry expands there is a pressing need to have a clearly and well-written textbook that focuses on desalination fundamentals and other industrial aspects. The book would serve a large sector of the desalination community that includes process engineers, designers, students, and researchers. Fundamentals of the desalination process are based on physical principles that include mass and energy conservation, mass, momentum, and heat transfer, and thermodynamics. The authors firmly believe that good understanding of these fundamentals is necessary to analyze or evaluate the performance for any of the existing and known desalination processes. Moreover, understanding the fundamentals would allow for critical evaluation of novel schemes or devising new schemes. Other aspects included in the book are the historical background of the desalination process, developments during the second half of the twentieth century, recent trends, and future challenges. The book focuses on the processes widely used in industry, which include the multistage flash desalination and reverse osmosis. Also, other desalination processes with attractive features and high potential are considered. The book includes a large number of solved examples, which are explained in simple and careful matter that allow the reader to follow and understand the development. The data used in development of the examples and case studies are extracted from existing desalination plants. Also, the examples represent practical situations for design and performance evaluation of desalination plants. The book also includes comparison of model predictions against results reported in literature as well as available experimental and industrial data. Although, this textbook will target the desalination community, which may include practicing engineers, designers, developers, graduate students, and researchers, however, the contents can be used by engineers in other industrial disciplines. Several industries include similar unit operation processes, i.e., evaporators, condensers, flashing units, membrane separation, and chemical treatment. Examples of such industries include wastewater treatment, food, petroleum, petrochemical, power generation, and pulp and paper. Process fundamentals and design procedures of such unit processes follow the same procedures given in this textbook.

It should be stressed that this is the first textbook on desalination that can be used for undergraduate and graduate instruction. Although, there are a number of books on desalination, most are of the editorial type with research and development articles, which are not suitable for educational purposes. Other books are descriptive and have introductory material.

This textbook includes sections on thermal desalination, membrane desalination, associated processes, and process economics. The thermal desalination part is focused on the single effect evaporation process, the multiple effect evaporation, and the multistage flash desalination. It should be noted that complete and full description and development is made for each thermal desalination process. Because of the similarities among various thermal desalination processes, this implied repeating some parts of the process elements and model equations. As a result, the reader would find it simple to study each process separately without having to search various sections for some common features or equations. This is except for the correlations and equations for water physical properties, thermodynamic losses, and heat transfer coefficients. Another major feature of the analysis provided in this textbook is the inclusion of the performance charts for each process. This data is found necessary to provide the reader with practical limits on process operation and performance.

In Chapter 2, the single effect evaporation with submerged evaporator is discussed. The development includes process description, mathematical model, and performance evaluation. This is followed by discussion of various types of evaporators. It should be stressed that coverage of this material is essential to develop the theoretical and physical backgrounds necessary to understand the thermal multiple effect processes. Chapter 3 includes analysis of the single effect evaporation process combined with various types of heat pumps. These include mechanical, thermal, adsorption, and absorption vapor compression. The analysis includes description of processes, mathematical models, and solved examples. In addition, comparison is made between model predictions and industrial data for the single effect mechanical vapor compression system.

Chapter 4 includes the multiple effect evaporation processes. The Chapter starts with description of different process layouts. This is followed by analysis of the forward feed system, which includes process developments, description, detailed models, and case studies. The last part of the Chapter includes similar treatment for the parallel feed configurations. The Chapter also includes performance evaluation of both configurations, comparison with field data, future challenges and trends.

In Chapter 6, the analysis of the multistage flash desalination starts with coverage of process developments. This is followed by process synthesis, which starts with the single stage flashing and proceeds to the multistage configuration. In Section 6.5, details of the conventional brine circulation multistage flashing process is given and it includes process description, model assumptions, equations, solution method, and a number of solved examples. The development then focuses on analysis of performance evaluation as a function of the parameters that have the strongest effect on the unit product cost. These parameters includes the heat transfer area for evaporators and condensers, dimensions of the flashing or evaporation stages, capacity and dimensions of the ejectors, size and load of pumping units, and other associated processes. The remainder of Chapter 6 covers other possible layouts for the multistage flash desalination process, which includes the once through system, the brine mixing process, and the thermal vapor compression configuration.

All of the thermal desalination processes can be combined with vapor compression heat pumps, which includes thermo-compressors or steam jet ejectors, mechanical vapor compression, adsorption vapor compression, absorption vapor compression, and chemical vapor compression. The main objective for using vapor compression heat pumps is to improve the process efficiency and utilization of low grade energy that commonly discarded into the environment. This approach conserves on the system energy and reduces the amount of fossil fuels that upon combustion generates green house gases and other pollutants. The analysis of the heat pumps focuses on description of the elements forming each heat pump, mathematical and empirical models, solution procedure, and performance analysis. The remaining sections on the combined systems of thermal desalination and vapor compression focus on development integrated mathematical models for the system, operating regimes, performance evaluation, and comparison against systems with no vapor compression and field data. The vapor compression desalination systems are given in Chapters 3 and 5, which covers the single and the multiple effect evaporation. Section 6.5 gives model and analysis of the multistage flash desalination combined with thermal vapor compression system.

Chapters 7 and 8 include the reverse osmosis membrane desalination process. Chapter 7 starts with a brief description of various membrane separation process, separation mechanism, and recent trends. Also, description is given for membrane modules, process layout, and staging of modules. The reverse osmosis models include simple and short cut techniques as well as more detailed models. The analysis of membrane systems focus on determination the required feed pressure and membrane area to provide the desired permeate or fresh water flow rate. Chapter 8 covers elements of feed pretreatment, membrane biofouling, and membrane cleaning.

In Chapter 9, a number of associated processes are analyzed, which includes the venting, steam jet ejectors, wire mesh mist eliminator, and orifices in the MSF process. The chapter includes description of various unit process, fundamental models as well as fitting correlations. In addition, a number of examples are solved for each system.

Desalination economics are covered in chapter 10 and it includes elements of economic analysis and a number of case studies. The necessary elements to perform economic analysis of desalination processes include process capital and amortization, operating and maintenance cost, energy cost, and inflation effects.

The case studies include examples for various thermal and membrane desalination systems. In addition, all case studies include comparison with available cost data from literature studies and existing desalination plants.

The book contents are suitable for instructing training courses for practicing engineers. The courses could be introductory, intermediate, or advanced. The introductory level training course would cover the first chapter, and sections in all other chapters on description of desalination process, associated unit processes, and performance evaluation. As for the intermediate or advanced training courses it should include a number of the solved examples and case studies. These may include design problems on the multistage flash desalination, the multiple effect evaporation with/without thermal or mechanical vapor compression, the single effect evaporation with mechanical vapor compression, and the reverse osmosis. The difference between the intermediate and advanced training courses comes in coverage of the descriptive part and the type of design problems and analysis.

Using the book for instruction of an undergraduate course depends on the level of instruction, where a simple freshmen or sophomore course would include most of the descriptive part and simple problems that include basic material and energy balances and evaluation of the stream physical properties. On the other hand, a desalination course being taught on the junior or senior levels would allow for more detailed design in addition to the process description section. Finally, the teaching material for a graduate course should involve a larger number of the design problems for various desalination processes. In the graduate course, many of the descriptive part should be assigned for reading ahead of the class period and its coverage should be instructed in an interactive discussion between students and instructor.

The book includes a comprehensive computer package. The computer package is written in visual basic. The package allows for flow sheet as well as design displays. The results of the package are displayed on the screen and are also written to text files. The screen displays as well as the text files can be printed. The flow sheet display includes the unit processes forming the desired desalination process. Upon completion of the calculations each unit process in the flow sheet becomes active, where pressing a specific unit process gives a small display with the stream properties of the unit. The design display gives more a comprehensive list of the design parameters, i.e., thermal conductivity of the evaporator/condenser tubes, fouling factors, etc. Completion of the design calculations gives the major design results, i.e., heat transfer areas, properties of outlet streams, and thermal performance ratio.

Hisham T. El-Dessouky (eldessouky@kuc01.kuniv.edu.kw)
Hisham M. Ettouney (Hisham@kuc01.kuniv.edu.kw)

Symbols

A	Area, m^2.
BPE	Boiling point elevation, °C
C_p	Specific heat at constant pressure, kJ/kg °C.
Cr	Compression ratio defined as pressure of compressed vapor to pressure of entrained vapor.
d	Vapor flow rate formed in flashing boxes, kg/s.
D	Vapor flow rate formed by flashing or boiling, kg/s.
Er	Expansion ratio defined as pressure of compressed vapor to pressure of entrained vapor.
h	Heat transfer coefficient, kW/m^2 °C.
H	Enthalpy of liquid water, kJ/kg.
H"	Enthalpy of vapor phase, kJ/kg.
k	Thermal conductivity, kW/m °C.
L	Length, m.
LMTD	Logarithmic temperature difference, °C.
M	Mass flow rate, kg/s.
M_z	Mass of adsorbing solids, kg.
P	Pressure, kPa.
PR	Performance ratio, dimensionless.
ΔP	Pressure drop, kPa.
Q	Heat transfer rate, kJ/s.
r	Tube radius, m.
R	Universal gas constant, kJ/kg°C.
R_f	Fouling resistance, m^2 °C/kW
Rs	Load ratio, mass flow rate of motive steam to mass flow rate of entrained vapor.
s	Salt concentration, mg/ℓ.
sA	Specific heat transfer area, m^2/(kg/s).
sM_{cw}	Specific cooling water flow rate, dimensionless.
T	Temperature, °C.
ΔT	Temperature drop, °C.
U	Overall heat transfer coefficient, kW/m^2 °C.
v	Vapor specific volume, m^3/kg.
V	Vapor velocity, m/s
w	Entrainment ratio, mass flow rate of entrained vapor to mass flow rate of motive steam.
X_{NC}	Mass fraction of non-condensable gases.
X	Salinity, ppm.

Z	Length, m.

Greek Letters

α	Adsorption capacity, kg water/kg zeolite.
η	efficiency.
λ	Latent heat, kJ/kg.
ρ	Density, kg/m^3
ε	Void fraction
γ	Compressibility ratio.

Subscripts

b	Brine.
c	Condenser.
cw	Cooling water.
d	Distillate product.
e	Evaporator
ev	Entrained vapor.
f	Feed seawater
i	Inlet stream or inner diameter
m	Motive steam.
n	Nozzle.
o	Outlet stream or outer diameter.
p	Demister.
s	Compressed vapor or heating steam.
t	Throat of the ejector nozzle.
v	Formed vapor.
w	Tube wall.
z	Solid bed.

Contents

CHAPTER 5
Multiple Effect Evaporation – Vapor Compression 211

CHAPTER 6
Multi Stage Flash Distillation 271

Chapter 1

Introduction

Objectives

The main objective of this chapter is to give an overview of various desalination processes, developments, and the needs for industrial desalination. This is made through discussion of the following topics:

- Water resources
- Classification of salt water
- History of industrial desalination
- Definition and classification of industrial desalination processes
- Market status for desalination processes

1.1 Resources and Need for Water Desalination

The earth contains about 1.4×10^9 km^3 of water, which covers approximately 70% of the planet surface area; the percentage of salt water in this large amount is 97.5%. The remaining 2.5% is fresh water with 80% of this amount frozen in the icecaps or combined as soil moisture. Both forms are not easily accessible for human use. The remaining quantity, about 0.5%, is believed to be adequate to support all life on Earth. Unfortunately, this water is not distributed evenly throughout the plant and it is not available in sufficient quantities either when or where it is needed. Table 1 gives a summary for distribution of various water resources across the globe. The global daily average of rainfall is 2×10^{11} m^3. This amount is poorly distributed across the globe.

The solar energy is the main driver for formation of fresh water from oceans. The thermal energy absorbed by the earth surface generates sufficient temperature gradients that drive water evaporation from the large surfaces of ocean water. The water vapor rises through the ambient air and forms a cloud cover at various elevations. The clouds are formed of fine water droplets with an average diameter of 10 μm. The clouds are transported over land, where precipitation takes place. The form of precipitation depends on the surrounding air temperature. Snow is formed in clod climates and higher elevations; while, rain is formed in warmer climates and lower elevations. On the other hand, mixtures of ice, snow, and rain are formed during spring time of clod climates.

Precipitation depends on the wind direction and speed, which have fixed patterns that varies subject to location and seasonal temperature variations. Also, precipitation is affected by geographical conditions, i.e., presence of mountains, flat land, as well as local ambient conditions, i.e., temperature, and humidity. The wind pattern, geographical forms, and ambient conditions generates zones of constant water precipitation, monsoon seasons, and areas of very low precipitation.

Precipitated water forms lakes, rivers, underground surface water, deep aquifers, or massive flood areas. These forms could be seasonal or permanent. For example, rapid increase in the ambient temperature during the spring season could result in the melting of large amounts of winter snow over a short period of time. Such events are experience in several regions in Europe, Russia, and the US. Also, the monsoon rain in the Indian content results in precipitation of large amounts of water over a very short period of time. Both forms of water precipitation give rise to destruction of property and loss of life. Permanent rivers form the life line to several regions around the globe, where it transports water from high precipitant area to dessert land. The best example for this situation is the river Nile, which originates in the high mountains of Ethiopia and Kenya and travels more than 2000 km. Through the ages the river Nile gave life to the Nile valley and supported the development of Egyptian civilization.

A major part of water precipitation ends up as ground moisture in the form of sub-surface water or deep aquifers. Deep aquifers proved to be viable source for bottled drinking water. This is because of regulated and limited rates of water use from these sources insure sufficient natural replenishment of the source. Also, the natural process through various rock formations provides the water with natural minerals and keeps its pH at acceptable levels.

Table 1
Distribution of water resources across the globe

Resource	Volume km^3	Percent of total water	Percent of Fresh Water
Atmospheric Water	12900	0.001	0.01
Glaciers	24064000	1.72	68.7
Ground Ice	300000	0.021	0.86
Rivers	2120	0.0002	0.006
Lakes	176400	0.013	0.26
Marshes	11470	0.0008	0.03
Soil Moisture	16500	0.0012	0.05
Aquifers	10530000	0.75	30.1
Lithosphere	23400000	1.68	
Oceans	1338000000	95.81	
Total	1396513390		

Classification of various types of water is based on the purpose for which the water is used. The first water grade is set for safe drinking, household purposes, and a number of industrial applications. This water category has a salinity range of 5 to 1000 ppm. This type of water is found in rivers and lakes and can be generated by industrial desalination processes. In large cities, various levels of water salinity are used, where water with salinity below 150 ppm is used for drinking while higher salinity water of up 1000 ppm is used for various household applications. This proved to be more effective, because the average per capita consumption of the low salinity drinking water (150 ppm) is limited to 2 liters/day. On the other hand, the per capita consumption rate for other household purposes is 200-400 liters/day, which is used for cooking, washing, cleaning, gardening, and other purposes. On industrial scale, the most stringent water quality is set by the makeup water for boilers and applications related to the electronic industry and pharmaceuticals. The water quality for this application is limited to a maximum salinity of 5 ppm. This high degree of purity is achieved through the use of ion exchangers, which operates on low salinity river water or industrially desalinated water. Other industrial applications call for less stringent water quality than those used for boilers. Applications include chemical reactions, dairy and food, washing and cleaning, and cooling

The second water category has a salinity range of 1000-3000 ppm. This type of water is suitable for irrigation purposes and industrial cooling. This applies for higher salinity water, which includes brackish and seawater. The salinity range for brackish water is 3000-10000 ppm. As for the seawater its average salinity is 34,000 ppm. Water with salinity above 10000 ppm is termed as high salinity water. The salinity of seawater varies subject to local conditions, where it is affected by ambient and topographical conditions. For example, enclosed seas have higher salinity than open seas and oceans. Also, seas, which are found in areas of high temperatures or that receive high drainage rates of saline water, would certainly have higher degree of salinity. For example, the salinity of the Gulf water near the shores lines of Kuwait, Saudi Arabia, and the United Arab Emirates may reach maximum values close to 50,000 ppm. On the other hand, the salinity of the Gulf water near the Western shores of Florida, USA, may reach low values of 30,000 ppm. This is because of the large amount of fresh water received from rivers and springs in that area.

The amount of fresh water resources is nearly constant since the start of life on earth. On the other hand, the world population has increased more rapidly over a period of less than 200 years. Figure 1 shows a bar chart for the population development over the past 200 years and forecast for the next 50 years. The figure shows the following:

- In 1804 the world population was 1 billion.
- It took 123 years to reach 2 billion in 1927.
- In 1960 or after 33 years the population increased to 3 billion.
- After 13 years and in 1987, the population increased to 5 billion.
- In 1999, the population has reached 6 billion.
- It is expected that a population of 7.5 billion will be reached in 2020 and about 9 billion in 2050.

At present, about 40% of the world's population is suffering from serious water shortages. By the year 2025, this percentage is expected to increase to more than 60%. This is because of the rapid increase of population, changes in the life-style, increased economic activities, and pollution that limit the use of fresh water resources. Moreover, common use of unhealthy water in developing countries causes 80-90% of all diseases and 30% of all deaths. Even in industrial countries, long spells of dry seasons and limited rainfall forces governments, states, and municipalities to adopt severe water restriction programs that affect the population at large. Such situations are reported on frequent basis in several countries around the globe. The water shortage extends to include underground water supplies, previously considered to be an unlimited resource in many countries. In this regard, several cases are reported for well failure, decline of the water table, and seawater intrusion into the fresh water aquifers. This situation has forced many countries, industrial and developing, to adopt active and efficient programs for reclamation of industrial and municipal wastewater.

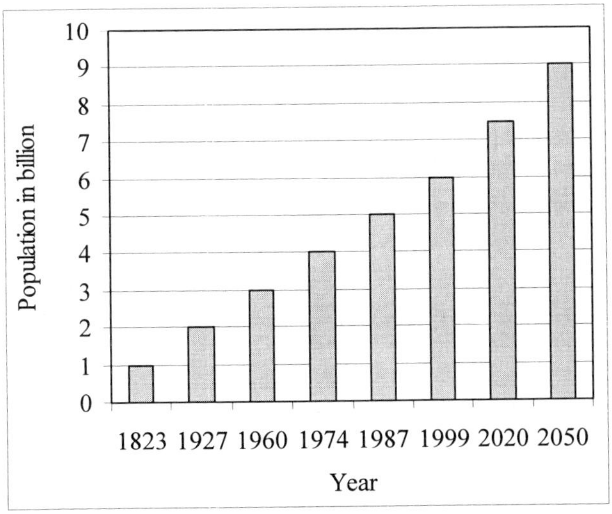

Fig. 1. Change in the world population since 1823 and until 2050

Inspection of the global map shows clearly the extent of desert and arid zones, which covers major portions of all continents. The most famous of these deserts is the great Sahara that encompasses all of the Arabian Peninsula and North Africa. The Great Sahara runs from the eastern shores of Saudi Arabia and for a distance of more than 4000 km to the western shores of Morocco. In particular, the Arabian Peninsula that includes Saudi Arabia, Kuwait, Qatar,

Bahrain, United Arab Emirates, and Oman does not have a single natural river. Other large desserts are found in China, the south west of the US, in most of the Australian continent, and in South America.

In addition, desertification occurs across the globe and at very rapid rate and it has strong effect on the weather pattern, rainfall, and the environment. Desertification is primarily caused by unregulated humane activities that result in the destruction of delicate habitats, such as woodlands, rainforests, swamps, and Savannah. Many of the flat lands used for farming purposes are being turned into dessert because of the continuous loss of the rich top soil and poor farming practices. Similarly, over-grazing activities has quickly turned many of the Savannah land into desserts. Examples for desertification of Savannah are found in Sudan, Kenya, Ethiopia, and China. Another bad practice is the unregulated harvesting of woodlands for the production of wood lumber and paper pulp. Also, rainforests in equatorial regions, which are considered an essential element of the global environment, is being replaced at a steady rate by farming and mining land. Such activities are driven by greed and search for quick profit. Unfortunately, such delicate habitats never recover and eventually become new dessert area.

The combined effect of the continuous increase in the world population, changes in life style, and the limited natural resources of fresh water makes industrial desalination of seawater a major contender for providing sustainable source of fresh water for arid zones and during drought periods. This solution is also supported by the fact that more than 70% of the world population live within 70 km of seas or oceans. During the second half of the twentieth century, desalination of seawater proved to be the most practical and many cases the only possible solution for many countries around the globe, i.e., the Gulf States, Mediterranean and Caribbean Islands. At the turn of century, desalination is being considered by a larger number of countries as the most viable and economical solution for providing fresh water.

1.2 Composition of Seawater

The main ions found in seawater include Na^-, Ca^{++}, K^+, Mg^{++}, $(SO_4)^-$, and Cl^-. Of course, all other ions found in nature are present in the seawater, but at a much smaller concentrations. The chemical composition of open sea is constant; however, the total dissolved amount of dissolved solids changes subject to local conditions. This is because the diffusion time for salts or the time required to obtain complete mixing of all seas and oceans is much smaller than the time required for complete filling or replenishment. Table 2 shows typical composition of seawater, which has a total salinity of 36000 ppm. In addition, to the dissolved ions found in seawater the seawater includes a wide variety of fine suspended matter that include sand, clay, microorganisms, viruses, and colloidal matter. The size of these compounds varies over a range of $5x10^{-2}$ to 0.15 μm.

Table 2
Typical composition of seawater with salinity of 36,000 ppm.

Compound	Composition	Mass Percent	ppm
Chloride	Cl^-	55.03	19810.8
Sodium	Na^+	30.61	11019.6
sulfate	$(SO_4)^{--}$	7.68	2764.8
Magnesium	Mg^{++}	3.69	1328.4
Calcium	Ca^{++}	1.16	417.6
Potassium	K^+	1.16	417.6
Carbonic Acid	$(CO_3)^{--}$	0.41	147.6
Bromine	Br^-	0.19	68.4
Boric Acid	$H_3BO_3^-$	0.07	25.2
Strontium	Sr^{++}	0.04	14.4
Total		100	36000

1.3 Historical Background

Up to the 1800 desalination was practiced on ship boards. The process involved using single stage stills operated in the batch mode. Energy is supplied from cock stoves or furnaces without recovering the heat of condensation. The equipment and product quality varied considerably and were dependent on the manufacturer and operator. Mist carryover was always a problem. The sugar industry established in the early 1800 resulted in considerable progress of evaporation processes. This involved development of more efficient and larger scale stills for production of syrup and sugar. The start of the desalination industry dates back to the early of part of the twentieth century. In 1912, a six effect desalination plant with a capacity of 75 m^3/d is installed in Egypt. The total production capacity of the desalination increased during the period 1929-1937 due to the start of the oil industry. However, exponential growth occurred during the period from 1935 to 1960 at an annual rate of 17%.

The recent history of thermal desalination processes shown in Table 3 is summarized in the following:

- In 1957, the landmark of the four-stage flash distillation plant by Westinghouse was installed in Kuwait. The plant did not have the standard MSF features defined by the Patent of Silver in 1957, where the number of flashing stage was close to three times the system performance ratio. The MSF patent by Silver gives a major advancement over the Westinghouse configuration because of the much smaller specific heat transfer area for the condenser tubing. This reduced considerably the capital cost since the high tubing cost in the Westinghouse system was replaced by inexpensive partitions in the MSF systems.

- The first two MSF plants were installed in Kuwait and Guernsey in UK. The Guernsey plant was operated during draught periods; however, it had severe corrosion problems and was taken out of service in the early seventies.
- The Point Loma MSF plant with a capacity of 1 migd was constructed in 1962.

Table 3
Historical developments in thermal and membrane desalination processes.

Year	Achievement
1957	First industrial scale flashing unit by Westinghouse in Kuwait. Four stage flashing system a performance ratio of 3.3.
1957	Silver patent for the MSF configuration.
1959	Shuwaikh mix (poly-phosphate based) allowed for increase in the plant factor to values between 70-90%.
1960	First MSF plants commissioned in Shuwaikh, Kuwait and in Guernsey, Channel Island. The MSF unit in Shuwaikh had 19 stages, a 4550 m^3/d capacity, and a performance ratio of 5.7. In Guernsey, the unit had 40 stages, a 2775 m^3/d capacity, and a performance ratio of 10.
1962	Point Loma MSF plant with a capacity of 1 migd
1965	Dearation of feed stream.
1966	Reduction in specific volume
1967	First on-line ball cleaning system by Weirwestgarth in the Bahamas.
1967	Acid cleaning
1969	Co-Generation, energy cost reduction by 50%
1969	Increase in load factor to 85%
1970	Development of commercial grade RO membranes
1973	Cladding of partition walls.
1973	Construction of the standard MSF units, 6 migd, 24 stages, and a performance ratio of 6-8.
1980	Design and operation of low temperature mechanical vapor compression units
1980	Design and operation of low temperature multiple effect evaporation units combined with thermal vapor compression
1985	Use of polymer antiscalent at top brine temperatures of 110 °C.
1996	Construction of the largest MSF units known to day with capacity of 57,735 m^3/d in UAE.
1999	Construction of large scale RO plant in Florida, USA
1999	Increase in unit capacity of multiple effect evaporation units
2000	Design and construction of high performance of MSF system with 43 stages, 17280, and a performance ratio of 13

- In 1959, the first attempt was made to develop antiscalent materials in thermal desalination plants. This attempt was made in Kuwait and the antiscalnet was known as "Shuwikh Mix" or "SALVAP". The polyphosphate based material suppressed effectively scale formation and its use was limited to a top brine temperature of 95 °C. Use of this material increased operation from 200-600 hours to an excess of 8000 hours, Temperley (1995). Early studies of the Shuwaikh mix by Butt and Bou-Hassan (1981) showed the need for further development of the antiscaling material, since operation show difficulties in maintaining high productivity and performance ratio.
- During the sixties several achievements have been made, which includes design of MSF plants with capacities up to 25,000 m^3/d or the 6 migd. In addition, a 12.5 performance ratio MSF plant, largest know for MSF, was constructed in the Channel Islands with a 6800 m^3/d or 1.5 migd capacity, in two decks, and 40 stages.
- Other developments in the sixties include the on-line ball cleaning systems, antiscalent chemical additives, acid cleaning, feed dearation, and construction of co-generation plants.
- During the seventies, specifications for plant construction, operation, chemical treatment, corrosion prevention, and control were compiled as a result of accumulated experience, Abu-Eid and Fakhoury (1974).
- The Japanese manufacturer has emerged as a major power in construction of MSF plants. Large number of 22,500-25,500 m^3/d or 6 migd MSF plants were constructed during the seventies in the Gulf.
- In the eighties, use of polymer antiscalent started to replace the polyphosphate, which was limited to a top brine temperature of 90 °C. The polymer antiscalent allowed for operation at higher temperatures of 110 °C, which resulted in increase of the performance ratio to 8.65 and the capacity to 7.2 migd.
- The eighties also included design and operation of the low temperature single and multiple effect evaporation processes. The single effect system is based on mechanical vapor compression and the multiple designs included the stand alone mode and the thermal vapor compression units. Operation at low temperature is characterized by low tendency for scaling and allowed for the use of inexpensive aluminum alloys.
- In the mid 1990's larger capacity MSF and MEE plants were constructed. Also, system operation was considerably improved to achieve plant factors close to 90% and continuous operation for periods varying between 2-5 years.

Use of membranes for desalination is thought to mimic functions of biological membranes, i.e., cell membranes, lungs, kidneys, skin, etc. The most simple separation form is the household sieves being used over many centuries for separation of fine grain ground from coarse particles and grain shells. Similarly, fine cloth was made from cotton fibers and used to manufacture cheese. Both forms of separation are based on differences in particle size.

However, developments in membrane technology have focused on adoption of other separation mechanisms, such differences in solution and diffusion rates of various species across the membrane material.

Historical developments in artificial membranes are summarized in the following points:

- In 1823, Dutrochet gave correct explanation of osmosis (passage of solvent across a membrane from low to high concentration) and dialysis (passage of solute across a membrane from high to low concentration).
- In 1867, Traube and Pfeffer performed one of the first quantitative studies on performance of artificial membranes.
- Moritz Taube, 1867, prepared the first synthetic membrane.
- In the late 1800's Graham discovered that arranging a membrane between a reservoir of pressurized air and another reservoir of unpressurized air could produce oxygen-enriched air.
- Early use of membranes was applied to recovery of NaOH by dialysis from wastewater containing hemicellulose from the viscose-rayon industry.
- Also, uranium isotopes (235 and 238) are separated in the vapor phase through porous membranes.
- Reid and Breton, 1959, at the University of Florida developed cellulose acetate RO membranes.
- Loeb and Sourirajan, 1963, from the University of California, Los Angeles developed the first asymmetric cellulose acetate membrane, with higher salt rejection and water flux.

As for commercialization of the RO membranes it is summarized in the following points:

- In the late 1960s, the Loeb-Sourirajan cellulose acetate membranes are used to construct spiral wound modules.
- In 1971, Dupont introduced the Permasep B-9 permeator for brackish water desalination. The permeator contains millions of asymmetric aromatic polyamide (aramid) hollow fine fibers.
- In late 1973, Dupont introduced the Permasep B-10 permeator, also using asymmetric aramid fibers, capable of producing potable water from seawater in a single pass.
- In the mid-1970s, cellulose triacetate hollow fiber permeators were introduced by Dow Chemical Company, followed by Toyobo of Japan
- During the same period, Fluid Systems and FilmTec introduced the spiral wound polyamide thin film composite membranes.
- Throughout the 1980s, improvements were made to these membranes to increase water flux and salt rejection with both brackish water and seawater.
- Today the predominate membrane materials are still aramids, polyamides, and cellulose acetate and triacetate in spiral wound and hollow fiber configurations.

- Applications of the RO membranes include potable water production, waste recovery, food applications, kidney dialysis, high-purity water for boiler feed, and ultrapure water electronics applications.
- In 2000, the RO technology was used to treat more than $9x10^6$ m^3/d of water per day, and this market is expected to continue growing during the first half of the 21st century.

1.4 Definition and Classification of Industrial Desalination Processes

The industrial desalination processes involve the separation of nearly salt-free fresh water from sea or brackish water, where the salts are concentrated in the rejected brine stream, Fig. 2. It is worth mentioning that in electrodialysis the positive and negative ions are separated from the salt water. The desalination processes can be based on thermal or membrane separation methods, Fig. 3. The thermal separation techniques include two main categories; the first is evaporation followed by condensation of the formed water vapor and the second involves freezing followed by melting of the formed water ice crystals. The former process is the most common in desalination and nearly at all cases it is coupled with power generation units, which may be based on steam or gas turbine systems. The evaporation process may take place over a heat transfer area and is termed as boiling or within the liquid bulk and is defined as flashing.

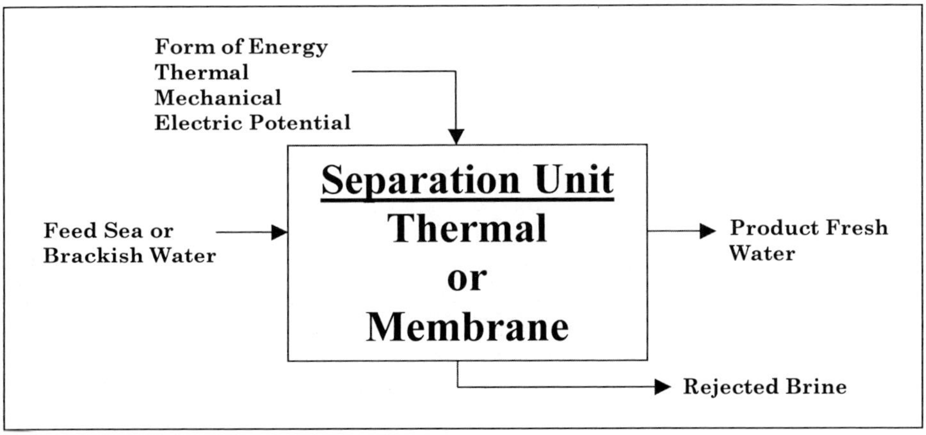

Fig. 2. Definition of desalination processes.

The evaporation processes include the multistage flash desalination (MSF), the multiple effect evaporation (MEE), the single effect vapor compression

(SEE), humidification-dehumidification (HDH), and solar stills. The HDH and solar stills are different from other evaporation processes by the following:
- Water is evaporated at temperatures lower than the boiling temperature.
- The main driving force for evaporation is the concentration difference of water vapor in the air stream.

The single effect vapor compression includes mechanical vapor compression (MVC), thermal vapor compression (TVC), absorption vapor compression (ABVC), adsorption vapor compression (ADVC), and chemical vapor compression (CVC). Vapor compression is combined with the single or multiple effect desalination units to improve the process thermal efficiency. In this technique, the low temperature vapor formed in the same effect or the last evaporation effect is compressed (upgraded) to a higher temperature and is then used to derive or initiate the evaporation process in the first or the same evaporation effect. The vapor compression devices include mechanical compressors, steam jet ejectors (which is known as thermal vapor compression), adsorption/desorption beds, and absorption/desorption columns. Solar energy can be used to desalinate water directly in solar stills or used as an energy source for other thermal processes.

The main membrane desalination process is reverse osmosis (RO), where fresh water permeates under high pressure through semi-permeable membranes leaving behind highly concentrated brine solution. The other membrane process is electrodialysis (ED) with very limited industrial applications. In this process the electrically charged salt ions are separated through selective ion exchange membranes leaving behind low salinity product water. Accordingly, a highly concentrated brine stream is formed on the other side of the membrane.

The desalination processes can also be classified according to the type of main energy form of energy used to drive the process. This classification is shown in Fig. 4. As is shown the thermal energy processes is divided into two categories, where energy is either added or removed. In the processes, where the energy is added, includes the MSF, MEE, HDH, and the processes combined with thermal, chemical, adsorption, or absorption heat pumps. The heating steam in these processes can be obtained from a co-generation power plant, a dedicated boiler unit, or from solar energy. Desalination through energy removal includes the freezing process. The RO and MVC systems are classified as mechanical energy desalination. In the RO process the mechanical energy (or the pressure difference across the membrane) drives water through the membrane and retains the salt in the brine stream. As for the MVC process, the mechanical energy of the vapor compressor increases the pressure and temperature of the distillate vapor, which is used to heat the feed seawater. The last category shown in Fig. 4 employs the electrical energy to separate water and salt. This is the electrodiaylsis process, where the electric energy drives the electrically charged ions through selective membranes.

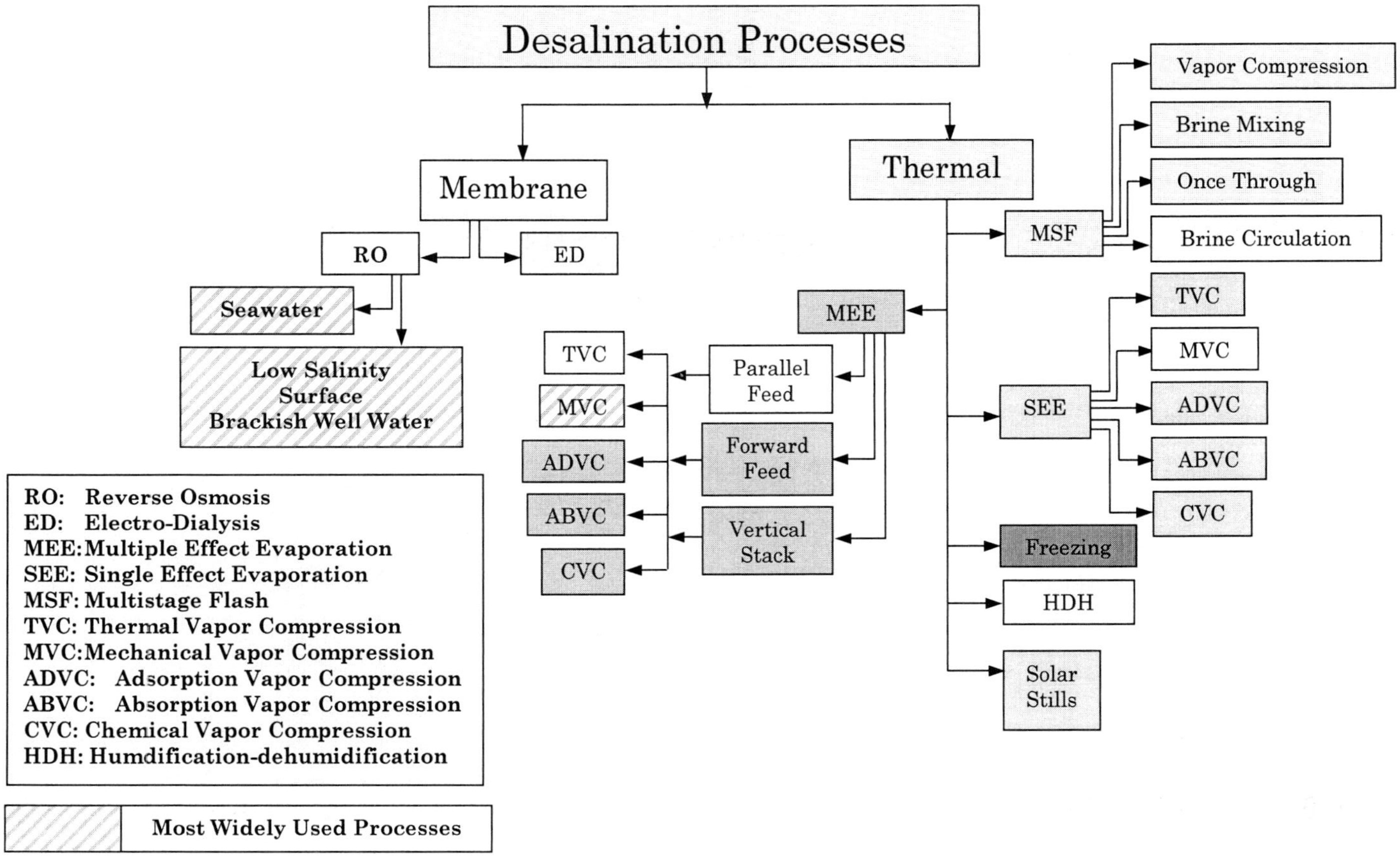

Fig. 3. Thermal and membrane desalination processes

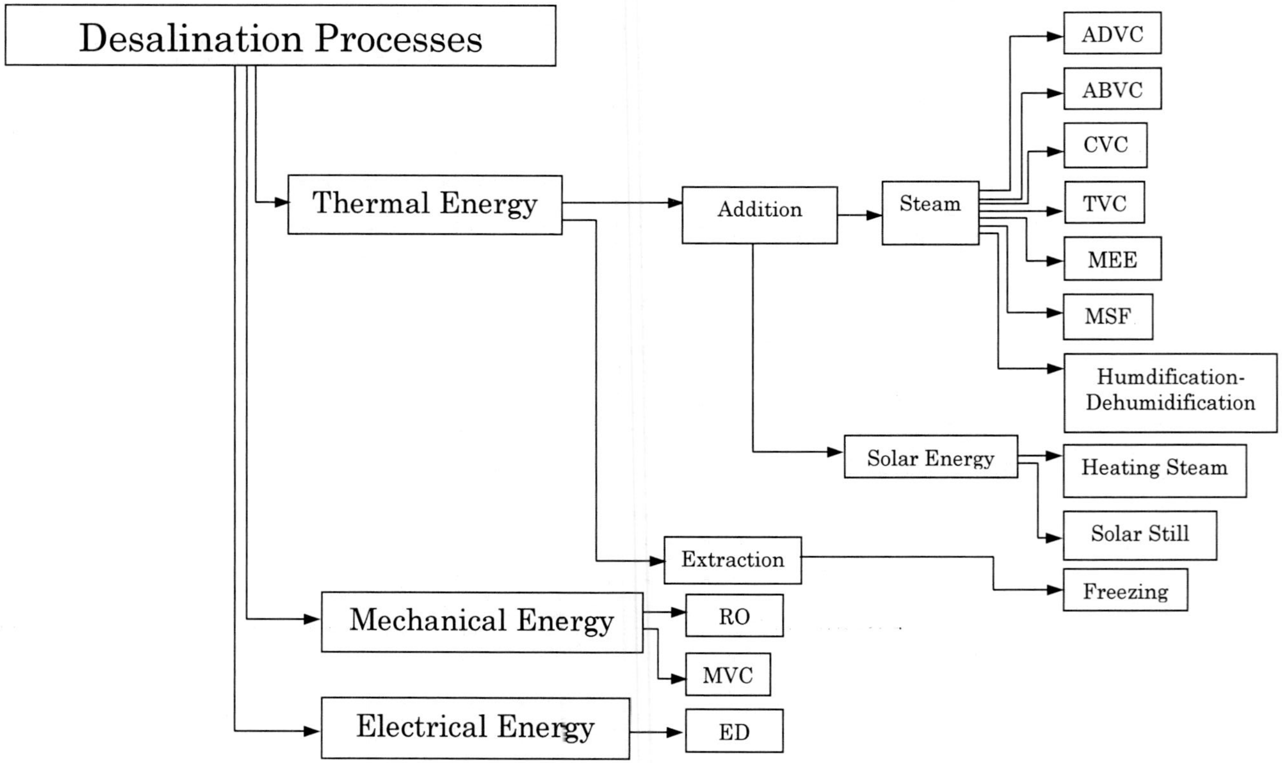

Fig. 4. Energy classification of desalination processes

1.5 Market Status for Desalination Processes

In 2000, thermal desalination processes remains to be a front runner in seawater desalination. In addition, the MSF process constitutes more than 54% of the operating capacity of all desalination processes and more than 93% of all thermal processes. Also, the RO represents more than 88% of all the membrane based process.

Table 4 shows developments in the production rates at the years 1996 (IDA 1996) and 2000 (IDA 2000) for major producer countries that include the Gulf States, the US, and others. The table also includes percentages for the production capacities of various methods, which includes MSF, RO, ED, MEE, and MVC. As is shown in Table 4, the RO method dominates the desalination markets in the US, Japan, and Spain. On the other hand, MSF process is the principal process in the Gulf countries. This is except for Bahrain, where the industry is divided equally between the MSF and the RO processes. However, most of the RO plants in Bahrain process brackish water. Limited use of the RO process in the Gulf countries is a result of the harsh conditions in the area. The confined water body of the Gulf has a high salinity range that varies between 42,000-51,000 ppm, which depends on the seasonal temperature. Also, the summer season extends over a long period from April to October with high temperature averages of 40 ^oC and 30 ^oC for the air and water, respectively. The RO process used in the US, Japan, and other countries experience a much milder conditions, where the open sea water salinity is much lower with a value of 35,000 ppm, and the summer temperature averages for air and water are below 25 ^oC.

Currently, the desalination industry is experiencing vast expansion around the globe. This is because of increase in water demand and the higher cost of fresh water from natural resources. An example for this expansion is found in Spain, where the production capacity is doubled over the past five years. Also, 720,000 m^3/d desalination capacity or approximately 20% of the current installed capacity is being constructed in Saudi Arabia at a total cost of $\$2x10^9$. In addition, the production capacity will be doubled over the next two decades by installing $4.4x10^6$ m^3/d at a total cost of $\$50x10^9$. Several other examples can be sited for other countries including India, Egypt, Syria, Indonesia, Singapore, USA, and China. Some of these countries are new to the desalination industry, such as Syria, Indonesia, and Singapore. While other are in the process of expanding their production to meet specific needs.

Table 4
Desalination production capacity and percentages of various processes in the Gulf countries, the US and other countries. In each category, the first number is for 1996 and second is for 2000.

Country	Total Capacity (m^3/d)	Percentage Relative to total world production	MSF(%)	MEE(%)	MVC(%)	RO(%)	ED(%)
Saudi	5253208	23.6	65.66	0.31	1.21	30.97	1.85
Arabia	5429334	20.96	64.22	0.329	1.39	32.254	1.8
USA	3092533	15.6	1.71	1.78	4.51	78.04	11.37
	4327596	16.7	1.32	4.49	6.35	74.63	13.56
UAE	2164507	9.8	89.80	0.38	2.97	6.49	0.24
	2890689	11.16	86.66	7.7	0.03	5.51	0.09
KUWAIT	1538426	6.8	95.47	0.68	0.00	3.39	0.33
	1614861	6.2	96.52	0.08	0.00	3.25	0.15
JAPAN	745318	3.67	4.72	1.97	0.00	86.41	6.78
	945163	3.65	3.86	2.34	0.00	84.32	7.35
LIBYA	683308	3.37	67.70	0.94	1.84	19.56	9.79
	701303	2.71	65.66	10.7	0	15.91	7.73
QATAR	566904	2.79	94.43	0.64	3.26	0.00	0.00
	572870	2.21	94.34	3.86	0	1.8	0.00
SPAIN	529891	2.61	10.62	0.90	8.65	68.91	10.90
	1233835	4.76	4.51	3.5	2.79	84.25	4.95
ITALY	518711	2.56	43.22	1.88	15.14	20.43	19.16
	581478	2.24	43.76	12.4	6.53	21.67	16.24
BAHRAIN	309158	1.52	52.02	0.00	1.46	41.73	4.50
	473391	1.83	62.74	9.67	0	26.88	0.71
OMAN	192586	0.95	84.06	2.18	0.00	11.73	0.00
	377879	1.21	87.31	1.111	3.7	7.63	0.237

Total world production (1996) $=20.3\times10^6$ m^3/d
Total world production (2000) $=25.909\times10^6$ m^3/d

References

Abu-Eid, Z.M., and Fakhoury, A.G., Some special design features of Kuwait MSF plants, Desalination, **23**(1974)263-284.

Butt, F.H., and Bou-Hassan, A.H., Performance of a polyphosphate-based scale control additive in an MSF plant of Kuwait, Desalination, **36**(1981)129-136.

IDA, The 1996 IDA worldwide desalting plants inventory, Report No. 15.

IDA, The 2000 IDA worldwide desalting plants inventory, Report No. 16.

Silver, R.S., Multi-stage flash distillation – The first 10 years, 3rd Int. Sym. On Fresh Water from the Sea, Athens, Greece, **1**(1970)191-206.

Temperley, T.G., The coming of age of desalination, IDA World Congress on Desalination and Water Sciences, Madrid, October, 1997.

Chapter 2

Single Effect Evaporation

Objectives

The objectives of this chapter include the following:

- Development of mathematical model for the single effect evaporation process.
- Discussing and analyzing the performance of single effect evaporation process.
- Outlining features of various evaporator configurations

2.1 Single Effect Evaporation

The single-effect evaporation desalination system has very limited industrial applications. The system is used in marine vessels. This is because the system has a thermal performance ratio less than one, i.e.; the amount of water produced is less than the amount of heating steam used to operate the system. However, understanding of this process is essential since it constitutes many of the elements forming other single-effect vapor compression systems as well as the multiple effect evaporation processes. This would facilitate understanding of these systems, which are more complex.

2.1.1 Process Description

Figure 1 shows a schematic diagram for the single effect evaporation system. The main components of the unit are the evaporator and the feed preheater or the condenser. The evaporator consists of an evaporator/condenser heat exchange tubes, a vapor space, un-evaporated water pool, a line for removal of non-condensable gases, a water distribution system, and a mist eliminator. The feed preheater has a shell and tube configuration and operates in a counter-current mode, where the latent heat of condensed vapor is transferred to the intake seawater, which includes the feed (M_f) and the cooling seawater (M_{cw}).

The intake seawater (M_{cw}+M_f) at a temperature (T_{cw}) and a salt concentration (X_f) is introduced into the tube side of the preheater where its temperature increases to (T_f). The cooling water (M_{cw}) is dumped back to the sea. The function of the cooling water in the condenser is the removal of the excess heat added to the system in the evaporator by the heating steam. This implies that the evaporator does not consume all the supplied heat, instead, it degrades its quality. The heating of the feed seawater (M_f) in the condenser tubes from (T_{cw}) to (T_f) is essential to increase the thermal performance of the process. The heat needed to warm the seawater inside the condenser tubes is supplied by condensing the vapor formed by boiling in the evaporator (M_d).

The vapor condensation temperature and consequently the pressure in the vapor space for both the evaporator and the condenser is controlled by

- The cooling water flow rate, M_{cw}.
- The feed water temperature, T_{cw}.
- The available heat transfer area in the condenser, A_c.
- The overall heat transfer coefficient between the condensing vapor and the circulating seawater, U_c.

Accordingly, the condenser has three functions:

- Removes the excess heat from the system.
- Improves the process performance ratio.
- Adjusts the boiling temperature inside the evaporator.

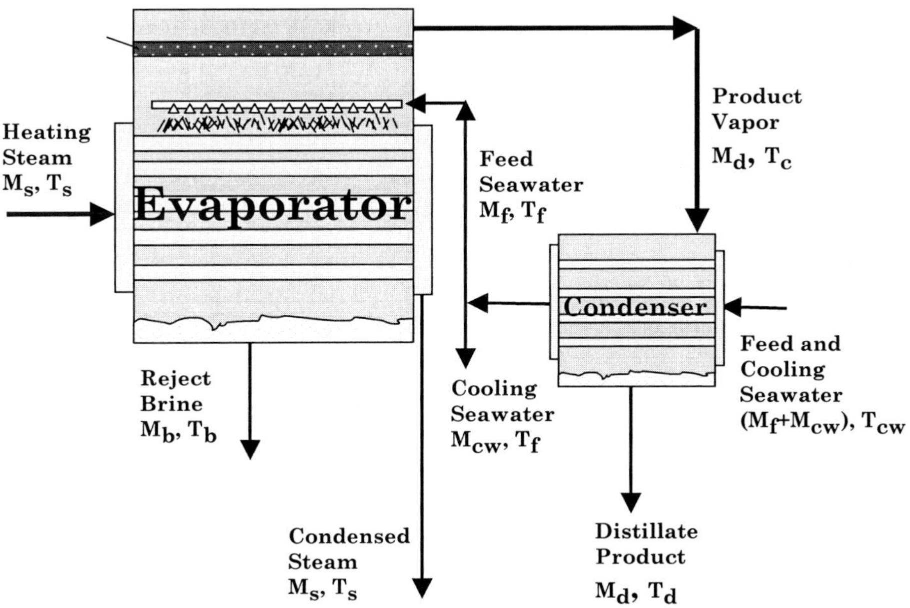

Fig. 1. Single effect evaporation desalination process

The feed seawater (M_f) is chemically treated and deaerated before being pumped to the evaporator. The chemical treatment is needed to prevent the foaming and the tendency for scale formation in the evaporator. Both factors may seriously impair unit operation. Within the evaporator, the feed water is sprayed at the top where it falls in the form of thin film down the succeeding rows of tubes arranged horizontally. Condensation of the saturated heating steam and release of its latent heat provides the required sensible and latent for water evaporation from the feed seawater. As a result, the feed water temperature (T_f) is raised to the boiling temperature (T_b). The magnitude of (T_b) is dictated mainly by the nature of chemicals used to control the scale formation and the state of the heating steam. The vapor formed by boiling with a rate of (M_d) is free of salts. Figure 2 shows that the temperature of the generated vapor (T_v) is less than the boiling temperature by the boiling point elevation (BPE). Similarly, the temperature of the condensed vapor (T_d) is lower than the temperature of the generated vapor by losses caused by the demister, the transmission lines, and condensation.

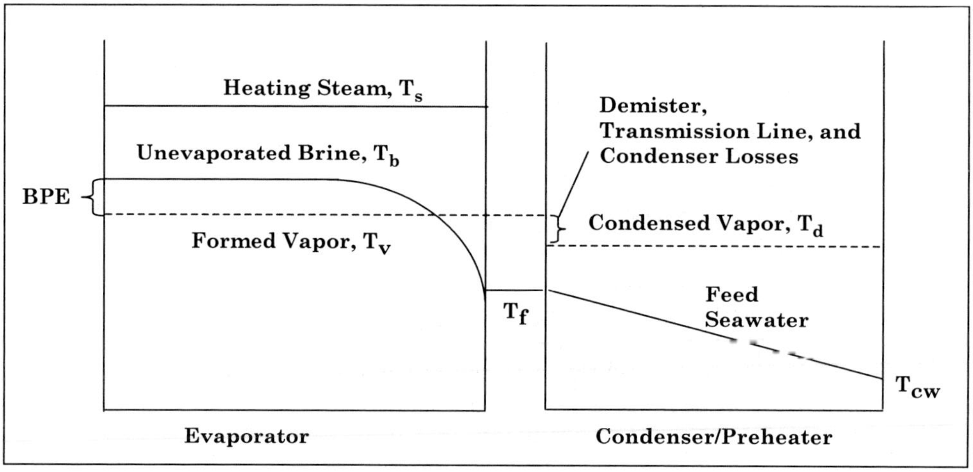

Fig. 2. Temperature profiles in evaporator and condenser of the single effect evaporation desalination process

The generated vapor flows through a knitted wire mist separator known as the wire mesh demister to remove the entrained brine droplets. The vapor should be completely freed from brine droplets to prevent the contamination of the product water. This also prevents exposure of the condenser tubes to the brine, which can result in scaling, surface corrosion, and reduction of the heat transfer rates. Also, in thermal vapor compression units presence of entrained water droplets in the vapor flowing into the steam jet ejector can result in erosion of the ejector nozzle and diffuser. The saturation temperature of the vapors departing the demister is lower than (T_v). This temperature depression is caused by the frictional pressure loss in the demister. Other pressure drop takes place during the vapor transfer between the evaporator and preheater; also pressure drop occurs during vapor condensation. This will further decrease the vapor condensation temperature.

The non-condensable gases in the vapor space of the condenser must be vented continuously to avoid downgrading of the heat transfer capacity of the condenser. The blanket of noncondensibles masks some of the heat transfer area from the condensing vapor. In addition, the non-condensable gases reduce the partial pressure of the condensing vapors. As a result, condensation takes place at a lower temperature. This reduces the process efficiency because of the decrease in the net driving force for heat transfer and consequently reduces the feed seawater temperature (T_f). Removal of the gases is made at points where the temperature approaches that of the feed water. This permit the cooling of the noncondensable gases to the minimum possible temperature, thereby, minimizing the amount of vapor that may escape with the gases and decreases the volume of pumped gases. In addition, it is possible to operate the counter-current condenser so that the exit water is within 3 to 5 °C of the condensation temperature of the saturated vapor. This improves the thermal performance of the unit and minimizes the mass flow rate of cooling water.

2.1.2 Process Modeling

The model for the single-effect evaporation system is divided into six parts:

- Material balances.
- Evaporator and condenser energy balances.
- Boiling point elevation and thermodynamic losses.
- Evaporator and condenser heat transfer area.
- Summary of performance parameters.

Material Balances

The overall mass and salt balances assume that the distillate water is salt free. The two balance equations are given by

$$M_f = M_d + M_b \tag{1}$$

$$M_f\, X_f = M_b\, X_b \tag{2}$$

where (M) is the mass flow rate, (X) is the salinity, and the subscripts b, d, and f denotes the rejected brine, distillate, and feed seawater. Equation 1 can be used to eliminate (M_f) from Eq. 2 and generate a relation between (M_b) and (M_d). This result is given by

$$M_b = M_d\, (X_f\,/(X_b - X_f)) \tag{3}$$

Similarly, M_b can be eliminated from Eq. 2 to generate a relation between M_f and M_d. This result is given by

$$M_f = M_d\ (X_b\ /(X_b - X_f)) \tag{4}$$

Evaporator and Condenser Energy Balances

The energy balance of the evaporator conserves the energies of the heating steam, vapors formed, feed seawater, and rejected brine. In the evaporator, saturated steam flowing from the steam boiler at a rate equal to M_s is used in to raise the temperature of the feed seawater M_f from the inlet temperature T_f to the boiling temperature T_b. In addition, it supplies the latent heat required to evaporate the specified mass of vapor, M_d, or

$$Q_e = M_f\ C_p\ (T_b - T_f) + M_d\ \lambda_v = M_s\ \lambda_s \tag{5}$$

where Q_e is the thermal load of the evaporator, C_p is the specific heat at constant pressure of the brine, and λ is the latent heat of evaporation. The reference temperature in Eq. 5 is T_b. The specific heat in Eq. 5 is calculated at an average temperature of $(T_f + T_b)/2$ and salinity of X_f of the feed seawater. As is shown in Eq. 5 and Fig. 2, the vapor temperature is equal to T_v, which is lower than boiling temperature by the boiling point elevation, BPE.

The condenser operates on the vapor formed in the evaporator, (M_d). The latent heat of condensation is transferred to feed seawater with a mass flow rate of (M_f+M_{cw}). The feed seawater (M_f) is introduced into the evaporator; while the remaining part (M_{cw}), which is known as the cooling water, is rejected. The vapor is assumed saturated at a temperature equal to (T_v).

The heat load of the condenser is given by

$$Q_c = (M_f + M_{cw})\ C_p\ (T_f - T_{cw}) = M_d\ \lambda_v \tag{6}$$

where Q_c is the thermal load of the condenser, C_p is the specific heat at constant pressure of the brine, M is the mass flow rate, T is the temperature, and λ is the latent heat of evaporation. The subscripts cw, f, d, and v denote the cooling seawater, feed seawater to the evaporator, distillate vapor, and condensing vapor. The seawater heat capacity, Cp, is calculated at an average temperature of $(T_f+T_{cw})/2$ and a salinity of X_f.

The overall energy balance for the system is given by

$$M_s\,\lambda_s = M_b\,C_p\,(T_b - T_{cw}) + M_d\,C_p\,(T_v - T_{cw}) + M_{cw}\,C_p\,(T_f - T_{cw}) \tag{7}$$

The reference temperature in Eq. 7 is T_{cw}. The heat capacities are calculated at the temperature average shown for each term and a salinity of X_b, 0, and X_f for the terms that include M_b, M_d, and M_{cw} terms, respectively. Equation 6 is used to eliminate the last term on the right hand side in Eq. 7. This substitution reduces the overall energy balance to

$$M_s\,\lambda_s = M_b\,C_p\,(T_b - T_f) + M_d\,C_p\,(T_v - T_f) + M_d\,\lambda_v \tag{8}$$

The vapor temperature T_v is then defined in terms of the boiling temperature (T_b) and the boiling point elevation (BPE)

$$T_b = T_v + BPE \tag{9}$$

Substitution of Eq. 9 into Eq. 8 gives

$$M_s\,\lambda_s = M_b\,C_p\,(T_v + BPE - T_f) + M_d\,C_p\,(T_v - T_f) + M_d\,\lambda_v \tag{10}$$

Equation 10 is arranged to give

$$M_s\,\lambda_s = M_b\,C_p\,(T_v - T_f) + M_b\,Cp\,BPE + M_d\,C_p\,(T_v - T_f) + M_d\,\lambda_v \tag{11}$$

The flow rate of rejected brine, M_b, is eliminated in Eq. 11 by using the relation given in Eq. 3. This gives

$$M_s\,\lambda_s = M_d\,C_p\,(T_v - T_f) + M_d\,(X_f/(X_b - X_f))\,C_p\,(T_v - T_f) + M_d\,(X_f/(X_b - X_f))\,Cp\,BPE + M_d\,\lambda_v \tag{12}$$

Equation 12 is then simplified to

$$M_s\,\lambda_s = M_d\,((1+X_f/(X_b - X_f))\,C_p\,(T_v - T_f) + (X_f/(X_b - X_f))\,Cp\,BPE + \lambda_v) \tag{13}$$

Equation 13 is then written in terms of the flow rates ratio of the distillate and the heating steam, or the performance ratio, PR. This gives

$$PR = \frac{M_d}{M_s} = \frac{\lambda_s}{(\lambda_v + C_p(T_v - T_f)\frac{X_b}{X_b - X_f} + \frac{X_f}{X_b - X_f}\,C_p BPE)} \tag{14}$$

Equation 14 is used to determine the system performance ratio as a function of the temperatures of the feed and condensed vapor, the salinity of feed and rejected brine, the boiling point elevation, the latent heats of heating steam and condensing vapor, and the heat capacity of water.

Equation 6 is arranged to obtain the specific cooling water flow rate. The derivation of this relation proceeds as follows:

$$M_{cw}\ C_p\ (T_f\ - T_{cw}) = M_d\ \lambda_d - M_f\ C_p\ (T_f - T_{cw}) \tag{15}$$

The seawater feed flow M_f is eliminated in the above equation by use of the relation given in Eq. 4. This gives

$$M_{cw}\ C_p\ (T_f - T_{cw}) = M_d\ \lambda_d - M_d\ (X_b/(X_b - X_f))\ C_p\ (T_f - T_{cw}) \tag{16}$$

Further arrangement of Eq. 15 gives the specific flow rate of water cooling, which is given by

$$sM_{cw} = \frac{M_{cw}}{M_d} = \frac{\lambda_d - (X_b/(X_b - X_f))Cp(T_f - T_{cw})}{C_p(T_f - T_{cw})} \tag{17}$$

Evaporator and Condenser Heat Transfer Area

The dimensions of the required heat transfer surface area in the evaporator A_e are obtained from:

- The amount of the heat to be transferred Q_e.
- The overall heat transfer coefficient U_e.
- The difference between the condensation temperature of the steam, T_s, and the boiling temperature of the seawater T_b.

This relation is given by

$$A_e = Q_e/(U_e\ (T_s - T_b)) \tag{19}$$

Substituting the value of Q_e from Eq. 5 into the above equation gives

$$A_e = \frac{M_f\ C_p\ (T_b - T_f) + M_d\ \lambda_v}{U_e(T_s - T_b)} \tag{20}$$

The flow rate of the feed seawater, M_f, is eliminated in Eq. 20 by the use of Eq. 4, which relates M_f, M_d, and salinity of the feed and rejected brine. This reduces Eq. 20 to the following

$$A_e = \frac{M_d\left(\frac{X_b}{X_b - X_f}\right)C_p\ (T_b - T_f) + M_d\ \lambda_v}{U_e(T_s - T_b)} \tag{21}$$

Equation 21 is arranged to obtain the specific heat transfer area for the evaporator, which is defined as the ratio of the heat transfer area to the distillate product flow rate. The resulting expression for the specific heat transfer area is given by

$$\frac{A_e}{M_d} = \frac{\left(\frac{X_b}{X_b - X_f}\right)C_p\ (T_b - T_f) + \lambda_v}{U_e(T_s - T_b)} \tag{22}$$

Equation (22) can also be written in terms of the boiling point elevation (BPE), where

$$\frac{A_e}{M_d} = \frac{\left(\frac{X_b}{X_b - X_f}\right)C_p\ (T_b - T_f) + \lambda_v}{U_e(T_s - T_v - \mathrm{BPE})}$$

Inspection of this equation shows that the increase in the (BPE) would reduce the temperature driving force and hence increases the specific heat transfer area. In other words, the (BPE) represents an extra resistance to heat transfer.

The heating surface area of the evaporators A_e is usually, but not always, taken as that in contact with the boiling liquid, whether on the inside or outside of the tubes. The overall heat transfer coefficient based on the outside surface area U_e is related to the individual thermal resistance by the following well-known expression:

$$\frac{1}{U_e} = \frac{1}{h_i}\frac{r_o}{r_i} + R_{f_i}\frac{r_o}{r_i} + \frac{r_o \ln(r_o / r_i)}{k_w} + R_{f_o} + \frac{1}{h_o} \tag{23}$$

where h is the heat transfer coefficient, R_f is the fouling resistance, k_w is the thermal conductivity of tube material and r is the radius. The subscripts i and o refer to the inner and outer tube surfaces, respectively.

The heat transfer between the condensing vapor and the feed water in the condenser can be written in terms of the condenser load, the overall heat transfer coefficient, U_c, the condenser heat transfer area, A_c, and the logarithmic mean temperature difference, $(LMTD)_c$, thus:

$$A_c = \frac{Q_c}{U_c (LMTD)_c} = \frac{M_d \lambda_d}{U_c (LMTD)_c} \quad (24)$$

Examining Fig. 2 show that the $(LMTD)_c$ is defined as

$$(LMTD)_c = \frac{(T_f - T_{cw})}{\ln \frac{(T_d - T_{cw})}{(T_d - T_f)}} \quad (25)$$

The specific heat transfer area in the condenser is then given by

$$\frac{A_c}{M_d} = \frac{\lambda_d}{U_c (LMTD)_c} \quad (26)$$

The overall heat transfer coefficient for the evaporator and the condenser are calculated from the correlations developed by El-Dessouky et al. (1997), see appendix C.

Summary of Performance Parameters

Performance of the single-effect system is developed in the previous sections. The performance parameters include the following:

- The amount of product fresh water per unit mass of heating steam, or the thermal performance ratio (PR) given by Eq. 14.
- The specific heat transfer surface area (sA) given by Eqs. 22 and 27.
- The specific cooling water flow rate (sM_{cw}) given by Eq. 17.

The thermal performance ratio and the specific flow rate of cooling give a measure for the system operating cost. Increase in the thermal performance ratio, implies reduction in energy consumption, which amounts for 30-50% of the unit product cost. On the other hand, the specific heat transfer area is a measure of the process capital.

The above system parameters are defined by the following relations:

$$PR = \frac{M_d}{M_s} = \frac{\lambda_s}{\left(\lambda_v + C_p (T_v - T_f) \frac{X_b}{X_b - X_f} + \frac{X_f}{X_b - X_f} C_p BPE \right)} \quad (27)$$

$$sA = \frac{A_e + A_c}{M_d} = \frac{\left(\frac{X_b}{X_b - X_f}\right) C_p\,(T_b - T_f) + \lambda_v}{U_e(T_s - T_b)} + \frac{\lambda_d}{U_c(LMTD)_c} \tag{28}$$

and

$$sM_{cw} = \frac{M_{cw}}{M_d} = \frac{\lambda_d - (X_b/(X_b - X_f))Cp(T_f - T_{cw})}{C_p(T_f - T_{cw})} \tag{29}$$

An approximate form of the performance ratio, Eq. 27, is obtained by neglecting the sensible heat effects, the second and the third terms in the dominator. This result is arrived at by comparing the order of magnitudes of the three terms in the dominator of Eq. 27. The magnitude of λ_v is higher than 2000 kJ/kg, while the order of the second or the third terms is in the range of 50 kJ/kg. This is obtained by setting Cp to 4 kJ/kg °C, T_v-T_f to 5 °C, and $X_b/(X_b$-$X_f)$ to 2.5. The result of this analysis, show that the performance ratio for a single effect configuration can be approximated by

$$PR = \lambda_s/\lambda_v \tag{30}$$

Equation 30 is useful in checking the model results.

2.1.3 System Performance

The following set of examples illustrates application of the single stage evaporator model. The first example is a non-iterative and direct solution case study in which the system temperatures are specified and it is required to determine the heat transfer area of the evaporator and down condenser, the thermal performance ratio, and the cooling water flow rate. In the second example, the evaporator thermal load is specified and it is required to determine the brine boiling temperature, the thermal performance ratio, the heat transfer areas, and the cooling water flow rate. In the third example, an existing system, where the heat transfer areas are known, is analyzed to determine the feed seawater temperature, the heating steam temperature, the steam flow rate, the thermal performance ratio, and cooling water flow rate.

The following set of specifications is used in solution of the system model:

- The seawater temperature, T_{cw}, varies over a range of 5 to 30 °C.
- The feed water temperature, T_f, is less than the brine boiling temperature by 4 to 15 °C.

- The steam temperature, T_s, is higher than the brine boiling temperature by 4 to 15 °C.
- The distillate flow rate, M_d, is always kept constant at 1 kg/s.
- The seawater salinity range is 32,000 to 42,000 ppm.
- The salinity of the rejected brine, X_b, is 70000 ppm.
- The boiling temperature, T_b, varies over a range of 55 to 100 °C.
- The heat capacity of seawater, distillate, and reject brine are assumed constant and equal to 4.2 kJ/kg °C.

Example 1:

A single-effect evaporator generates a distillate product at a flow rate of 1 kg/s. The system operating temperatures are as follows:

- The boiling temperature, T_b, is 75 °C.
- The intake seawater temperature, T_{cw}, is 25 °C.
- The feed temperature, T_f, is 70 °C.
- The steam temperature, T_s, is 82 °C.

Determine the heat transfer areas in the evaporator and the condenser, the thermal performance ratio, the flow rates of feed seawater and reject brine, and the flow rate of cooling seawater.

Solution: The solution proceeds with evaluation of the vapor temperature. This requires calculation of the boiling point elevation (BPE) using the correlation given in appendix B.

$$\begin{aligned} BPE = & (0.0825431+0.0001883\,(75)+0.00000402\,(75)^2)\,(7) \\ & +(-0.0007625+0.0000902\,(75)-0.00000052\,(75)^2)\,(7)^2 \\ & +(0.0001522-0.000003\,(75)-0.00000003\,(75)^2)\,(7)^3 \\ = & 0.903\ ^{o}C \end{aligned}$$

The resulting temperature of the vapor formed in the evaporator (T_v) is calculated from Eq. 9,

$$T_v = T_b - BPE = 75 - 0.903 = 74.097\ ^{o}C$$

The temperatures of the heating steam and vapor are used to calculate the latent heat for the steam and distillate vapor, λ_s and λ_v, are calculated from the correlations given in appendix A. The resulting values are:

$$\begin{aligned} \lambda_s & = 2501.897149 - 2.407064037\ T_s + 1.192217\text{x}10^{-3}\ T_s^2 - 1.5863\text{x}10^{-5}\ T_s^3 \\ & = 2501.897149 - 2.407064037\ (82) + 1.192217\text{x}10^{-3}\ (82)^2 \\ & \quad - 1.5863\text{x}10^{-5}\ (82)^3 = 2303.788\ \text{kJ/kg} \end{aligned}$$

$$\lambda_v = 2501.897149 - 2.407064037\ T_v + 1.192217x10^{-3}\ (T_v)^2 - 1.5863x10^{-5}\ (T_v)^3$$
$$= 2501.897149 - 2.407064037\ (74.097)$$
$$+ 1.192217x10^{-3}\ (74.097)^2 - 1.5863x10^{-5}\ (74.097)^3 = 2323.6\ kJ/kg$$

The overall heat transfer coefficients in the evaporator and condenser are calculated using the correlations given in appendix C. The resulting values for the two coefficients are:

$$U_e = 1.9695{+}1.2057x10^{-2}T_b{-}8.5989x10^{-5}(T_b)^2{+}2.5651x10^{-7}(T_b)^3$$
$$= 1.9695{+}1.2057x10^{-2}(75){-}8.5989x10^{-5}(75)^2{+}2.5651x10^{-7}(75)^3$$
$$= 2.4983\ kJ/s\ m^2\ {}^oC$$

$$U_c = 1.7194{+}3.2063x10^{-3}T_v{+}1.5971x10^{-5}(T_v)^2{-}1.9918x10^{-7}(T_v)^3$$
$$= 1.7194{+}3.2063x10^{-2}(74.097){+}1.5971x10^{-5}(74.097)^2$$
$$-1.9918x10^{-7}(74.097)^3$$
$$= 1.961\ kJ/s\ m^2\ {}^oC$$

The system performance parameters are calculated from Eqs. 28-30. The thermal performance ratio is given by

$$PR = \frac{M_d}{M_s} = \frac{\lambda_s}{\left(\lambda_v + C_p(T_v - T_f)\frac{X_b}{X_b - X_f} + \frac{X_f}{X_b - X_f}C_p BPE\right)}$$

$$= 2303.788/(2323.6 + 4.2\ (74.097 - 70)\frac{70000}{70000 - 42000}$$
$$+ \frac{42000}{70000 - 42000}(4.2)(0.903))$$
$$= 0.97$$

The specific heat transfer area is given by

$$sA = \frac{A_e + A_c}{M_d}$$

$$= \frac{\left(\frac{X_b}{X_b - X_f}\right)C_p\ (T_b - T_f) + \lambda_v}{U_e(T_s - T_b)} + \frac{\lambda_v}{U_c(LMTD)_c}$$

$$= \frac{\left(\frac{70000}{70000-42000}\right)4.2\ (75 - 70) + 2323.6}{2.4983(82 - 75)} + \frac{2323.6}{(1.961)(18.12)}$$
$$= 135.9 + 65.4 = 201.3\ m^2/(kg/s)$$

The specific cooling water flow rate is given by

$$sM_{cw} = \frac{M_{cw}}{M_d} = \frac{\lambda_v - (X_b/(X_b - X_f))\,Cp\,(T_f - T_{cw})}{C_p(T_f - T_{cw})}$$

$$= \frac{2323.6 - (70000/(70000 - 42000))4.2\,(70 - 25)}{4.2\,(70 - 25)}$$

$$= 9.8$$

To complete the analysis, other system variables are calculated below. The flow rates of feed seawater and rejected brine area calculated from Eqs. 3 and 4. Substituting for $X_f = 42000$ ppm, $X_b = 70000$ ppm, and $M_d = 1$ kg/s in Eq. 3 and 4 results in

$M_b = X_f /(X_b\text{-}X_f) = 42000/(70000\text{-}42000) = 1.5$ kg/s

$M_f = X_b /(X_b\text{-}X_f) = 70000/(70000\text{-}42000) = 2.5$ kg/s

The steam flow rate is obtained from the performance ratio result, where

$M_s = M_d / PR = 1/0.97 = 1.03$ kg/s

The evaporator and condenser loads are obtained from Eqs. 5 and 6, respectively. The resulting values are:

$Q_e = M_s\,\lambda_s = (1.03)(2303.788) = 2372.9$ kJ/s

$Q_c = M_d\,\lambda_v = (1)(2323.6) = 2323.6$ kJ/s

The condenser load, Eq. 6, is used to obtain the cooling seawater flow rate, M_{cw},

$M_{cw} = M_d\,\lambda_v/(C_p(T_f - T_{cw}) - M_f$

$= (1)\,(2326.34)/(4.2(70\text{-}25)) - 2.5 = 9.8$ kg/s

As for the actual heat transfer areas in the evaporator and the condenser, their values are identical to the specific value, because M_d is equal to 1.

Example 2:

A single effect evaporator has a thermal load, Q_e, of 2355 kJ/s and the heating steam temperature is 115 °C. The seawater temperature, T_{cw}, is 30 °C and the feed seawater temperature, T_f, is less than the boiling temperature, T_b, by 10 °C. If the distillate product flow rate is 1 kg/s calculate the boiling temperature, the heat transfer areas, and the flow rate of the cooling seawater. Use the same salinity and flow rates given in Example 1 for the feed seawater and reject brine.

Solution: The thermal load of the evaporator is used to calculate the heating steam flow rate. This requires calculation of the steam latent heat, λ_s, at 115 °C. From the latent heat correlation this value is equal to 2216.73 kJ/kg. Therefore, the flow rate of the heating steam is determined from the Eq. 5, where

$$Q_e = M_s \lambda_s$$

$$2355 = M_s (2216.73)$$

This gives, M_s = 1.0624 kg/s. Since, the distillate flow rate is known, then, the system thermal performance ratio is determined from Eq. 28, where,

$$PR = M_d/M_s = 1/1.0624 = 0.94$$

The brine boiling temperature can be obtained from the evaporator thermal load, Eq. 5. This is

$$Q_e = M_d \lambda_v + M_f C_p (T_b - T_f)$$

The iteration sequence is simple and is based on evaluation of the right hand side of the above equation at an assumed value for brine boiling temperature, T_b. The iteration error is then set equal to the difference of the calculated value for thermal load, Q_e, and its actual value of 2355 kJ/s. The iterations are terminated when the iteration error changes sign. Good initial guesses for the brine boiling temperature is within a range of (T_s – 20) °C. Results for two iterations are given below to illustrate the above solution sequence.

In the first iteration, the brine boiling temperature is assumed equal to 102 °C. At this condition, the feed temperature is equal 92 °C. The values of T_v and λ_v are then calculated at T_b = 102 °C and X_b = 70000 ppm. First the boiling point elevation is calculated from the correlation given in appendix B. The values of B and C in this correlation are first determined

$$BPE = (0.0825431+0.0001883\,(102)+0.00000402\,(102)^2)\,(7)$$
$$+(-0.0007625+0.0000902\,(102)-0.00000052\,(102)^2)\,(7)^2$$
$$+(0.0001522-0.000003\,(102)-0.00000003\,(102)^2)\,(7)^3$$
$$= 0.994\ ^{o}C$$

The resulting temperature of the vapor formed in the evaporator, T_v, is calculated from Eq. 9,

$$T_v = T_b - BPE = 102 - 0.994 = 101.006\ ^{o}C$$

The vapor latent heat at 101.006 °C is 2254.6 kJ/kg. The above values are substituted in Eq. E.1 to calculate the evaporator thermal load

$$Q_e = M_d\,\lambda_v + M_f\,C_p\,(T_b - T_f)$$
$$= (1)\,(2254.6) + (2.5)\,(4.2)\,(10) = 2359.6\ kJ/s$$

The iteration error is then calculated

$$Error = Q_{e_{actual}} - Q_{e_{calculated}}$$
$$= 2355 - 2359.6 = -4.6\ kJ/s$$

The above error is small enough and further iterations are not necessary.

The calculated values of T_b and T_v are used to obtain the overall heat transfer coefficient in the evaporator and condenser,

$$U_e = 1.9695+1.2057x10^{-2}T_b-8.5989x10^{-5}(T_b)^2+2.5651x10^{-7}(T_b)^3$$
$$= 1.9695+1.2057x10^{-2}(102)-8.5989x10^{-5}(102)^2$$
$$+2.5651x10^{-7}(102)^3$$
$$= 2.58\ kJ/s\ m^2\ ^{o}C$$

$$U_c = 1.7194+3.2063x10^{-3}T_v+1.5971x10^{-5}(T_v)^2-1.9918x10^{-7}(T_v)^3$$
$$= 1.7194+3.2063x10^{-3}(101.006)+1.5971x10^{-5}(101.006)^2$$
$$-1.9918x10-7(101.006)3$$
$$= 2.001\ kJ/s\ m^2\ ^{o}C$$

The above results allow for calculations of the heat transfer areas in the evaporator and condenser. The specific heat transfer area is given by

$$
\begin{aligned}
sA &= \frac{A_e + A_c}{M_d} \\
&= \frac{\left(\frac{X_b}{X_b - X_f}\right) C_p\,(T_b - T_f) + \lambda_v}{U_e (T_s - T_b)} + \frac{\lambda_v}{U_c (LMTD)_c} \\
&= \frac{\left(\frac{70000}{70000 - 42000}\right) 4.2\,(10) + 2254.6}{2.58 (115 - 102)} + \frac{2254.6}{2.001 (30.02)} \\
&= 70.35 + 37.5 \\
&= 107.85\ m^2/(kg/s)
\end{aligned}
$$

The specific cooling water flow rate is given by

$$
\begin{aligned}
sM_{cw} &= \frac{M_{cw}}{M_d} = \frac{\lambda_v - (X_b/(X_b - X_f))\,Cp\,(T_f - T_{cw})}{C_p (T_f - T_{cw})} \\
&= \frac{2254.6 - (70000/(70000 - 42000))4.2\,(92 - 30)}{4.2\,(92 - 30)} \\
&= 6.16
\end{aligned}
$$

Example 3:

The heat transfer area in the evaporator and condenser of a single stage evaporator are 85 and 40 m^2. The intake seawater temperature is equal to 10 °C and the brine boiling temperature is 65 °C. If the distillate product flow rate is 1 kg/s calculate the feed seawater temperature, the temperature and flow rate of heating steam, the flow rate of the cooling seawater, and the thermal performance ratio. Use the same feed and brine salinity given in the Example 1 as well as the results for the flow rates of the intake seawater and reject brine.

Solution: The boiling point elevation, vapor temperature, and vapor latent are calculated at a brine boiling temperature of 65 °C and a brine salinity of X_b = 70000 ppm. First the boiling point elevation is calculated from the correlation given in appendix B. The values of B and C in this correlation are first determined

$$
\begin{aligned}
BPE &= (0.0825431 + 0.0001883\,(65) + 0.00000402\,(65)^2)\,(7) \\
&\quad + (-\,0.0007625 + 0.0000902\,(65) - 0.00000052\,(65)^2)\,(7)^2 \\
&\quad + (0.0001522 - 0.000003\,(65) - 0.00000003\,(65)^2)\,(7)^3 \\
&= 0.87\ ^{o}C
\end{aligned}
$$

This gives a vapor temperature of

$T_v = T_b - BPE = 65 - 0.87 = 64.13\ ^oC$

The vapor latent heat at this temperature is 2348.3 kJ/kg. The overall heat transfer coefficient in the condenser and evaporator are obtained at T_v and T_b, where

$$U_c = 1.7194+3.2063x10^{-3}T_v+1.5971x10^{-5}(T_v)^2-1.9918x10^{-7}(T_v)^3$$
$$= 1.7194+3.2063x10^{-3}(64.13)+1.5971x10^{-5}(64.13)^2 - 1.9918x10^{-7}(64.13)^3$$
$$= 1.94\ kJ/s\ m^2\ ^oC$$

$$U_e = 1.9695+1.2057x10^{-2}T_b-8.5989x10^{-5}(T_b)^2+2.5651x10^{-7}(T_b)^3$$
$$= 1.9695+1.2057x10^{-2}(65)-8.5989x10^{-5}(65)^2+2.5651x10^{-7}(65)^3$$
$$= 2.46\ kJ/s\ m^2\ ^oC$$

The heat rate design equations (Eqs. 22 and 24) for the condenser and evaporator are used to calculate the temperature of the intake seawater and the heating steam flow rate, where,

$$A_c = (M_d\ \lambda_v)/(U_c\ (LMTD)_c)$$

$$40 = (1)\ (2348.3)/((1.94)(LMTD)_c)$$

The above equation gives $(LMTD)_c = 30.3\ ^oC$, which is used to calculate the feed seawater temperature, or

$$(LMTD)_c = (T_f - T_{cw})/ln((T_v - T_{cw})/(T_v - T_f))$$

$$30.3 = (T_f - 10)/ln((64.13 - 10)/(64.13 - T_f))$$

which gives $T_f = 49.4\ ^oC$. This result is used to solve Eq. 22 for T_s, where,

$$A_e = (M_d\ \lambda_v + M_f\ C_p\ (T_b - T_f))/(U_e\ (T_s - T_b))$$

$$85 = ((1)(2348.3) + (2.5)(4.2)(65 - 49.4))/(2.46(T_s - 65))$$

This gives $T_s = 77.01\ ^oC$. The steam latent heat at this temperature is equal to 2316.4 kJ/kg. The thermal load of the evaporator is used to determine the flow rate of the heating steam. This is

$$M_s\ \lambda_s = M_d\ \lambda_v + M_f\ C_p\ (T_b - T_f)$$

M_s (2316.4) = (1) (2348.3) + (2.5) (4.2) (65 – 49.4)

The resulting steam flow rate is 1.084 kg/s. The thermal performance ratio of the system is then calculated,

$PR = M_d/M_s = 1/1.084 = 0.92$

Finally, the flow rate of the cooling seawater is obtained from the down condenser balance, where,

$(M_f+M_{cw})\ C_p\ (T_f - T_{cw}) = M_d\ \lambda_v$

$(2.5+M_{cw})$ (4.2) (49.4 – 10) = (1) (2348.3)

The resulting flow is then calculated, M_{cw} = 11.69 kg/s.

Performance Charts

Characteristics of the single-effect system are presented as a function of variations in the boiling temperature (T_b). Other system variables that can be varied include the seawater temperature and salinity, which are affected by plant location and seasonal conditions. Results are presented in terms of variations in the performance ratio (PR), the specific heat transfer area (sA), and the specific cooling water flow rate (sM_{cw}). In all calculations, it is assumed that the heating steam temperature is higher than the brine boiling temperature by 7 °C and the brine boiling temperature is higher than the feed seawater temperature by 5 °C.

Variations in the system performance ratio, specific heat transfer area, and specific cooling water flow rate as a function of the seawater temperature and the boiling temperature are shown in Figs. 3, 4, and 5, respectively. As is shown in Fig. 3, the system performance ratio is independent of the intake temperature of the seawater (T_{cw}). This is because the seawater feed temperature (T_f), is specified in terms of the boiling temperature (T_b). Therefore, variations in the intake temperature of the seawater will only affect the cooling water flow rate as well as the condenser heat transfer area. Another important result shown in Fig. 3 is the very low sensitivity of the system performance ratio on the boiling temperature. Variations in the ratio of the latent heat of tho steam and formed vapor are almost constant over the range of boiling temperatures used in the calculations.

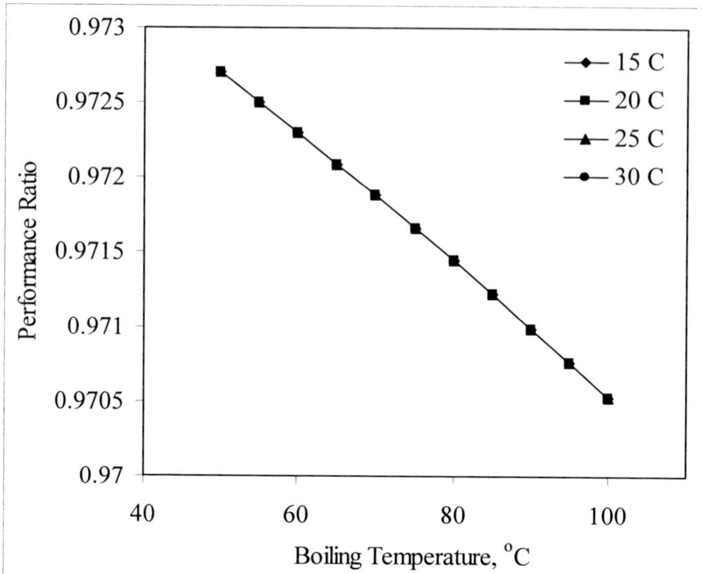

Figure 3: Effect of sewater temperature and boiling temperature on the performance ratio in single-effect evaporation deslaintion process

Effects of the boiling temperature and the intake seawater temperature on the specific heat transfer area are shown in Fig. 4. As is shown, the increase of the boiling temperature (T_b) decreases the specific heat transfer area in the evaporator and the condenser. A similar result is also obtained upon the decrease of the intake seawater temperature. The decrease of the evaporator specific heat transfer area with the increase of the boiling temperature is caused by enhancement of the heat transfer coefficient in the evaporator. At higher boiling temperatures, the heat-transfer coefficient in the evaporator increases because of reduction in the thermal resistance through the condensing vapor film, the brine film, and the metal wall. The heat transfer coefficient in the condenser also increases at higher boiling temperature, because of the increase in the temperature of the condensing vapors. The decrease of the intake temperature increases the driving force for heat transfer in the condenser. This results in reduction of the specific heat transfer area.

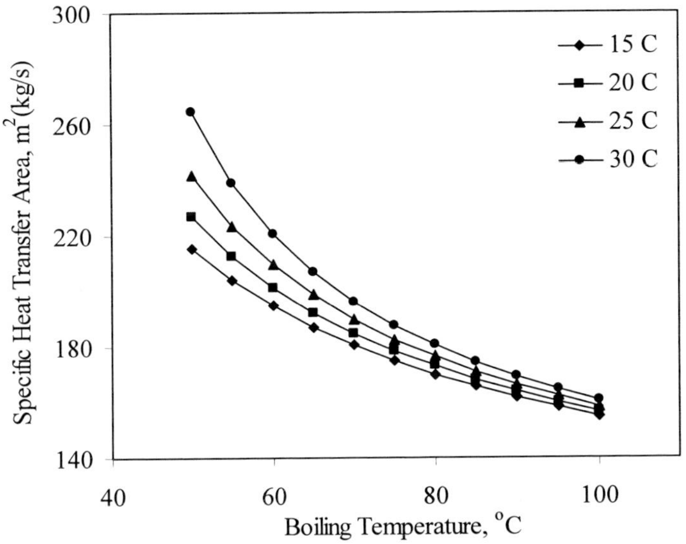

Figure 4: Effect of sewater temperature and boiling temperature on the specific heat transfer area in single-effect evaporation deslaintion process

Variations in the specific cooling water flow rate as a function of the boiling temperature and the intake seawater temperature are shown in Fig. 5. At higher boiling temperatures, the amount of heat absorbed per unit mass of feed seawater (Mf) increases because of the increase in the feed seawater temperature (Tf). This decreases the excess heat which must be removed by the cooling seawater. Also, the condenser load is lower because of the decrease in the latent heat at higher vapor temperatures. In addition, decrease of the intake seawater temperature increases the thermal load per unit mass of cooling seawater.

System performance as a function of the seawater temperature and salinity are shown in Figs. 6, 7, 8. These results are obtained at a boiling temperature (T_b) of 75 °C and a salinity ratio for rejected brine and feed seawater, X_b/X_f, of 1.667. All other system parameters are kept constant at the values specified in the previous sections.

As is shown in Fig. 6, the system performance ratio is less than one. As discussed before, the performance ratio is independent of variations in the seawater temperature. Also, it is insensitive to variations in the seawater salinity. This is because it is only dependent on the latent heat ratio of the steam and the condensing vapor and the difference between (T_b) and (T_f), which is kept constant.

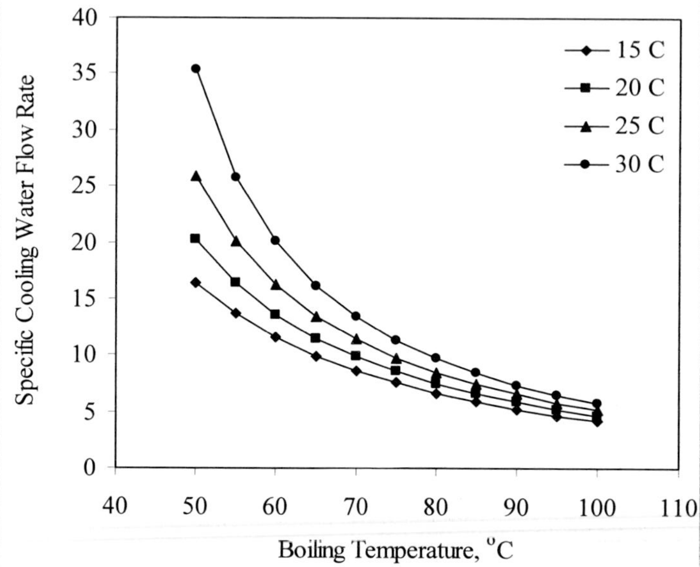

Figure 5: Effect of sewater temperature and boiling temperature on specific cooling water flow rate in single-effect evaporation deslaintion process

Variations in the specific heat transfer area and the specific cooling water flow rate are shown in Figs. 7 and 8, respectively. Both design and operating parameters are virtually independent of the seawater salinity. The seawater salinity has a limited effect on the system, which is measured by the degree of the boiling point elevation. This effect is limited to 2 °C at higher salinity. On the other hand, the specific heat-transfer area and the cooling water flow rate are affected by variations in the seawater temperature. This is because of the variations in the driving force for heat transfer in the condenser.

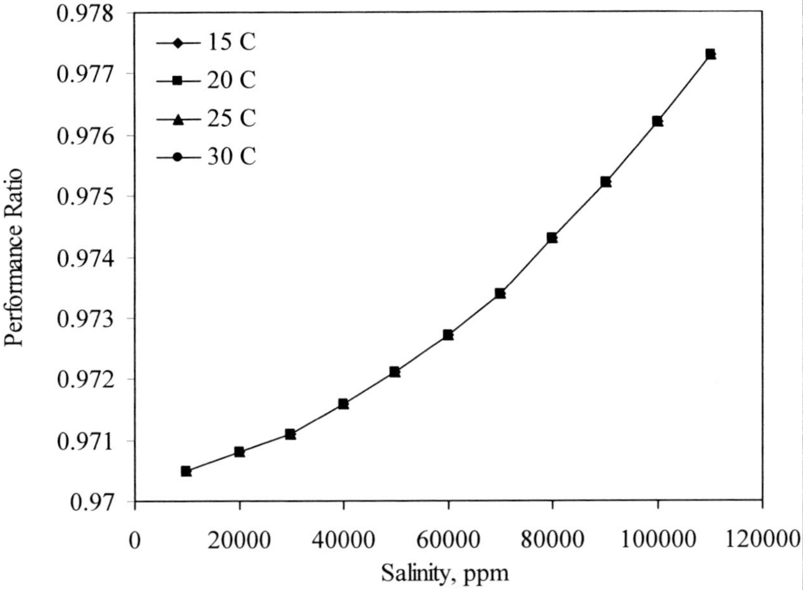

Figure 6: Variation in the system performance ratio as a function of seawater salinity and temperature

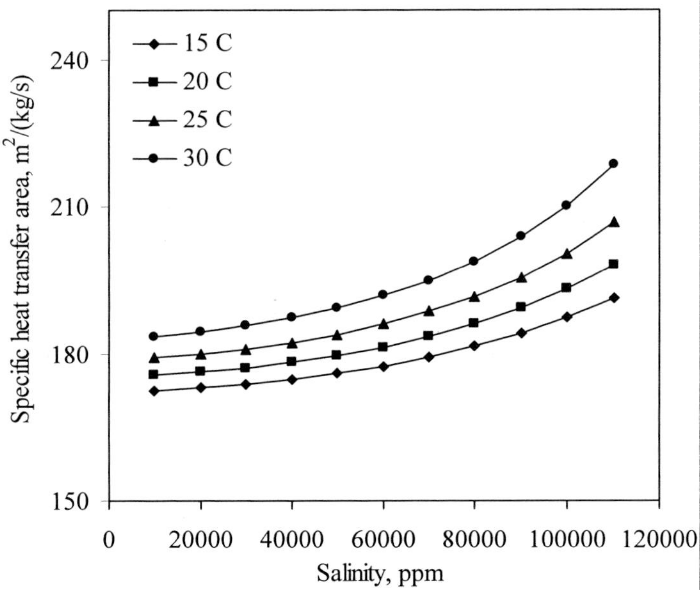

Figure 7: Variation in the specific heat transfer area as a function of the seawater salinity and temperature

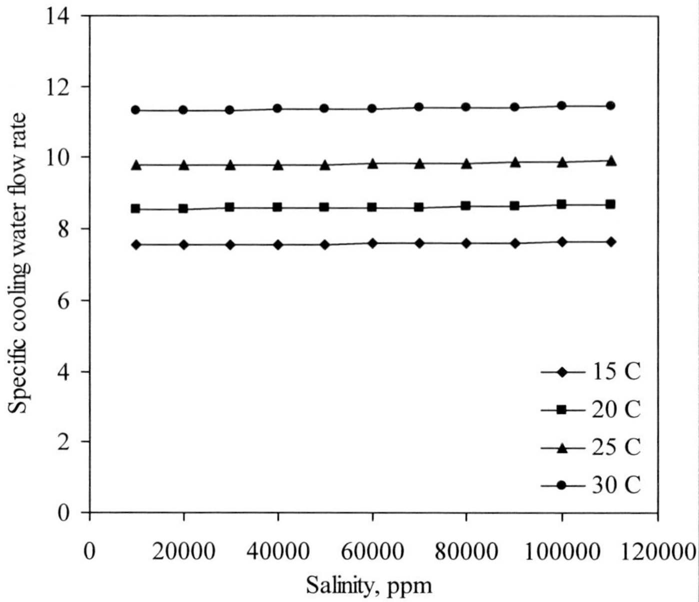

Figure 8: Variation in the specific cooling water flow rate as a function of the seawater salinity and temperature

2.1.4 Summary

Modeling and analysis is presented for the single-effect evaporation desalination system. Although, this system is of very limited use in the desalination industry, it constitutes basic elements found in industrial desalination systems. Modeling and analysis of this simple system is necessary to understand basics and fundamentals of the desalination process, which are also found in actual desalination systems. Detailed results are presented to show the dependence of the factors controlling the fresh water cost, which are the thermal performance ratio, specific heat transfer area, and specific cooling water flow rate, on design and operating variables. These variables are the brine boiling temperature, the intake seawater temperature, and water salinity.

The following conclusions are made in the light of the results and discussion given in the previous section:

- The performance ratio of the single-effect evaporation desalination process is always less than one.
- Performance ratios below one exist as a result of rejecting large amounts of energy in the brine and distillate product.

- The performance ratio is insensitive to variations in the boiling temperature and the intake seawater temperature. This is because the performance ratio depends on the ratio of latent heats of the steam and vapor formed in the evaporator. This ratio varies slightly upon the decrease or increase of the boiling temperature, because of simultaneous adjustment in the steam temperature, i.e., $T_s=(T_b+7)$ °C.
- The specific heat transfer area and the specific cooling water flow rate are sensitive to variations in the boiling temperature and the intake seawater temperature. Both parameters decrease with the increase of the boiling temperature and the decrease of the intake seawater temperature. This is because of the increase in the overall heat transfer coefficient in the evaporator and condenser, which causes large enhancement in the heat transfer rate. Also, the decrease of the intake seawater temperature increases the driving force for heat-transfer in the condenser and increases the amount of heat removed per unit mass of cooling water.
- The system performance parameters are insensitive to variations in the seawater salinity. This is because of the limited effect of seawater salinity on the system parameters. This effect is limited by the boiling point elevation, which is less than 2 °C at high temperatures and seawater salinity.

In summary, analysis of the single-effect evaporation desalination system shows the need for more efficient management of the system energy. Also, system operation is recommended at higher boiling temperatures. Proper energy management will result in higher system performance ratios. This will be found in other single-effect systems, which utilize vapor compression, or in multi-effect configurations. System operation at higher boiling temperature results in reduction of the specific heat transfer area and the specific cooling water flow rate. This reduction lowers the first cost, i.e., construction cost of the evaporator, condenser, and seawater pump. In addition, the operating cost is lower as a result of reduction in the energy required to operate the seawater-pumping unit.

2.2 Evaporators

Evaporators are the heart of any evaporation desalination process; moreover they are an essential element in several industrial applications that include chemical, petroleum, and food processes. The evaporator configuration is based on creating a hot surface, where heating steam condenses on one side and vapor is formed on the other side. Evaporators include the following types:

- Submerged tube
- Falling film

2.2.1 Submerged Evaporators

The basic form of evaporators is the submerged type, Fig. 9. As is shown, the system is a combined unit of an evaporator and a condenser. In this system, the evaporator tubes are submerged in a liquid pool. The liquid surrounding the submerged tubes reaches saturation temperature and evaporation proceeds as the heating steam condenses inside the tubes. The formed vapor flows through a demister pad that removes the entrained liquid droplets. The vapor flows to the condenser, where it condenses on the outside surface of the condenser tubes. As condensation takes place, the latent heat of condensation preheats the feed liquid in the condenser before entering the evaporator unit.

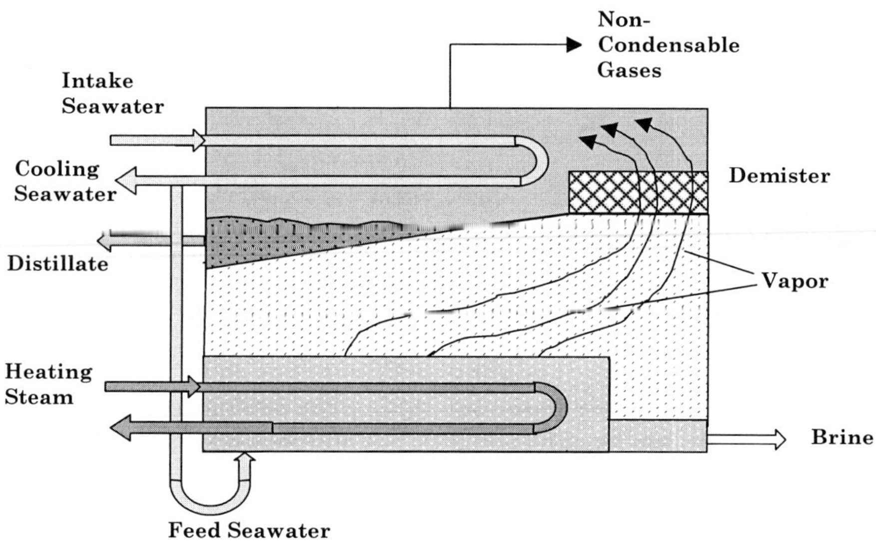

Fig. 9. Submerged evaporator

Features of the submerged evaporator include the following:

- This design is used as a still or kettle reboiler. One of the most common uses is the household humidifier and electric kettle. The submerged evaporator was the most common design during the first half of the twentieth century and was used in various types of applications including desalination.
- The system is suitable if there is no danger of scaling or fouling.
- If the system is prone to scaling or fouling, such as in desalination, use of antiscalent or adjustment of operating conditions is necessary to control the scaling rate. If scaling or fouling is not controlled then system operation may

not be efficient. As mentioned before, operation of the submerged evaporator at these conditions would require prolonged cleaning time equal to or more than the actual production time.
- The heat transfer coefficient for the submerged evaporator is much lower than that for the more common falling film design. This is because the hydrostatic head imposed by the liquid on the tube surface hinders formation, growth, and release of vapor bubbles on the hot surface.
- Development of submerged evaporators is an active research area, especially in absence of fouling and scaling potential. The research focus on use of additive materials that enhance the evaporation process and limits the resistance caused by the liquid hydrostatic head.

2.2.2 Falling Film Evaporators

The falling film evaporator has two main configurations, which includes the horizontal and the vertical tube, Figs. 1 and 10. The horizontal tube is the most common design used in the desalination industry. Features of this configuration include the following:
- The heating steam flows inside the tubes and the liquid is sprayed on the outside surface of the tubes.
- The liquid spray forms a thin falling film on the outside surface of the tube bundles. As a result, formation, growth, and escape of vapor bubbles meet smaller resistance than that found in the submerged tube configuration.
- The tubes are arranged in several rows with a square pitch to simplify the cleaning process.
- The main advantage of the horizontal tube configuration is the complete wetting of the tube surface area. This is achieved by proper selection of the tube pitch, tube diameter, spray pattern, and liquid flow rate. The details of the evaporation process within the system are rather complex, since it involves simultaneous evaporation and condensation. This is caused as the ascending vapor is contacted by the un-saturated water droplets. Accordingly, the part of the vapor releases its latent heat to the liquid droplet causing increase in its temperature to saturation conditions. Subsequently, the saturated water droplets evaporate as they fall on the hot surface of the tubes.

Features of the vertical tube falling film configuration (Fig. 10) include the following:
- The liquid is introduced at the top part of the vertical tubes, where it form a falling film on the inside surface of thc tubes.
- The heating steam flows on the outside surface of the tubes, where it condenses and releases its latent heat to the falling film.
- The heat exchange process result in steam condensation and evaporate ion of the liquid on the tube side.

- Design and operation of the vertical tube configuration is more complex than the horizontal system. This is because the falling film may collapse during its down pass and result in partial wetting of the surface. This result in dry patches or areas covered with very small amount of liquid. As a result, complete evaporation of the liquid, which would leave salt scaling on the tube surface. Also, formation of dry patches would result in increase in the surface temperature of the tube. This would result in increase of the thermal stresses as a result increase in the tube expansion in the hot spots. Ultimately, the tube operation life is reduced because of bucking, scaling, and increase in corrosion rates.

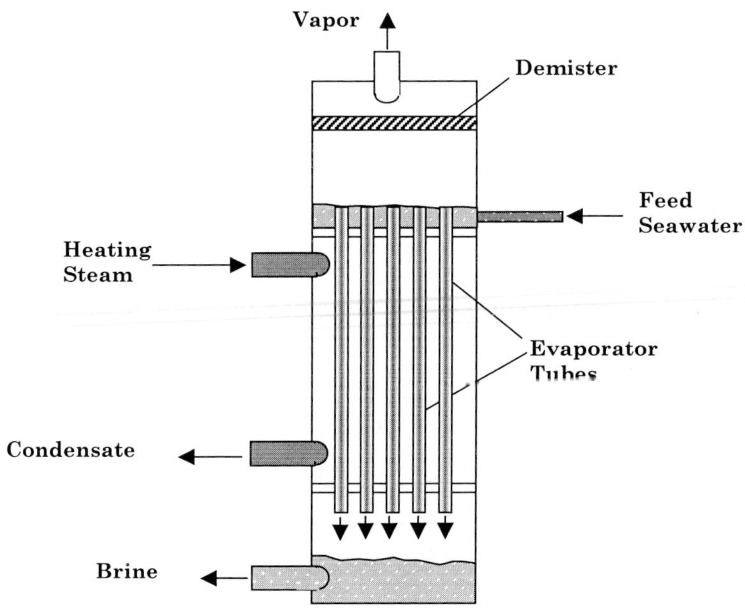

Fig. 10. Vertical tube falling film evaporator

2.2.3 Plate Evaporators

One of the very attractive developments in the desalination industry is the use of plate evaporators instead of the shell and tube configuration. Although, it have been reported that scale formation in plate evaporators requires frequent and time consuming unit cleaning and dismantling. Such operational difficulties necessitate further investigative research to reduce the scale formation problem, where plate design is configured to allow for an on-line cleaning processes with sponge balls or other devices.

The plate evaporators/condensers can be manufactured of metals, plastics, or polymer coated metals. The plate heat exchangers have lower hold-up volume, closer temperature approach, smaller weight, smaller space requirement, higher heat transfer coefficient, and lower fouling resistance. Such enhancements vary over a range of 5-70%. The plate evaporators/condensers operate over the following ranges: temperatures of 35-150 °C, pressures of 10-15 bar, plate areas of 0.02-4.45 m^2/plate, flow rates of 3500 m^3/hr, and approach temperatures of 1 °C. On commercial scale, the most common material used for plate evaporators/condensers is stainless steel, however, plastic units are available with heat transfer areas between 5-100 m^2, operating pressures and temperatures up to 10 bars and 100 °C.

Review Questions

1. The thermal performance ratio for a single stage evaporator is always less one. Analyze the model equation for the thermal performance ratio to determine reasons causing this special characteristic.
2. Give reasons for the increase in the system performance ratio as the temperature of the heating steam is increased.
3. Why the heat transfer area in the evaporator decreases with the increase in the brine boiling temperature.
4. What factors affect the heat transfer area of the down condenser?
5. What is the effect of seasonal variations on the intake seawater temperature on the performance of the down condenser and the desalination unit? What should be done during operation to keep the unit operation unaffected by such variations?

Problems

1. A single-effect evaporator generates a distillate product at a flow rate of 1 kg/s. The system operating temperatures are as follows:
   - The boiling temperature, T_b, is 90 °C.
   - The feed temperature, T_f, is 85 °C.
   - The steam temperature, T_s, is 102 °C.

   Determine the heat transfer areas in the evaporator and the condenser, the thermal performance ratio, the flow rates of feed seawater and reject brine, and the flow rate of cooling seawater. Assume that the specific heat of seawater is constant and equal to 4.2 kJ/kg °C.

2. Repeat problem 1 and use the specific heat correlation given in appendix C for seawater. Compare the results against those obtained in problem 1. Comparison should include the relative percentage difference of the two sets of results.
3. A single stage evaporator has a thermal load, Q_e, of 2355 kJ/s and the heating steam temperature is 105 °C. The seawater temperature, T_{cw}, is 15 °C and the feed seawater temperature, T_f, is less than the boiling temperature, T_b, by 5 °C. If the distillate product flow rate is 1 kg/s calculate the boiling temperature, the heat transfer areas, and the flow rate of the cooling seawater. Assume the specific heat for the brine is constant and equal to 4.2 kJ/kg °C.
4. The heat transfer area in the evaporator and condenser for a single stage evaporator is 90 and 30 m^2, respectively. The system is designed to operate at temperature of 85 °C for the boiling brine and an intake seawater temperature of 15 °C. Calculate the feed seawater temperature, the heating steam temperature, the steam flow rate, the cooling seawater flow rate, and the system performance ratio.
5. If the seawater temperature drops to 5 °C in problem 4, determine this effect on the system thermal performance ratio. Note that the heat transfer area remains constant as well as the heating steam temperature and the brine boiling temperature.
6. Reanalyze problem 5 by varying the steam temperature to achieve the same performance ratio in problem 4. Also, introduce a mixer unit, in which the cooling seawater stream, M_{cw}, is mixed with the intake seawater in order to raise its temperature to the design value of 20 °C. Which of the two schemes is more feasible in actual operation, i.e., increasing Ts or use of a mixer recycle unit?

Chapter 3

Single Effect Evaporation – Vapor Compression

Objectives

This chapter focuses on evaluation of the single effect evaporation system combined with various types o f heat pumps. The evaluation in the following:

- Process description.
- Model development.
- Performance evaluation.

The systems considered in this chapter include thermal, mechanical, absorption, and adsorption vapor compression.

3.1 Single Effect Thermal Vapor Compression

The single-effect thermal vapor-compression desalination process is of very limited use on industrial scale. However, thermal vapor compression is used with the MEE system, which is known as MEE–TVC. The thermal vapor compression method is attractive due to its simple operation, inexpensive maintenance, simple geometry, and absence of moving parts. Modeling, simulation, and analysis of the single-effect evaporation unit forms the basis for studying of the MEE system and the MEE combined with vapor compression. The following sections include description of the process elements, the steady-state mathematical model for the TVC system, solution method, examples, and system performance as a function of the design and operating parameters. The mathematical model for the process is previously developed by El-Dessouky, 1997

3.1.1 Process Description

Single effect thermal vapor compression (TVC) seawater desalination process in its simple form is illustrated schematically in Fig. 1. The main components of the unit are the evaporator, the steam jet ejector, and the feed heater or the condenser. The evaporator consists of an evaporator/condenser heat exchanger, a vapor space, a water distribution system, and a mist eliminator. On the other hand, the steam jet ejector is composed of a steam nozzle, a suction chamber, a mixing nozzle, and a diffuser. The feed heater or the heat sink unit is usually a counter-current surface condenser in which the non-condensable gases leave at a temperature approaching the temperature of the feed water. This permit the cooling of the non-condensable gases to the minimum possible temperature, thereby, minimizing the amount of vapor that may escape with the gases and decreases the volume of pumped gases. In addition, it is possible to operate the counter-current condenser so that the exit water is within 3 to 5 °C of the condensation temperature of the saturated vapor. This improves the thermal performance of the unit and minimizes the mass flow rate of cooling water.

The intake seawater at a flow rate of (M_{cw}+M_f) at temperature T_{cw} and salt concentration X_f is introduced into the tube side of the condenser where its temperature increases to T_f. The cooling water M_{cw} is dumped back to the sea. The function of circulating the cooling water in the condenser is the removal of the excess heat added to the system in the form of motive steam necessary to drive the steam jet ejector. It is important to emphasize that the evaporator does not consume the supplied heat, instead, it simply degrades its quality. The heating of the feed seawater M_f in the condenser from T_{cw} to T_f is essential to increase the thermal performance of the process. The heat needed to warm the seawater inside the condenser is supplied by condensing a controlled portion of vapor formed by boiling in the evaporator M_c. The vapor condensation temperature and consequently the pressure in the vapor space for both the evaporator and the condenser is controlled by

- The cooling water flow rate, M_{cw}.
- Feed water temperature, T_{cw}.
- The available heat transfer area in the condenser, A_c.
- The overall heat transfer coefficient between the condensing vapor and the circulating seawater, U_c.

Accordingly, the condenser has three functions: (1) remove excess heat from the system, (2) improve the process performance ratio, PR, and (3) adjust the boiling temperature inside the evaporator.

The feed seawater M_f is chemically treated and deaerated before being pumped to the evaporator. The chemical treatment is needed to prevent the foaming and the tendency for scale formation in the evaporator. Both factors may seriously impair unit operation. Within, the evaporator, the feed water at T_f is sprayed at the top where it falls in the form of thin film down the succeeding rows of tubes arranged horizontally. The feed water temperature is raised from T_f to the boiling temperature T_b. The magnitude of T_b is dictated by the nature of chemicals used to control the scale formation and the state of the heating steam. This temperature is mastered through settling the pressure in the vapor space of the evaporator. The vapor formed by boiling with a rate of M_d is free of salts. The temperature of the generated vapor T_v is less than the boiling temperature T_b by the boiling point elevation (BPE). The vapor generated therein flows through a knitted wire mist separator known as wire mesh demister to remove the entrained brine droplets. The vapor should be completely freed from brine droplets to prevent the contamination of both the product water and the heat transfer surfaces on which it condenses. Also, the presence of entrained water droplets with the vapor flowing into the steam jet ejector will erode the ejector nozzle and diffuser. The saturation temperature of the vapor departing the demister is lower than T_v due to the temperature depression because of the frictional pressure loss in the demister. The vapor flows from the demister flows to the condenser where it splits into two portions, the first part M_c condenses outside the tubes of the condenser while the rest M_{ev} is entrained by the steam jet ejector. Although the two streams are drawn separately in the flow diagram, to show the process, they flow from the evaporator to the condenser in the same pipeline. The non-condensable gases accumulated in the vapor space of the condenser must be vented to avoid the downgrading of the heat transfer capacity of the condenser. The blanket of non-condensable gases masks some of the heat transfer area from condensing operation. If the condenser operates at a pressure less than the atmospheric pressure, a pumping device such as an ejector or a vacuum pump is needed to draw off the vent gases from the system. It is worth mentioning that parts of the process description are similar to those given in the Chapter 2. However and as discussed in the preface, the repetition provides the reader with a complete picture for each pocess.

The schematic diagram for the steam jet thermo-compressor or steam booster with its corresponding state points and the variation in both the velocity and the pressure for the motive and entrained vapor through the ejector are shown in Fig. 2. The ejector is used to increase the pressure of the entrained vapor M_{ev} from pressure P_{ev} to a higher pressure P_s. This process takes place through converting the pressure energy of motive steam M_m to generate vacuum and compress the entrained vapor to the required pressure. As the motive steam at flow rate of M_m expands in the nozzle from state 1 to state 2, its static pressure energy is converted to kinetic energy. The nozzle is a converging/diverging shape to expand the steam to velocities greater than the speed of sound (supersonic). The suction chamber is used to keep the nozzle properly positioned with respect to the diffuser and to direct the entrained vapor. The entrained vapor M_{ev} enters the suction chamber at pressure P_{ev} where it mixes with the motive steam. The mixing process is violent and rapid. The two streams mix together as they pass through the converging section of the venturi diffuser. The mixture enters the throat section of the diffuser, completely mixed, at the sonic velocity of the mixture. The mixed stream is self compressed through the diverging section of the venturi diffuser, where the cross sectional area increases and the velocity decreases, converting the kinetic energy of the mixture to static pressure energy. The mixture leaves the ejector at a pressure P_s that is intermediate to the motive (P_m) and suction (P_{ev}) pressures.

The steam jet ejector must be designed and operated at critical conditions to allow normal and stable operation. This condition is associated with absence of violent fluctuations in the suction pressure. If the ejector is designed to operate with a full stable range, it will have a constant mass flow rate of the entrained vapor for different discharge pressures when the upstream conditions remain constant. The ejector is critical when the compression ratio is greater than or equal to the critical pressure ratio of the suction vapor. For water vapor this ratio is 1.81. That is, the suction pressure must be less than 0.55 times the discharge pressure to obtain critical or stable conditions in the steam jet ejector.

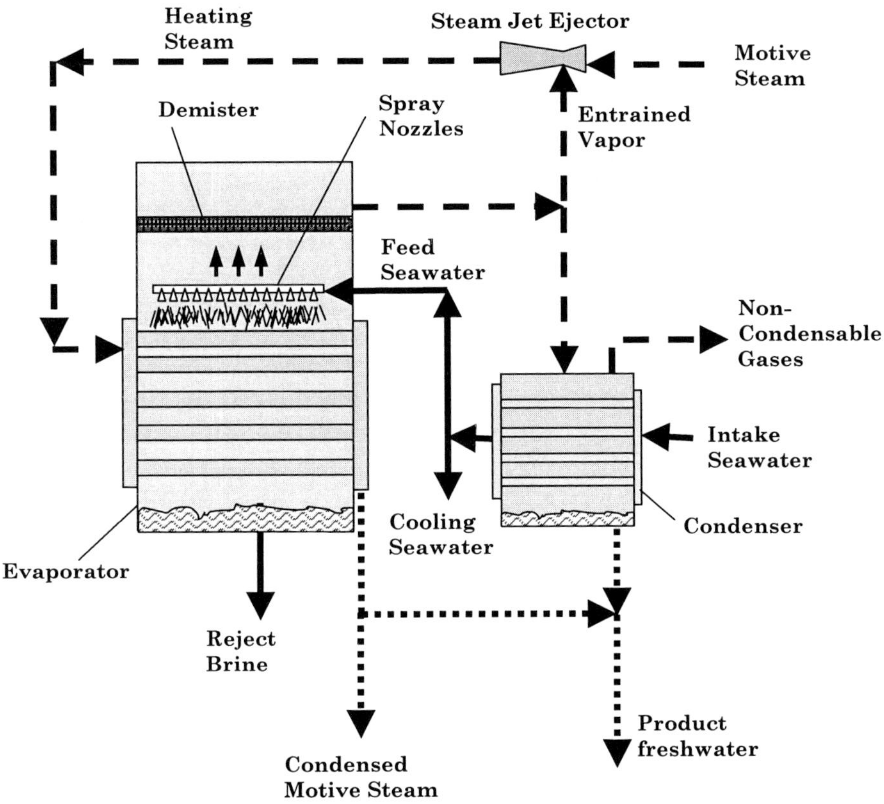

Fig. 1. Single effect thermal vapor compression evaporator-desalination process

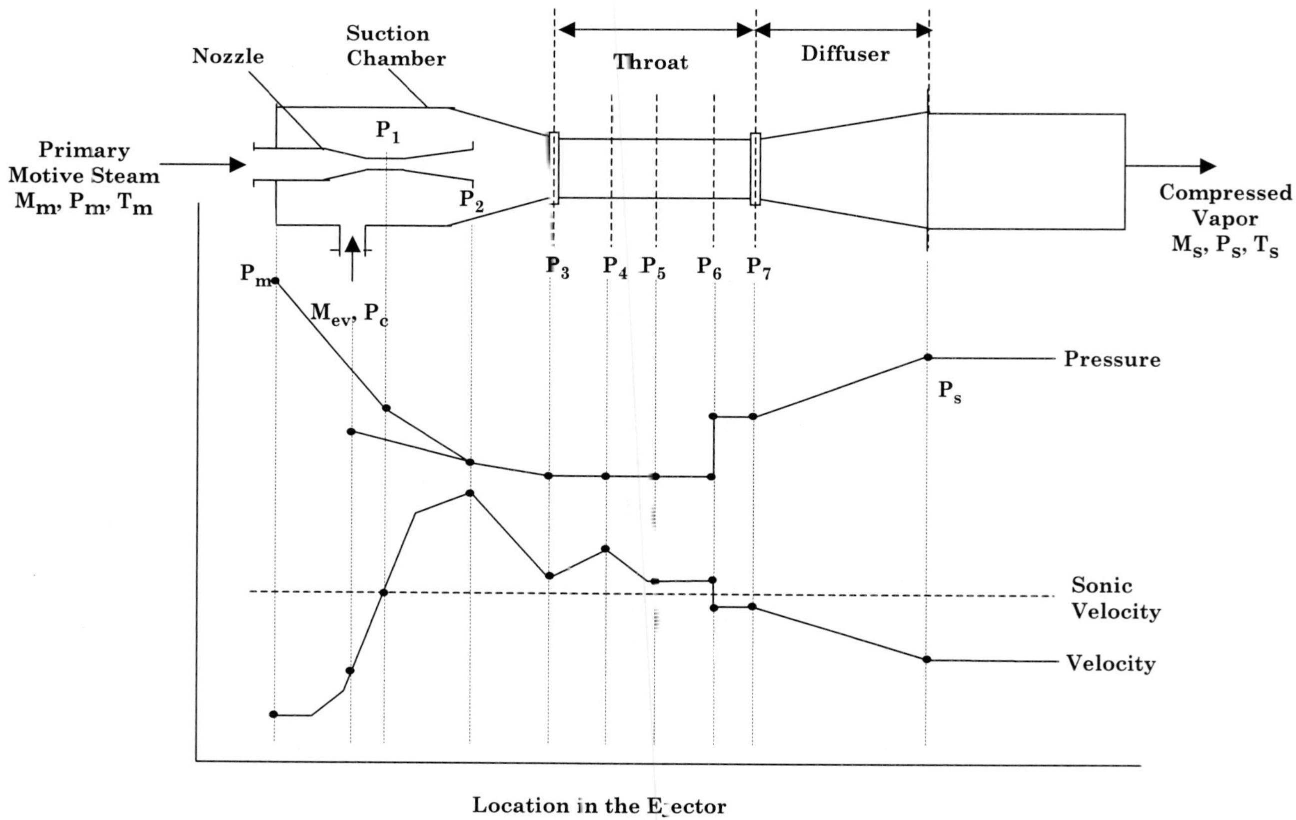

Fig. 2. Variation in stream pressure and velocity as function of location along the ejector.

3.1.2 Process Modeling

Development of the TVC model is divided into seven sections:

- Performance parameters.
- Material balance.
- Evaporator and condenser energy balances.
- Boiling point elevation and thermodynamic losses.
- Evaporator and condenser heat transfer areas.
- Steam jet ejector design equations.

Performance Parameters

Performance of the TVC is determined in terms of the following variables:

- The amount of product fresh water per unit mass of motive steam, or the performance ratio, PR.
- The specific heat transfer surface area, sA.
- The specific cooling water flow rate, sM_{cw}.

The above system parameters are defined by the following relations:

$$PR = \frac{M_d}{M_m} \tag{1}$$

$$sA = \frac{A_e + A_c}{M_d} \tag{2}$$

$$sM_{cw} = \frac{M_{cw}}{M_d} \tag{3}$$

where M is the mass flow rate and the subscript c, cw, d, e, and m denotes the condenser, cooling water, distillate product, evaporator, and motive steam, respectively. The variables A_e and A_c are the heat transfer area in the evaporator and condenser, respectively.

Material Balance

The distillate and rejected brine flow rates are obtained by solution of the overall mass and salt balances. The two balance equations assume that the distillate water is salt free. The two balance equations are given by

$$M_f = M_d + M_b \tag{4}$$

$$\frac{M_d}{M_f} = \frac{X_b - X_f}{X_b} \tag{5}$$

where M is the mass flow rate, X is the salinity, and the subscripts b, d, and f denote the rejected brine, distillate, and feed seawater.

Evaporator and Condenser Energy Balances

In the evaporator, the dry saturated steam flowing from the steam jet ejector and admitted into the evaporator (M_m + M_{ev}) is used in to raise the temperature of the feed seawater M_f from the inlet temperature T_f to the boiling temperature T_b. In addition, it supplies the latent heat required to evaporate the specified mass of vapor, M_d, or:

$$Q_e = M_f\, C_p\, (T_b - T_f) + M_d\, \lambda_v = (M_m + M_{ev})\, \lambda_s \tag{6}$$

where Q_e is the thermal load of the evaporator, C_p is the specific heat at constant pressure of the brine, and λ is the latent heat of evaporation. Correlations for C_p and λ are given in Appendix A.

The condenser operates on the remaining fraction of vapor formed in the evaporator, M_c, which is not entrained by the steam jet ejector. The condensation latent heat is transferred to the feed seawater with a mass flow rate of M_f + M_{cw}. The fraction M_f of the seawater feed is introduced into the evaporator, while the remaining part, M_{cw}, which is known as the cooling water, is rejected back to the sea. The feed seawater temperature is assumed equal to 25 °C. As for the feed vapor it is assumed saturated at a temperature equal to T_c, which is lower than the boiling temperature, T_b, by the boiling point elevation and thermodynamic losses.

The heat load of the condenser is given by

$$Q_c = (M_f + M_{cw})\, C_p\, (T_f - T_{cw}) = M_c \lambda_c \tag{7}$$

where Q_c is the thermal load of the condenser. The subscript c, cw, and f denote the condenser, cooling seawater, and un-entrained vapor.

Boiling Point Elevation and Thermodynamic Losses

The generated vapor is at the saturation temperature, T_v, which corresponds to the pressure in the evaporator vapor space. This temperature is less than the boiling temperature T_b by the boiling point elevation BPE, where,

$$T_b = T_v + BPE \tag{8}$$

The boiling point elevation (BPE), at a given pressure, is the increase in the boiling temperature due to the salts dissolved in the water. Correlation for the boiling point elevation of seawater is given Appendix B.

The condensation temperature of vapor outside the tube bundle of the condenser T_c is less than the boiling temperature in the evaporator T_b by the boiling point elevation (BPE) and the saturation temperature depression associated with pressure losses in the demister (ΔT_p) and inside the condenser horizontal tubes (ΔT_c). Thus:

$$T_c = T_b - (BPE + \Delta T_p + \Delta T_c) \tag{9}$$

The correlation for the pressure drop in the demister is given in Appendix B. As for the pressure drop of the vapor flowing over the condenser tubes it is assumed has a negligible value. This is the pressure recovery due to flow deceleration compensates the pressure drop caused by friction. Therefore, the net pressure fall and consequently the saturation temperature depression in the condensation process can be neglected, Marto (1991), Muller (1991), Sinnott (1996).

Evaporator and Condenser Heat Transfer Area

The dimensions of the required heat transfer surface area in the evaporator A_e are obtained from:

- The amount of the heat to be transferred Q_e.
- The overall heat transfer coefficient U_e.
- The difference between the condensation temperature of the steam, T_s, and the boiling temperature of the seawater T_b.

This relation is given by

$$A_e = \frac{Q_e}{U_e(T_s - T_b)} \tag{10}$$

The heating surface area of the evaporators A_e is usually, but not always, taken as that in contact with the boiling liquid, whether on the inside or outside of the tubes.

The heat transfer between the condensing vapor and the feed water in the condenser can be written in terms of an overall heat transfer coefficient (U_c), condenser heat transfer area (A_c), and the logarithmic mean temperature difference $(LMTD)_c$, thus:

$$A_c = \frac{Q_c}{U_c (LMTD)_c} \tag{11}$$

The $(LMTD)_c$ is defined as:

$$(LMTD)_c = \frac{T_f - T_{cw}}{\ln \dfrac{T_c - T_{cw}}{T_c - T_f}} \tag{12}$$

In Eqs. 11 and 12 the overall heat transfer coefficient is based on the outside surface area and is related to the individual thermal resistance by the following well-known expression:

$$\frac{1}{U_e} = \frac{1}{h_i}\frac{r_o}{r_i} + R_{f_i}\frac{r_o}{r_i} + \frac{r_o \ln(r_o / r_i)}{k_w} + R_{f_o} + \frac{1}{h_o} \tag{13}$$

where h is the heat transfer coefficient, R_f is the fouling resistance, k_w is the thermal conductivity of tube material and r is the radius. The subscripts i and o refer to the inner and outer tube surfaces, respectively. Correlations for the evaporator and condenser heat transfer coefficients are given in the Appendix C.

Steam Jet Ejector

The most important and critical step in modeling the TVC desalination system is the evaluation of the performance of the steam jet ejector. The main data required from analyzing the steam jet ejector is the determination of the mass of motive steam required per unit mass of the entrained vapor (Ra), given the pressure of the motive steam (P_m), discharge pressure (P_s) and the suction pressure (P_{ev}). There are a limited number of methods available in the literature to analysis the steam jet ejector. However, these methods require tedious and lengthy calculation procedures. Additionally, most of these methods are based on using many correction factors that depend heavily on the detail design of the ejector. The technique developed here is established on the data and method presented by Power, 1994. Power found that none of procurable ways were superior to his simple method. The method is most accurate for motive steam pressures above 5.1 bar and low compression ratios associated with (Ra) values less than 4. The curves used in the calculations represent smoothed data from several sources and agree with manufacturer's data within the about 10% over the best-fit range. El-Dessouky, 1997, developed the following relationships to evaluate the performance of the steam jet ejector. The entertainment ratio is defined by:

$$Ra = 0.296 \frac{(P_s)^{1.19}}{(P_{ev})^{1.04}} \left(\frac{P_m}{P_{ev}}\right)^{0.015} \left(\frac{PCF}{TCF}\right) \qquad (14)$$

where Ra is the entrainment ratio and defined as the mass of motive steam per unit mass of entrained vapor, P_m, P_s and P_{ev} are the pressures of the motive steam, discharge mixture and entrained vapor respectively, PCF is motive steam pressure correction factor and TCF is the entrained vapor temperature correction factor. The following two equations are developed to calculate both PCF and TCF.

$$PCF = 3x10^{-7} (P_m)^2 - 0.0009 (P_m) + 1.6101 \qquad (15)$$

$$TCF = 2x10^{-8} (T_{ev})^2 - 0.0006 (T_{ev}) + 1.0047 \qquad (16)$$

where P_m is in kPa and T_{ev} is in °C. The previous equations are valid only for ejector operating with steam as the motive fluid and the entrained gas is water vapor. These equations are valid in the following ranges: Ra $\leq$ 4, 500 $\geq T_{ev} \geq$ 10 °C, 3500 $\geq P_m \geq$ 100 kPa, and $\frac{P_s}{P_{ev}} \geq 1.81$.

It is interesting to realize that the consideration of the thermodynamic losses such as BPE, and temperature depression corresponding to the pressure drops in the demister increases in the energy demand for the jet ejector. This is because the vapor must be compressed, not simply through the working temperature drop ($T_s - T_b$), but through the working temperature drop plus the thermodynamic losses, i.e., $\{T_s - [T_b - (BPE + \Delta T_p)]\}$, or ($T_s - T_{ev}$).

Solution Procedure

The following set of specifications are used in solution of the TVC system:

- The seawater temperature.
- The feed water temperature.
- The seawater salinity.
- The salinity of the rejected brine.
- The thickness of the demister pad.

- The vapor velocity in the demister.
- The density of the demister pad.
- Boiling temperature.
- Ejector compression ratio.
- Motive steam pressure.

The solution procedure is shown in Fig. 3 and proceeds as given below:

- The mass flow rates of the reject brine and feed seawater, M_b and M_f, for a specified distillate water flow rate, M_d, of 1 kg/s, are calculated from Eqs. 4 and 5.
- The boiling point elevation, BPE, is calculated from the correlation in Appendix B.
- The saturation temperature T_v is calculated from Eq. 8 and the corresponding saturation pressure, P_v, is obtained from the steam tables or calculated from the correlation given in Appendix (A).
- The pressure drop in the demister (ΔP_p) is calculated from the correlation given in Appendix (B). This value is used to calculate the vapor pressure past the demister, P_{ev}, which is equal to $P_v - \Delta P_p$.
- The saturation vapor temperature, T_{ev}, is calculated at the saturation vapor pressure, P_{ev}, from the steam tables or the saturation temperature correlation given in Appendix A.
- The compressed vapor pressure, P_s, is obtained from the specification of the compression ratio, Cr, and the entrained vapor pressure, P_{ev}. This is followed by calculation of the saturation temperature, T_s, at the corresponding vapor pressure, P_s, from the steam tables or the correlation given in Appendix A.
- The evaporator thermal load, Q_e, is calculated from Eqs. 6.
- The overall heat transfer coefficient in the evaporator, U_e, is calculated from the correlation given in Appendix C.
- The overall heat transfer coefficient in the condenser, U_c, is calculated from the correlation given in Appendix C.
- The entrainment ratio, Ra, is calculated from Eqs. 14-16.
- The mass flow rates of the motive steam and entrained vapor, M_s and M_{ev}, are obtained by substituting the values of the entrainment ratio, Ra, and the evaporator load, Q_e, in Eq. 6.
- The condenser load, Q_c, and the cooling water flow rate, M_{cw}, are obtained from the condenser energy balance, Eq. 7.
- The evaporator and condenser areas, A_e and A_c, are obtained from Eqs. 10 and 11.
- The system performance parameters, PR, sA, and sM_{cw}, are calculated from Eqs. 1-3.

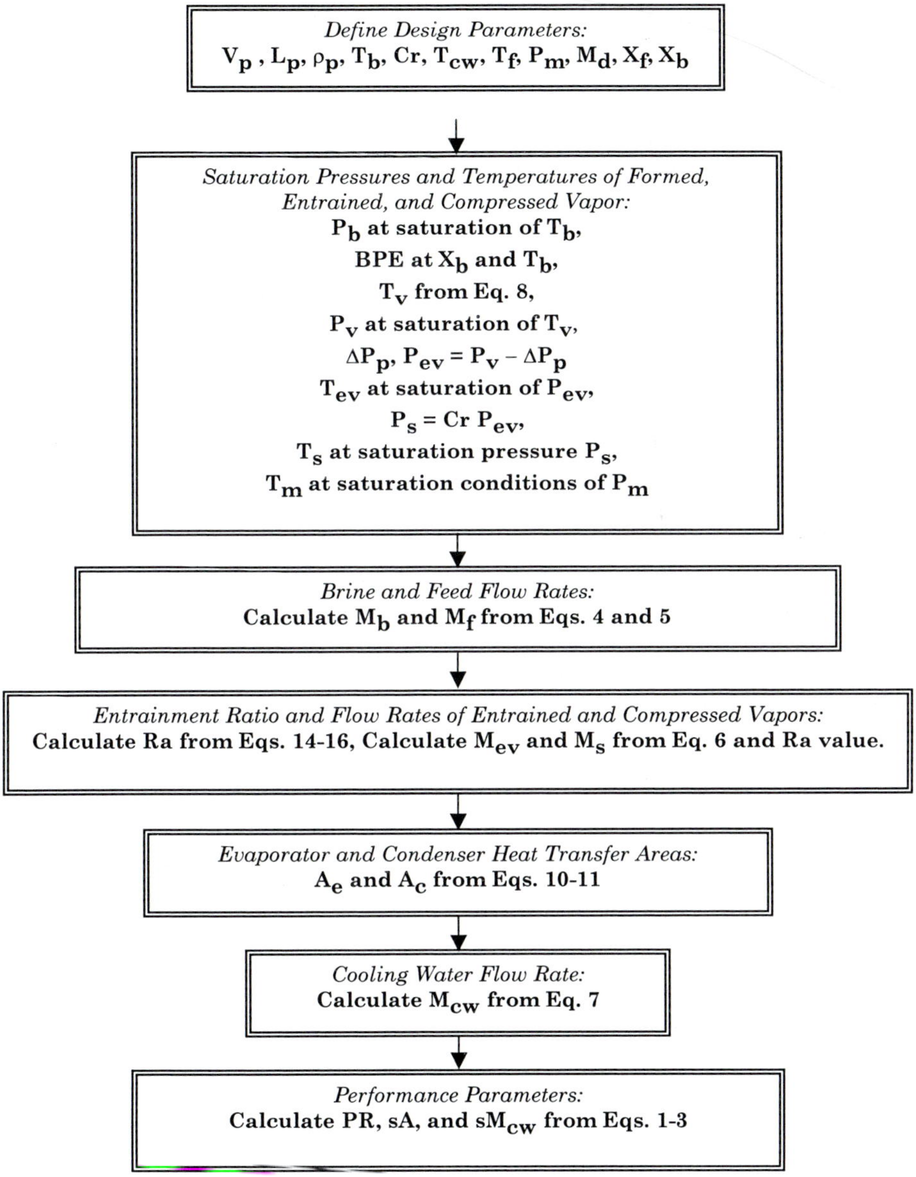

Fig. 3. Solution procedure of the TVC mathematical model

3.1.3 System Performance

Evaluation of the TVC system is illustrated in the following examples. The first includes a design problem to determine the specific heat transfer area, the flow rate of the cooling water, and the performance ratio. The second example rates the performance of an existing system, where the heat transfer area of the evaporator and the condenser are known.

Example 1:

A single-effect thermal vapor-compression system is designed at the following operating conditions:

- Boiling temperature, T_b, of 75 °C.
- Compression ratio, Cr, of 2.5.
- Motive steam pressure, P_m, of 750 kPa.
- Brine reject concentration, X_b = 70000 ppm
- Intake seawater salinity, X_f = 42000 ppm
- Intake seawater temperature, T_{cw} = 25 °C
- System capacity, M_d = 1 kg/s
- Boiling temperature, T_b = 75 °C
- Feed seawater temperature, $T_f = (T_b - 5) = 70$ °C
- Condenser efficiency, $\eta = 0.9$.

Solution

Substituting for X_f = 42000 ppm, X_b = 70000 ppm, and M_d = 1 kg/s in Eq. 4 results in

$$M_f = X_b/(X_b - X_f) = 70000/(70000 - 42000) = 2.5 \text{ kg/s}$$

Equation 4 is then used to calculate M_b,

$$M_b = M_f - M_d = 2.5 - 1 = 1.5 \text{ kg/s}$$

The boiling point elevation, BPE, is calculated from the correlation given in Appendix B. The values of B and C are evaluated from

$$\text{BPE} = (0.0825431 + 0.0001883\,(75) + 0.00000402\,(75)^2)\,(7) \\ +(-\,0.0007625 + 0.0000902\,(75) - 0.00000052\,(75)^2)\,(7)^2$$

$+(0.0001522\text{-}0.000003\ (75) - 0.00000003\ (75)^2)\ (7)^3$
$= 0.903\ ^{o}C$

The resulting value of T_v is calculated from Eq. 8,

$$T_v = T_b - BPE = 75 - 0.903 = 74.097\ ^{o}C$$

The corresponding saturation vapor pressure, P_v, is obtained from the correlation given in Appendix A,

$$\begin{aligned}P_v = EXP((&-7.419242 + (0.29721)\\&((0.01)(74.097+273.15-338.15))\\&-0.1155286\ ((0.01)\ (74.097+273.15-338.15))^2\\&+0.008685635\ ((0.01)\ (74.097+273.15-338.15))^3\\&+0.001094098\ ((0.01)\ (74.097+273.15-338.15))^4\\&-0.00439993\ ((0.01)\ (74.097+273.15-338.15))^5\\&+0.002520658\ ((0.01)\ (73.01+273.15-338.15))^6\\&-0.0005218684\ ((0.01)\ (74.097+273.15-338.15))^7)\\&((647.286/(74.097+273.15))-1))(22089000/1000)\\= 37.1\ kPa&\end{aligned}$$

The pressure drop in the demister is evaluated from the correlation given in Appendix B. In this equation ρ_p, V, L, and δ_w are set equal to 300 kg/m^3, 1.8 m/s, 0.1 m, and 0.28 mm. This results in

$$\Delta P_p = 3.88178\ (\rho_p)^{0.375798}\ (V)^{0.81317}\ (\delta_w)^{-1.56114147}$$
$$= 390\ Pa/m$$

Which gives a total pressure drop of 0.039 kPa through the demister. The vapor pressure past the demister is then calculated

$$P_{ev} = P_v - \Delta P_p = 37.1 - 0.039 = 37.061\ kPa$$

Therefore, the vapor saturation temperature past the demister, T_{ev}, is assumed equal to the saturation temperature, T_v. Another assumption applied here is the equality of the vapor condensation temperature, T_c, in the condenser and the vapor temperature in the evaporator, T_v.

The specified value for the compression ratio, Cr, and the entrained pressure value, P_{ev}, are used to calculated pressure of the compressed vapor, P_s, which is

$P_s = (Cr)\ (P_{ev}) = (2.5)\ (37.1) = 92.75\ kPa$

The corresponding saturation temperature, T_s, is calculated from the correlation for the saturation temperature given Appendix A,

$$T_s = \left(42.6776 - \frac{3892.7}{(\ln(P_s/1000) - 9.48654)}\right) - 273.15$$

$$= \left(42.6776 - \frac{3892.7}{(\ln(92.75/1000) - 9.48654)}\right) - 273.15$$

$$= 97.6\ ^oC$$

The overall heat transfer coefficients in the evaporator and condenser are calculated from the correlations given in Appendix C

$$U_e = \begin{pmatrix} 1969.5 + 12.057\,(T_b) - 0.85989x10^{-1}\,(T_b)^2 \\ + 0.25651x10^{-3}\,(T_b)^3 \end{pmatrix} x10^{-3}$$

$$= \begin{pmatrix} 1969.5\ + 12.057\,(75) - 0.85989x10^{-1}\,(75)^2 \\ + 0.25651x10^{-3}\,(75)^3 \end{pmatrix} x10^{-3}$$

$$= 2.62\ kW/m^2\ ^oC$$

$$U_c = \begin{pmatrix} 1719.4 + 3.2063\,(T_c) + 1.5971x10^{-2}\,(T_c)^2 \\ -1.9918x10^{-4}\,(T_c)^3 \end{pmatrix} x10^{-3}$$

$$= \begin{pmatrix} 1719.4 + 3.2063\,(74.097) + 1.5971x10^{-2}\,(74.097)^2 \\ -1.9918x10^{-4}\,(74.097)^3 \end{pmatrix} x10^{-3}$$

$$= 1.96\ kW/m^2\ ^oC$$

The entrainment ratio, Ra, is obtained from Eq. 14. This requires calculations of the correction factors, PCF and TCF, from Eqs. 15 and 16. These results are

$$PCF = 3x10^{-7}\ (P_m)^2 - 0.0009\ (P_m)\ + 1.6101$$
$$= 3x10^{-7}\ (750)^2 - 0.0009\ (750)\ + 1.6101$$
$$= 1.104$$

$$TCF = 2x10^{-8}\ (T_c)^2 - 0.0006\ (T_c) + 1.0047$$
$$= 2x10^{-8}\ (74.097)^2 - 0.0006\ (74.097) + 1.0047$$

$= 0.96$

$$Ra = 0.296 \frac{(P_s)^{1.19}}{(P_{ev})^{1.04}} \left(\frac{P_m}{P_{ev}}\right)^{0.015} \left(\frac{PCF}{TCF}\right)$$

$$= 0.296 \frac{(92.75)^{1.19}}{(37.1)^{1.04}} \left(\frac{750}{37.1}\right)^{0.015} \left(\frac{1.104}{0.96}\right) = 1.82$$

The amount of motive steam is obtained by solution of the evaporator balance, Eq. 6. This gives

$$M_f\,Cp\,(T_b\text{-}T_f)+M_d\,\lambda_v = (M_s+M_{ev})\,\lambda_s$$

In the above equation, λ_v and λ_s, are calculated from the correlation given in Appendix A. T_s (96.46)

$$\lambda_v = 2501.897149 - 2.407064037\,T_v + 1.192217\text{x}10^{-3}\,T_v^{\,2} - 1.5863\text{x}10^{-5}\,T_v^{\,3}$$
$$= 2501.897149 - 2.407064037(74.097) + 1.192217\text{x}10^{-3}\,(74.097)^2 - 1.5863\text{x}10^{-5}\,(74.097)^{\,3}$$
$$= 2323.6\ \text{kJ/kg}$$

$$\lambda_s = 2501.897149 - 2.407064037\,T_s + 1.192217\text{x}10^{-3}\,T_s^{\,2} - 1.5863\text{x}10^{-5}\,T_s^{\,3}$$
$$= 2501.897149 - 2.407064037(97.6) + 1.192217\text{x}10^{-3}\,(97.6)^2 - 1.5863\text{x}10^{-5}\,(97.6)^{\,3}$$
$$= 2263.6\ \text{kJ/kg}$$

The heat capacity is calculated at T_b and X_f from the correlation given Appendix A and its value is equal to 3.86 kJ/kg °C. Substitution of the values for T_b, T_f, M_d, M_f, Cp, λ_v and λ_s in Eq. 6 gives

$$(2.5)(3.86)\,(75-70) + (1)(2326.37) = (M_m + M_{ev})(2266.76)$$

M_m/Ra replaces the amount of entrained vapor, M_{ev}, which results in

$$(2.5)(3.99)(75-70) + (1)(2323.6) = (M_m + M_m/1.82)\,(2263.6)$$

Solving the above equation gives M_m = 0.67 kg/s. The amount of entrained vapor, M_{ev}, is then calculated from the entrainment ratio value,

$M_{ev} = M_m/Ra = 0.67/1.82 = 0.37$ kg/s

The cooling water flow rate is obtained from the condenser balance, Eq. 7.

$(M_f + M_{cw})\,(Cp)\,(T_f - T_{cw}) = (\eta)\,(M_d - M_{ev})\,(\lambda_c)$

$(2.5 + M_{cw})\,(3.97)\,(70-25) = (0.9)\,(1-0.37)\,(2323.6)$

Solution of the above equation gives M_{cw} = 4.83 kg/s. The evaporator and condenser loads are obtained from Eqs. 6 and 7, respectively. The resulting values are:

$Q_e = (M_m + M_{ev})(\lambda_s) = (0.67 + 0.37)(2263.6) = 2354.1$ kW

$Q_c = \eta\,(M_c)(\lambda_c) = (0.9)(0.63)(2323.6) = 1317.5$ kW

In the condenser load the value of λ_c is identical to λ_v. This is because of neglecting various forms of thermodynamic losses caused by pressure drop and during condensation. The evaporator and condenser areas are then calculated from Eqs. 10 and 11.

$$A_e = \frac{Q_e}{U_e(T_s - T_b)} = \frac{2354.1}{(2.62)(97.6-75)} = 39.8 \text{ m}^2$$

$$A_c = \frac{Q_c}{(U_c)(LMTD)_c} = \frac{1317.5}{(1.96)(16.25)} = 41.18 \text{ m}^2$$

The $(LMTD)_c$ value in the condenser is calculated from Eq. 12

$$(LMTD)_c = \frac{(T_f - T_{cw})}{\ln \frac{T_c - T_{cw}}{T_c - T_f}} = \frac{70-25}{\ln \frac{73.01-25}{73.01-70}} = 16.25\ ^{o}\text{C}$$

Since the distillate flow rate is set at 1 kg/s, the above values for M_{cw}, A_c, and A_e are the specific values. The performance ratio is calculated from Eq. 1, which gives

$PR = M_d/M_m = 1/0.675 = 1.48$

and the resulting specific cooling water flow rate, M_{cw}, and the specific heat transfer area, sA, given by Eqs. 2 and 3, $sM_{cw} = M_{cw}/M_d = 4.83$, and $sA = (A_e+A_c)/M_d = 44.47 + 41.18 = 85.65\ m^2/(kg/s)$.

Example 2:

The heat transfer areas for the evaporator and condenser of in a single-effect thermal vapor-compression system are 37.1 m^2 and 54.8 m^2, respectively. The boiling temperature in the evaporator is 65 °C. The steam jet ejector operates at a compression ratio of 3.2 and a motive steam pressure of 550 kPa. Other data includes a salinity of 70,000 ppm for the brine reject, a feed seawater salinity of 42,000 ppm, and a seawater temperature of 25 °C. The feed seawater temperature to the evaporator is less than the boiling temperature by 5 °C. Assume that thermodynamic losses in the demister, transmission lines, and during condensation have negligible effects on the system. Also, assume that the condenser efficiency is equal to 90%. Evaluate the performance ratio of the system, the specific flow rate of cooling water, and the production capacity.

Solution:

The boiling point elevation is evaluated at T_b = 65 °C and X_b = 70,000 ppm, where

$$\begin{aligned} BPE &= (0.0825431+0.0001883\ (65)+0.00000402\ (65)^2)\ (7) \\ &+(-\ 0.0007625+0.0000902\ (65) - 0.00000052\ (65)^2)\ (7)^2 \\ &+(0.0001522-0.000003\ (65) - 0.00000003\ (65)^2)\ (7)^3 \\ &= 0.87\ ^oC \end{aligned}$$

This gives a vapor temperature of

$$T_v = T_b - BPE = 65 - 0.87 = 64.13\ ^oC$$

Invoking the negligible effect of thermodynamic losses gives the following equality

$$T_v = T_{ev} = T_c = 63.074\ ^oC$$

The correlations in Appendix A are used to calculate the corresponding saturation pressure and the latent heat, with values of 22.95 kPa and 2350.86 kJ/kg, respectively. The pressure of the compressed vapor is obtained from

$$Cr = P_s/P_{ev}$$
$$3.2 = P_s/22.95$$

$$P_s = 73.45 \text{ kPa}$$

which gives a saturation temperature of 91.29 °C. Accordingly, the expansion ratio for the ejector is obtained from

$$Er = P_m/P_{ev}$$
$$Er = 550/22.95 = 23.9$$

The overall heat transfer coefficients in the evaporator and condenser are calculated from the correlations given in Appendix C

$$U_e = \begin{pmatrix} 1969.5 + 12.057\,(T_b) - 0.85989\text{x}10^{-1}\,(T_b)^2 \\ + 0.25651\text{x}10^{-3}\,(T_b)^3 \end{pmatrix} \text{x}10^{-3}$$
$$= \begin{pmatrix} 1969.5 + 12.057\,(65) - 0.85989\text{x}10^{-1}\,(65)^2 \\ + 0.25651\text{x}10^{-3}\,(65)^3 \end{pmatrix} \text{x}10^{-3}$$
$$= 2.46 \text{ kJ/s m}^2 \text{ °C}$$

$$U_c = \begin{pmatrix} 1719.4 + 3.2063\,(T_c) + 1.5971\text{x}10^{-2}\,(T_c)^2 \\ - 1.9918\text{x}10^{-4}\,(T_c)^3 \end{pmatrix} \text{x}10^{-3}$$
$$= \begin{pmatrix} 1719.4 + 3.2063\,(63.074) + 1.5971\text{x}10^{-2}\,(63.074)^2 \\ - 1.9918\text{x}10^{-4}\,(63.074)^3 \end{pmatrix} \text{x}10^{-3}$$
$$= 1.935 \text{ kJ/s m}^2 \text{ °C}$$

The entrainment ratio, Ra, is then calculated, where

$$PCF = 3\text{x}10^{-7}\ (P_m)^2 - 0.0009\,(P_m) + 1.6101$$
$$= 3\text{x}10^{-7}\ (550)^2 - 0.0009\,(550) + 1.6101$$
$$= 1.2058$$

3.1.3 System Performance

$$TCF = 2x10^{-8}\ (T_c)^2 - 0.0006\ (T_c) + 1.0047$$
$$= 2x10^{-8}\ (63.074)^2 - 0.0006\ (63.074) + 1.0047 = 0.966$$

$$Ra = 0.296 \frac{(P_s)^{1.19}}{(P_{ev})^{1.04}} \left(\frac{P_m}{P_{ev}}\right)^{0.015} \left(\frac{PCF}{TCF}\right)$$
$$= 0.296 \frac{(73.45)^{1.19}}{(22.95)^{1.04}} \left(\frac{550}{22.95}\right)^{0.015} \left(\frac{1.20585}{0.966}\right) = 2.47$$

The condenser rating gives the following thermal load

$$A_c = \frac{Q_c}{(U_c)(LMTD)_c}$$
$$54.8 = \frac{Q_c}{(1.935)(13.9)}$$
$$Q_c = 1474.24 \text{ kW}$$

In the above equation $(LMTD)_c$ value is obtained from

$$(LMTD)_c = \frac{(T_f - T_{cw})}{\ln \dfrac{T_c - T_{cw}}{T_c - T_f}} = \frac{60 - 25}{\ln \dfrac{63.074 - 25}{63.074 - 60}} = 13.9\ ^{o}C$$

The thermal load of the condenser is then used to calculate the flow rate of the condensed vapor, or,

$$Q_c = \eta\ (M_c)(\lambda_c)$$
$$1474.24 = (0.9)(M_c)(2350.86)$$

This gives M_c = 0.69 kg/s. Similarly, the thermal load of the evaporator is calculated from the rate equation, where,

$$A_e = \frac{Q_e}{U_e(T_s - T_b)}$$
$$37.1 = \frac{Q_e}{(2.46)(91.29 - 65)}$$
$$Q_e = 2400 \text{ kW}$$

The evaporator thermal load together with the latent heat for condensation of the compressed are used to calculate the flow rate of the compressed vapor flow rate:

$Q_e = (M_s)(\lambda_s)$
$2400 = (M_s)\ (2280)$
$M_s = 1.05$ kg/s

Recalling that the entrainment ratio (Ra) is defined as

$Ra = M_m/M_{ev}$

where $M_m = M_s - M_{ev}$. Substitution for the M_m expression in the above equation together with the value of Ra gives the value M_m, or,

$2.47 = (M_s - M_{ev})/M_{ev} = (1.05 - M_{ev})/M_{ev}$

This gives $M_{ev} = 0.303$ kg/s. Therefore

$M_m = 1.05 - 0.303 = 0.747$ kg/s

Therefore, the total system capacity is equal to the sum of M_c and M_{ev}, or, 0.99 kg/s for M_d. Substituting for X_f = 42000 ppm, X_b = 70000 ppm, and M_d = 0.99 kg/s in the balance equations gives

$$M_f = \frac{M_d X_b}{X_b - X_f} = \frac{(0.99)(70000)}{(70000-42000)} = 2.475 \text{ kg/s}$$

$$M_b = M_f - M_d = 2.475 - 0.99 = 1.485 \text{ kg/s}$$

The cooling water flow rate is obtained from the condenser balance, Eq. 7.

$(M_f + M_{cw})\ (Cp)\ (T_f\text{–}T_{cw}) = (\eta)\ (M_d - M_{ev})\ (\lambda_c)$
$(2.475+M_{cw})\ (3.97)\ (60\text{–}25) = (0.9)\ (0.99\text{–}0.303)\ (2350.86)$

Solution of the above equation gives M_{cw} = 7.98 kg/s.

Performance Charts

System performance is presented in terms of variations in the system design parameters as a function of the boiling temperature, T_b, the compression ratio, Cr, and the pressure of the motive steam, P_m. The system parameters include variations in the overall heat transfer coefficient in the evaporator (U_e)

and the condenser (U_c), the performance ratio, PR, the specific heat transfer area, sA, and the specific cooling water flow rate, sM_{cw}.

Variations in the system performance ratio as a function of the boiling temperature, motive steam pressure, and compression ratio are shown in Figs. 4-6. As is shown the system performance ratio varies over a range of 1-2. The higher performance ratio values are obtained at low boiling temperatures, low compression ratios, and high motive steam pressures. At low boiling temperatures, the amount of motive steam consumed to compress the entrained vapor is low. This is because of the small increase in the vapor pressure at low temperatures. For example, the vapor pressure between 55 and 60 °C increases from 15.8 to 19.9 kPa is 26.5%. On the other hand, the vapor pressure increases from 70.14 to 84.55 kPa as the temperature increases from 90 to 95 °C.

At low compression ratios, the amount of motive steam consumed to compress the entrained vapor is small. Therefore, the system performance ratio is higher. The same result also applies at high motive steam pressures. Irrespective of this, the sensitivity of the performance ratio to variations in the motive steam pressure is less pronounced than those found as a function of the boiling temperature and the compression. This result is shown in Fig. 6 with limited variations in the system performance ratio as the motive steam pressure is increased over a range of 250-1750 kPa. For each set of data in Fig. 6, the boiling temperature and the compression ratio are kept constant. At such conditions, the amount of latent heat consumed by the boiling brine is constant, which implies a constant temperature for the compressed vapor. As the pressure of the motive steam is increased its latent is lower. Therefore, to maintain constant latent heat in the compressed vapor, it is necessary to entrain larger amounts of the vapor leaving the evaporator. This reduces the consumed amount of motive steam.

Variations in the specific heat transfer area are shown in Figs. 7-9. The results are shown as function of the boiling temperature, the motive steam pressure, and the compression ratio. As is shown in Fig. 7, the specific heat-transfer area decreases drastically as the boiling temperature is increased. This result is caused by the increase in the overall heat transfer coefficient in the evaporator and the condenser at high boiling temperatures. As the overall heat transfer coefficient increases, because of the decrease in the resistance to heat transfer, the area for heat transfer is decreased. The specific heat transfer area is also decreased at higher boiling temperatures. This is because the amount of distillate formed only depends on the salinity of the feed seawater and the rejected brine.

Similar results are shown in Fig. 8 for variations in the specific heat transfer area as a function of the compression ratio. At constant boiling temperatures and higher compression ratios, the pressure of the compressed vapor is larger. This is because the pressure of the entrained vapor does not change at constant boiling temperatures. Simultaneously, the temperature of the compressed vapor is also increased as the compression ratio is elevated. The increase in the temperature of the compressed vapor enhances the rates of heat transfer. This is caused by the increase of the driving force for heat transfer across the evaporator, which is measured by the difference of T_s-T_b. As a result, the evaporator heat transfer area is reduced at higher compression ratios. Irrespective of this, the heat-transfer area increases in the condenser. This is because of the increase in the condenser load, which is caused by the reduction in the amount of entrained vapor at higher compression ratios. However, the decrease in the evaporator area is more pronounced than the increase in the condenser area. The net result of the above is the decrease in the specific heat transfer area upon the increase of the compression ratio.

Effect of the motive steam pressure on the specific heat transfer area is shown in Fig. 9. The results are obtained at a compression ratio of 1.895 and boiling temperature range of 55-82 °C. These results are similar to those obtained for the variations in the system performance ratio, Fig. 4. As is shown in Fig. 9, the specific heat transfer area is insensitive to variations in the motive steam pressure. This is because of limited variations in the overall heat transfer coefficient in the evaporator and condenser as well as the amount of entrained vapor.

Variations in the specific cooling water flow rate are shown in Figs. 10-12. The results are obtained over the same parameter range as discussed before. As is shown, the specific cooling water flow rate is highly sensitive to variations in the boiling temperature and the compression ratio, Figs. 10 and 11. However, it is insensitive to variations in the motive steam pressure, Fig. 12. This result is consistent with the discussion given for variations in other design parameters, i.e., and performance ratio and specific heat transfer area. Sensitivity of the specific cooling water flow rate with respect to the boiling temperature and the compression ratio is caused by large increase in the overall heat transfer coefficient in the evaporator and condenser. An opposite behavior is observed upon the increase in the motive steam pressure. In this regard, limited sensitivity in the specific cooling water flow rate is found upon the increase in the motive steam pressure.

3.1.3 System Performance

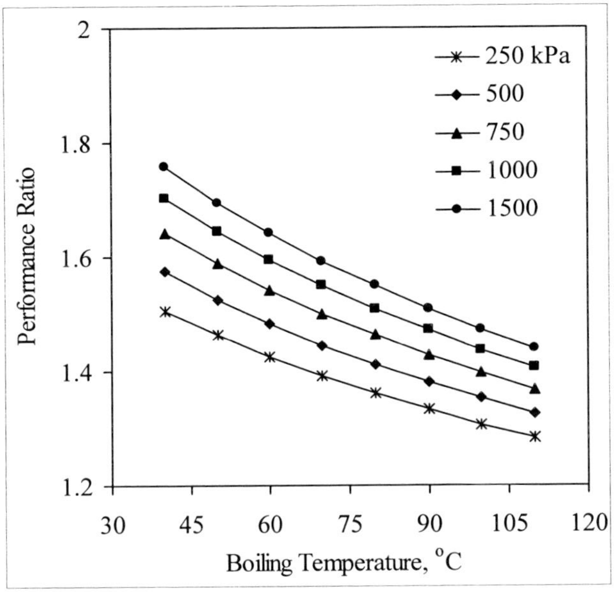

Fig. 4. Variations in the performance ratio as a function of the boiling temperature and the motive steam pressure.

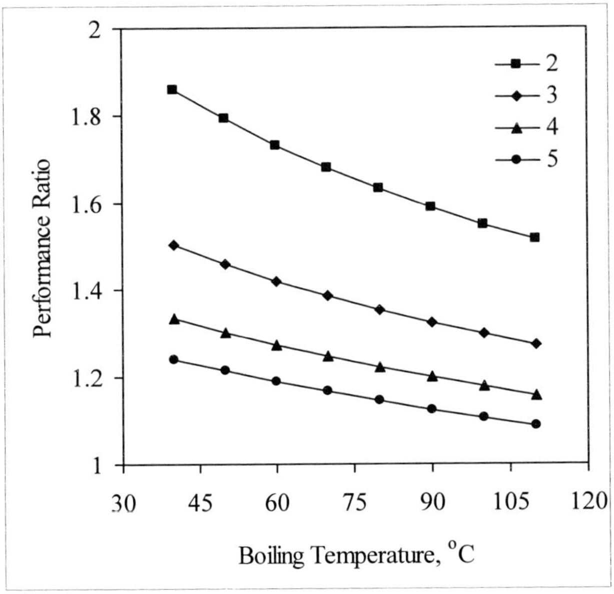

Fig. 5. Variations in the performance ratio as a function of the boiling temperature and the compression ratio.

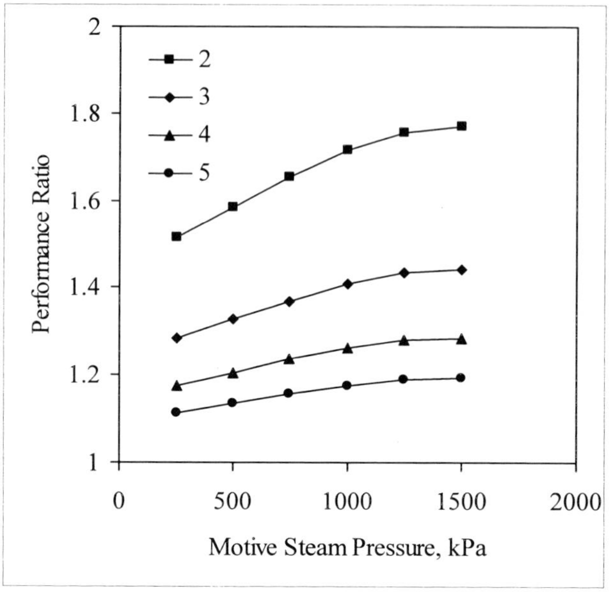

Fig. 6. Variations in the performance ratio as a function of the motive steam pressure and the compression ratio.

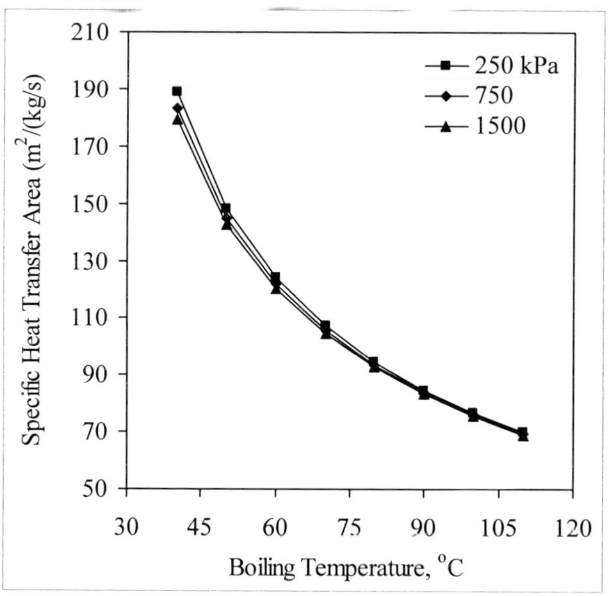

Fig. 7. Variations in the specific heat transfer area as a function of the boiling temperature and the motive steam pressure.

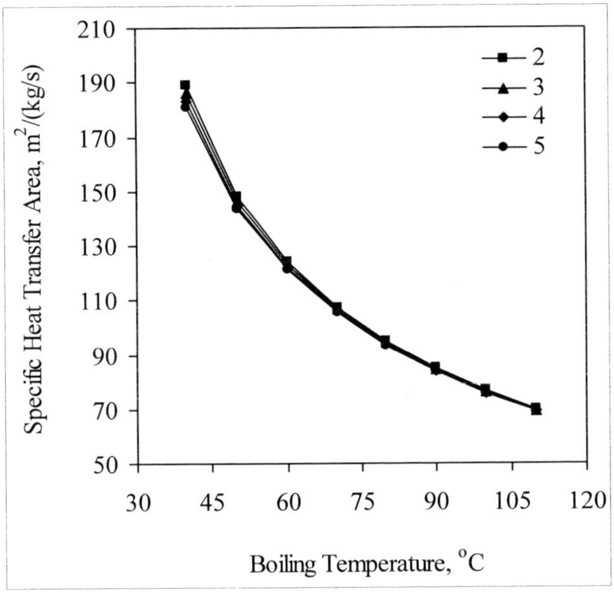

Fig. 8. Variations in the specific heat transfer area as a function of the boiling temperature and the compression ratio.

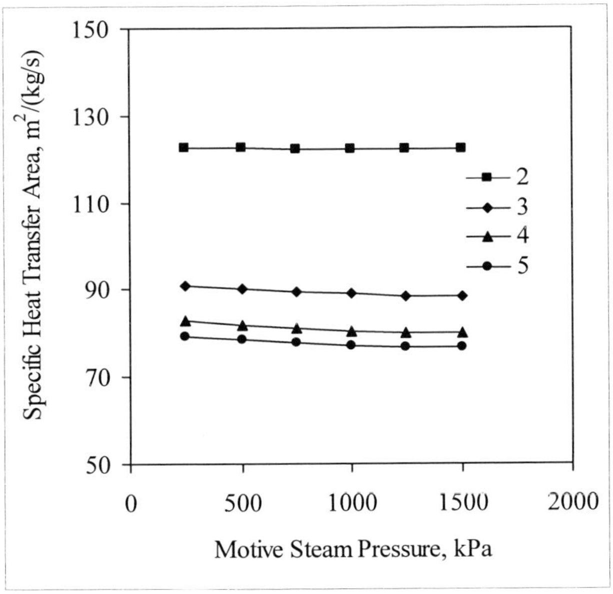

Fig. 9. Variations in the specific heat transfer area as a function of the motive steam pressure and the compression ratio.

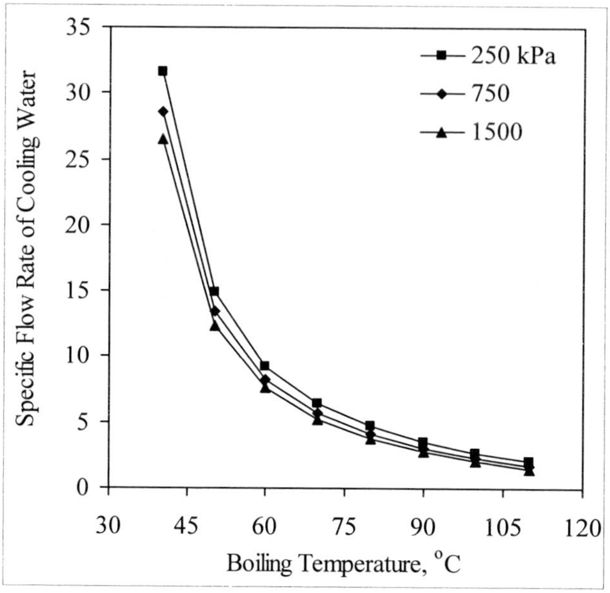

Fig. 10. Variations in the specific flow rate of cooling water as a function of the boiling temperature and the motive steam pressure.

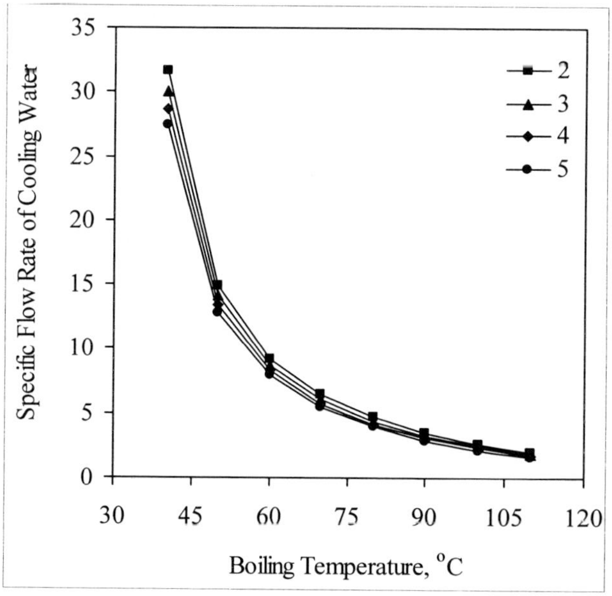

Fig. 11. Variations in the specific flow rate of cooling water as a function of boiling temperature and the compression ratio.

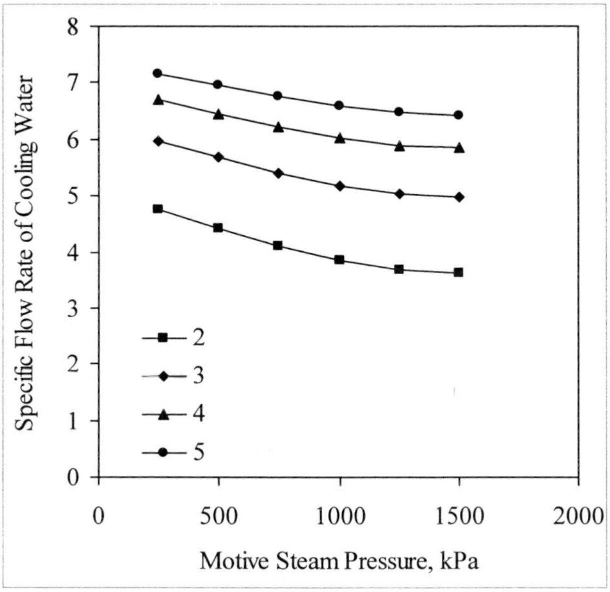

Fig. 12. Variations in the specific flow rate of cooling water as a function of the motive steam pressure and the compression ratio.

3.1.4 Summary

The TVC system is not found on industrial scale, however, its modeling, design, and analysis is considered because it provides the basis for the more complex system of multiple effect evaporation with thermal vapor compression. The mathematical for the TVC system includes material and energy balance equations for the condenser and evaporator. Also, the model includes the heat transfer equations for the condenser and evaporator as well as an empirical equation for the steam jet ejector. The analysis of the system is made as a function of variations in the thermal performance ratio, the specific heat transfer area, and the specific flow rate of cooling water. The analysis is performed over a range of the boiling temperature, the motive steam pressure, and the compression ratio. The following conclusions are made in the light of the results and discussion given in the previous section:

- The performance ratio decreases with the increase of the boiling temperature and the compression ratio. This is because of the increase in the motive steam consumption. This increase is necessary in order to achieve the required level of vapor compression.

- The performance ratio increases, but with limited sensitivity, upon the increase in the motive steam pressure. This result is caused by small increase in the amount of entrained vapor at higher motive steam pressures. In turn, this reduces the amount of consumed motive steam.
- The specific heat transfer area and the specific cooling water flow rate are sensitive to variations in the boiling temperature and the compression ratio. Both design parameters decrease with the increase of the boiling temperature and the compression ratio. This is because of the increase in the overall heat transfer coefficient in the evaporator and condenser, which causes large enhancement in the heat transfer rate.
- The specific heat transfer area and the specific cooling water flow rate have limited sensitivity with variations in the motive steam pressure.

In summary, it is recommended to operate of the single-effect vapor-compression desalination unit at intermediate values for the boiling temperature, i.e., 70-80 ^{o}C, and low compression ratios, i.e., values close to 2. This is necessary to have performance ratios close to or higher than 1.5. In addition, at such conditions high reduction is observed in the specific heat transfer area and the specific cooling water flow rate. This reduction will lower the first cost, i.e., construction cost of the evaporator, condenser, and seawater pump. In addition, the operating cost will be lower as a result of reduction in the energy required to operate the seawater-pumping unit.

References

El-Dessouky, H., Modeling and simulation of thermal vapor compression desalination plant, Symposium on Desalination of Seawater with Nuclear Energy, Taejon, Republic of Korea, 26-30 May, 1997.

Marto, P.J. Heat transfer in condensers, in Boilers, Evaporators and Condensers, Kakac, S., Editor, John Wiley, N.Y., 1991.

Muller, A.C., Condensers, in Hemisphere Handbook of Heat Exchanger Design, Hewitt, G.F., Editor, Hemisphere, N.Y., 1991.

Power, B.R., Steam jet ejectors for process industries, McGraw-Hill, Inc., N.Y., 1994.

Sinnott, R.K., Coulson & Richardson's Chemical Engineering, Vol. 6, 2nd Ed., Butterworth Heinemann, Oxford, 1996.

Problems

1. Use the ejector model (Eqs. 14-16) to develop a performance diagram for the steam jet ejector as a function of the entrainment ratio (Ra = M_m/M_{ev}), compression ratio (Cr = P_s/P_{ev}), and expansion ratio (Er = P_m/P_{ev}). The chart will cover the following ranges ($0.2 \le w \le 10$), ($0.2 \le Cr \le 5$), and ($1 \le Er \le 1000$). Discuss variations in the entrainment ratio as a function of the expansion and compression ratios.
2. A TVC system is used to desalinate seawater at 35 °C with 42000 ppm salinity. The maximum allowable brine temperature is 100 °C. The heat transfer coefficient for the evaporator and the two preheaters is constant and equal to 5.016 kW/m^2 °C. The specific heat transfer area is 109.46 m^2 per (kg/s) of fresh water and the heat transfer area of the distillate preheater is 200 m^2. The flow rates of the hot and cold stream in the preheaters are equal. The temperatures of the distillate and rejected brine flowing from the preheaters are 45 °C and 40 °C, respectively. Calculate the thermal performance ratio.
3. Calculate the thermal performance ratio and the specific heat transfer area for a TVC system operating the following conditions:
   - Motive steam pressure = 845.4 kPa
   - Distillate product temperature = 100 °C
   - Boiling temperature = 95 °C
   - Feed salinity = 42000 ppm
   - Feed temperature = 30.44 °C
4. A TVC system generates a distillate product at a flow rate of 1 kg/s. The system operating temperatures are as follows:
   - The boiling temperature, T_b, is 90 °C.
   - The feed temperature, T_f, is 85 °C.
   - The compressed vapor temperature, T_s, is 102 °C.
   - The motive steam pressure, P_m, is 15 bar.

   Determine the heat transfer areas in the evaporator and the condenser, the thermal performance ratio, the flow rates of feed seawater and reject brine, and the flow rate of cooling seawater. Assume the following:
   - The specific heat of seawater and brine streams is constant and equal to 4.2 kJ/kg °C.
   - Use the correlations for the overall heat transfer coefficient given in Appendix C.
   - Neglect thermodynamic losses in the demister and during condensation.
5. A single effect evaporator with a thermal load, Q_e, of 26500 kJ/s and the heating steam temperature is 105 °C is converted into a TVC system. The seawater temperature, T_{cw}, is 15 °C and the feed seawater temperature, T_f, is less than the boiling temperature, T_b, by 5 °C. The motive steam pressure

is 10 bars. Calculate the performance ratio before and after turning the system into the TVC configuration. Also, calculate the heat transfer area of the evaporator and condenser. Note that turning the system from the single effect configuration to the TVC mode has no effect on effect on the heat transfer area.

6. The heat transfer area in the evaporator and condenser for a TVC system is 90 and 30 m^2, respectively. The system is designed to operate at temperature of 85 °C for the boiling brine and an intake seawater temperature of 15 °C. Calculate the feed seawater temperature, the heating steam temperature, the steam flow rate, the cooling seawater flow rate, and the system performance ratio. Use a motive steam pressure of 5 bars.
7. A thermal vapor compression system operates at the following conditions:
   - Product flow rate = 10 kg/s
   - Feed water salinity = 42000 ppm
   - Feed water temperature = 14.4 °C
   - Pressure of motive steam = 4.6 kWm2 °C

   Calculate the following:
   - The evaporator heat transfer area.
   - The thermal performance ratio.
   - The change in the thermal performance ratio for the following conditions:
     - The feed water temperature increases to 30 °C.
     - The evaporation increases to 100 °C.
     - The motive steam pressure decreases to 500 kPa.
8. If the seawater temperature drops to 5 °C in problem 4, determine this effect on the system thermal performance ratio. Note that the heat transfer area remains constant as well as the heating steam temperature and the brine boiling temperature. What would be your recommendation to restore the system performance ratio to its original value.

3.2 Single Effect Mechanical Vapor Compression

The single-effect mechanical vapor-compression desalination process (MVC) is the most attractive among various single stage desalination processes. The MVC system is compact, confined, and does not require external heating source, which is opposite to thermal, absorption, or adsorption vapor compression. The system is driven by electric power; therefore, it is suitable for remote population areas with access to power grid lines. Another advantage of the MVC system is the absence of the down condenser and the cooling water requirements. This is because the compressor operates on the entire vapor formed within the system. Other advantages of the system include:

- Moderate investment cost.
- Proven industrial reliability to long lifetime operation.
- Simple seawater intake and pretreatment.
- The system adopts the horizontal falling film tube configuration, which allows for high heat transfer coefficient.
- The low temperature operation, 60 °C, allows for reduced scaling and heat losses and minimum requirement of thermal insulation.
- The system is modular type and it is simple to enlarge production volume by adopting additional modules.
- High product purity.
- Simple system adjustment to load variations, through temperature manipulation.

The major part of literature studies of the MVC system is focused on description of system characteristics and performance. Literature studies concerning modeling and analysis are limited to a small number, which includes:

- In 1981, Matz and Fisher compared the economics of the MVC system to the reverse osmosis (RO) desalination processes. The analysis was motivated by the need for integrated and compact desalination systems for remote resort areas. The MVC and RO systems are operated by electric current and do not require energy from external steam boilers. Comparison, based on power consumption, show that the specific power consumption for the RO system is 10-8 kWh/m^3 and for the MVC is 18.5–10 kWh/m^3. Further, the study predicted that future development of either system is expected to reduce the power consumption down to 5 kWh/m^3 for the RO system and to 8 kWh/m^3 for the MVC system. The main conclusion made by Matz and Fisher, 1981, is that neither system has a definitive edge, regarding the total production cost. This is because other cost elements in the RO system, which include membrane replacement and intensive chemical treatment, result in comparable total product cost for both systems.
- Lucas and Tabourier, 1985, reported performance data for single, two, four, and six effect MVC systems. The capacities for these systems vary from 300-

2500 m^3/d for the single and the six effect systems, respectively. They reported a specific power consumption of 11 kWh/m^3, which lay within the range reported by Matz and Fisher, 1981. In the single-effect configuration, the compressor increases the vapor temperature from 58 °C to 63 °C, which gives a compression ratio of 1.3. This ratio is 1.85 in the four-effect system, because the vapor temperature is increased from 49.5 °C to 62.5 °C. A range of 2-4 °C is reported for the temperature difference between the hot and cold streams in the feed preheaters.

- Matz and Zimerman (1985) reported similar performance and economic data for single and two effect vapor compression systems. The system operate at capacities between 50-1000 m^3/d, low top brine temperature between 50-70 °C, and specific power consumption slightly below 10 kWh/m^3.
- A decade later, Zimerman (1994) reported expansion of the MVC industry to more than 200 units operating in single or multi-effect modes. Although, the number is much larger than the few units found in 1985, it represents a small fraction in the global desalination industry, more than 12,000 operating units dominated by MSF and RO.
- Veza, 1995, reported on reliability of two MVC units installed in Canary Islands in 1987 and 1989. Over this period, the plant factor for both units vary between 87 and 90% with specific power consumption of 10.4-11.2 kWh/m^3 and a production capacity of 500 m^3/d/unit. The high plant factor is caused by low temperature operation, 60 °C, which reduces the scale formation rate.
- Comparison of the MVC versus other single effect desalination processes is studied by Al-Juwayhel et al., 1997. The study includes mathematical models for MVC system as well as other systems. Analysis of the MVC system focused on determination of the specific power consumption as a function of the top brine temperature. Model results are found consistent with literature data, where the specific power consumption varied over a range of 8-16 kWh/m^3.

In summary, it can be concluded that the MVC system remains to be used on a limited scale, however, it has high operation reliability, its specific power consumption is comparable to the RO system, and its production capacity suits either small or large consumption rates. Simulation studies are focused on economic comparison of MVC, RO, and other desalination systems. Other simulation analysis includes simplified mathematical models for the system or models for analyzing the plant energy consumption.

3.2.1 Process Description

The MVC system contains five major elements, which include mechanical vapor compressor, evaporator, preheaters for the intake seawater, brine and product pumps, and venting system. Figure 13 shows a schematic diagram for the system. As is shown, the compressor and evaporator form one single unit. The evaporator contains falling film horizontal tubes, spray nozzles, suction vapor tube, and wire-mesh mist eliminator. The feed preheaters are plate type heat exchanger, which operates on the intake seawater and the hot liquid streams leaving the evaporator.

The feed seawater enters the evaporator at a flow rate of M_f and a temperature of T_f. The feed seawater is sprayed over the horizontal tubes. The spray forms a falling film over succeeding tube rows. Formation of the thin film enhances the heat transfer rate and makes the evaporation process more efficient. The seawater temperature increases from T_f to T_b before evaporation commences. The formed vapors, M_d, are at a temperature of T_b. The vapors transfer from the evaporator section to the compressor through the vapor suction tube, which is guarded by a wire-mesh mist eliminator. This is necessary to avoid entrainment of brine droplets in the vapor stream, which would result in damage of the compressor blades. Limited temperature depression occurs as the vapors flow through the demister. The vapors flow tangentially through the compressor, where it is superheated from T_b to T_s. Upon compression, the vapors are forced inside the horizontal tubes, where it loses the superheat energy and its temperature drops from T_s to the saturation temperature T_d. Condensation takes place at T_d and the released latent heat is transferred to the brine film. The temperature difference T_s-T_b affects the compressor power consumption and is dictated by the temperature of the feed seawater.

The balance of energy within the system is maintained by recovery of the thermal load in the rejected brine and product streams. This is achieved in the feed preheater, which a plate type heat exchanger. In this unit, the intake seawater is at a low temperature, t_{cw}, and a flow rate M_f. The rejected brine and product streams leaving the evaporator are at higher temperatures of T_b and T_d, respectively. As heat is exchanged between the three streams the temperature of the seawater is increased to T_f and the temperature of the rejected brine and product streams is reduced to T_o.

Temperature profiles of the system are shown in Fig. 14. As is shown, the temperature of the feed seawater increases from T_{cw} to T_f in the preheater unit. Simultaneously, the temperatures of the rejected brine and the product stream decrease from T_b and T_d, respectively, to the same temperature T_o. Inside the evaporator, the temperature of the feed seawater increases from T_f to the boiling temperature T_b. The formed vapor is at the same boiling temperature, which is higher than the saturation temperature, T_v, by the boiling point elevation, T_b-T_v.

The formed vapor is compressed and superheated to a temperature T_s. Condensation of the compressed vapor takes place at a temperature of T_d.

Ranges of the temperature difference of various streams are:

- The difference (T_s-T_d) varies from 4 to 10 °C,
- The difference (T_b-T_f) varies from 1-5 °C,
- The difference (T_d-T_b) varies from 1-5 °C, and
- The difference (T_o-T_{cw}) varies from 1-5 °C.

Maintaining the temperature differences within these ranges is essential to achieve the following:

- Keep the compressor power consumption within practical limits.
- Avoid excessive increase in the evaporator heat transfer area.
- Operate the preheater units at reasonable LMTD values to minimize the heat transfer area.

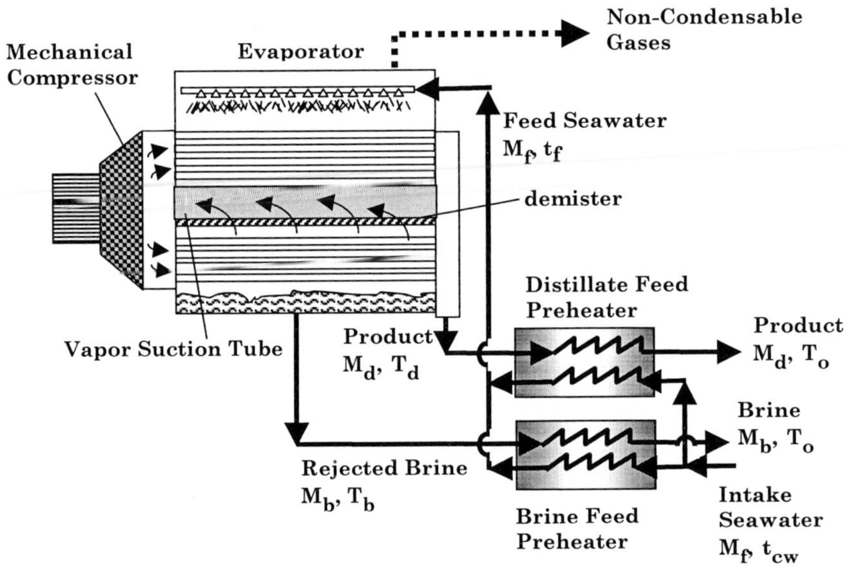

Fig. 13. Single effect mechanical vapor compression (MVC).

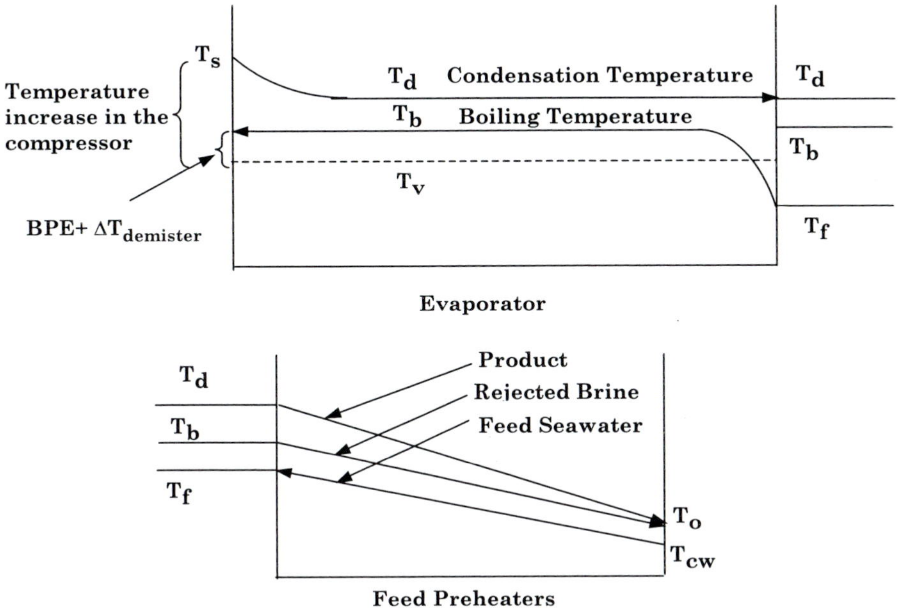

Fig. 14. Temperature profiles in MVC.

3.2.2 Process Modeling

Two models are developed for the MVC system; the first model includes several simplifying assumptions. This results in closed form solution that does not require iterative solution method. Such model is simple to use and generates useful results, which can be used for quick assessment of design and operating data. The second model negates the simplifying assumptions adopted in the first model. This makes the equations non-linear; therefore iterative solution is needed to determine the system characteristics.

Simplified MVC Model

The assumptions invoked in development of this model include the following:

- Different heat transfer areas for the preheaters,
- Constant, but not equal, overall heat transfer coefficients in the evaporator and preheaters.
- Equal temperature for the effluent heating streams in the preheaters.
- The heat capacity for all streams is constant and equal to 4.2 kJ/kg °C.

- The latent heat of formed vapor and condensing steam are temperature dependent.
- Effect of the boiling point elevation is neglected.
- The distillate is salt free.
- The driving force for heat transfer in the evaporator is assumed constant and equal to the difference between the condensation and evaporation temperatures.

The model is into four parts, which include the following:

- Material balances.
- Evaporator and preheaters energy balances.
- Evaporator and preheaters heat transfer area.
- Performance parameters.

Material Balances

The overall mass and salt balances are given by

$$M_f = M_d + M_b \tag{17}$$

$$M_f X_f = M_b X_b \tag{18}$$

where M is the mass flow rate, X is the salinity, and the subscripts b, d, and f denotes the rejected brine, distillate, and feed seawater. Equation 1 can be used to eliminate M_f from Eq. 18 and to generate a relation between M_b and M_d. This result is given by

$$M_b = M_d (X_f /(X_b - X_f)) \tag{19}$$

Similarly, M_b can be eliminated from Eq. 18 to generate a relation between M_f and M_d. This result is given by

$$M_f = M_d (X_b /(X_b - X_f)) \tag{20}$$

Evaporator and Feed Preheaters Energy Balances

Two preheaters are used to increase the feed temperature from the intake seawater temperature T_{cw} to T_f. The intake seawater is divided into two portions, αM_f and $(1-\alpha)M_f$. In the first preheater, heat is exchanged between αM_f and the condensed vapors, and in the second preheater, heat is exchanged between $(1-\alpha)M_f$ and the rejected brine. The thermal load of the two heat exchangers is given in terms of the intake seawater

$$Q_h = M_f\, C_p\, (T_f - T_{cw}) \tag{21}$$

The seawater feed flow rate given by Eq. (20) is substitute in Eq. (21). This gives

$$Q_h = M_d\, C_p\, (X_b/(X_b - X_f))\, (T_f - T_{cw}) \tag{22}$$

Equation (21) can be also written in terms of the heat load of the condensed vapor and the rejected brine, which gives

$$Q_h = M_d\, C_p\, (T_d - T_o) + M_b\, C_p\, (T_b - T_o) \tag{23}$$

The brine flow rate, M_b, given by Eq. (19) is substituted in Eq. (23). This reduces Eq. (23) into

$$Q_h = M_d\, C_p\, (T_d - T_o) + M_d\, (X_f/(X_b - X_f))\, C_p\, (T_b\text{-}T_o) \tag{24}$$

Equating Eqs. (22) and (24) gives

$$\begin{aligned}(M_d\, C_p\, (T_d - T_o) &+ M_d\, (X_f/(X_b - X_f))\, C_p\, (T_b\text{-}T_o)) \\ &= M_d\, C_p\, (X_b/(X_b - X_f))\, (T_f - T_{cw})\end{aligned} \tag{25}$$

Equation (25) is simplified to determine the outlet temperature of the heating streams, T_o, which gives

$$T_o = (T_{cw} - T_f) + (X_f / X_b)\, T_b + ((X_b - X_f)/X_b)\, T_d \tag{26}$$

In the evaporator, heat is supplied to the feed seawater, where its temperature increases from T_f to T_b. Also, latent heat is consumed by the formed vapor at a temperature of T_b. This energy is supplied by the latent heat of condensation for the compressed vapors at T_d and by the superheat of the compressed vapors, T_s-T_d. The evaporator thermal load is given by

$$Q_e = M_f\, C_p\, (T_b - T_f) + M_d\, \lambda_b = M_d\, \lambda_d + M_d\, C_{p_v}\, (T_s - T_d) \tag{27}$$

In the above equation λ_b and λ_d are the latent heat of formed vapor at T_b and condensing vapor at T_d. The feed flow rate given by Eq. (20) is substituted in Eq. (27). The resulting equation is

$$M_d\, (X_b/(X_b - X_f))\, C_p\, (T_b - T_f) + M_d\, \lambda_b = M_d\, \lambda_d + M_d\, C_{p_v}\, (T_s - T_d) \tag{28}$$

Equation (28) is then simplified to determine the seawater feed temperature, T_f. This is given by

$$T_f = ((X_b - X_f)/X_b)\,((\lambda_b - \lambda_d)/C_p - (C_{p_v}/C_p)\,(T_s - T_d)) + T_b \tag{29}$$

Evaporator and Condenser Heat Transfer Area

The heat transfer area for the evaporator is determined in terms of thermal load, the driving force for heat transfer, and the overall heat transfer coefficient. The thermal load for the evaporator is equal to the sum of the latent heat of formed vapor at T_b and the sensible heat added to the feed seawater to increase its temperature from T_f to T_b. The driving force for heat transfer is assumed equal to the difference between the condensation and evaporation temperatures, T_d-T_b. As for the overall heat transfer coefficient it is calculated from the correlation given in Appendix C. This evaporator heat transfer area is then given by

$$A_e = \frac{M_d\lambda_b + M_f C_p (T_b - T_f)}{U_e (T_d - T_b)} = \frac{M_d\lambda_d + M_d C_{p_v}(T_s - T_d)}{U_e (T_d - T_b)} \tag{30}$$

The heat transfer area for the two preheaters obtained in similar manner, however, the driving force for heat transfer is taken as the logarithmic mean at the preheater ends. These equations are given by

$$A_d = \frac{M_d C_p (T_d - T_o)}{U_d (LMTD)_d} = \frac{\alpha M_f C_{p_\ell} (T_f - T_{cw})}{U_d (LMTD)_d} \tag{31}$$

$$\begin{aligned} A_b &= \frac{M_b C_p (T_b - T_o)}{U_b (LMTD)_b} \\ &= \frac{M_d (X_f/(X_b - X_f)) C_p (T_b - T_o)}{U_b (LMTD)_b} \\ &= \frac{(1-\alpha) M_f C_p (T_f - T_{cw})}{U_b (LMTD)_b} \end{aligned} \tag{32}$$

The $(LMTD)_d$ is defined as:

$$(LMTD)_d = \frac{(T_d - T_f) - (T_o - T_{cw})}{\ln \dfrac{T_d - T_f}{T_o - T_{cw}}} \tag{33}$$

The $(LMTD)_b$ is defined as:

$$(LMTD)_b = \frac{(T_b - T_f) - (T_o - T_{cw})}{\ln \dfrac{T_b - T_f}{T_o - T_{cw}}} \tag{34}$$

Performance Parameters

Performance of the MVC is determined in terms of the following variables:

- The specific power consumption, kWhr/m^3.
- The specific heat transfer surface area, sA.

The compressor mechanical energy is given

$$W = \frac{\gamma}{\eta(\gamma - 1)} P_v v_v \left(\left(\frac{P_s}{P_v} \right)^{\left(\frac{\gamma-1}{\gamma}\right)} - 1 \right) \tag{35}$$

where W is specific power consumption, P is the pressure, v is the specific volume, η is the compressor efficiency, and γ is the isentropic efficiency. It should be noted that the inlet pressure (P_v) is equal to the vapor pressure of the formed vapor at (T_b) and the outlet pressure (P_s) is the compressed vapor pressure at (T_s).

The specific heat transfer area is obtained by summing Eqs. (30-32). The resulting summation is divided by M_d, which results in

$$\begin{aligned} sA &= \frac{A_e + A_d + A_b}{M_d} \\ &= \frac{\lambda_d + C_{p_v}(T_s - T_d)}{U_e (T_d - T_b)} + \frac{C_p (T_d - T_o)}{U_d (LMTD)_d} + \frac{(X_f/(X_b - X_f)) C_p (T_b - T_o)}{U_b (LMTD)_b} \end{aligned} \tag{36}$$

Solution of MVC Simple Model

Solution of the above model is sequential and requires no iterations. Solution proceeds as follows:

- The mass flow rates of the reject brine and feed seawater, M_b and M_f, are calculated from Eqs. (20) and (21).
- The seawater feed temperature, T_f, is obtained from Eq. (29).
- The effluent temperature of heat streams, T_o, is obtained from Eq. (26).

- The areas for evaporator, brine preheater, and product preheater, are calculated from Eqs. (30-32).
- The specific power consumption, W, is obtained from Eq. (35).
- The specific heat transfer area, sA, is obtained from Eq. (36).

Example 1:

A single-effect mechanical vapor-compression system is to be designed at the following conditions:

- The distillate flow rate, $M_d = 1$ kg/s.
- The heat capacity of the vapor is constant, $C_{p_v} = 1.884$ kJ/kg °C.
- The heat capacity of all liquid streams is constant, $C_p = 4.2$ kJ/kg °C.
- The overall heat transfer coefficient in the evaporator, $U_e = 2.4$ kJ/s m^2 °C.
- The overall heat transfer coefficient in the brine preheater, $U_b = 1.5$ kJ/s m^2 °C.
- The overall heat transfer coefficient in the product preheater, $U_d = 1.8$ kJ/s m^2 °C.
- The intake seawater temperature, $T_{cw} = 25$°C.
- The condensed vapor temperature, $T_d = 62$ °C.
- The compressed vapor temperature, $T_s = T_d + 3 = 65$ °C.
- The evaporation temperature, $T_b = T_d - 2 = 60$ °C.
- The feed seawater salinity, $X_f = 42000$ ppm.
- The salinity of the rejected brine, $X_b = 70000$ ppm.
- Compressor efficiency, $\eta = 58.9\%$

Calculate the specific power consumption, the heat transfer areas for the evaporator and preheaters, and the specific heat transfer area.

Solution

Substituting for $X_f = 42000$ ppm, $X_b = 70000$ ppm, and $M_d = 1$ kg/s in Eq. (20) results in

$$M_f = X_b/(X_b - X_f) = 70000/(70000 - 42000) = 2.5 \text{ kg/s}$$

Equation (17) is then used to calculate M_b,

$$M_b = M_f - M_d = 2.5 - 1 = 1.5 \text{ kg/s}$$

The latent heats of condensation and evaporation, λ_d and λ_b, are then calculated from the correlation given in Appendix (A).

$$\lambda_d = 2499.5698 - 2.204864\ T_d - 2.304\text{x}10^{-3} T_d^2$$
$$= 2499.5698 - 2.204864\ (62) - 2.304\text{x}10^{-3}\ (62)^2 = 2354.01\ \text{kJ/kg}$$
$$\lambda_b = 2499.5698 - 2.204864\ T_b - 2.304\text{x}10^{-3} T_b^2$$
$$= 2499.5698 - 2.204864\ (60) - 2.304\text{x}10^{-3}\ (60)^2 = 2358.98\ \text{kJ/kg}$$

Equation (29) is used to calculate T_f

$$T_f = ((X_b - X_f)/X_b)\ ((\lambda_b - \lambda_d)/Cp - (Cp_v/Cp)\ (T_s - T_d)) + T_b$$
$$= ((70000\text{-}42000)/70000)\ ((2358.98 - 2354.01)/4.2 - (1.884/4.2)\ (65 - 62)) + 60$$
$$= 59.39\ ^oC$$

Equation (26) is used to calculate T_o

$$T_o = (T_{cw} - T_f) + (X_f / X_b)\ T_b + ((X_b - X_f)/X_b)\ T_d$$
$$= (25 - 58.73) + (42000/70000)\ (60) + ((70000 - 42000)/70000)\ (62)$$
$$= 26.4\ ^oC$$

The evaporator area is calculated from Eq. (30)

$$A_e = \frac{M_d \lambda_d + M_d C_{p_v} (T_s - T_d)}{U_e (T_d - T_b)}$$
$$= \frac{(1)(2354.01) + (1)(1.884)(65 - 62)}{(2.4)(62 - 60)}$$
$$= 492.77\ \text{m}^2$$

The value of $(LMTD)_d$ is obtained from Eq. (33)

$$(LMTD)_d = \frac{(T_d - T_f) - (T_o - T_{cw})}{\ln \dfrac{T_d - T_f}{T_o - T_{cw}}}$$
$$= \frac{(62 - 59.39) - (26.4 - 25)}{\ln \dfrac{62 - 59.39}{26.4 - 25}} = 1.94\ ^oC$$

Similarly the value of $(LMTD)_b$ is determined from Eq. (34)

$$(\mathrm{LMTD})_b = \frac{(T_b - T_f) - (T_o - T_{cw})}{\ln \dfrac{T_b - T_f}{T_o - T_{cw}}}$$

$$= \frac{(60 - 59.39) - (26.4 - 25)}{\ln \dfrac{60 - 59.39}{26.4 - 25}} = 0.95\ ^{o}\mathrm{C}$$

The heat transfer area for the two preheaters are determined from Eqs. (31-32)

$$A_d = \frac{M_d C_p (T_d - T_o)}{U_d (\mathrm{LMTD})_d} = \frac{(1)(4.2)(62 - 26.4)}{(1.8)(1.94)} = 42.78\ \mathrm{m}^2$$

$$A_b = \frac{M_b C_p (T_b - T_o)}{U_b (\mathrm{LMTD})_b} = \frac{(1.5)(4.2)(60 - 26.4)}{(1.5)(0.95)} = 148.94\ \mathrm{m}^2$$

The specific power consumption, W, is calculated from Eq. (35). This requires determination of P_o, P_i, v_i, and γ. The compressor inlet and outlet pressures, P_i and P_o, is equal to the saturation pressure of compressed vapor at T_s and the formed vapor at T_d, respectively. These values can be obtained from the steam tables or from the correlation given in Appendix A:

$$P_o = 10.17246 - 0.6167302\,(T_s) + 1.832249\mathrm{x}10^{-2}(T_s)^2$$
$$-1.77376\mathrm{x}10^{-4}(T_s)^3 + 1.47068\mathrm{x}10^{-6}(T_s)^4$$
$$= 10.17246 - 0.6167302\,(65) + 1.832249\mathrm{x}10^{-2}(65)^2$$
$$-1.77376\mathrm{x}10^{-4}(65)^3 + 1.47068\mathrm{x}10^{-6}(65)^4 = 25.03\ \mathrm{kPa}$$

$$P_i = 10.17246 - 0.6167302\,(T_b) + 1.832249\mathrm{x}10^{-2}(T_b)^2$$
$$-1.77376\mathrm{x}10^{-4}(T_b)^3 + 1.47068\mathrm{x}10^{-6}(T_b)^4$$
$$= 10.17246 - 0.6167302\,(60) + 1.832249\mathrm{x}10^{-2}(60)^2$$
$$-1.77376\mathrm{x}10^{-4}(60)^3 + 1.47068\mathrm{x}10^{-6}(60)^4 = 19.88\ \mathrm{kPa}$$

The specific volume of inlet vapor at T_b can be obtained from steam tables or the correlation given in the Appendix (A). This given by

$$
\begin{aligned}
v_i &= 163.3453 - 8.04142\,(T_b) + 0.17102\,(T_b)^2 \\
&\quad -1.87812\text{x}10^{-3}(T_b)^3 + 1.03842\text{x}10^{-5}(T_b)^4 - 2.28215\text{x}10^{-8}(T_b)^5 \\
&= 163.3453 - 8.04142\,(60) + 0.17102\,(60)^2 \\
&\quad -1.87812\text{x}10^{-3}(60)^3 + 1.03842\text{x}10^{-5}(60)^4 - 2.28215\text{x}10^{-8}(60)^5 \\
&= 4.836\ \text{m}^3/\text{kg}
\end{aligned}
$$

The value of the compression ratio γ is 1.32. The specific power consumption is then calculated from Eq. (35)

$$
\begin{aligned}
W &= \frac{\gamma}{\eta(\gamma-1)} P_i v_i \left(\left(\frac{P_o}{P_i} \right)^{\left(\frac{\gamma-1}{\gamma}\right)} - 1 \right) \\
&= \frac{1.32}{0.589\,(1.32-1)} (19.88)(7.69) \left(\left(\frac{25.03}{19.88} \right)^{\left(\frac{1.32-1}{1.32}\right)} - 1 \right) \left(\frac{1000}{3600} \right) \\
&= 17.13\ \text{kWhr/m}^3
\end{aligned}
$$

The specific heat transfer area is obtained directly by summing the values of A_e, A_b, and A_b. This is because the distillate flow-rate is set at 1 kg/s. The value of sA is equal to 492.78 + 42.78 + 148.94 = 684.49 m^2.

Example 2:

A single-effect mechanical vapor-compression system has the following design data:

- Evaporator heat transfer area = 400 m^2.
- Distillate feed preheater heat transfer area = 7 m^2.
- Brine feed preheater heat transfer area = 15 m^2.
- The heat capacity of the vapor is constant, C_{p_v} = 1.884 kJ/kg °C.
- The heat capacity of all liquid streams is constant, C_p = 4.2 kJ/kg °C.
- The overall heat transfer coefficient in the evaporator, U_e = 2.4 kW/m^2 °C.
- The overall heat transfer coefficient in the brine preheater, U_b = 6.3 kW/ m^2 °C.
- The overall heat transfer coefficient in the distillate preheater, U_d = 6.7 kW/m^2 °C.
- The intake seawater temperature, T_{cw} = 25°C.

- The compressed vapor temperature, $T_s = (T_d + 7)$ °C.
- The feed seawater salinity, $X_f = 42000$ ppm.
- The salinity of the rejected brine, $X_b = 70000$ ppm.
- Compressor efficiency, $\eta = 58.9\%$

Calculate the following:

- Flow rate of the distillate product.
- Flow rate of the brine reject
- Flow rate of the feed seawater.
- Temperature of the feed seawater.
- Temperature of the outlet brine and product streams.
- Temperature of the brine stream leaving the evaporator.
- Temperature of the condensate product stream.

Solution of this problem is iterative, where the following equations are solved iteratively. Equation solution can be simultaneous or sequential. The equations include the following:

$$M_f = (M_d\, X_b)/(X_b - X_f)$$

$$M_b = M_f - M_d$$

$$T_f = ((X_b - X_f)/X_b)\,((\lambda_b - \lambda_d)/Cp - (Cp_v/Cp)\,(T_s - T_d)) + T_b$$

$$T_o = (T_{cw} - T_f) + (X_f / X_b)\,T_b + ((X_b - X_f)/X_b)\,T_d$$

$$A_e = \frac{M_d \lambda_d + M_d C_{p_v}(T_s - T_d)}{U_e (T_d - T_b)}$$

$$A_d = \frac{M_d C_p (T_d - T_o)}{U_d (LMTD)_d}$$

$$A_b = \frac{M_b C_p (T_b - T_o)}{U_b (LMTD)_b}$$

The assumed initial guess include the following:

- $M_d = 0.8$ kg/s
- $T_d = 74$ °C
- $T_b = 72$ °C

Solution of the first two equations gives M_f and M_b

$$M_f = M_d X_b / (X_b - X_f) = (0.8)(70000)/(70000 - 42000) = 2 \text{ kg/s}$$

$M_b = M_f - M_d = 2 - 0.8 = 1.2$ kg/s

The latent heats of condensation and evaporation, λ_d and λ_b, are calculated from correlation given in Appendix (A)

$$\lambda_d = 2499.5698 - 2.204864\ T_d - 2.304x10^{-3} T_d^2$$
$$= 2499.5698 - 2.204864\,(74) - 2.304x10^{-3}\,(74)^2 = 2323.79\ \text{kJ/kg}$$

$$\lambda_b = 2499.5698 - 2.204864\ T_b - 2.304x10^{-3} T_b^2$$
$$= 2499.5698 - 2.204864\,(72) - 2.304x10^{-3}\,(72)^2 = 2328.88\ \text{kJ/kg}$$

The third equation is used to calculate T_f

$$T_f = ((X_b - X_f)/X_b)\ ((\lambda_b - \lambda_d)/Cp - (Cp_v/Cp)\ (T_s - T_d)) + T_b$$
$$= ((70000\text{-}42000)/70000)\ ((2328.88\text{-}2323.79)/4.2 - (1.884/4.2)\ (7)) + 72$$
$$= 71.22\ ^oC$$

The fourth equation is used to calculate T_o

$$T_o = (T_{cw} - T_f) + (X_f / X_b)\ T_b + ((X_b - X_f)/X_b)\ T_d$$
$$= (25 - 71.22) + (42000/70000)\ (72) + ((70000 - 42000)/70000)\ (74)$$
$$= 26.58\ ^oC$$

The heat transfer equations are then used to update the initial guess, where the evaporator area is used to calculate a new value for M_d

$$A_e = \frac{M_d \lambda_d + M_d C_{p_v} (T_s - T_d)}{U_e (T_d - T_b)}$$
$$400 = \frac{(M_d)(2323.79) + (M_d)(1.884)(7)}{(2.4)(74 - 72)}$$

which gives $M_d = 0.82$ kg/s

The value of $(LMTD)_d$ for distillate product preheater is calculated

$$(LMTD)_d = \frac{(T_d - T_f) - (T_o - T_{cw})}{\ln \dfrac{T_d - T_f}{T_o - T_{cw}}} = \frac{(74 - 71.22) - (26.58 - 25)}{\ln \dfrac{74 - 71.22}{26.58 - 25}} = 2.124\ ^oC$$

Similarly the value of $(LMTD)_b$ is calculated for the brine preheater

$$(LMTD)_b = \frac{(T_b - T_f) - (T_o - T_{cw})}{\ln \frac{T_b - T_f}{T_o - T_{cw}}} = \frac{(72 - 71.22) - (26.58 - 25)}{\ln \frac{72 - 71.22}{26.58 - 25}} = 1.133\ ^oC$$

The heat transfer area for the two preheaters are then used to update the condensate and brine streams leaving the evaporator

$$A_d = \frac{M_d C_p (T_d - T_o)}{U_d (LMTD)_d}$$

$$7 = \frac{(0.82)(4.2)(T_d - 26.58)}{(6.7)(2.124)}$$

which gives a new value for $T_d = 55.5\ ^oC$

$$A_b = \frac{M_b C_p (T_b - T_o)}{U_b (LMTD)_b}$$

$$15 = \frac{(0.82)(4.2)(T_b - 26.58)}{(6.3)(1.133)}$$

which gives a new value for $T_b = 57.6\ ^oC$. The iterations continue to give the following final solution:

- $M_d = 0.8$ kg/s
- $M_b = 1.2$ kg/s
- $M_f = 2$ kg/s
- $T_d = 74.8\ ^oC$
- $T_b = 72.8\ ^oC$
- $T_o = 27.8\ ^oC$
- $T_f = 70.8\ ^oC$

The specific power consumption (W) is calculated from Eq. (35). This requires determination of P_s, P_v, and v_v The compressor inlet and outlet pressures, P_v and P_s, is equal to the saturation pressure of compressed vapor at T_s and the formed vapor at T_d, respectively. These values can be obtained from the steam tables or from the correlation given Appendix (A):

- $P_v = 35.2$ kPa
- $P_s = 38.3$ kPa

- $v_v = 4.49\ m^3/kg$

The value of the compression ratio γ is 1.32. The specific power consumption is then calculated from Eq. (35)

$$W = \frac{\gamma}{\eta(\gamma-1)} P_v v_v \left(\left(\frac{P_s}{P_v} \right)^{\left(\frac{\gamma-1}{\gamma}\right)} - 1 \right)$$

$$= \frac{(0.8)(1.32)}{0.589\,(1.32-1)} (35.26)(4.49) \left(\left(\frac{38.311}{35.26} \right)^{\left(\frac{1.32-1}{1.32}\right)} - 1 \right) \left(\frac{1000}{3600} \right)$$

$$= 5\ \mathrm{kWhr/m^3}$$

Detailed MVC Model

The assumptions used to develop the second model include the following:

- Different heat transfer areas for the preheaters
- Equal temperature for the effluent heating streams.
- The heat capacities for brine and product streams depend on temperature and composition.
- The Overall heat transfer coefficient in the preheaters is constant, but not equal.
- The latent heat of formed vapor and condensing steam are temperature dependent.
- The specific heat of the vapor is constant.
- The effect of the boiling point elevation, BPE, is included in the calculations.
- The distillate is salt free.
- The driving force for heat transfer in the evaporator is assumed constant and equal to the difference between the condensation and evaporation temperatures.

The basic model equations are similar to those given for the simplified model. However, mathematical manipulations of the energy balances cannot be made because of the nonlinear nature of the equations. Correlations for the boiling point elevation, saturation pressure, saturation volume, evaporator heat transfer coefficient, enthalpies of fresh water vapor and liquid, and the specific heat of the seawater and brine are given in the appendices. The following is a list of the equations used in the detailed model.

Material and salt balances

$$M_b = M_d\,(X_f/(X_b - X_f)) \quad (37)$$

$$M_f = M_d + M_b \quad (38)$$

Preheaters energy balances

$$M_f\,C_p\,(T_f - T_{cw}) = M_d\,C_p\,(T_b - T_o) + M_b\,C_p\,(T_b - T_o) \quad (39)$$

Evaporator energy balances

$$M_f\,C_p\,(T_b - T_f) + M_d\,\lambda_v = M_d\,\lambda_d + M_d\,C_{p_v}\,(T_s - T_d) \quad (40)$$

Evaporator heat transfer area

$$A_e = \frac{M_d\lambda_d + M_d C_{p_v}(T_s - T_d)}{U_e(T_d - T_b)} \quad (41)$$

Distillate/feed preheater heat transfer area

$$A_d = \frac{M_d C_p (T_d - T_o)}{U_d (LMTD)_d} = \frac{\alpha M_f C_p (T_f - T_{cw})}{U_d (LMTD)_d} \quad (42)$$

$$(LMTD)_d = \frac{(T_d - T_f) - (T_o - T_{cw})}{\ln\dfrac{T_d - T_f}{T_o - T_{cw}}} \quad (43)$$

Brine/feed preheater heat transfer area

$$A_b = \frac{M_b C_p (T_b - T_o)}{U_b (LMTD)_b} = \frac{(1-\alpha) M_f C_p (T_f - T_{cw})}{U_b (LMTD)_b} \quad (44)$$

$$(LMTD)_b = \frac{(T_b - T_f) - (T_o - T_{cw})}{\ln\dfrac{T_b - T_f}{T_o - T_{cw}}} \quad (45)$$

Performance parameters

$$W = \frac{\gamma}{\eta(\gamma-1)} P_i v_i \left(\left(\frac{P_o}{P_i} \right)^{\left(\frac{\gamma-1}{\gamma}\right)} - 1 \right) \qquad (46)$$

$$sA = \frac{A_e + A_d + A_b}{M_d}$$
$$= \frac{\lambda_d + C_{p_v}(T_s - T_d)}{U_e(T_d - T_b)} + \frac{(H(T_d) - H(T_o))}{U_d(LMTD)_d} + \frac{(X_f/(X_b - X_f))\, C_p (T_b - T_o)}{U_b (LMTD)_b} \qquad (47)$$

Solution of MVC detailed Model

Specification made to solve the above equation system include:

- The distillate flow rate, M_d = 1 kg/s.
- The intake seawater temperature, T_{cw} = 25°C.
- The condensed vapor temperature, T_d = 62 °C.
- The range for compressed vapor temperature, $T_s = T_d + 3 = 65$ °C.
- The evaporation temperature, $T_b = T_d - 2 = 60$ °C.
- The heat capacity of the vapor is constant, C_{p_v} = 1.884 kJ/kg °C.
- The overall heat transfer coefficient in the brine preheater, U_b = 1.5 kJ/s m^2 °C.
- The overall heat transfer coefficient in the product preheater, U_d = 1.8 kJ/s m^2 °C.
- The feed seawater salinity, X_f = 42000 ppm.
- The salinity of the rejected brine, X_b = 70000 ppm.
- Compressor efficiency, η = 58.9%.

Solution proceeds as follows:

- The mass flow rates of the reject brine and feed seawater, M_b and M_f, are calculated from Eqs. (37-38).
- The temperatures of the seawater feed and the effluent heating stream, T_f and T_o, are obtained by iterative solution of Eqs. (39-40).
- The areas for evaporator, brine preheater, and product preheater, are calculated from Eqs. (41,42, 44).
- The specific power consumption, W, is obtained from Eq. (46).
- The specific heat transfer area, sA, is obtained from Eq. (47).

Example 3:

As presented in the simple model, substituting X_f = 42000 ppm, X_b = 70000 ppm, and M_d = 1 kg/s in Eq. 4 gives M_f = 2.5 kg/s. This value is substituted in Eq. (37) giving M_b = 1.5 kg/s. The overall heat transfer coefficients in the evaporator is then calculated from the following correlation

$$\begin{aligned} U_e &= \left(1961.9 + 3.2\,(T_d) + 12.6\text{x}10^{-2}\,(T_b)^2 - 3.16\text{x}10^{-4}\,(T_b)^3\right)\text{x}10^{-3} \\ &= \left(1961.9 + 3.2\,(60) + 12.6\text{x}10^{-2}\,(60)^2 - 3.16\text{x}10^{-4}\,(60)^3\right)\text{x}10^{-3} \\ &= 2.438 \text{ kJ/s m}^2\ {}^{\circ}\text{C} \end{aligned}$$

Iterations are made to solve Eqs. (39) and (40), which proceeds as follows:

Iteration	T_f	T_o
Initial Guess	27.00000	59.00000
Iteration 1	27.00000	59.02030
Iteration 2	27.00932	59.00000
Iteration 3	26.54499	59.33199
Iteration 4	26.54495	59.33193

The evaporator area is calculated from Eq. (41)

$$\begin{aligned} A_e &= \frac{M_d \lambda_d + M_d C_{p_v}(T_s - T_d)}{U_e (T_d - T_b)} \\ &= \frac{(1)(2354.01) + (1)(1.884)(65-62)}{(2.438)(62-60)} \\ &= 483.78 \text{ m}^2 \end{aligned}$$

The value of $(LMTD)_d$ is obtained from Eq. (43)

$$\begin{aligned} (LMTD)_d &= \frac{(T_d - T_f) - (T_o - T_{cw})}{\ln \dfrac{T_d - T_f}{T_o - T_{cw}}} \\ &= \frac{(62-59.33) - (26.54-25)}{\ln \dfrac{(62-59.33)}{(26.54-25)}} = 2.055\ {}^{\circ}\text{C} \end{aligned}$$

Similarly the value of $(LMTD)_b$ is determined from Eq. (45)

$$(LMTD)_b = \frac{(T_b - T_f) - (T_o - T_{cw})}{\ln \dfrac{T_b - T_f}{T_o - T_{cw}}}$$

$$= \frac{(60 - 59.33) - (26.54 - 25)}{\ln \dfrac{60 - 59.33}{26.54 - 25}} = 1.046 \ ^\circ C$$

The heat transfer area for the two preheaters are determined from Eqs. (42-44)

$$A_d = \frac{M_d (H(T_d) - H(T_o))}{U_d (LMTD)_d} = \frac{(1)\ (259.35 - 111.04)}{(1.8)(2.055)} = 40.08 \ m^2$$

$$A_b = \frac{M_b C_p (T_b - T_o)}{U_b (LMTD)_b} = \frac{(1.5)(3.845)(60 - 26.54)}{(1.5)(1.045)} = 123.01 \ m^2$$

The specific power consumption, W, and the specific heat transfer area are obtained from Eqs. (46) and (47), respectively. It should be noted that the value of W is identical to that obtained in the simplified model, because T_s and T_b have the same values. Therefore, values of P_s, P_v, and v_v, are the same and have values of 25.038 kPa, 19.78 kPa, and 7.69 m^3/kg, respectively. The resulting specific power consumption is 17.127 kWh/m^3. As for the specific heat transfer area, it is equal to the sum of A_e, A_d, and A_b; this gives a total value of 646.87 $m^2/(kg/s)$.

3.2.3 System Performance

The detailed model is used to simulate and analyze the performance of the MVC desalination process. Analysis is made as a function of the brine boiling temperature, T_b, and the temperature difference of the condensing vapor and the boiling brine, T_d -T_b. The brine boiling temperature is varied between 60 and 105 °C and temperature difference between 1 and 4 °C. All calculations are made for a distillate flow rate of 1 kg/s and a temperature difference of 3 °C between the compressed and condensing vapors, T_s-T_d. Results include the specific power consumption and the specific heat transfer areas for the evaporator and the two preheaters.

The specific power consumption for the system is shown in Fig. 15. As is shown, the specific power consumption increases at lower boiling temperatures and upon the increase of the temperature difference between the condensing vapor and the boiling brine. The decrease in the specific power consumption at higher boiling temperatures is caused by the decrease in the vapor specific volume. Similarly, increase of the temperature difference between the condensing and formed vapors gives a larger compression ratio. Either effect increases the compressor power consumption. Levels of the specific power consumption shown in Fig. 15 vary between high values of 25 kWh/m^3 to low values below 10 kWh/m^3. In practice, the specific power consumption for the MVC system is close to 15 kWh/m^3 at a boiling temperature of 60 °C. Superposing these values on Fig. 15, indicate that the system is operated at temperature difference of 3 °C for (T_s-T_d) and between 1-2 °C for (T_d-T_b).

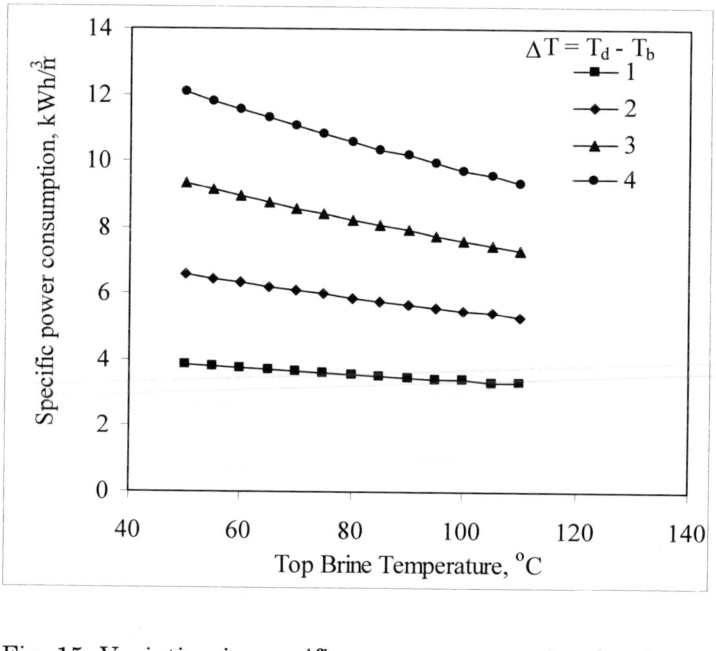

Fig. 15. Variation in specific power consumption for the evaporator as a function of top brine temperature and temperature difference of condensing steam and boiling brine.

Variations for the evaporator specific heat transfer area are shown in Fig. 16. The specific heat transfer area decreases at higher boiling temperatures and increases upon the decrease of the temperature difference between condensing vapor and boiling brine. Higher boiling temperatures enhance the heat transfer rates; this is because of the decrease in the liquid density and viscosity and the increase of the thermal conductivity of the liquid and metal walls. As a result, the overall heat transfer coefficient increases and results in reduction of the heat transfer area. The temperature difference of the condensing vapor and boiling brine is the driving force for heat transfer across the evaporator tubes. Lowering this difference decreases the driving force for heat transfer and in turn increases the heat transfer area. As is shown in Fig. 16, the evaporator heat transfer area is more sensitive to variations in the temperature difference of the condensing vapor and boiling brine. Variations in the evaporator heat transfer area are limited to a low value of 8% as the boiling temperature is increased from 60 to 105 °C. On the other hand, a four folds increase occurs in the evaporator heat transfer area as the temperature difference of the condensing vapors and boiling brine is increased from 1 to °C. In actual practice, the evaporator specific heat transfer area varies between 400-600 $m^2/(kg/s)$ at boiling temperature of 60 °C. Applying this value to the data shown in Fig. 16 gives a 2 °C for the operating difference between the temperatures of the condensing vapor and boiling brine. This is an interesting result because it is consistent with the previous superposition result obtained for the specific power consumption data.

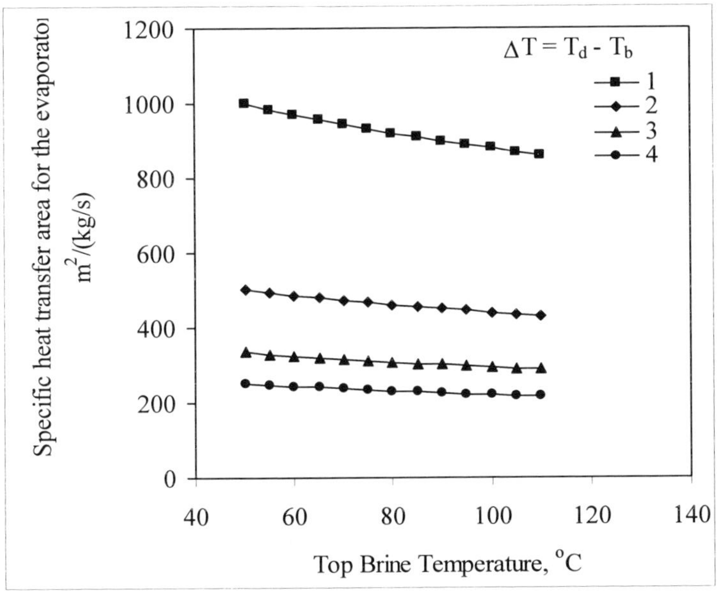

Fig. 16. Variation in specific heat transfer area for the evaporator as a function of top brine temperature and temperature difference of condensing steam and boiling brine.

Figure 17 shows the results for the specific heat transfer area for the distillate-feed preheater. The area increases at higher boiling temperatures, T_b, and lower difference for the condensing vapor and boiling brine (T_d-T_b). The heat load of the preheater is given in terms of the temperature difference (T_d-T_o). This difference increases at higher boiling temperatures, T_b, since T_d is kept higher than T_b by 1-4 °C. Decrease of the temperature difference for the condensing vapor and the boiling brine dictates the decrease of the feed seawater temperature, T_f. This decrease is necessary to supply the evaporator with a smaller amount of energy in order to provide for the smaller enthalpy difference of the condensing and forming vapors. It is expected that the decrease in the T_f value should increase the value of $(LMTD)_d$. However, this decrease is also associated with simultaneous decrease in the T_d, which occurs at a larger rate. As a result, the $(LMTD)_d$ value decreases as the temperature difference of the condensing vapor and the boiling brine is decreased. The decrease in the value of $(LMTD)_d$ results in a larger in heat transfer area.

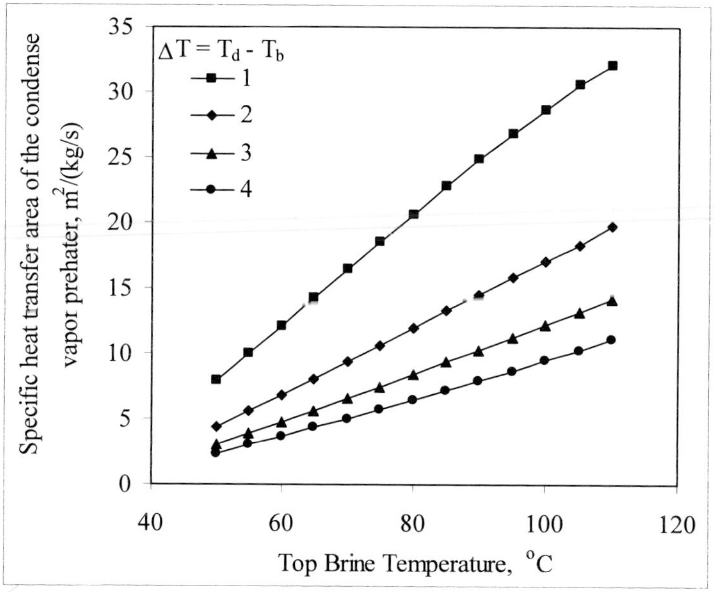

Fig. 17. Variation in specific heat transfer area of condensed vapor preheater as a function of top brine temperature and temperature difference of condensing steam and boiling brine.

The specific heat transfer area for the brine-feed preheater is shown in Fig. 18. As is shown, the heat transfer area increases upon the increase of the brine boiling temperature, T_b, and the temperature difference of the condensing vapor and boiling brine, T_d -T_b. The increase in the brine boiling temperature increases the preheater thermal load, which is defined in terms of the difference, T_b-T_o. As for the increase in the preheater area with the increase in the temperature difference of the condensing vapor and the boiling brine, it is caused by the decrease in the $(LMTD)_b$ value. As discussed above increase in the value of (T_d - T_b) at constant T_b increases the feed seawater temperature, T_f. This increase reduces the temperature difference at the hot end of the preheater, because the brine temperature remains constant. The decrease in the temperature difference at the hot end of the preheater reduces the $(LMTD)_b$ value and in turn increases the preheater area.

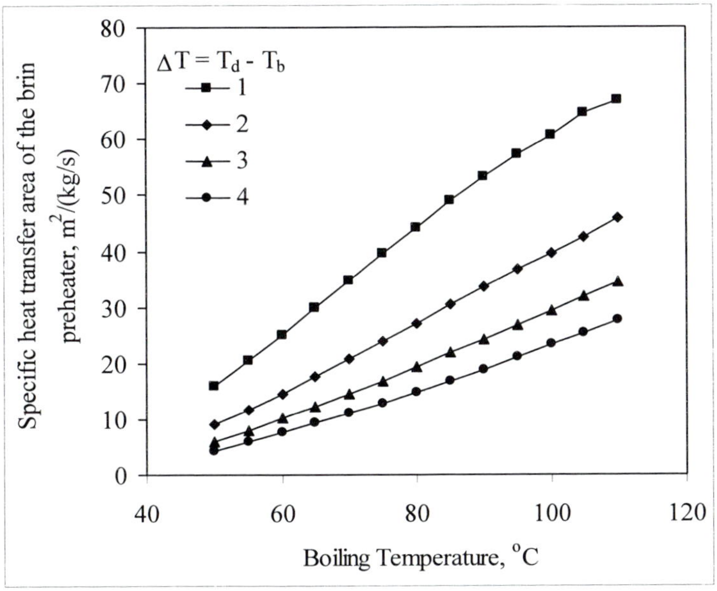

Fig. 18. Variation in specific heat transfer area of brine preheater as a function of top brine temperature and temperature difference of condensing steam and boiling brine.

Comparison of the specific heat transfer area for the two preheaters show that the brine preheater area is 3 to 5 times larger than the distillate preheater area. This is because of the higher thermal load found in the brine preheater, where the brine mass flow rate, M_b, is 1.5 kg/s, while the distillate flow rate, M_d, is only 1 kg/s. Difference dependence is found for variation in the area of the two preheaters as a function of the temperature difference of the condensing vapor and boiling brine. The area for the distillate-feed preheater is found to increase at lower differences and the opposite behavior is found for the brine-feed preheater. This result indicate that selection of the temperature difference between the condensing vapor and the boiling brine should be optimized to minimize the heat transfer areas for the two preheaters as well as the evaporator preheater and the specific power consumption.

Results for the total specific heat transfer area is illustrated in Fig. 19. As is shown the total specific increases with the increase in the brine boiling temperature and the temperature difference of the condensing vapor and boiling brine. The evaporator specific heat transfer area dictates this behavior, since its specific area is larger than the specific areas for the two preheaters.

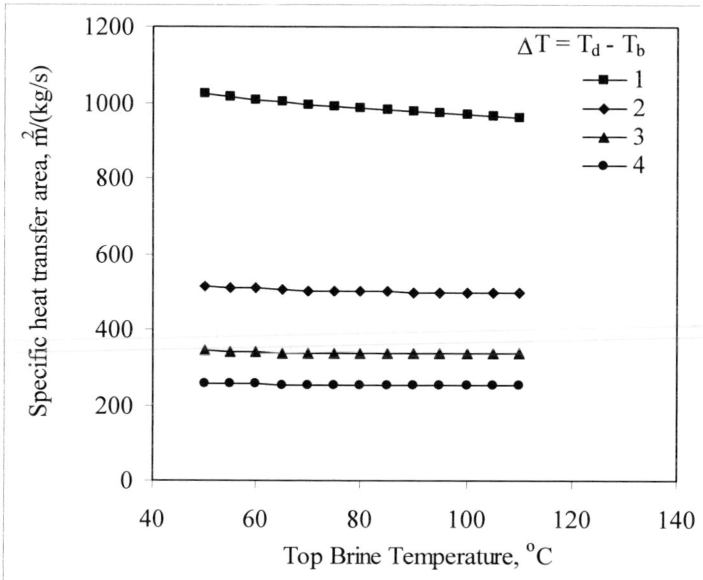

Fig. 19. Variation in the specific heat transfer area as a function of top brine temperature and temperature difference of condensing steam and boiling brine.

3.2.4 Industrial Data and Practice

Limited amount of literature data or industrial technical reports can be found on characteristics of the single-effect MVC system. The majority of available data reports specific power consumption, capacity, product purity, plant factor, material of construction, overall dimensions, and the brine boiling temperature. It is common that specific data on the heat transfer areas or temperatures of various streams are not reported. Moreover and as discussed the introduction, the number of literature studies on the MVC system is small and the majority of the articles focus on the process main characteristics, i.e., capacity, plant factor, and specific power consumption.

Table 1 includes a summary of data extracted from literature on single-effect MVC system. The last column in the table includes predictions of the detailed model. The model predictions are made at a brine boiling temperature of 60 °C, which consistent with industrial practice. The model predicts a specific power consumption of 10.24 kWh/m^3, which is consistent literature data. This value is obtained for a temperature difference of 3 °C for (T_s-T_d). The evaporator specific heat transfer area is also consistent with value reported by Veza, 1995. This value is obtained at a temperature difference of 2 oC for (T_d-T_b), which is lower than the reported value, Lucas and Tabourier, 1985. Although, no values are reported for the specific heat transfer area of the feed preheaters, the predicted values are consistent with the thermal load of each unit.

Table 1
Comparison of industrial data and model predictions

	Matz and Fisher 1981	Matz and Zimerman 1985	Lucas and Tabourier 1985	Veza 1995	Model
Specific Power Consumption kWh/m^3	17-18	10	10	10-11	10.24
Capacity, m^3/d	50-500	250-450	25-300	500	-
Boiling Temperature, °C	40-50	50-70	-	59	60
Evaporator area m^2/(kg/s)	-	-	-	448.9	483
Brine-Feed Preheater specific area, m^2/(kg/s)	-	-	-	-	206
Distillate-Feed Preheater specific area, m^2/(kg/s)	-	-	-	-	50
T_d-T_b, °C	-	-	1	-	2
T_o-T_{cw}, °C	-	-	2-4	-	1.17
T_h-T_f and T_d-T_f, °C	-	-	2-3	-	0.3 and 2.3
T_s-T_b, °C	-	-	5	-	3

3.2.5 Summary

Analysis of the system performance by the mathematical models shows consistency of predictions and industrial practice. The specific power consumption is found to vary over a similar range, 9-17 kWh/m^3 at 60 °C. In addition, the predicted evaporator specific heat transfer area is close to the industrial practice, with values between 400-600 m^2/(kg/s) at 60 °C. The temperature values predicted by the model are also found consistent with reported industrial data.

References

Al-Juwayhel, F., El-Dessouky, H., and Ettouney, H., Analysis of single-effect evaporator desalination systems combined with vapor compression heat pumps, **114**(1997)253-275.

IDA Worldwide Desalting Plants Inventory, Wangnick, Report No. 13, December 1995, Int. Desalination Association (IDA), Gnarrenburg.

Lucas, M., and Tabourier, B., The mechanical vapour compression process applied to seawater desalination: A 1500 ton/day unit installed in the nuclear power plant of Flamanville, France, Desalination, **52**(1985)123-133.

Matz, R., and Fisher, U., A comparison of the relative economics of sea water desalination by vapor compression and reverse osmosis for small to medium capacity plants, Desalination, **36**(1981)137-151.

Matz, R., and Zimerman, Z., Low-temperature vapour compression and multi-effect distillation of seawater. Effects of design on operation and economics. Desalination, **52**(1985)201-216.

Veza, J.M., Mechanical vapour compression desalination plants – A case study, Desalination, **101**(1995)1-10.

Zimerman, Z., Development of large capacity high efficiency mechanical vapor compression (MVC) units, Desalination, **96**(1994)51-58.

Problems

1. An MVC system is to be designed to produce 5000 m^3/d of fresh water. The boiling temperature is 70 °C and the temperatures of the compressed vapor and condensate are higher by 8 °C and 3 °C, respectively. The salinity of the

feed seawater is 38000 ppm and the rejected brine salinity is 70000 ppm. For preliminary design considerations neglect thermodynamic losses, assume constant specific heat for all liquid stream (4.2 kJ/kg °C), constant specific heat for the vapor streams (1.884 kJ/kg °C), and constant overall heat transfer coefficients of 2.7, 7.2, and 7.8 for the evaporator, distillate preheater, and brine preheater, respectively. Calculate the following: M_b, M_f, T_f, T_o, A_b, A_d, A_e, and W.

2. Determine the effect of dependence of the specific heat of liquid streams, boiling point rise, latent heat, pressure, enthalpy, overall heat transfer coefficients, and specific volume on temperature and concentration on the design values obtained in problem 1. Use the correlations given in the appendices to calculate the physical properties and thermodynamic losses.
3. An MVC system is used to desalinate seawater at 35 °C with 42000 ppm salinity. The maximum allowable brine temperature is 100 °C. The heat transfer coefficient for the evaporator and the two preheaters is constant and equal to 5.016 kW/m^2 °C. The specific heat transfer area is 109.46 m^2 per (kg/s) of fresh water and the heat transfer area of the distillate preheater is 200 m^2. The flow rates of the hot and cold stream in the preheaters are equal. The temperatures of the distillate and rejected brine flowing from the preheaters are 45 °C and 40 °C, respectively. Calculate the specific power consumption.
4. An MVC system has the following design data:

   M_d = 1 kg/s, A_d = 10 m^2.

   A_b = 20 m^2, A_e = 500 m^2.

   U_e = 2.4 kW/m^2 °C, U_d = 6.7 kW/m^2 °C.

   U_b = 6.3 kW/m^2 °C, X_f = 42000 ppm.

   X_b = 70000 ppm.

   Determine T_b, T_d, T_o, T_f, and T_s if T_{cw} = 25 °C, Cp = 4.2 kJ/kg °C, and Cp_v = 1.884 kJ/kg °C. Also, determine the specific power consumption of the compressor and the flow rates of the brine and feed seawater.
5. For the same conditions in the previous problem determine T_b, T_d, T_o, T_f and T_s if T_{cw} drops to 15 °C. Also, determine the specific power consumption and the flow rates of the brine and feed seawater.
6. If fouling conditions arise in the system described in problem 2, where the overall heat transfer coefficient in the evaporator is drop to 1.8 kW/m^2 °C. If all other conditions are kept the same, determine T_b, T_d, T_o, T_f, and T_s at the new fouling conditions. Also, calculate the specific power consumption and the flow rates of brine and feed streams.

3.3 Single Effect Absorption Vapor Compression

The absorption vapor compression (ABVC) single or multiple effect desalination processes are not found on full commercial or industrial scale. However, the literature includes a large number of studies on development, innovation, and performance evaluation of ABVC systems for refrigeration and air conditioning processes, Kuehn et al. (1998). On the other hand, evaluation of the combined systems of these heat pumps and various thermal desalination processes is limited to a small number of studies in the literature. Weinberg et al. (1980) evaluated the performance of a coupled system of multiple effect evaporation, vacuum freezing, and lithium bromide absorption heat pump. The system is thought to enhance the performance ratio of the multiple effect evaporation to high values of 18-20 and operating temperatures between 0 to 60 ^oC. The low temperature operation minimizes corrosion and scaling problems. Alefeld and Ziegler (1985) proposed a fully integrated desalination system combined with LiBr-H_2O absorption heat pump. The system includes three stages, which process seawater and generates fresh water. Aly (1988) and Fathalah and Aly (1991) analyzed a solar powered LiBr-H_2O heat pump, which generates high grade steam to operate MEE desalination system. More emphasis, in their analysis, was given to the performance of the solar power unit and air conditioning in the evaporator unit. Yanniotis and Pilavachi (1996) modeled the performance of sodium hydroxide heat pumps in MEE systems. Model results are validated against experimental measurements and were found to have reasonable agreement. Al-Juwayhel et al. (1997) studied the performance of single effect evaporation desalination systems combined with various types of heat pumps. As discussed before, results for the ABVC gave thermal performance ratios close to three times higher than the single effect thermal vapor compression system. On the other hand, El-Dessouky and Ettouney (1997) showed 50% increase in the thermal performance ratio for the MEE-ABVC and MEE-ADVC systems over the MEE-TVC with values close to 20.

3.3.1 Process Description

Elements of the ABVC system are shown in Fig. 20. In this system, the evaporator constitutes horizontal falling film tubes, brine spray nozzles, demister, and the brine pool. The down condenser has a similar shell and tube configuration, where condensation takes place on the shell side. The absorber is also a shell and tube falling film configuration. Absorption of water vapor by the LiBr-H_2O solution occurs on the shell side of the absorber, while heating of the feed seawater and vapor formation takes place on the tube side of the absorber. The generator has a similar layout to the evaporator, where dilute LiBr-H_2O solution forms a falling film on the outside surface of the tubes and the motive steam is condensed inside the tubes. The heat exchange unit between the concentrated and diluted $LiBr-H_2O$ solution is used to exchange heat from the concentrated to the dilute solution. This improves the overall process efficiency. The process is described in the following points:

- The intake seawater stream flows through the down condenser, where it condenses part of the vapor formed in the evaporator. The remaining part of the vapor is fed to the shell side of the absorber in the heat pump.
- The intake seawater temperature increases from (T_{cw}) to (T_f) as it absorbs the latent heat of condensation of the condensing vapor. Part of the feed seawater is rejected back to the sea (M_{cw}) which is known as the cooling seawater. The remaining portion of the intake seawater is the feed seawater stream (M_f), which is chemically treated and deaerated before being fed to the tube side of the absorber.
- The concentrated $LiBr-H_2O$ solution absorbs the vapor stream entering the absorber. The absorption process is exothermic and releases sufficient amount of heat that sustains increase of the feed seawater temperature to the saturation temperature. Also, vapor is formed from the feed seawater within the absorber. This vapor forms part of the heating steam in the evaporator.
- The temperature of the absorbed vapor and the concentration of the outlet dilute $LiBr-H_2O$ solution define the equilibrium conditions in the absorber. It should be noted that boiling point elevation for the $LiBr-H_2O$, or the temperature difference between the dilute $LiBr-H_2O$ solution and the absorbed vapor, varies over a range of 10-50 °C as the mass fraction of $LiBr-H_2O$ in the dilute solution is increased from 0.25-0.45.
- The dilute $LiBr-H_2O$ solution enters the generator, where it is sprayed on the outside surface of the tubes. The solution absorbs the latent heat of motive steam that condenses on the tube side of the generator. The heating process increases the temperature of the $LiBr-H_2O$ solution to saturation and results in evaporating the same amount of water absorbed by the solution in the absorber. The concentration of the concentrated $LiBr-H_2O$ solution and the temperature of the formed vapor define the equilibrium conditions in the generator.
- The combined vapor formed in the generator and absorber derives the evaporation process in the evaporator. The brine stream leaving the absorber is sprayed on the outside surface of the evaporator tubes, where it absorbs the latent heat of condensation from the condensing steam on the tube side of the evaporator.
- The concentrated brine leaving the evaporator is rejected back to the sea and the formed vapor is routed to the down condenser. The sum of the condensate of the heating steam and the condensate in the down condenser forms the distillate product stream.
- Demisters in various units prevent droplet entrainment of brine and LiBr.

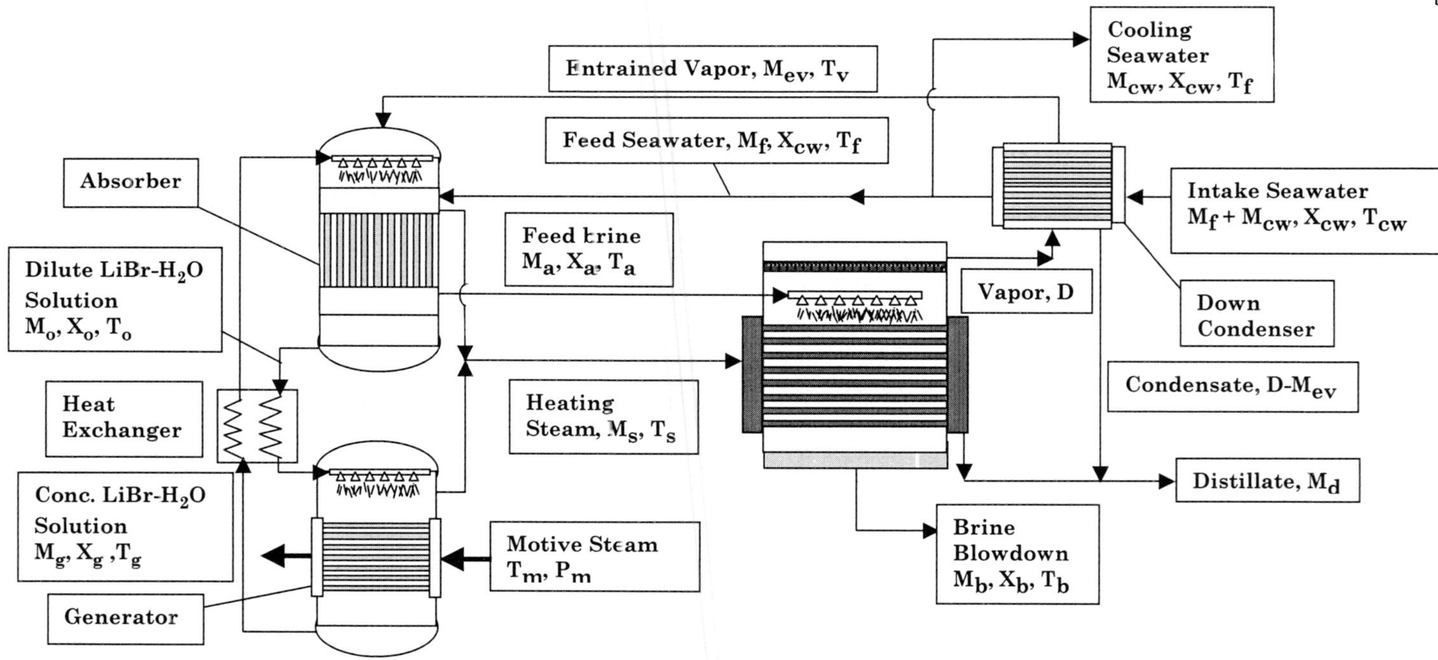

Fig. 20. Single effect absorption vapor compression desalination process (ABVC)

3.3.2 Process Modeling

The steady-state model includes a set of material and energy balances, heat transfer equations, and thermodynamic relations. Assumptions used in the model include:

- The vapor formed in the evaporator, absorber and generator is salt free. This assumes that the entrainment of brine droplets by the vapor stream is negligible and has no effect on salinity of the distillate product.
- Energy losses from the effects to the surroundings are negligible. This is because of operation at relatively low temperatures, between 100-40 °C, and the effects are well insulated.
- The physical properties of various streams are calculated at the average temperature of influent and effluent streams.

The overall material and salt balances are given by

$$M_f = M_d + M_b \tag{48}$$

$$M_b = M_f (X_f / X_b) \tag{49}$$

where M is the mass flow rate, X is the salt concentration, and the subscript b, d, and f denotes the brine, the distillate, and the feed seawater. In Eq. (49) the brine blow down salinity (X_b) is set at 90% of the saturation salinity of the $CaSO_4$ solution

$$X_b=0.9(457628.5-11304.11T_b+107.5781T_b^2-0.360747T_b^3) \tag{50}$$

This equation is obtained by curve fitting of the salinity/temperature relation for the solubility of $CaSO_4$, El-Dessouky et al. (2000b).

In the evaporator, the saturated falling brine film absorbs the latent heat of the condensing steam. This evaporates a controlled mass of vapor, D at T_v, where

$$M_s \lambda_s = D \lambda_v \tag{51}$$

where λ is the latent heat. The subscripts s and v denote the heating steam and the vapor formed, respectively. In the evaporator, absorber, and generator the boiling temperature are higher than the corresponding vapor saturation temperature by the boiling point elevation, $(BPE(T_b,X_b))$, and the temperature rise caused by the hydrostatic pressure head, ΔT_y. This is

$$T_b = T_v + BPE(T_b,X_b) + \Delta T_y \quad (52)$$

The term, ΔT_y, is negligible in horizontal falling films, because of the very small thickness of the boiling film.

The condensation temperature in the down condenser, T_c, is lower than the evaporation temperature, T_v, by the boiling point elevation, $(BPE(T_b,X_b))$, and the saturation temperature depressions associated with pressure losses in the demister, (ΔP_p), transmission lines between the effects, (ΔP_t), and vapor condensation inside the tubes, (ΔP_c). The resulting condensation temperature is

$$T_c = T_b - (BPE(T_b,X_b) + \Delta T_p + \Delta T_t + \Delta T_c) \quad (53)$$

The pressure drop during condensation, ΔP_c, is defined as the algebraic sum of the decrease caused by friction (ΔP_r) and the increase caused by gravity (ΔP_g) and vapor deceleration (ΔP_a). This relation is given by

$$\Delta P_c = (\Delta P_r - \Delta P_g - \Delta P_a) \quad (54)$$

Correlations for the pressure drop components ΔP_p, ΔP_t, ΔP_r, ΔP_g, and ΔP_a are given in the study by El Dessouky et al. (1998).

In the down condenser, the temperature of the intake seawater, $M_{cw}+M_f$, is increased from T_{cw} to T_f. Condensing part of the vapors formed in the evaporator provides the required heating energy. This energy balance is given by

$$(D - M_{ev})\lambda_c = (M_{cw} + M_f)\,C_p\,(T_f - T_{cw}) \quad (55)$$

where the subscripts c, cw, and ev denote the condensing vapors, the intake seawater, and the entrained vapor in the absorber.

The following relation gives the flow rate of the heating steam

$$M_s = M_{ev} + M_{ab} \quad (56)$$

where M_{ev} is the amount of entrained vapor in the absorber, subsequently released in the generator as a part of the heating steam. M_{ab} is remaining part of heating steam generated in the absorber. Inspection of Fig. 20 shows that the total distillate flow rate is given by

$M_d = D + M_{ab}$ (57)

The energy balance around the absorber is given by

$M_g H_g + M_{ev} H''_{ev} + M_f H_f = (M_g + M_{ev}) H_o + M_{ab} H''_s + M_a H_a$ (58)

where M_g is the flow rate of the concentrated LiBr-H_2O solution entering the absorber, M_a is the brine mass flow leaving the absorber. The water vapor saturation enthalpies H''_{ev} and H''_s are obtained at the condensation temperature in the down condenser (T_c) and the heating steam temperature in the evaporator (T_s), respectively. In Eq. (57) H_g and H_o are the enthalpies of the concentrated and diluted LiBr-H_2O solution evaluated at (C_g,T_g) and (C_o,T_o). It should be noted that T_g and T_o are obtained from the equilibrium relation in Appendix A at the water vapor saturation temperatures of T_s and T_c, respectively. The enthalpies of the feed seawater and the feed brine H_f and H_a are calculated at (T_f, X_{cw}) and (T_a, X_a), respectively. The heating steam temperature is related to the feed brine temperature by the boiling point elevation, or,

$T_s = T_a - BPE(X_a,T_a)$ (59)

The energy equation for the generator balances the amount of input energy in the motive steam and the dilute LiBr-H_2O solution against the amount of output energy in the concentrated LiBr-H_2O solution and the heating steam. This relation is given by

$(M_g + M_{ev}) H_o + M_m H''_m = M_g H_g + M_{ev} H''_s$ (60)

where M_m and H''_m are the mass flow rate and enthalpy of motive steam.

The material and salt balance around the absorber for the feed seawater and the feed brine are given by

$M_f = M_a + (M_s - M_{ev})$ (61)

$X_f M_f = M_a X_a$ (62)

Similarly, the following relations give the total mass and salt balance for the concentrated and diluted LiBr-H_2O solutions

$M_o = M_g + M_{ev}$ (63)

$$M_o\,C_o = M_g\,C_g \tag{64}$$

The design equations for the heat transfer area are developed for the evaporators, the preheaters, and the down condenser. For the evaporators, the heat transfer area, A_e, is

$$A_e = \frac{M_s \lambda_s}{U_e (T_s - T_b)} \tag{65}$$

where U is the overall heat transfer coefficient, and the subscript e refers to the evaporator.

The heat transfer area of the down condenser is given by

$$A_c = \frac{(D - M_{ev})\,\lambda_c}{U_c A_c (LMTD)_c} \tag{66}$$

$$(LMTD)_c = \frac{T_f - T_{cw}}{\ln \dfrac{T_c - T_{cw}}{T_c - T_f}} \tag{67}$$

The overall heat transfer coefficient in the evaporator and condenser are:

$$U_e = 1.9394 + 1.40562\text{x}10^{-3} T_b - 2.0752\text{x}10^{-4} (T_b)^2 + 2.3186\text{x}10^{-6} (T_b)^3 \tag{68}$$

$$U_c = 1.6175 + 0.1537\text{x}10^{-3} T_v + 0.1825\text{x}10^{-3} (T_v)^2 - 8.026\text{x}10^{-8} (T_v)^3 \tag{69}$$

where U_e and U_c are the overall heat transfer coefficient in the evaporator and down condenser in kW/m^2 ^oC, T_b is the brine boiling temperature, and T_v is the vapor condensation temperature in the condenser. All temperatures in the above correlations are in ^oC. The standard deviations for the above correlations are 2.03% and 1.76%. These correlations are tested and proved to be reliable through comparison against other correlations in the literature and available experimental and design data.

The system performance is defined in terms of the following parameters:

- Performance ratio, which is defined as the amount of the distillate product per unit mass of the motive steam

$$PR = M_d/M_m \tag{70}$$

- Specific flow rate of cooling water, which is defined as the amount of the cooling water per unit mass of distillate product

$sM_{cw} = M_{cw}/M_d$ (71)

- Specific heat transfer area, which is defined as the ratio of the total heat transfer area of the evaporator and condenser to the total flow rate of distillate product

$sA = (A_e+A_c)/M_d$ (72)

- Conversion ratio, which is defined as the amount of distillate product per unit mass of feed seawater

$CR = M_d/M_f$ (73)

Solution Method

The solution procedure is shown in Fig. 21. Solution of the model equations requires definition of the following system variables:

- The distillate flow rate, M_d, is 1 kg/s.
- The intake seawater temperature, T_{cw}, is 25°C.
- The heating steam temperature, T_s, is higher than the brine boiling temperature T_b by 2-10 °C.
- The feed seawater temperature, T_f, is lower than the vapor condensation temperature T_c by 5 °C.
- The feed brine temperature, T_a, is lower than the temperature of the dilute LiBr-H_2O solution T_b by 5 °C.
- The motive steam temperature, T_m, is higher than the temperature of the concentrated LiBr-H_2O solution T_g by 5 °C.
- The range for the heating steam temperature, T_s, is 50-100 °C.
- The feed seawater salinity, X_f, is 36000 ppm.

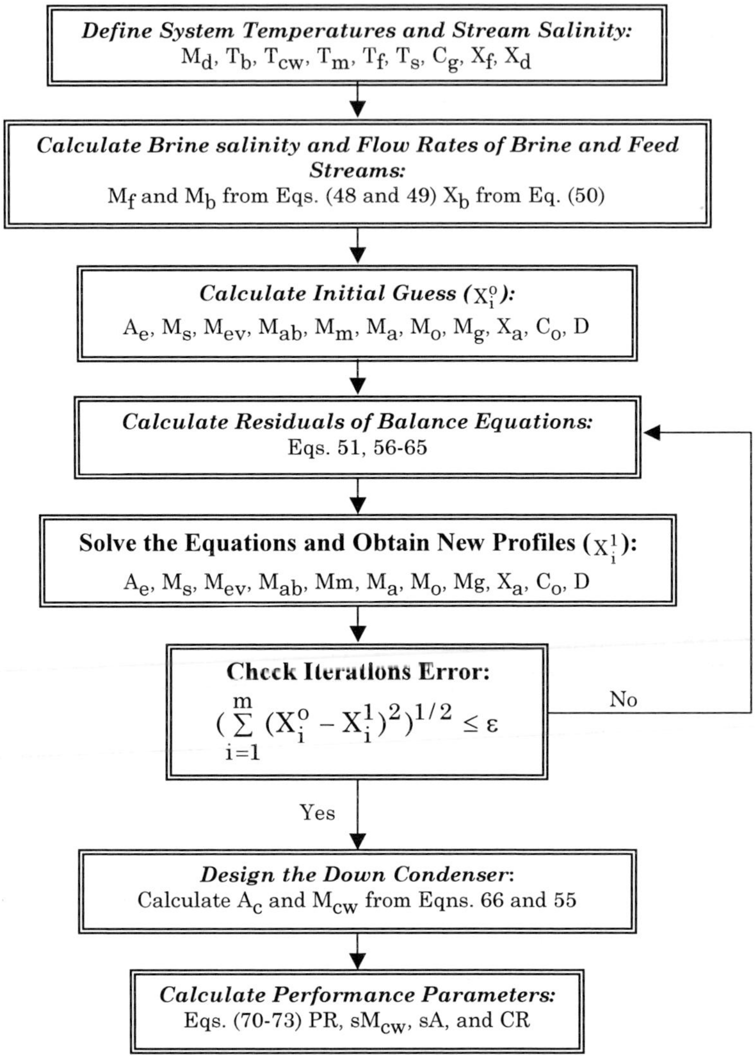

Fig. 21. Solution algorithm of the absorption heat pump and the single effect evaporation desalination system.

As is shown in Fig. 21 solution sequence proceeds as follows:

- The system capacity, stream temperatures, and stream salinity are defined as specified above.
- Eqs. 48-50 are solved to determine the feed and brine flow rates and the salinity of the brine blowdown.
- An initial guess is assumed for the following:
  - Evaporator area.
  - The flow rates of the heating steam, motive steam, entrained vapor, vapor formed in the absorber, dilute LiBr-H_2O, and the concentrated LiBr-H_2O.
  - Concentrations of the dilute LiBr-H_2O solution and brine leaving the absorber.
- The above variables are calculated by solution of Eqs. 51, 56-65. Solution proceeds iteratively using Newton's method. Iterations continue until the tolerance criterion is achieved with a value of $1x10^{-4}$ for ε.
- The flow rate of the cooling seawater and the heat transfer area of the condenser are calculated from Eqs. 55 and 66, respectively.
- The performance parameters are calculated from Eqs. 70-73.

Example 1:

Design a single effect evaporation desalination unit combined with absorption vapor compression. The system operates at the following conditions:

- The system capacity is 1 kg/s.
- Compressed vapor temperature is 80 °C.
- The brine boiling temperature is 77 °C.
- The mass fraction of LiBr in the concentrated solution is 0.7.
- The intake seawater temperature is 25 °C.
- The intake seawater salinity is 36000 ppm.
- The temperature of the seawater stream leaving the condenser is lower than the temperature of the condensing vapor by 5 °C.
- The motive steam temperature is higher than the boiling temperature of the concentrated LiBr solution by 5 °C.
- The temperature of the seawater stream leaving the absorber is lower than the boiling temperature of the dilute LiBr solution by 5 °C.
- The temperature approach for the hot and cold ends in the LiBr heat exchanger is 3 °C.
- Seawater velocity inside the condenser tubes is 1.5 m/s.
- Salinity of product fresh water is 0 ppm.
- Outer diameter of the condenser and evaporator tubes is 0.015 m.
- Wall thickness of the condenser and evaporator tubes is 0.005 m.
- Thermal conductivity of the condenser and evaporator tubes is 0.042 kW/m °C.
- Total fouling resistance inside/outside the condenser and evaporator tubes is 0.001 kW.

- Velocity of steam condensate inside the evaporator tubes is 1.5 m/s.
- Velocity of falling film on the outside surface of the evaporator tubes is 1.5 m/s.
- Thickness of falling film on the outside surface of the evaporator tubes is 0.001 m.

Solution:

The model solution starts with evaluation of some the system parameters, which includes X_b, $BPE(X_b,T_b)$, T_v, T_f, M_f, and M_b. The rejected brine salinity is calculated from the saturation correlation, where

$$X_b = 0.9(457628.5-(11304.11)T_b+ (107.5781)T_b^2-(0.36074702)T_b^3)$$

$$X_b = 54314.7 \text{ ppm}$$

The boiling point elevation in the evaporator is obtained from the correlation in Appendix B

$$BPE(X_b,T_b) = 1.36\ ^oC$$

Therefore, the vapor temperature in the evaporator is obtained from

$$T_v = T_b - BPE(T_b,X_b) = 77 - 1.36 = 75.64\ ^oC$$

This gives a feed seawater temperature of

$$T_f = T_v - 5 = 75.64 - 5 = 70.64\ ^oC$$

The flow rates of the feed seawater and rejected brine are obtained from the following equations

$$M_f = (M_d)(X_b)/(X_b - X_f) = (1)(54314.7)/(54314.7-36000) = 2.97 \text{ kg/s}$$

$$M_b = M_f - M_d = 2.97 - 1. = 1.97 \text{ kg/s}$$

The second part of the model solution involves iterative and simultaneous solution of the mass and energy balances and the heat transfer equations to determine the following system variables:

- Flow rate of compressed vapor (M_s) = 0.666 kg/s
- Heat transfer area of the evaporator (A_e) = 257.8 m^2
- Flow rate of entrained vapor (M_{ev}) = 0.342 kg/s

- Flow rate of vapor formed in the absorber (M_{ab}) = 0.324 kg/s
- Flow rate of motive steam (M_m) = 0.37 kg/s
- Flow rate of concentrated LiBr-H_2O solution = 0.34 kg/s
- Flow rate of seawater leaving absorber (M_a) = 2.64 kg/s
- Salinity of seawater leaving absorber (X_a) = 40413.1 ppm
- Mass fraction of LiBr in dilute solution (X_o) = 0.349
- Boiling vapor flow rate from evaporator (D) = 0.663 kg/s
- Flashing vapor flow rate from evaporator (D_f) = 0.0132 kg/s

Other system parameters obtained after the final iteration include the following:

- Performance ratio (PR) = $\frac{1}{M_m} = \frac{1}{0.37} = 2.7$
- Boiling temperature of concentrated LiBr-H_2O (T_g) = 158.7 °C
- Boiling temperature of diluted LiBr-H_2O (T_a) = 85.9 °C
- Temperature of compressed vapor (T_s) = 80.04 °C
- Temperature of seawater leaving absorber (T_a) = 80.93 °C
- Motive steam temperature (T_m) = 163.7 °C
- Pressure of motive steam (P_m) = 677.4 kPa

Enthalpy data at the above conditions include the following:

- Enthalpy of compressed vapor = 2636.27 kJ/kg
- Enthalpy of concentrated LiBr-H_2O = 4851.4 kJ/kg
- Enthalpy of diluted LiBr-H_2O = 1387.6 kJ/kg
- Enthalpy of seawater leaving condenser = 283.7 kJ/kg
- Enthalpy of seawater leaving absorber = 323.8 kJ/kg
- Enthalpy of motive steam = 2077.03 kJ/kg

Analysis of the condenser unit gives the following results:

- Flow rate of cooling seawater (M_{cw}) = 1.28 kg/s
- Condenser heat transfer coefficient (U_c) = 3.48 kJ/m^2 °C
- Condenser heat transfer area (A_c) = 11.3 m^2

3.3.3 System Performance

System performance is evaluated as a function of the heating steam temperature, the temperature difference of the heating steam and the boiling brine, and the mass fraction of LiBr-H_2O in the concentrated solution. Effects of the heating steam temperature and the mass fraction of the LiBr-H_2O in the concentrated solution are shown in Figs. 22-24 for the variations in the performance ratio, the specific heat transfer area, and the specific flow rate of cooling water. The three figures show insensitive and independent behavior of the system parameters on the mass fraction of LiBr-H_2O in the concentrated solution. The following causes this behavior:

- Various system temperatures, which includes the heating steam, the boiling temperature, the feed seawater, and the feed brine are independent of the LiBr-H_2O mass fraction in the concentrated solution. These temperatures affect the system variables, which are used to calculate the system performance parameters.
- Increase in the LiBr-H_2O mass fraction in the concentrated solution affects only the flow rate of the concentrated solution. At higher concentrations, the solution enthalpy increases and results in reduction of the concentrated solution flow rate. This is necessary to balance the system energy in the absorber and generator.

The main effect of increasing the LiBr-H_2O mass fraction in the concentrated solution is the need for higher pressure motive steam. The is illustrated in the following data are, which are obtained for a heating steam temperature of 100 °C:

- C_g = 0.75, Tg = 201.87 °C, T_m = 206.87, P_m = 1789 kPa (17.89 bar), PR = 2.4, sA = 220.9, sM_{cw} = 0, CR = 0.092.
- C_g = 0.45, T_g = 122.93 °C, T_m = 127.93, P_m = 253 kPa (2.53 bar), PR = 2.3, sA = 212.9, sM_{cw} = 0, CR = 0.092.

Selection between the two points depends on the following factors:

- Availability of high pressure steam, i.e., 17 bar versus 3 bar.
- Increase in the system second law efficiency upon operation at low steam pressures, Hamed et al. (1996), Darwish and El-Dessouky (1996).
- Use of smaller tube diameter for higher pressure steam, El-Dessouky et al. (2000a).
- Elimination of the control loops required for reduction of the steam pressure to lower values of 3 bar, Alatiqi et al. (1999).

As is shown in Figs. 22-24 effects of the heating steam temperature on the system performance are more pronounced than the mass fraction of the LiBr-H_2O in the concentrated solution. This behavior is dramatic concerning the specific heat transfer area and the specific flow rate of cooling water. As is shown in Fig. 23, the specific heat transfer area has values above 400 $m^2/(kg/s)$ at heating steam temperatures close to 50 °C. On the other hand, the specific heat transfer area decreases to values between 200-250 $m^2/(kg/s)$, which are considered the industrial practice, as the temperature is increased to values between 80-100 °C. This behavior is primarily caused by increase in the overall heat transfer coefficient at higher temperatures. This enhances the heat transfer rate and results in reduction of the required heat transfer area. A lesser factor is the reduction in the latent heat of evaporation or condensation at higher temperatures, which results in reduction in the thermal load of the evaporator and the down condenser. It should be stressed that the temperature difference driving force between heating steam and boiling brine has no effect because it is kept constant in the calculations.

At higher heating steam temperatures, the specific flow rate of cooling water is zero, Fig. 24. This is because of the limitations imposed on the salinity of the brine blow down stream, Fig. 25, where at higher temperatures the difference in the salinity of the feed seawater and the brine blow down is less than 1000-2000 ppm. As a result of the constant production capacity, the feed flow rate of seawater increases to higher values and reduces the flow rate of the cooling seawater. The opposite behavior occurs at lower temperature, where higher conversion ratios are achieved. This reduces the feed flow rate and results in the increase in the cooling seawater flow rate.

Variations in the system performance as a function of the heating steam temperature and the temperature difference between the heating steam and the boiling brine are shown in Figs. 26-28. As is shown the two parameters have strong effect on the specific heat transfer area and the specific flow rate of cooling water. As is shown in Fig. 27, the increase in the temperature reduces the specific heat transfer area. At larger temperature differences, the driving force for heat transfer increases and results in reduction in the heat transfer area. Simultaneously, this effect increases the amount of distillate product, which increases conversion ratio. As discussed before, increase in the conversion ratio is associated with simultaneous decrease in the feed seawater flow rate and increase in the flow rate of cooling seawater, Fig. 28.

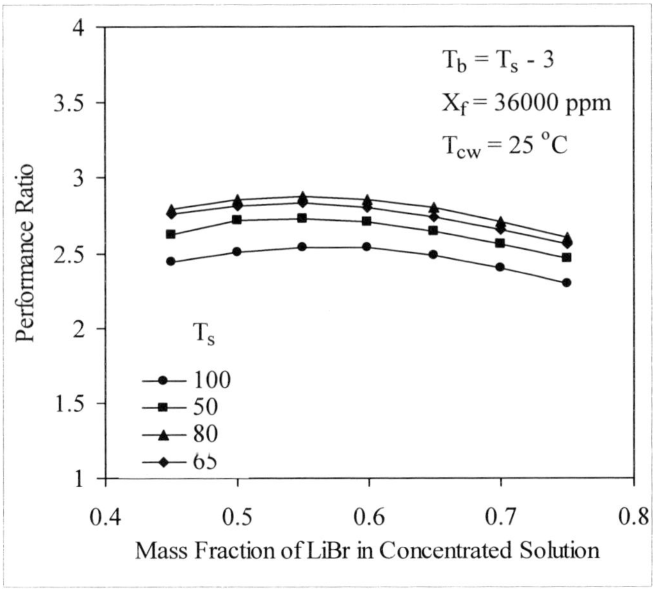

Fig. 22. Variation in the performance ratio as a function of the mass fraction of LiBr in concentrated solution and the heating steam temperature

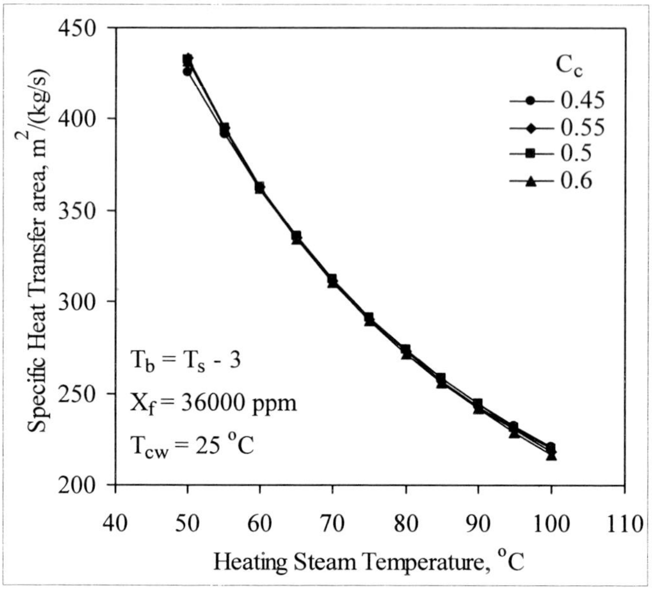

Fig. 23. Variation in the specific heat transfer as a function of the mass fraction of LiBr in concentrated solution and the heating steam temperature

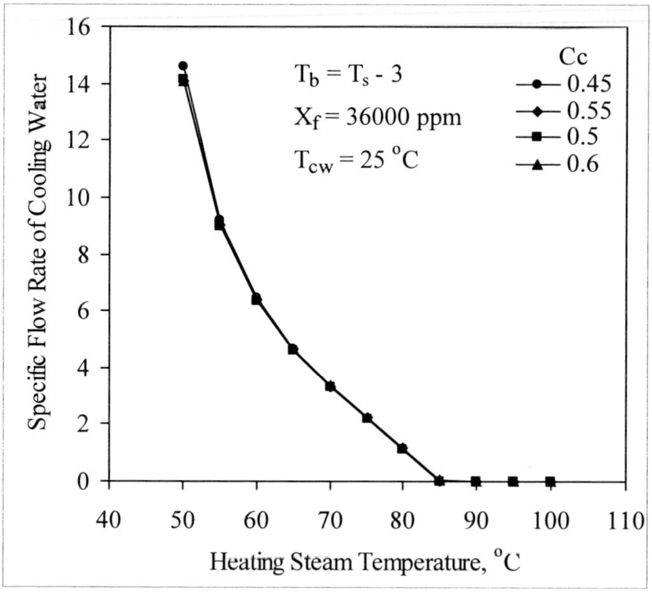

Fig. 24. Variation in the specific flow rate of cooling as a function of the mass fraction of LiBr in concentrated solution and the heating steam temperature

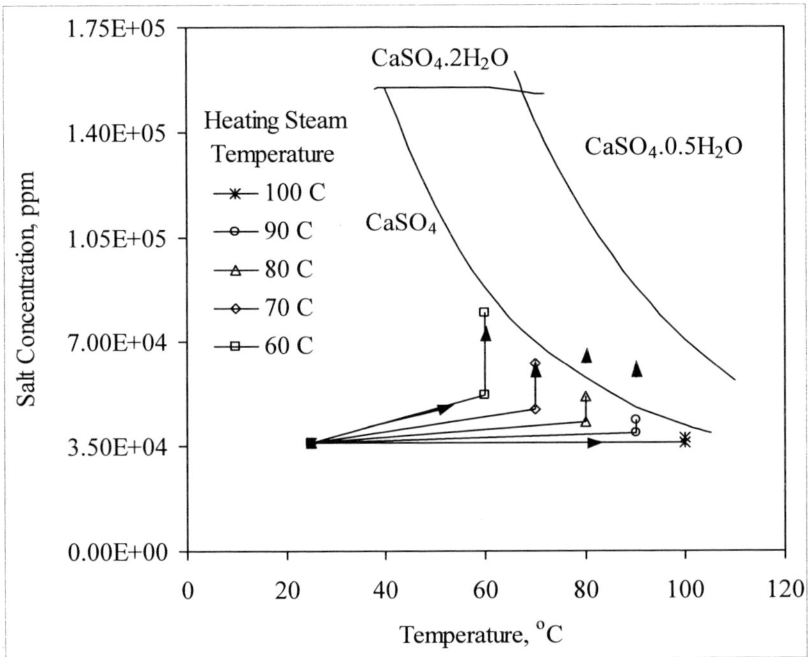

Fig. 25. Calcium sulfate solubility and top brine temperature for ABVC system.

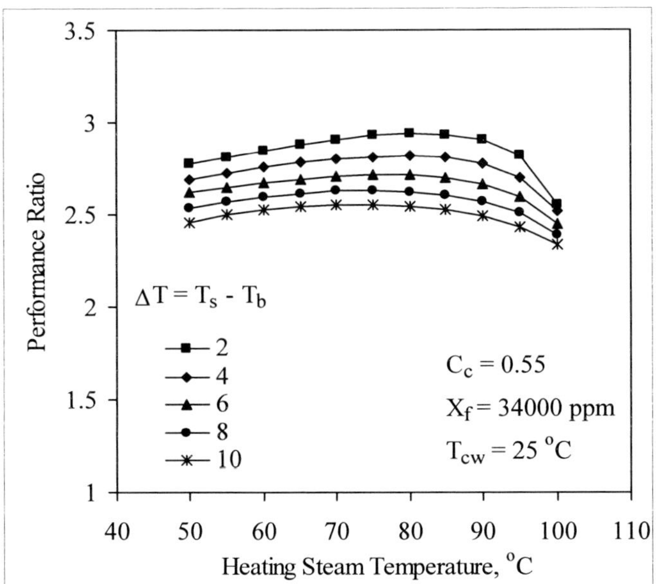

Fig. 26. Variation in the performance ratio as a function of the temperature difference of heating steam and top brine temperature and the heating steam temperature

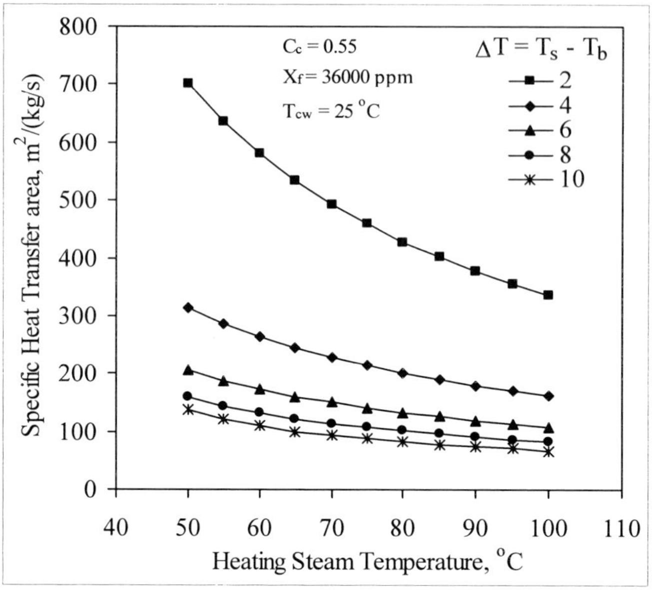

Figure 27: Variation in the specific heat transfer area as a function of the temperature difference of heating steam and top brine temperature and the heating steam temperature

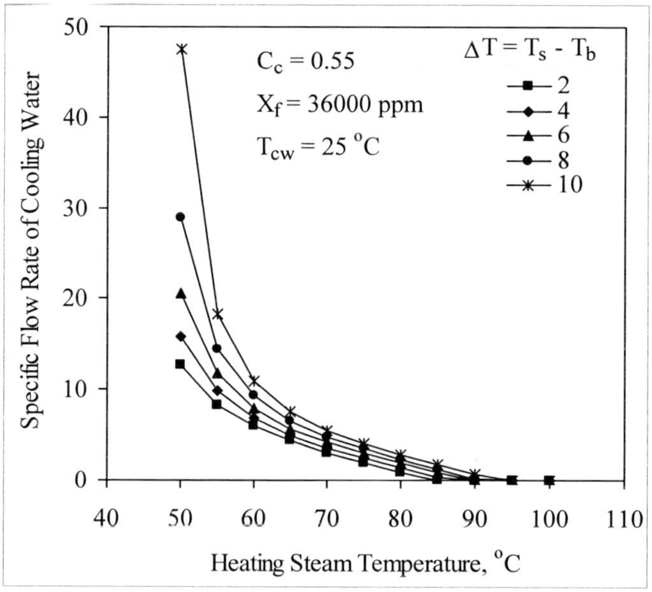

Fig. 28. Variation in the cooling water specific flow rate as a function of the temperature difference of heating steam and top brine temperature and the heating steam temperature

3.3.4 Summary

The absorption heat pump combined with the single effect evaporation desalination process is analyzed as a function of the design and operating parameters. The following conclusions are made in light of results and analysis:

- The thermal performance ratio varies over a range of 2.4-2.8 and is close to 50-70% higher than that of the single effect thermal vapor compression, El-Dessouky and Ettouney (1999).
- Effects of the LiBr mass fraction in the concentrated solution are minimal on the system performance. However, choice of this parameter is dependent on steam availability.
- The specific heat transfer area decreases with the increase in the heating steam temperature and the temperature difference of the heating steam and boiling brine.
- The specific flow rate of cooling water decreases at higher heating steam temperatures and lower temperature difference between the heating steam and boiling brine.

In summary, selection of the optimum design and operating conditions should take into considerations attractive features for system operation at higher temperatures. At these conditions, drastic reduction occurs in the specific flow rate of cooling water and the specific heat transfer area. Both factors result in considerable savings in the capital and production cost.

References

Alatiqi, I., Ettouney, H.M., and El-Dessouky, H.T., Process control in water desalination industry: An overview, Desalination, **126**(1999)33-40.

Alefeld, G., Ziegler, F., Advanced heat pump and air-conditioning cycles for the working pair H_2O/LiBr: Industrial applications, ASHRAE Tech. Data Bull., June 1985, pp 11-24.

Al-Juwayhel, F., El-Dessouky, H.T., and Ettouney, H.M., Analysis of single-effect evaporator desalination systems driven by vapor compression heat pumps, Desalination, **114**(1997)253-275.

Aly, S.E., vapour compression distillation using waste heat absorption systems, Desalination, **68**(1988)57-68.

Darwish M.A., and El-Dessouky, H., The heat recovery thermal vapour-compression desalting system: A Comparison with other thermal desalination Processes, Applied Thermal Engineering, **18**(1996)523-537.

El-Dessouky, H.T., and Ettouney, H.M., Simulation of Combined Multiple Effect Evaporation-Vapor Compression Desalination Processes, 1st IDA Int. Desalination Conference in Egypt, Cairo, Egypt, September, 1997.

El-Dessouky, H.T., Alatiqi, I., Bingulac, S., and Ettouney, H.M., Steady-state analysis of the multiple effect evaporation desalination process, Chem. Eng. Tech., **21**(1998)15-29.

El-Dessouky, H.T., and Ettouney, H.M., Single effect thermal vapor compression desalination process: Thermal analysis, Heat Transfer Eng., **20**(1999)52-68.

El-Dessouky, H.T., Ettouney, H.M., Al-Fulaij, H., and Mandani, F., Multistage flash desalination combined with thermal vapor compression, Chem. Eng. Proc., in print, 2000a.

El-Dessouky, H.T., Ettouney, H.M., and Al-Juwayhel, F., Parallel feed multiple effect evaporation with thermal vapor compression, Trans. I. Chem. E., in print, 2000b.

Fathalah, K., and Aly, S.E., Theoretical study of a solar powered absorption/MED combined system, Energy Convers., **31**(1991)529-544.

Hamed, O.A., Zamamiri, A.M., Aly, S., and Lior, N., Thermal performance and exergy analysis of thermal vapor compression desalination system, Energy Convers. Mgmt, **37**(1996)379-387.

Kuehn, T.H., Ramsey, J.W., Threlkeld, J.L., Thermal environmental engineering, 3rd ed., Prentice Hall, New York, 1998.

Weinberg, J., Ophir, A., and Fisher, U., Coupling of multi-effect distillation with vacuum freezing to reduce energy cost in sea water desalination, Proceedings of the 7th Int. Symposium on Fresh Water from the Sea, 1980, Vol. I, 283-292.

Yanniotis, S., and Pilavachi, P.A., Mathematical modeling and experimental validation of an absorption-driven multiple –effect evaporator, Chem. Eng. Technol., **19**(1996)448-455.

3.4 Single Effect Adsorption Vapor Compression

The adsorption-desorption heat pump is environmentally friendly. The pump uses benign fluids, which does not contribute in the destruction of the ozone layer. Their role in the greenhouse effect is negligible because they can be driven by renewable or waste energy sources and also due to their high thermal efficiency. Moreover, the process is simple, does not include moving parts, has a long life, and is vibration free. For these reasons, in recent years, the adsorption-desorption heat pump has attracted increasing attention in the concern of replace the traditional compressor-based systems, which utilize ozone harmful fluids. Applications of the adsorption-desorption heat pumps are found in air conditioning and in ice making.

3.4.1 Process Description

The ADVC system is shown diagrammatically in Fig. 29. The system includes the evaporator/condenser unit, two adsorption beds, feed preheaters, and a heat exchanger for the thermal fluid circulating between the adsorption and desorption beds. It is interesting to note that the evaporator and condenser form a single unit in this configuration, which replaces the individual condenser and evaporator in conventional adsorption heat pumps. Also, the feed preheaters are plate type and are used to exchange heat between the feed seawater and the condensed vapor and the rejected brine. The adsorber plays the role of the bottom condenser in the TVC system. That is, this adsorber absorbs or rejects the excess heat added to the system in the second adsorber.

The closed cycle of the heat pump is composed of the following steps:

1. Initially, bed I is assumed to be cold and saturated with water. The mass of the bed is the mass of the adsorbent M_z plus the associated water M_d. The temperature of the bed is T_a. The second bed is dry and hot at T_c. The temperature of the cold bed T_a must be less than the temperature of the water adsorbed in the bed. This temperature is fixed by the equilibrium relationship for the zeolite-water pair. On the other hand, the temperature of the hot bed T_c is equal to the temperature of heating steam flowing to the first effect. The first step commences, when the circulating fluid starts to transfer heat between the two beds. Thus, heating the first bed and cooling the second bed occurs simultaneously. During this phase, no heat is exchanged between the adsorbers and any external heat source or sink. The heat flowing into the first adsorber, Q_{2-1}, is represented by the path abe$_1$ on the Clapeyron diagram (Fig. 30), while, the heat transferred from the second bed, Q_{2-1}, is described by the route cde$_2$ on the same diagram. The process is terminated when the

first bed is heated to T_{e_1} and second bed is cooled to T_{e_2}. For the heat transfer to take place T_{e_2} should be higher than T_{e_1}.

2. The second step starts when the first bed is connected to the external source of heating steam (boiler), where its temperature is increased from T_{e_1} to T_c. At the same time, a stream of cooling-water is used to reduce the second bed temperature from T_{e_2} to T_a.
3. During the heating process and once the pressure inside the first bed becomes higher than the condenser pressure, the bed is opened to the tube side of the evaporator where the generated steam condenses.
4. At the same time, when the pressure in the second bed becomes less than the evaporator pressure, the bed is opened to the shell side of the evaporator where the vapor formed in the evaporator flows to the bed where it is adsorbed.

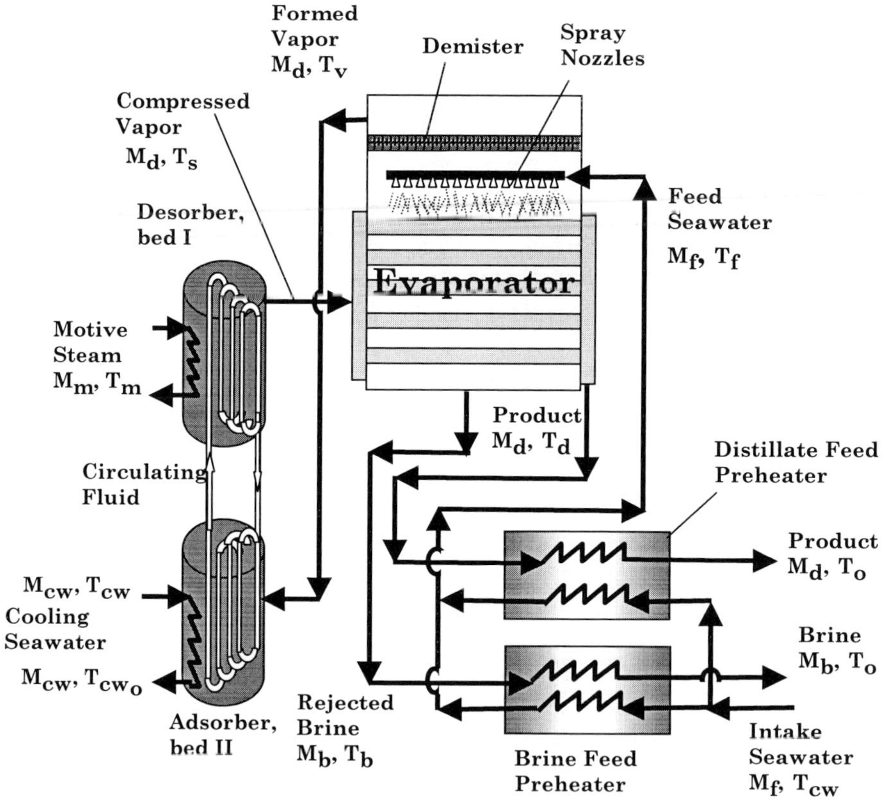

Fig. 29. Single effect-evaporator driven by adsorption heat pump

The previously described four steps represent the first half of the heat pump cycle. The second half of the cycle originates by circulating the heat transfer fluid in the reverse direction. During this second half of the cycle, bed I is cooled and adsorbs vapor from the evaporator. Simultaneously, bed II is heated and generates the heating steam, which condenses inside the evaporator tubes.

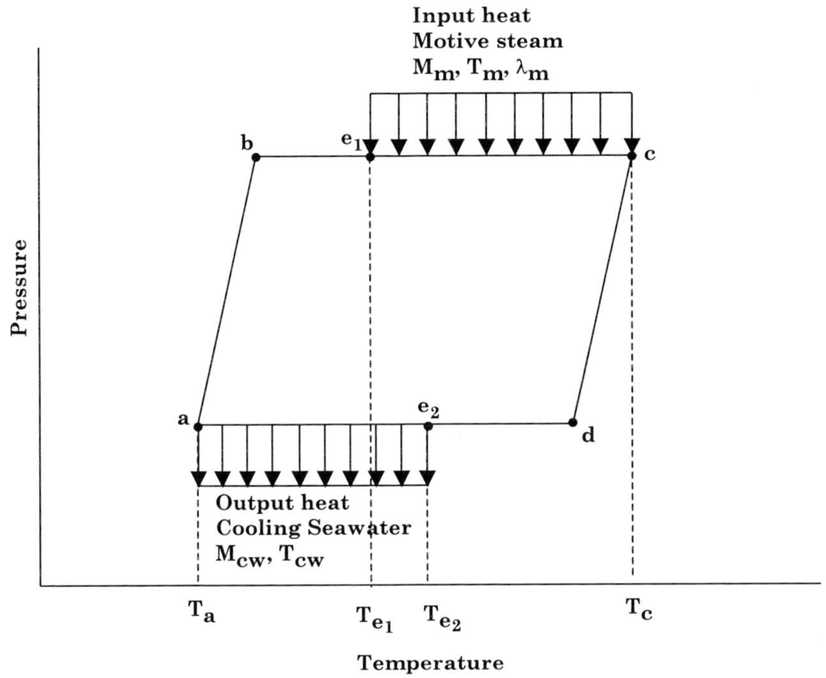

Fig. 30. Clapeyron diagram for the adsorption/desorption vapor compression cycle

3.4.2 Process Modeling

The mathematical model for the single effect adsorption vapor compression desalination system includes balance equations for the evaporator, feed preheaters, adsorption bed, and desorption bed. The model assumptions used in development include the following:

- Steady state conditions. This implies use of a minimum of two adsorption/desorption units. Therefore, as one of the two units go through the process of circulating the thermal fluid between the two beds the other unit is used for simultaneous absorption of vapor from the evaporator and generation of heating steam.

- The adsorber pressure is uniform. Therefore, the vapor pressure and the adsorbent temperature are related by the adsorption equilibrium equation.
- The bed contents are in thermal equilibrium. Therefore, the adsorbent and the adsorbate have the same temperature.
- No heat losses to the surroundings.
- Model parameters, such as the fluid density, heat transfer coefficients, and velocity are assumed constant.
- The mass of vapor adsorbed in the second bed is equal to the amount of steam generated in the first bed,
- Constant and equal rates for adsorption and desorption, and
- Constant rate of heat exchange between the two beds.

The model equations include the following:

- Overall material and salt balances

$$M_f = M_d + M_b \tag{74}$$

$$M_b = M_f (X_f / X_b) \tag{75}$$

- Preheaters energy balance

$$M_f C_p (T_f - T_{cw}) = M_d (H(T_d) - H(T_o)) + M_b C_p (T_b - T_o) \tag{76}$$

- Evaporator energy balance

$$M_f C_p (T_b - T_f) + M_d \lambda_v = M_d \lambda_d + M_d C_{p_v} (T_s - T_d) \tag{77}$$

- Boiling point elevation

$$T_b = T_v + BPE(T_b, X_b) + \Delta T_y \tag{78}$$

- Evaporator heat transfer area

$$A_e = \frac{M_d \lambda_d + M_d C_{p_v} (T_s - T_d)}{U_e (T_d - T_b)} \tag{79}$$

- Feed/distillate preheater heat transfer area

$$A_d = \frac{M_d (H(T_d) - H(T_o))}{U_d (LMTD)_d} = \frac{\alpha M_f C_p (T_f - T_{cw})}{U_d (LMTD)_d} \tag{80}$$

$$(\mathrm{LMTD})_d = \frac{(T_d - T_f) - (T_o - T_{cw})}{\ln \dfrac{T_d - T_f}{T_o - T_{cw}}} \tag{81}$$

– Feed/brine preheaters heat transfer area

$$A_b = \frac{M_b C_p (T_b - T_o)}{U_b (\mathrm{LMTD})_b} = \frac{(1-\alpha) M_f C_p (T_f - T_{cw})}{U_b (\mathrm{LMTD})_b} \tag{82}$$

$$(\mathrm{LMTD})_b = \frac{(T_b - T_f) - (T_o - T_{cw})}{\ln \dfrac{T_b - T_f}{T_o - T_{cw}}} \tag{83}$$

– Correlations for the overall heat transfer coefficient in the evaporator

$$U_e = 1.9394 + 1.40562\mathrm{x}10^{-3} T_b - 2.0752\mathrm{x}10^{-4} (T_b)^2 + 2.3186\mathrm{x}10^{-6} (T_b)^3 \tag{84}$$

where U_e is the overall heat transfer coefficient in the evaporator in kW/m^2 °C and T_b is the brine boiling temperature in °C.

– Energy balance during cooling of the second bed from T_{e_2} to T_a

$$M_{cw}\, C_p\, (T_{cw_o} - T_{cw_i}) = M_z\, C_{p_z}\, (T_{e_2} - T_a) \tag{85}$$

– Heat transferred from the second to the first bed

$$Q_{21} = M_d\, \lambda_v + M_z\, C_{p_z}\, (T_c - T_{e_2}) + M_d\, (H(T_v) - H(T_a)) \tag{86}$$

– Energy required to heat the first bed

$$Q_{21} + M_m \lambda_m = M_d\, \lambda_s + M_z\, C_{p_z}\, (T_c - T_a) + M_d\, (H(T_c) - H(T_a)) \tag{87}$$

– Combined energy balance (Eqs. 86 and 87)

$$M_m\, \lambda_m = M_d (\lambda_d - \lambda_v) + M_z C_{p_z} (T_{e_2} - T_a) + M_d\, (H(T_c) - H(T_v)) \tag{88}$$

– Combined energy balance (Eqs. 86 and 88)

$$M_m\, \lambda_m = M_d\, (\lambda_d - \lambda_v) + M_{cw}\, C_p\, (T_{cw_o} - T_{cw_i}) + M_d\, C_p\, (T_c - T_v) \tag{89}$$

– Efficiency of the circulating fluid heat exchanger

$$\eta = M_z\, C_{p_z}\, (T_{e_2} - T_a)/(M_{cw}\, C_p\, (T_{e_2} - T_{cw_i})) \tag{90}$$

– Energy balance of the circulating fluid heat exchanger

$$M_z\, C_{p_z}\, (T_{e_2} - T_a) = M_{cw}\, C_p\, (T_{cw_o} - T_{cw_i}) \tag{91}$$

– Combined energy balance and heat exchanger efficiency for circulating fluid (Eqs. (90) and (91))

$$T_{e_2} = (T_{cw_o} - T_{cw_i}\, (1 - \eta))/\eta \tag{92}$$

– Constraint on the temperature of inlet/outlet cooling water

$$T_{cw_i} = T_{cw} \tag{93}$$

$$T_{cw_o} = T_{e_2} - \Delta T \tag{94}$$

– Equilibrium relations for adsorber and desorber

$$\ln(P) = a + b/T \tag{95}$$

where

$$a = 20.49 - 60.4\, \alpha + 787\, \alpha^2 - 2.14\text{x}10^3 \alpha^3$$

$$b = -8013 + 33.83\text{x}10^3 \alpha - 3\text{x}10^5 \alpha^2 + 7.9\text{x}10^5 \alpha^3$$

– Water balance in adsorber between points a and c

$$M_z = M_m/(\alpha_g - \alpha_a) \tag{96}$$

In the above equations P and T are the equilibrium pressure and temperature of the adsorber and desorber. In the above relation T is in K and P is in mbar. For the absorber P is equal to vapor pressure in the evaporator and for the desorber P is equal to the heating steam vapor pressure. Also, T equals to T_a for the absorber and equals to T_c for the desorber. The constant α is in kg of water per kg of zeolite, Karagiorgas and Meunier (1987).

Other system constraints include the following:

- ΔT varies between 3 and 5 °C, Van Benthem, et al., 1995,
- η varies between 0.85 and 0.9,
- α_a varies between 0.06 and 0.15 kg H_2O/kg zeolite,
- T_c is higher than T_s by 3 to 10 °C, and
- T_m is higher than T_c by 3 to 10 °C.
- T_f is lower than T_b by 2 to 5 °C.
- T_d is higher than T_b by 2 to 5 °C.

Solution Method

The solution procedure is shown in Fig. 31 and it includes the following steps:

- The system capacity, brine temperature, intake seawater temperature, the water content in the adsorber at point (a), the heat exchanger efficiency, and the temperature difference in the heat exchanger, the equilibrium water content at point (a), and salinity of intake seawater and rejected brine are specified.
- The system constraints are defined, which includes the saturation temperature of the condensate and the feed seawater temperature.
- Eqs. 74-75 are solved to determine the feed and brine flow rates.
- The boiling point elevation and the vapor temperature in the evaporator are calculated from the correlation given in Appendix B and Eq. 78.
- An initial guess is assumed for T_s and T_o. This is followed by iterative solution of Eqs. 76 and 77. Newton's method is used with an iteration error of $1x10^{-4}$.
- The evaporator and preheaters heat transfer areas are determined from Eqs. 79, 80, and 82.
- The constraints on the desorber temperature at point g and the motive steam temperature are used to determine both temperatures.
- The absorber temperature, T_a, is evaluated from Eq. 95.
- The temperatures of inlet and outlet cooling seawater, T_{cw_i} and T_{cw_o}, and the desorber temperature at point (e_2), T_{e_2}, are obtained from Eqs. 92-94.
- The desorber water content, α_g, is obtained from Eq. 95.
- The solid mass in the adsorber is determined from Eq. 96.
- The motive steam and the cooling seawater flow rates are obtained from Eqs. 88 and 89.

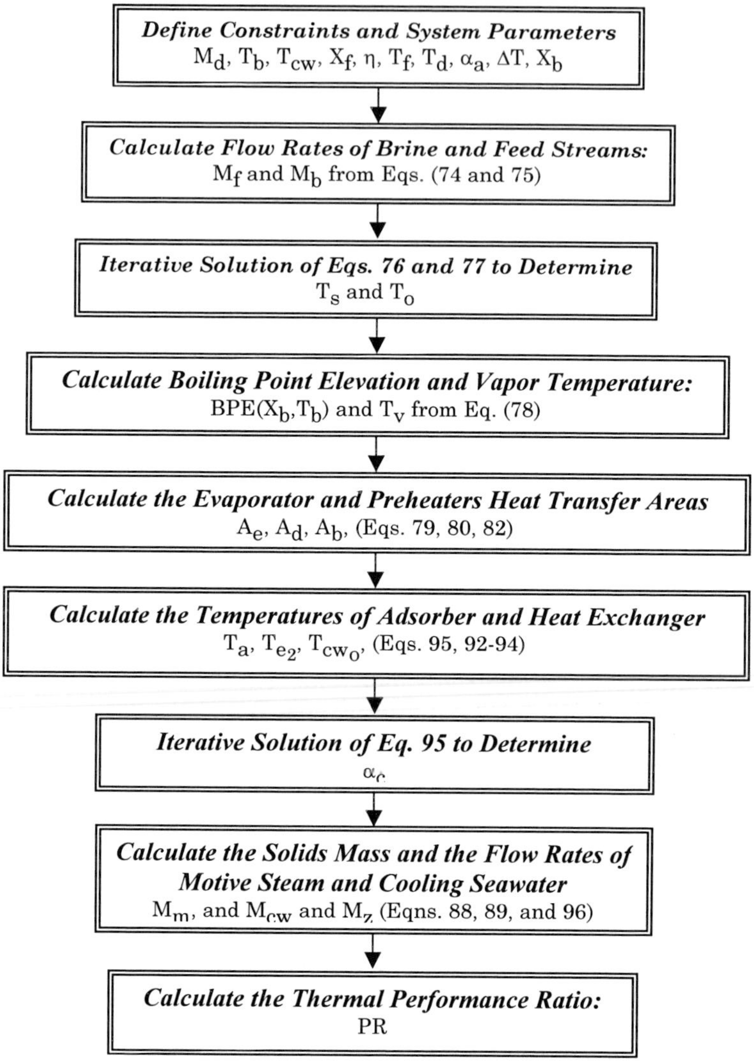

Fig. 31. Solution algorithm of the adsorption heat pump and the single effect evaporation desalination system.

Example 1:

A single effect adsorption vapor compression system is to be designed at the following conditions:

- Brine reject concentration (X_b) = 70000 ppm
- Intake seawater salinity (X_f) = 42000 ppm
- Intake seawater temperature (T_{cw}) = 25 °C
- System capacity (M_d) = 1 kg/s
- Boiling temperature (T_b) = 65 °C
- Specific heat of the vapor at constant pressure, Cp_v = 1.884 kJ/kg °C.
- Specific heat of zeolite, Cp_z = 0.9 kJ/kg K.
- Efficiency of heat exchanger in adsorber (η) = 0.9.
- Water content in adsorber (α_a) = 0.14 kg H_2O/kg solids
- Temperature difference of heat exchanger in adsorber (ΔT) = 5 °C

Determine the evaporator heat transfer area, thermal performance ratio, and flow rate of cooling water in the adsorber.

Solution:

The design procedure requires specification of the following parameters:

- Feed seawater temperature (T_f) = T_b – 2 = 63 °C
- Condensing vapor temperature (T_d) = T_b + 2 = 67 °C
- Desorber temperature (T_c) = (T_s + 5) °C
- Motive steam temperature (T_m) = (T_c + 5) °C

The vapor temperature in the evaporator is calculated by determining the boiling point elevation. The values of B and C are evaluated from

$$B = \left(6.71 + 6.34\text{x}10^{-2}(T_b) + 9.74\text{x}10^{-5}(T_b)^2\right)10^{-3}$$
$$= \left(6.71 + 6.34\text{x}10^{-2}(65) + 9.74\text{x}10^{-5}(65)^2\right)10^{-3} = 0.0112425$$

$$C = \left(22.238 + 9.59\text{x}10^{-3}(T_b) + 9.42\text{x}10^{-5}(T_b)^2\right)10^{-8}$$
$$= \left(22.238 + 9.59\text{x}10^{-3}(65) + 9.42\text{x}10^{-5}(65)^2\right)10^{-8}$$
$$= 2.3259345\text{x}10^{-7}$$

Substituting the values of B and C in the BPE equation gives

$$BPE = X_b\ (B + (X_b)(C))10^{-3}$$
$$= 70000\left(0.0112425 + (70000)\left(2.3259345\text{x}10^{-7}\right)\right)10^{-3} = 1.927\ ^\circ\text{C}$$

Therefore the vapor temperature (T_v) = 65 – 1.927 = 63.073 °C and the latent heat of the formed vapor (λ_v) = 2350.8 kJ/kg.

Solution of the overall material and salt balance for the feed seawater, distillate and rejected brine (Eqs. 74 and 75) gives:

- Flow rate of feed seawater (M_f) = 2.5 kg/s
- Flow rate of the rejected brine (M_b) = 1.5 kg/s

Iterative solution then proceeds for Eq. 76 to determine T_o, which gives a value of 27.9. The iteration sequence is shown in the following table:

Iteration	T_o	Error
Initial Guess	27	–
1	27.00932312	0.0009323
2	27.90120506	0.89188
3	27.90138435	0.00017929

Similarly Eq. 77 is solved iteratively to determine the compressed vapor temperature. The iteration sequence is shown in the following table:

Iteration	T_s	Error
Initial Guess	77	-
1	77.02658844	0.0265
2	83.06096649	6.034
3	83.06159973	0.000633

The compressed vapor temperature (T_s) is then used to calculate the

- Adsorber temperature (T_c) = T_s + 5 = 83.06 + 5 = 88.06 °C
- Motive steam temperature (T_m) = T_c + 5 = 88.06 + 5 = 93.06 °C
- Latent heat of motive steam (λ_m) = 2275.43 kJ/kg.

The heat transfer area of the evaporator is obtained by calculating the following:

$$\lambda_d = 2501.897149 - 2.407064037\, T_d + 1.192217\times10^{-3}\, T_d^2 - 1.5863\times10^{-5}\, T_d^3$$
$$= 2501.897149 - 2.407064037\,(67) + 1.192217\times10^{-3}\,(67)^2 - 1.5863\times10^{-5}\,(67)^3$$
$$= 2341.2 \text{ kJ/kg}$$

$$U_e = 1.9695+1.2057\times10^{-2}T_b-8.5989\times10^{-5}(T_b)^2+2.5651\times10^{-7}(T_b)^3$$
$$= 1.9695+1.2057\times10^{-2}(65)-8.5989\times10^{-5}(65)^2+2.5651\times10^{-7}(65)^3$$
$$= 2.4603 \text{ kJ/s m}^2\,^\circ\text{C}$$

$$A_e = \frac{M_d(\lambda_d + Cp_v(T_s - T_d))}{U_e(T_d - T_b)}$$
$$= \frac{(1)(2341.2 + 1.84(83.06 - 67))}{4.2\,(67 - 65)}$$
$$= 282.3 \text{ m}^2$$

The adsorber pressure and temperature at point (a) are then calculated

$$P_a = P(T_v) = 22.95 \text{ kPa}$$

$$a = 20.49 - 60.4\,\alpha_a + 787\,(\alpha_a)^2 - 2.14\text{x}10^3\,(\alpha_a)^3$$
$$= 20.49 - 60.4\,(0.14) + 787\,(0.14)^2 - 2.14\text{x}10^3\,(0.14)^3$$
$$= 21.587$$

$$b = -8013 + 33.83\text{x}10^3\,(0.14) - 3\text{x}10^5\,(\alpha_a)^2 - 7.9\text{x}10^5\,(\alpha_a)^3$$
$$= -8013 + 33.83\text{x}10^3\,(0.14) - 3\text{x}10^5\,(0.14)^2 - 7.9\text{x}10^5\,(0.14)^3$$
$$= -6989.04$$

$$T_a = b/(\log(P_a/100) - a)$$
$$= -6989.04/(\log(22.95/100) - 21.587) - 273$$
$$= 30.09\ ^{\circ}\text{C}$$

$$T_{cw_o} = (-(T_{cw_i}\,(1-\eta))/\eta - \Delta T)/(1 - 1/\eta)$$
$$= (-(25\,(1-0.9))/0.9 - 5)/(1 - 1/0.9)$$
$$= 70\ ^{\circ}\text{C}$$

$$T_{e_2} = (T_{cw_o} - T_{cw_i}(1-\eta))/\eta$$
$$= (70 - 25(1-0.9))/0.9$$
$$= 75.097\ ^{\circ}\text{C}$$

Iterative solution then proceeds for Eq. 95 to determine α_g, which gives a value of 0.0124. The iteration sequence is shown in the following table

Iteration	α_g	Error
Initial Guess	9.99999978E-03	–
1	1.00034522E-02	3.45242E-06
2	1.23926941E-02	0.002389242
3	1.24159195E-02	2.32254E-05
4	1.24160266E-02	1.071E-07

The solid mass is obtained from Eq. 96

$M_z = M_d/(\alpha_a - \alpha_g)$
$= 1/(0.14\text{-}0.0124) = 7.84$ kg dry zeolite

The motive steam flow rate is then calculated from Eqs. 88

$$M_m = (M_d (\lambda_d - \lambda_v) + M_z\, Cp_z\, (T_{e_2} - T_a) + M_d\, (H(T_c) - H(T_v)))/\lambda_m$$
$$= ((2341.2-2350.8) + 7.84\,(0.9)(75.097 - 30.09) +(1)(368.73 - 264))/2275.43$$
$$= 0.181 \text{ kg/s}$$

The cooling water flow (M_{cw}) is obtained from Eq. 89

$$M_m\lambda_m = M_d(\lambda_d - \lambda_v) + M_{cw}Cp(T_{cw_o} - T_{cw_i}) + M_d\,(H(T_c) - H(T_v))$$

which gives M_{cw} = 1.666 kg/s. Finally the system thermal performance ratio is obtained, where

$PR = M_d/M_m = 1/0.181 = 5.52$

3.4.3 System Performance

The system performance is evaluated as a function of the following parameters:

- The brine boiling temperature, T_b.
- The temperature difference between the condensing vapor temperature and boiling brine, T_d-T_b.
- The water content in the cold adsorber bed, α_a.

The results are shown in Figs. 32-37 and it includes variations in the thermal performance ratio, the specific heat transfer area, and the specific flow rate of cooling water. These results are obtained at the following system parameters:

- Brine reject concentration (X_b) = 70000 ppm
- Intake seawater salinity (X_f) = 42000 ppm
- Intake seawater temperature (T_{cw}) = 25 °C
- System capacity (M_d) = 1 kg/s
- Boiling temperature (T_b) = 65 °C
- Specific heat of the vapor at constant pressure, Cp_v = 1.884 kJ/kg °C.
- Specific heat of zeolite, Cp_z = 0.9 kJ/kg K.

- Efficiency of heat exchanger in adsorber (η) = 0.9.
- Water content in adsorber (α_a) = 0.14 kg H_2O/kg solids
- Temperature difference of heat exchanger in adsorber (ΔT) = 5 °C
- Feed seawater temperature (T_f) = (T_b – 2) °C
- Desorber temperature (T_c) = (T_s + 5) °C
- Motive steam temperature (T_m) = (T_c + 5) °C

Variations in the thermal performance ratio are shown in Fig. 32 as a function of the brine boiling temperature, T_b, and the temperature difference between the condensing vapor and the boiling brine, T_d-T_b. As is shown the thermal performance ratio increases at higher boiling temperatures and larger difference in the temperature of the condensing vapor and boiling brine. As is shown a thermal performance ratio close to 10 can be reached as the brine boiling temperature increases to 110 °C. However, it should be noted that achieving such higher thermal performance is subject to reducing the water content in the adsorber at point (g) to values between zero and 0.01 kg H_2O/kg zeolite. On the other hand, the thermal performance ratio varies around a value of 4-5 for brine boiling temperatures between 40-60 °C. The superior performance of the ADVC is certainly pronounced in comparison with other single effect systems.

Irrespective of the high thermal performance ratio, the ADVC system has similar design features to other single effect vapor compression systems. As is shown in Fig. 33 the evaporator heat transfer area decreases drastically upon the increase in the temperature difference of the condensing vapor and the boiling brine. This is because of the increase in the temperature driving force between the condensing vapor and the boiling brine. A similar effect takes place in the cooling seawater heat exchanger, Fig. 34, where increase in the system temperature increases the driving force between the bed and the cooling seawater stream. This in turn reduces the flow rate of the cooling seawater stream.

System performance as a function of the water content in the adsorber at point (a) and the brine boiling temperature are shown in Figs. 35-37. As is shown in Fig. 35 the thermal performance ratio varies between 2 and 7. As discussed before, the high performance ratio of 13 can only be achieved if the water content of the adsorber at point (g) is reduced to values below 0.01 kg H_2O/kg zeolite. As is shown in Fig. 36, the evaporator heat transfer area has no dependence on the water content in the adsorber bed at point (a) and it only depends on the brine boiling temperature. As for the specific flow rat of cooling water it depends on both parameters, where it decreases with the increase of the brine boiling temperature. Effect of the water content in the adsorber varies, where at low boiling temperatures its increase reduces the specific flow rate of cooling water. The opposite effect is obtained at higher boiling temperatures.

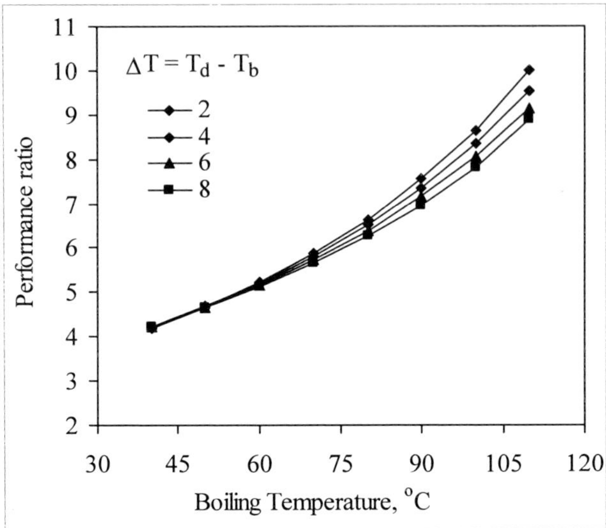

Fig. 32. Effect of boiling temperature and the temperature difference of condensed vapor and boiling brine on the thermal performance ratio of the ADVC system.

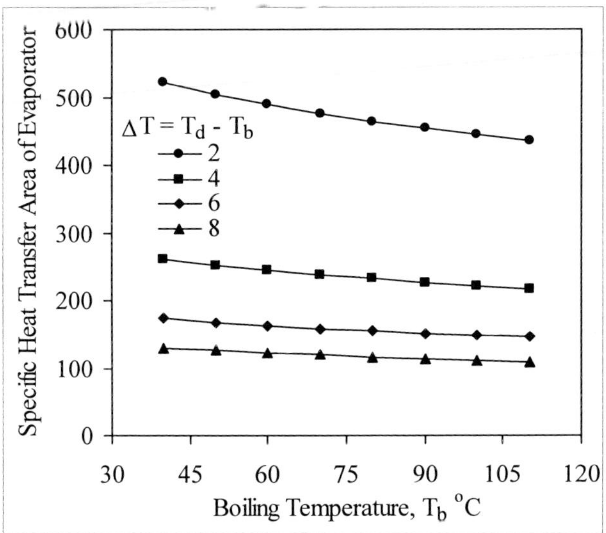

Fig. 33. Effect of boiling temperature and the temperature difference of condensed vapor and boiling brine on the evaporator specific heat transfer area for the ADVC system.

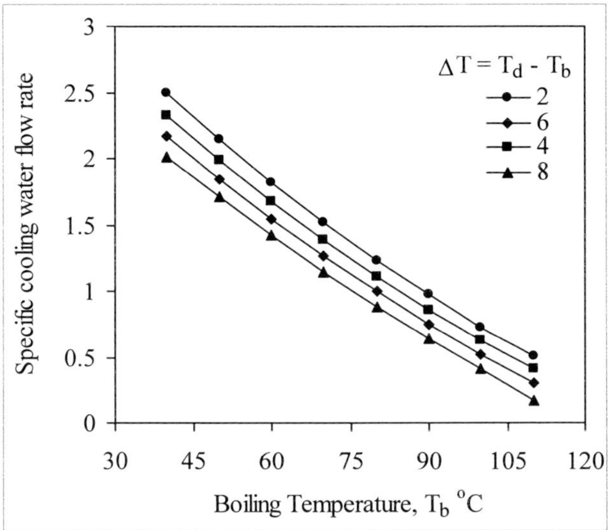

Fig. 34. Effect of boiling temperature and the temperature difference of condensed vapor and boiling brine on the specific flow rate of cooling water for the ADVC system.

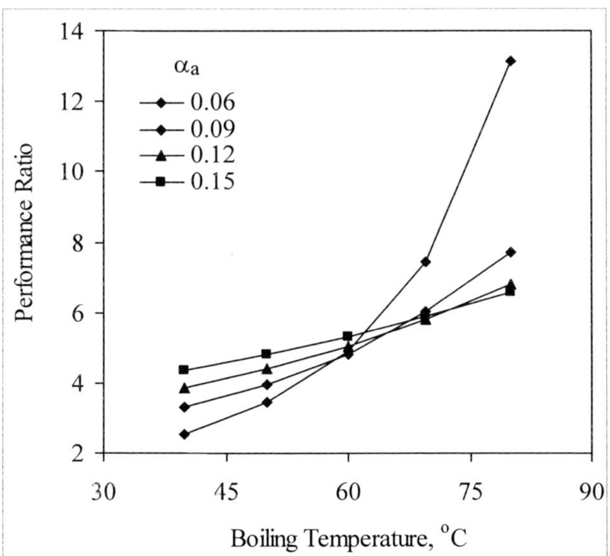

Fig. 35. Effect of boiling temperature and the water content in adsorber at point (a) on the thermal performance ratio of the ADVC system.

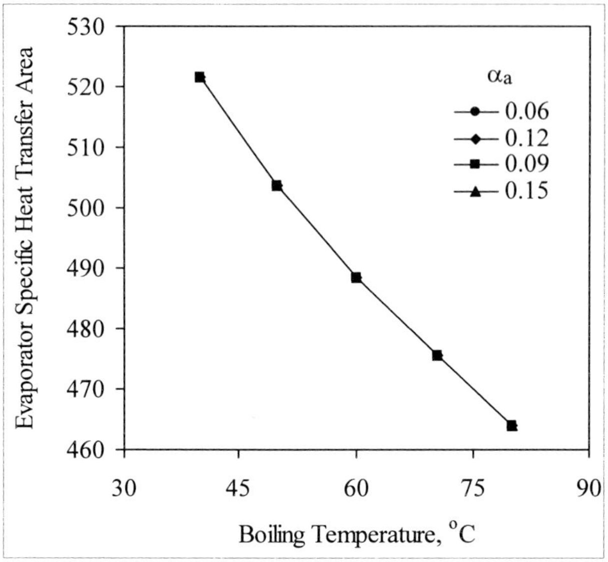

Fig. 36. Effect of boiling temperature and the water content in adsorber at point (a) on the evaporator specific heat transfer area for the ADVC system.

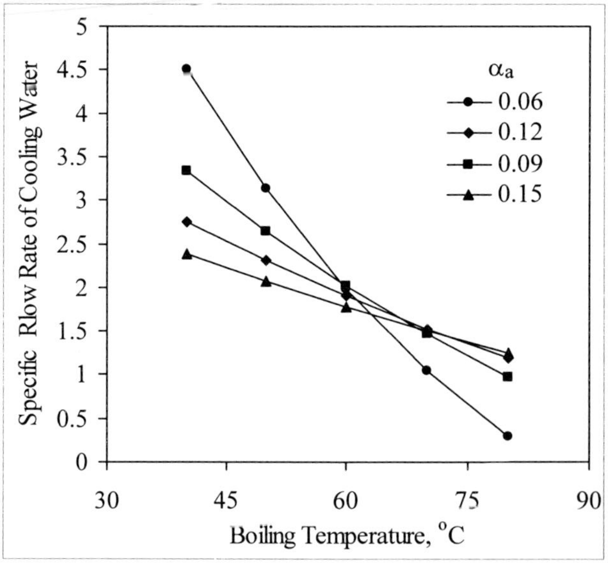

Fig. 37. Effect of boiling temperature and the water content in adsorber at point (a) on the specific flow rate of cooling water for the ADVC system.

3.4.4 Summary

The ADVC system is one of the most efficient single effect vapor compression desalination system. The system includes conventional unit processes found in other single effect configuration, i.e., evaporator and feed preheaters. In addition, its heat pump is rather simple and it includes two zeolite beds for adsorption and desorption. Operation of these beds is controlled by the design pressure and temperature for vapor adsorption and generation of the compressed vapor. A steady state mathematical model is presented to design and evaluate the system performance. The model is used to present step-by-step design calculations for the ADVC system. In addition, overall system performance is presented as a function main design and operating parameters. Results are presented in terms of variations in the thermal performance ratio, specific heat transfer area for the evaporator, and the specific flow rate of the cooling water. The system performance ratio is the highest among all other single effect vapor compression configurations. Also, the specific heat transfer area for the evaporator and the specific flow rate of the cooling water are similar to systems.

References

Karagiorgas, M., and Meunier, F., The dynamics of solid-adsorption heat pump connected with outside heat sources of finite capcity, Heat Recovery Systems & CHP, **7**(1987)285-299.

Van Benthem, G.H.W., Cacciola, G., and Restuccia, G., Regenerative adsorption heat pumps: Optimization of the design, Heat Recovery Systems & CHP, **15**(1995)531-544.

Problems

1. An ADVC system is to be designed to produce 2500 m^3/d of fresh water. The boiling temperature is 75 °C and the temperature of the saturated condensate is higher by 8 °C. The feed temperature is less than the brine boiling temperature by 3 °C and the water content in the adsorber at point (a) is 0.12 kg H_2O/kg zeolite. The salinity of the feed seawater is 39000 ppm and the rejected brine salinity is 70000 ppm. For preliminary design considerations assume the following:
   - All thermodynamic losses (including BPE) are negligible.
   - Constant specific heat for all liquid stream (4.2 kJ/kg °C).
   - Constant latent heat for all vapor streams (2500 kJ/kg).

- Constant specific heat for the vapor streams (1.884 kJ/kg °C).
- Constant overall heat transfer coefficients of 3.2, 7.5, and 8.2 kW/m^2 °C for the evaporator, distillate preheater, and brine preheater, respectively.

Calculate the following: M_b, M_f, T_s, T_o, A_b, A_d, A_e, PR, M_z, and M_{cw}. Assume a heat exchanger efficiency of 0.85.

2. Repeat the previous problem by considering the simultaneous effects of the following parameters:
   - Dependence of the specific heat on temperature and composition.
   - Effect of the boiling point rise.
   - Effect of demister losses.
   - Dependence of the latent on temperature.
   - Dependence of overall heat transfer coefficient on temperature.

   Compare results with the example solved in the Section 3.3.3.
3. An ADVC system is used to desalinate seawater at 37 °C with 42000 ppm salinity. The maximum allowable brine temperature is 90 °C. The heat transfer coefficient for the evaporator and the two preheaters is constant and equals to 6 kW/m^2 °C. The specific heat transfer area is 250 m^2 per (kg/s) of fresh water and the heat transfer area of the distillate preheater is 200 m^2. The flow rates of the hot and cold stream in the preheaters are equal. The temperatures of the distillate and rejected brine flowing from the preheaters are 45 °C and 40 °C, respectively. Calculate the thermal performance ratio.
4. An ADVC system has the following design data:

   M_d = 1 kg/s.

   A_d = 15 m^2.

   A_b = 25 m^2.

   A_e = 400 m^2.

   U_e = 2.4 kW/m^2 °C.

   U_d = 6.7 kW/m^2 °C.

   U_b = 6.3 kW/m^2 °C.

   X_f = 42000 ppm.

   X_b = 70000 ppm.

   Determine T_b, T_d, T_o, T_f, and T_s if T_{cw} = 28 °C, Cp = 4.2 kJ/kg °C, and Cp_v = 1.884 kJ/kg °C. Also, determine the thermal performance ratio and the flow rates of the brine and feed seawater.

Chapter 4

Multiple Effect Evaporation

Objectives

The main objective of this chapter is to evaluate the performance of the multiple effect evaporation desalination processes. This is achieved through discussion of the following:
- Process developments.
- Mathematical models and case studies.
- Detailed models and system performance.

4.1 Developments in Multiple Effect Evaporation

The multiple effect evaporation system is formed a sequence of single effect evaporators, where the vapor formed in one effect is used in the next effect. The vapor reuse in the multiple effect system allows reduction of the brine and the temperature to low values and prevent rejection of large amount of energy to the surrounding, which was the main drawback of the single effect system. In addition to the desalination industry, the main bulk of the multiple effect evaporation processes is found in the food, pulp and paper, petroleum, and petrochemical industries. As discussed in chapter 1 the origins of the multiple effect evaporation dates back to the 19th century with the growth of the sugar industry, where it was necessary to devise an efficient evaporation process to produce good quality sugar crystal at low prices.

Although, the first desalination plants were of the evaporation type their use was not expanded to full industrial scale because of limited design and operating experience. Such systems were plagued with excessive fouling, scaling, and corrosion. However, accumulated experiences during the 2nd half of the past century in thermal desalination processes, headed by the MSF process, have resulted in rapid progress and development of efficient and inexpensive chemical treatment for reduction and prevention of fouling, scaling, and corrosion. Such advances made it possible to maintain plant factors as high as 90% and to keep plants on-line for more than 2 years of operation. As a result, recent research, development, pilot plant operation, and field results show superior performance and the many attractive features of the multiple effect evaporation in comparison with the predominant MSF process.

The multiple effect evaporation process can be configured in forward, backward, or parallel feed, Fig. 1. The three configurations differ in the flow directions of the heating-steam and the evaporating brine. Selection among the three configurations relies on variation in the salt solubility as a function of the top brine temperature and the maximum brine concentration. At higher temperatures or higher brine concentrations, scale formation takes place inside and outside the tube surfaces. This results in the following:

- Decrease of the available flow area inside the tubes, which causes increase in the pressure drop and pumping energy, and
- Increase of the thermal resistance for heat transfer. This reduces the heat transfer efficiency, which results in a lower product flow rate.

Figure 2 shows variation in the solubility of calcium sulfate as a function of concentration and temperature. The diagrams illustrate solubility limits of calcium sulfate compounds as well as variations in the temperature-concentration profiles in the three MEE configurations. These profiles are given for the seawater and brine during their flow in the system preheaters and evaporators.

In the backward feed, the seawater is introduced into the last effect, which has the lowest temperature and pressure within the system. The brine flows through successive effects towards the first effect. The increase in the pressure and temperature across the effects dictates the use of brine pumping units between the effects.

This feature is a major disadvantage in the backward system; because of the increase in the pumping power, maintenance cost, and the increase in air leakage point through pump connections. The second disadvantage of the system is shown on Fig. 2c, where the brine with the highest concentration is subjected to the highest temperature in the system. As is shown, the temperature-concentration profile crosses the solubility limits for the calcium sulfate. The above two factors make the backward feed configuration inapplicable to seawater desalination.

Some examples for industrial applications of the parallel feed MEE can be found in literature, Temstet et al. (1995) and Temstet et al. (1996). Figures 1c shows a system schematic and Figs. 2a and 2b show the temperature-concentration profile in the parallel feed system. In this configuration, the feed seawater is divided into a set of parallel streams, which are fed into individual effects. In each effect the feed seawater is heated to the effect saturation temperature, before evaporation commences. The main advantage of the parallel feed configuration is the simplicity of its configuration in comparison with the other two layouts.

The main feature of the forward feed system is the ability to operate at high top brine temperatures, El-Dessouky et al. (1998). Detailed evaluation of this system is given in the next sections.

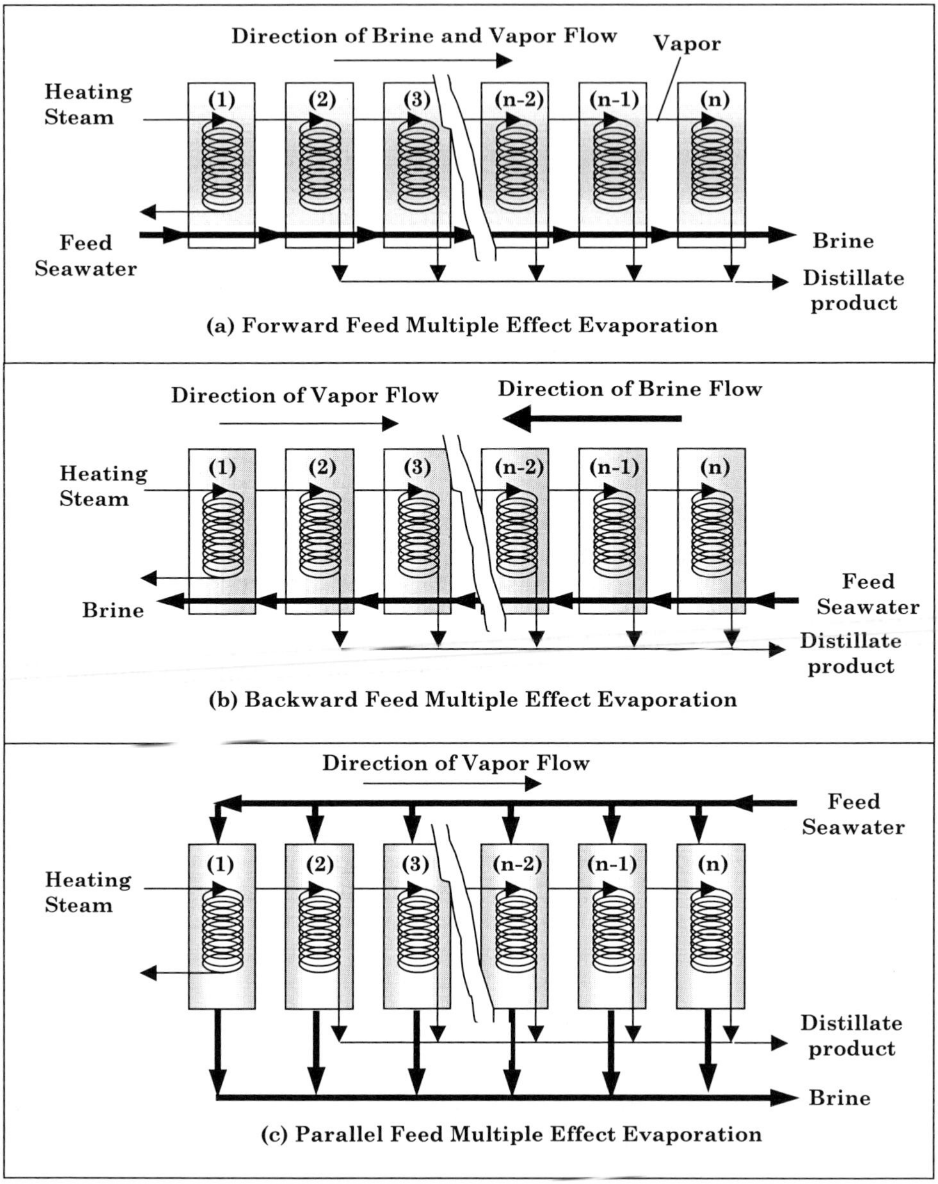

Fig. 1. Configurations of Multiple Effect Evaporation

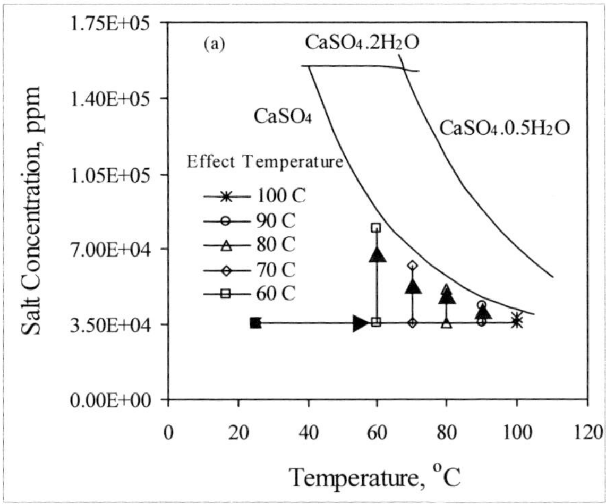

Fig. 2a. Calcium sulfate solubility and top brine temperature for parallel feed multiple effect evaporation.

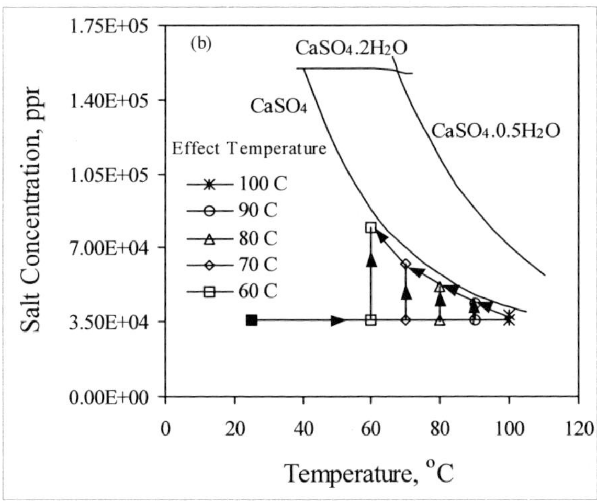

Fig. 2b. Calcium sulfate solubility and top brine temperature for parallel/cross flow multiple effect evaporation.

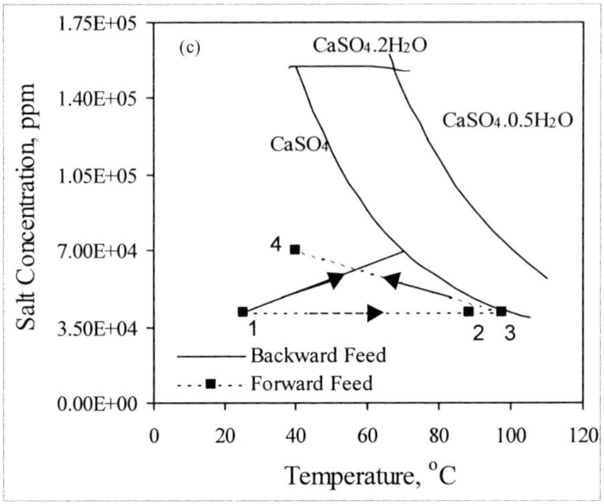

Fig. 2. Calcium sulfate solubility and top brine temperature for forward and backward feed multiple effect evaporation.

4.2 Forward Feed Multiple Effect Evaporation

Although the forward feed multiple effect evaporation system is not found on industrial scale for the desalination industry, it is widely used in the sugar and paper industries. The forward feed configuration was not used in the desalination industry because it has a more complex layout than the parallel feed configuration. In addition, the first multiple effect that were designed and constructed were of the parallel type. Field results of the parallel effect units proved their reliability; therefore, subsequent units remained to be of this design.

4.2.1 Process Description

Figure 3 shows a schematic diagram for the forward-feed Multiple-Effect Evaporation (MEE-FF) seawater desalination process. The system includes the evaporators, equal to n, a series of feed water preheaters, equal to n-2, a train of flashing boxes, equal to n-1, a down condenser, and a venting system. In the forward-feed configuration, the direction of heat flow as well as the flow direction of the brine and vapor is from left to right, i.e., from effect 1 to effect n. The pressure in the effects decreases in the flow direction. Each effect contains heat exchange tubes, vapor space, brine spray nozzles, mist eliminator, and brine collecting box. The horizontal falling film evaporator is the most widely used in the MEE desalination processes. The advantages of the horizontal falling film system are:

- Efficient water distribution and tube wetting,
- High heat-transfer rates,
- Absence of dry patches,
- Low scale formation and tube damage,
- Efficient disengagement of vapors and non-condensable gases,
- Proper venting of the non-condensable gases, and
- Simple monitoring of scaling and fouling.

The main drawback of the horizontal falling film MEE is scale and fouling on the outer surface of the tubes. This does not allow for the use of ball cleaning system, common in seawater internal flow. Such system proved to reduce internal scaling and fouling by 50% of the design value, Rautenbach and Schafer, 1997.

The intake seawater flows into the condenser of the last effect at a flow rate of $M_{cw}+M_f$. This stream absorbs the latent heat of vapors formed in the last effect and flashing box. The seawater stream is heated from the intake temperature, T_{cw}, to a higher temperature, T_f. The function of the cooling seawater, M_{cw}, is to remove the excess heat added to the system in the first effect by the motive steam. In the last effect, this heat is equivalent to the latent heat of the boiled off vapors. On the other hand, the feed seawater, M_f, is heated by the flashed off vapors formed in the last effect and the associated water flash box. The cooling seawater, M_{cw}, is rejected back to the sea. The feed seawater, M_f, is chemically treated, deaerated, and pumped through a series of preheaters. The temperature of the feed water increases from T_f to t_2 as it flows inside the tubes of the preheaters. Heating of the feed seawater is made by condensing the flashed off vapors from the effects, d_j, and the flash boxes, $\bar{d}_j$. The feed water, M_f, leaves the last preheater (associated with the second effect) and is sprayed inside the first effect. It is interesting to not that the preheater of the first effect is integrated in the heat exchanger of the effect. This is because there is no flash box in the first effect or flashed off vapors within the effect. The brine spray forms a thin film around the succeeding rows of horizontal tubes. The brine temperature rises to the boiling temperature, T_1, which corresponds to the pressure of the vapor space. The saturation temperature of the formed vapor, T_{v1}, is less than the brine boiling temperature by the boiling point elevation, $(BPE)_1$.

A small portion of vapor, D_1, is formed by boiling in the first effect. The remaining brine, M_f - D_1, flows into the second effect, which operates at a lower temperature and pressure. Vapor is formed in effects 2 to n by two different mechanisms, boiling and flashing. The amount vapor formed by boiling is D_j and the amount formed by flashing is d_j. Flashing occurs in effects 2 to n because the brine temperature flowing from the previous effect, T_{j-1}, is higher than the saturation temperature of the next effect, T_{v_j}. Therefore, vapor flashing is dictated by the effect equilibrium. In effects 2 to n, the temperature of the vapor formed by flashing, T'_{v_j}, is lower than the effect boiling temperature, T_j, by the boiling point elevation $(BPE)_j$ and the non-equilibrium allowance (NEA)j. In the flash boxes, a small quantity of flashing vapors, $\bar{d}_j$, is formed with a temperature equal to T''_{v_j}. This temperature is lower than the vapor condensation temperature in effect j, T_{c_j}, by the non-equilibrium allowance $(NEA)'_j$.

Motive steam, M_s, extracted from an external boiler drives vapor formation in the first effect. The vapor formed by boiling in the first effect, D_1, is used to drive the second effect, which operates at a lower saturation temperature, T_2. Reduction in the vapor temperature is caused by boiling point elevation, non-equilibrium allowance, and losses caused by depression in the vapor saturation pressure by frictional losses in the demister, transmission lines, and during condensation. These losses can be represented as an extra resistance to the flow of heat between condensing vapor and boiling brine. Therefore, it is necessary to increase the heat transfer area to account for these losses. The amount of vapor formed in effect j is less than the amount formed in the previous effect. This is because of the increase in the latent heat of vaporization with the decrease in the evaporation temperature.

The condenser and the brine heaters must be provided with good vents, first for purging during start-up and then for removing non-condensable gases, which may have been introduced with the feed or drawn in through leaks to the system. The presence of the non-condensable gases not only impedes the heat transfer process but also reduces the temperature at which steam condenses at the given pressure. This occurs partially because of the reduced partial pressure of vapor in a film of poorly conducting gas at the interface. To help conserve steam economy venting is usually cascaded from the steam chest of one preheater to the steam chest of the adjacent one. The effects operate above atmospheric pressure are usually vented to the atmosphere. The non-condensable gases are always saturated with vapor. The vent for the last condenser must be connected to vacuum-producing equipment to compress the non-condensable gases to atmosphere. This is usually a steam jet ejector if high-pressure steam is available. Steam jet ejectors are relatively inexpensive but also quite inefficient. Since the vacuum is maintained on the last effect, the unevaporated brine flows by itself from effect to effect and only a blow down pump is required on the last effect.

Summary of different processes that takes place in each effect, the associated flash box and feed preheater is shown in Fig. 4. As is shown the brine leaving the effect decreases by the amount of vapor formed by boiling, D_j, and by flashing, d_j. The distillate flow rate leaving the flash box increases by the amount of condensing vapors from the previous effect, D_{j-1} and d_{j-1}. The brine concentration increases from X_{j-1} to X_j upon vapor formation. The effect and flash box temperatures decrease from T_{j-1} to T_j and from T'_{j-1} to T'_j, respectively.

Comparison of the process layout for MSF and MEE, show that MSF is a special case of the MEE process. This occurs when the entire vapor formed in the effects is used to preheat the feed in the preheaters and non-is left for the evaporator tubes. In this case, the first effect, the flashing boxes, and the bottom condenser of the MEE replace the brine heater, the distillate collecting trays, and the heat rejection section of the MSF, respectively.

4.2.2 Process Modeling

Two models are presented in this section. The first is the simplified mathematical model, which gives a very efficient and simple tool for system design and evaluation. The model is solved through a simple sequence of manual calculations. Iterations are not exhaustive and do not require computer programming. Also, the assumptions invoked in model development do not sacrifice process fundamentals, specifically, equal heat transfer area in all effects.

The data generated by the model is limited to the following effect properties:

- Brine and distillate flow rates.
- Brine concentration.
- Temperature.
- Heat transfer area.

The model equations exclude the flash boxes and preheaters. The governing equation for the down condenser can be included and its solution is made upon completion of the effect iterations. The following assumptions are made to develop the MEE-FF simplified model:

- Constant specific heat, Cp, for the seawater at different temperature and concentration.
- Constant thermodynamic losses in all effects.
- Constant heat transfer area in all effects.
- No vapor flashing takes place inside the effects.
- Feed seawater is at the saturation temperature of the first effect.
- Equal thermal loads in all effects.

- The formed vapors are salt free.
- The driving force for heat transfer in the effect is equal to the difference of the condensation and evaporation temperatures.
- Energy losses to the surroundings are negligible.

Taking these assumptions into consideration, the mathematical model is developed below. The number of material and energy balance equations, which can be written for each effect, is three. This assumes that the seawater is modeled as a binary mixture of fresh water and salt. In addition, there are n equation for the heat transfer rate in each effect, which relates the effect thermal load to the area, overall heat transfer coefficient, and temperature driving force. Therefore, a total of 4xn equations are used to obtain the profiles of the flow rates, concentration, and temperature across the effects as well as the heat transfer area. The unknown values are as follow:

Brine flow rates, B_1, B_2, ..., B_{n-1}, B_n	(n unknown)
Brine concentration, X_1, X_2, ... , X_{n-1}	(n-1 unknown)
Distillate flow rate, D_1, D_2, ..., D_{n-1}, D_n	(n unknown)
Effect temperature, T_1, T_2, ..., T_{n-1}	(n-1 unknown)
Steam flow rate	(1 unknown)
Heat transfer area	(1 unknown)
	Total = (4 n) unknowns

Solution of the model equations to determine the variables, requires specification of the following system parameters:
- Temperature of the motive steam, T_s.
- Vapor temperature in effect n, T_n.
- Salt concentration in the brine stream leaving effect n, X_n.
- Salt concentration in the feed stream, X_f.
- Total distillate flow rate, M_d.

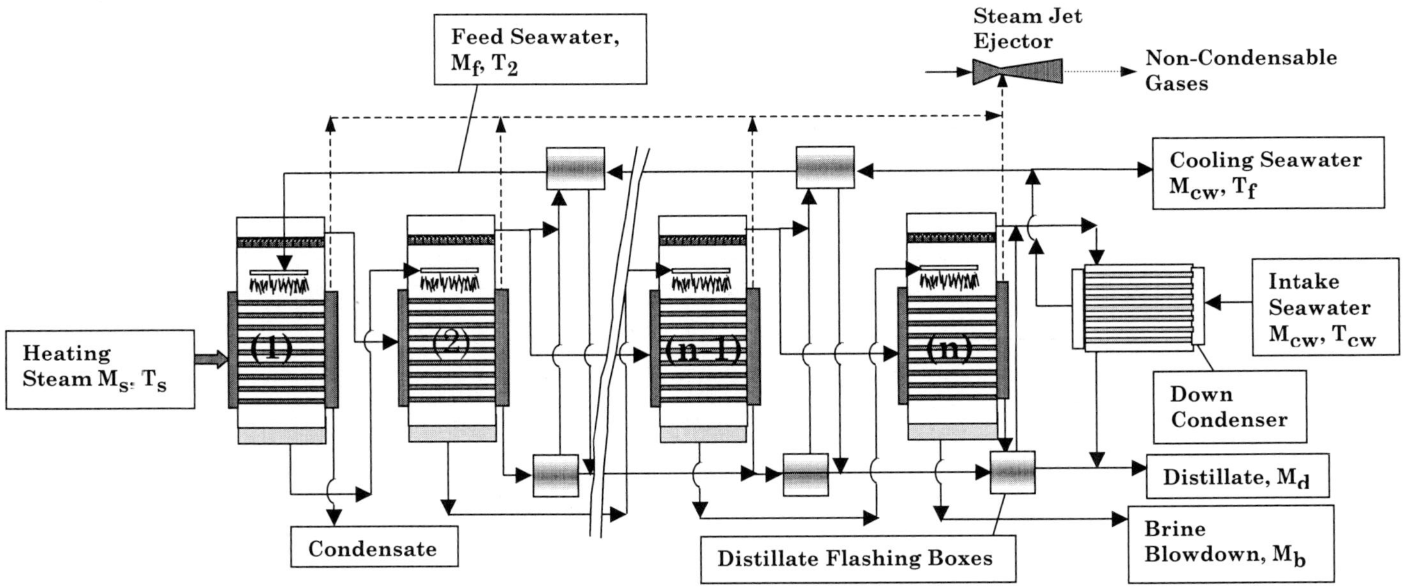

Fig. 3. Schematic of MEE-FF desalination process

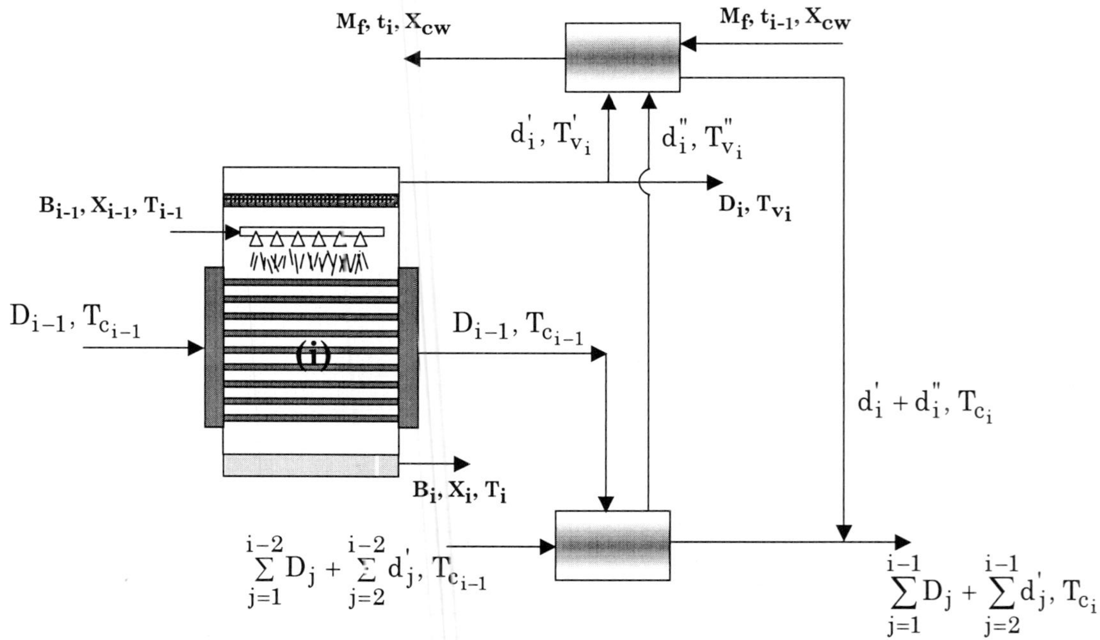

Fig. 4. Variables in evaporator, preheater, and flash box of effect i.

The overall material and salt balance equations are written to determine the brine flow rate leaving effect n, B_n, and the feed flow rate, M_f. These equations are

$$M_f = M_d + B_n \quad (1)$$

$$X_f M_f = X_n B_n \quad (2)$$

Substituting 1 in 2 and eliminating M_f gives

$$B_n = (X_f /(X_n - X_f)) M_d \quad (3)$$

All variables on the right hand side of Eq. 3 are previously specified; therefore, the value of B_n can be calculated. The overall balance, Eq. 1, is then used to determine M_f. Calculations of B_n and M_f are only made once are not included in the following iteration sequence.

Temperature Profile

The thermal load in all effects is assumed constant, thus

$$Q_1 = Q_2 = \ldots = Q_{n-1} = Q_n \quad (4)$$

with

$$Q_1 = M_s \lambda_s, \text{ for the first effect} \quad (5)$$

$$Q_i = D_i \lambda_{v_i}, \text{ for effects 2 to n} \quad (6)$$

where Q is the thermal load, M_s is the mass flow rate of motive steam, D_i is the distillate flow rate in effect i, λ_s is the steam latent heat at T_s, and λ_{v_i} is the latent heat of formed vapors at (T_i - ΔT_{loss}), and the subscript i, s, and v defines effect i, the steam, and the formed vapor. The thermal load in each effect can also be defined in terms of the heat transfer area in the effect, A, the temperature driving force, ΔT, and the overall heat transfer coefficient, U. This is

$$Q_i = A_i U_i \Delta T_i \quad (7)$$

Since the heat transfer area and thermal load are equal in all effects, then

$$Q_1/A_1 = Q_2/A_2 = \ldots = Q_{n-1}/A_{n-1} = Q_n/A_n \quad (8)$$

From 7 and 8, the following identity also applies

$$U_1 \Delta T_1 = U_2 \Delta T_2 = \ldots = U_{n-1} \Delta T_{n-1} = U_n \Delta T_n \quad (9)$$

The total temperature drop across the effects is defined as

$$\Delta T = T_s - T_n \quad (10)$$

where T_s and T_n are the temperatures of the motive steam and the vapor formed in the last effect, n. This drop is also equal to the sum of temperature drop per effect, or

$$\Delta T = \Delta T_1 + \Delta T_2 + \ldots + \Delta T_{n-1} + \Delta T_n \quad (11)$$

Equations 9 and 11 can be used to define ΔT_1 in terms of the overall heat transfer coefficient and the total temperature drop in all effects. From 9, ΔT_2 can be expressed in terms of ΔT_1 by

$$\Delta T_2 = \Delta T_1 \, U_1/U_2 \quad (12)$$

Also ΔT_3 can be expressed in terms of ΔT_2 by

$$\Delta T_3 = \Delta T_2 \, U_2/U_3 \quad (13)$$

Substituting 12 in 13 gives

$$\Delta T_3 = \Delta T_1 \, U_1/U_2 \, U_2/U_1$$

which simplifies to $\Delta T_3 = \Delta T_1 \, U_1/U_3$. The same applies for all other effects and this general relation is arrived at

$$\Delta T_i = \Delta T_1 \, U_1/U_i \quad (14)$$

Substituting the result given in Eq. 14 in Eq. 11 gives

$$\Delta T = \Delta T_1 \, U_1 \, (1/U_1 + 1/U_2 + \ldots + 1/U_{n-1} + 1/U_n) \quad (15)$$

Equation 15 is rearranged into the following form

$$\Delta T_1 = \frac{\Delta T_t}{U_1 \sum_{i=1}^{n} \frac{1}{U_i}} \qquad (16)$$

If estimates for U_i are made, then, temperature drop in all effects can be obtained from Eqs. 16 and 14. The actual temperature profile is then calculated from the following relations. In the first effect

$$T_1 = T_s - \Delta T_1 \qquad (17)$$

and in effects 2 to n

$$T_i = T_{i-1} - \Delta T_1 \, U_1/U_i \qquad (18)$$

Calculation of the temperature profile from Eqs. 17 and 18 requires specification of the overall heat transfer coefficients, U_i.

Profiles of salt concentration and flow rates of brine and distillate

The distillate flow rates are obtained from the following balance and the thermal loads, Eq. 6,

$$M_d = D_1 + D_2 + \ldots\ldots + D_{i-1} + D_n \qquad (19)$$

$$D_i \lambda_{v_i} = D_{i-1} \lambda_{v_{i-1}}, \text{ for effects 2 to n} \qquad (20)$$

Eq. 20 is used to express the values of D_i (for i = 2 to n) in terms of D_1, where

$D_2 = D_1 \lambda_{v_1}/\lambda_{v_2}$, and

$D_3 = D_2 \lambda_{v_2}/\lambda_{v_3} = D_1 (\lambda_{v_1}/\lambda_{v_2}) (\lambda_{v_2}/\lambda_{v_3}) = D_1 \lambda_{v_1}/\lambda_{v_3}$

A general recursive formula is then arrived at

$$D_i - D_1 \lambda_{v_1}/ \lambda_{v_i}, \text{ with i = 2 to n,} \qquad (21)$$

Substituting Eq. 21 in Eq. 19 gives

$$M_d = D_1 + D_1 \lambda_{v_1}/\lambda_{v_2} + \ldots\ldots + D_1 \lambda_{v_1}/ \lambda_{v_{n-1}} + D_1 \lambda_{v_1}/ \lambda_{v_n} \qquad (22)$$

Eq. 21 is then rearranged to obtain an expression for D_1

$$D_1 = M_d/(\lambda_{v_1}(1/\lambda_{v_1} + 1/\lambda_{v_2} + \ldots\ldots + 1/\lambda_{v_{n-1}} + 1/\lambda_{v_n})) \tag{23}$$

The recursive formula of 21 is then used to obtain the distillate flow rates in other effects

$$D_2 = D_1\, \lambda_{v_1}/\lambda_{v_2}$$

$$D_3 = D_1\, \lambda_{v_1}/\lambda_{v_3}$$

$$D_n = D_1\, \lambda_{v_1}/\lambda_{v_n}$$

The brine flow rate in the first effect can be obtained from

$$B_1 = M_f - D_1 \tag{24}$$

In effects 2 to n, this is given by

$$B_i = B_{i-1} - Di \tag{25}$$

Similar salt balances on the first effect and effect 2 to n are written to obtain X_1 and X_2 to X_n.

$$X_1 = X_f\ M_f/B_1 \tag{26}$$

$$X_i = X_{i-1}\ B_{i-1}/B_i \tag{27}$$

Heat Transfer Area

The heat transfer areas in effects 1 to n must be calculated to check the basic assumption of the model, i.e., equal heat transfer areas. The heat transfer area in the first effect is given by

$$A_1 = \frac{D_1\lambda_{v1}}{U_1(T_s - T_1)} \tag{28}$$

and for effects 2 to n it is defined as

$$A_i = \frac{D_i\lambda_i}{U_i(T_i - \Delta T_{loss})} \tag{29}$$

The ΔT_{loss} in the above equation corresponds to the thermodynamic losses in each effect and its value may vary from 0.5-3 °C.

Convergence Criterion and Setting up for New Iteration

The convergence criterion is based on the maximum difference in the heat transfer areas. This is given by

$$\Delta A_{max} = Max(A_{i+1} - A_i), \text{ with } i = 1,n\text{-}1 \quad (30)$$

If ΔA_{max} is greater than the specified iteration tolerance then the iterations continue. The iteration tolerance may be specified as a large number, i.e., 1 m^2, if a small number of iterations (1 or 2) are needed. However, if higher accuracy is required, then a smaller tolerance is specified, i.e., 0.1 or 0.01 m^2.

If the error is higher than the tolerance, then a new estimate for ΔT_i is made

$$\Delta T'_i = \Delta T_i \, A_i/Am \quad (31)$$

where Am is the average heat transfer area and is obtained from

$$A_m = \frac{\sum_{i=1}^{n} A_i}{n} \quad (32)$$

Iterations continue by calculating

- The temperature profile, T_i, in effects 1 to n from Eqs. 17 and 19.
- The distillate flow rate in the first effect, D_1, Eq. 23.
- The distillate flow rates in effects 2 to n, D_i, Eq. 21.
- The brine flow rate in the first effect, B_1, Eq. 24.
- The brine flow rates in effects 2 to n, B_i, Eq. 25.
- The salt concentration in the first effect, X_1, Eq. 26.
- The salt concentration effects 2 to n, X_i, Eq. 27.
- The heat transfer area in effects 1 to n, A_i, Eqs. 28 and 29.

The convergence criterion, Eq. 30, is then checked and iterations continue until it is achieved. Reaching the final solution is followed by calculation of the system performance characteristics, i.e., performance ratio, specific heat transfer area, and specific cooling water flow rate.

Performance Parameters

The performance ratio, PR, is defined as the flow rate ratio of distillate (M_d) and motive steam (M_s). This is

$$PR = M_d/M_s \tag{33}$$

The value of the steam flow rate, M_s, is obtained from the assumption of equal thermal loads, where

$$M_s = D_1 \lambda_{v_1} / \lambda_s \tag{34}$$

The specific heat transfer area is

$$sA = \frac{\sum_{i=1}^{n} A_i + A_c}{M_d} \tag{35}$$

where A_i is the heat transfer area in effect i and A_c is the down condenser heat transfer area, which is obtained from

$$A_c = \frac{Q_c}{U_c (LMTD)_c} \tag{36}$$

The $(LMTD)_c$ is defined as:

$$(LMTD)_c = \frac{T_f - T_{cw}}{\ln \dfrac{T_n - T_{cw}}{T_n - T_f}} \tag{37}$$

where T_{cw} is the intake seawater temperature, T_f, is the outlet seawater temperature, and T_n is the condensation temperature of the vapor formed in the last effect. The thermal load of the condenser is calculated from

$$Q_c = D_n \lambda_{v_n} \tag{38}$$

The specific cooling water flow rate is defined as

$$sM_{cw} = M_d/M_{cw} \tag{39}$$

where M_{cw} is the cooling water flow rate and is obtained from the condenser energy balance

$$D_n \lambda_{v_n} = (M_f + M_{cw}) \, Cp \, (T_f - T_{cw}) \quad (40)$$

It should be noted that T_f is the feed seawater temperature entering the preheater associated with effect, n-1.

Example 1:

The above model is used to determine performance of six effects MEE system. The following specifications are made to solve the simplified MEE model:
Number of effect, n = 6,
Motive steam temperature, T_s = 100 °C,
Total product flow rate, M_d = 1 kg/s,
Salt concentration in feed seawater, X_f = 42000 ppm,
Salt concentration in rejected brine, X_6 = 70000 ppm
Vapor temperature in last effect, T_6 = 40 °C.
Thermodynamic losses in all effects, ΔT_{loss} = 2 °C.
Seawater temperature leaving the condenser, T_f = 35 °C.
Intake seawater temperature, T_{cw} = 25 °C.

Before starting the iterations, the latent heat of the motive steam and the vapor formed in the last effect are obtained from the steam tables or the correlation given in Appendix A. This gives

$$\lambda_s = 2499.5698 - 2.204864 \, T_s - 2.304\text{x}10^{-3} \, T_s^2$$
$$= 2499.5698 - 2.204864 \, (100) - 2.304\text{x}10^{-3} \, (100)^2$$
$$= 2256.043 \text{ kJ/kg}$$

$$\lambda_{v6} = 2499.5698 - 2.204864 \, T_{v6} - 2.304\text{x}10^{-3} \, T_{v6}^2$$
$$= 2499.5698 - 2.204864 \, (40\text{-}2) - 2.304\text{x}10^{-3} \, (40\text{-}2)^2$$
$$= 2412.45 \text{ kJ/kg}$$

The flow rates of the brine leaving effect number 6 and the feed seawater are obtained from Eqs. 1 and 2. The brine flow rate in effect number 6 is

$$B_6 = (X_f / (X_6\text{-}X_f)) \, M_d = (42000/(70000\text{-}42000)) \, (1) = 1.5 \text{ kg/s}$$

Then the feed flow rate, M_f, is equal to the sum of M_d and B_6

$B_6 = 1 + 1.5 = 2.5$ kg/s

The total temperature drop across the effects, $T_s - T_6$, is equal to 100 – 40 = 60 °C. The overall heat transfer coefficients in effects 1 to 6 are specified and are assumed to remain constant throughout the iterations. The overall heat transfer coefficient in the first effect, U_1, is set equal to 2.4 kW/m² °C. Values in subsequent effects are obtained from

$$U_{i+1} = 0.95\, U_i$$

Values of the overall heat transfer coefficient in all effects are summarized in the following table

U_1	U_2	U_3	U_4	U_5	U_6
2.4	2.28	2.16	2.0577	1.9548	1.8571

The summation of the inverse for the overall heat transfer coefficients is required to calculate the temperature drop per effect. This summation is

$$\frac{1}{\sum_{i=1}^{6} U_i} = 1/U_1 + 1/U_2 + 1/U_3 + 1/U_4 + 1/U_5 + 1/U_6$$

$$= 1/2.4 + 1/2.28 + 1/2.16 + 1/2.0577 + 1/1.9548 + 1/1.8571$$
$$= 2.8529 \text{ m}^2 \text{ °C/kW}$$

The temperature drop in the first effect is then calculated

$$\Delta T_1 = \frac{\Delta T_t}{U_1 \sum_{i=1}^{n} \frac{1}{U_i}} = \frac{60}{(2.4)(2.8529)} = 8.7628 \text{ °C}$$

The values of ΔT_i are calculated for effects 2 to 6

$$\Delta T_2 = \Delta T_1 (U_1/U_2) = (8.7628)(2.4)/(2.28) = 9.224 \text{ °C}$$

$$\Delta T_3 = \Delta T_1 (U_1/U_3) = (8.7628)(2.4)/(2.166) = 9.7095 \text{ °C}$$

$$\Delta T_4 = \Delta T_1 (U_1/U_4) = (8.7628)(2.4)/(2.0577) = 10.2205 \text{ °C}$$

$$\Delta T_5 = \Delta T_1 (U_1/U_5) = (8.7628)(2.4)/(1.9548) = 10.7584 \text{ °C}$$

$\Delta T_6 = \Delta T_1\ (U_1/U_6) = (8.7628)(2.4)/(1.8571) = 11.3247\ ^\circ C$

The following table summarizes the above values

ΔT_1	ΔT_2	ΔT_3	ΔT_4	ΔT_5	ΔT_6
8.7628	9.224	9.7095	10.2205	10.7584	11.3247

It should be noted that the temperature drop per effect increases as the effect temperature is reduced, i.e., $\Delta T_1 > \Delta T_2 > \Delta T_3 > \Delta T_4 > \Delta T_5 > \Delta T_6$. This is dictated by

- Constant heat transfer area,
- Lower overall heat transfer coefficients at lower temperatures, and
- Constant thermal loads in all effects.

Therefore, the increase of the temperature drop at lower temperatures compensates the decrease in the overall heat transfer coefficient.

The temperature profile in effects 1 to 6 is then calculated from Eqs. 17 and 18.

$T_1 = T_s - \Delta T_1 = 100 - 8.762 = 91.2372\ ^\circ C$

$T_2 = T_1 - \Delta T_1\ (U_1/U_2) = 91.2372 - 8.762\ (2.4/2.28) = 82.0132\ ^\circ C$

$T_3 = T_2 - \Delta T_1\ (U_1/U_3) = 82.0132 – 8.762\ (2.4/2.166) = 72.3037\ ^\circ C$

$T_4 = T_3 - \Delta T_1\ (U_1/U_4) = 72.3037 - 8.762\ (2.4/2.28) = 62.0831\ ^\circ C$

$T_5 = T_4 - \Delta T_1\ (U_1/U_5) = 62.0831 - 8.762\ (2.4/2.28) = 51.3247\ ^\circ C$

To check the above values T_6 is calculated on

$T_6 = T_5 - \Delta T_1\ (U_1/U_6) = 51.3247 - 8.762\ (2.4/1.8571) = 40\ ^\circ C$

This value checks with the initial specification of 40 °C.

The following table includes summary of calculated temperatures as well as the temperature of the motive steam.

T_s	T_1	T_2	T_3	T_4	T_5	T_6
100	91.2	82.01	72.3	62.1	51.3	40

The latent heat values in all effects are calculated using the correlation given in Appendix A

λ_{v_1} = 2499.5698 – 2.204864 T_{v_1} - 2.304x10^{-3} $T_{v_1}^2$
= 2499.5698 - 2.204864 (91.2372-2) - 2.304x10^{-3} (91.2372-2)^2
= 2284.47 kJ/kg

λ_{v_2} = 2499.5698 – 2.204864 T_{v_2} - 2.304x10^{-3} $T_{v_2}^2$
= 2499.5698 - 2.204864 (82.0132-2) - 2.304x10^{-3} (82.0132-2)^2
= 2308.4 kJ/kg

λ_{v_3} = 2499.5698 – 2.204864 T_{v_3} - 2.304x10^{-3} $T_{v_3}^2$
= 2499.5698 - 2.204864 (72.3037-2) - 2.304x10^{-3} (72.3037-2)^2
= 2333.17 kJ/kg

λ_{v_4} = 2499.5698 – 2.204864 T_{v_4} - 2.304x10^{-3} $T_{v_4}^2$
= 2499.5698 - 2.204864 (62.0831-2) - 2.304x10^{-3} (62.0831-2)^2
= 2358.78 kJ/kg

λ_{v_5} = 2499.5698 – 2.204864 T_{v_5} - 2.304x10^{-3} $T_{v_5}^2$
= 2499.5698 - 2.204864 (51.3247-2) - 2.304x10^{-3} (51.3247-2)^2
= 2385.21 kJ/kg

The latent heat value in the effect number 6 is calculated previously, and its value is equal to 2412.46 kJ/kg. Summary of the latent heat values is given in the following table, which includes the latent heat of motive steam.

λ_s	λ_{v1}	λ_{v2}	λ_{v3}	λ_{v4}	λ_{v5}	λ_{v6}
2256.043	2284.47	2308.4	2333.17	2358.78	2385.21	2412.46

The flow rate profiles of the distillate and brine as well as the brine concentrations are calculated from Eqs. 21 and 23-27. The distillate flow rate in the first effect is calculated from Eq. 23

$D_1 = M_d /(1 + \lambda_{v_1}/\lambda_{v_2} + \lambda_{v_1}/\lambda_{v_3} + \lambda_{v_1}/\lambda_{v_4} + \lambda_{v_1}/\lambda_{v_5} + \lambda_{v_1}/\lambda_{v_6})$
= (1)/(1 + (2284.47/2308.4) + (2284.47/2333.17)
+ (2284.47/2358.78) + (2284.47/2385.21)
+ (2284.47/2412.46))
= 0.1712 kg/s

Subsequently, the distillate flow rates in effects 2 to n are calculated

$D_2 = D_1\ \lambda_{v_1}/\lambda_{v_2} = 0.1712\ (2284.47/2308.4) = 0.1694$ kg/s

$D_3 = D_1\ \lambda_{v_1}/\lambda_{v_3} = 0.1712\ (2284.47/2333.17) = 0.1676$ kg/s

$D_4 = D_1\ \lambda_{v_1}/\lambda_{v_4} = 0.1712\ (2284.47/2358.78) = 0.1658$ kg/s

$D_5 = D_1\ \lambda_{v_1}/\lambda_{v_5} = 0.1712\ (2284.47/2385.21) = 0.1639$ kg/s

$D_6 = D_1\ \lambda_{v_1}/\lambda_{v_6} = 0.1712\ (2284.47/2412.46) = 0.1621$ kg/s

The brine flow rates are obtained from Eqs. 24 and 25

$B_1 = M_f - D_1 = 2.5 - 0.1712 = 2.3288$ kg/s

$B_2 = B_1 - D_2 = 2.3288 - 0.1694 = 2.1594$ kg/s

$B_3 = B_2 - D_3 = 2.1594 - 0.1676 = 1.9918$ kg/s

$B_4 = B_3 - D_4 = 1.9918 - 0.1658 = 1.826$ kg/s

$B_5 = B_4 - D_5 = 1.826 - 0.1639 = 1.6621$ kg/s

The above calculations are checked by determining the value of B_6

$B_6 = B_5 - D_6 = 1.6621 - 0.1621 = 1.5$ kg/s

This value checks with the initial material balance calculations. The salt concentration profile is calculated from Eqs. 26 and 27.

$X_1 = X_f\ M_f\ /B_1 = 42000\ (2.5/2.3288) = 45087.6$ ppm

$X_2 = X_1\ B_1/B_2 = 45087.6\ (2.3288/2.1594) = 48625$ ppm

$X_3 = X_2\ B_2/B_3 = 48625\ \ (2.1594/1.9918) = 52716.8$ ppm

$X_4 = X_3\ B_3/B_4 = 52716.8\ (1.9918/1.826) = 57502.8$ ppm

$X_5 = X_4\ B_4/B_5 = 57502.8\ (1.826/1.6621) = 63174.3$ ppm

The value of X_6 is specified in the problem statement at 70,000 ppm. Summary for the values of distillate and brine flow rates and brine concentration are given in the following table.

Effect	1	2	3	4	5	6
D (kg/s)	0.1712	0.1694	0.1676	0.1658	0.1639	0.1621
B (kg/s)	2.3288	2.1594	1.9918	1.826	1.6621	1.5
X (ppm)	45087.6	48625	52716.8	57502.8	63174.3	70000

The heat transfer areas are calculated in effects 1 to 6. These values are

$$A_1 = D_1 \lambda_{v_1}/(U_1(T_s - T_1)) = (0.1712)(2284.47)/(2.4(100 - 91.24))$$
$$= 18.59 \text{ m}^2$$

$$A_2 = D_2 \lambda_{v_2}/(U_2(\Delta T_2 - \Delta T_{loss})) = (0.1694)(2308.4)/(2.28(9.224\text{-}2))$$
$$= 23.74 \text{ m}^2$$

$$A_3 = D_3 \lambda_{v_3}/(U_3(\Delta T_3 - \Delta T_{loss})) = (0.1676)(2333.17)/(2.166(9.7095\text{-}2))$$
$$= 23.41 \text{ m}^2$$

$$A_4 = D_4 \lambda_{v_4}/(U_4(\Delta T_4 - \Delta T_{loss})) = (0.1658)(2358.78)/(2.0577(10.2205\text{-}2))$$
$$= 23.12 \text{ m}^2$$

$$A_5 = D_5 \lambda_{v_5}/(U_5(\Delta T_5 - \Delta T_{loss})) = (0.1639)(2385.21)/(1.9548(10.7584\text{-}2))$$
$$= 22.83 \text{ m}^2$$

$$A_6 = D_6 \lambda_{v_6}/(U_6(\Delta T_6 - \Delta T_{loss})) = (0.1621)(2412.46)/(1.8571(11.3247\text{-}2))$$
$$= 22.58 \text{ m}^2$$

The maximum difference in effect areas is equal to 0.35 m^2. Assuming an error criterion of less than 0.0001 m^2 is required, therefore, a new iteration sequence has to be initiated. The second iteration starts with calculations of the new heat transfer area

$$A_m = \frac{\sum_{i=1}^{n} A_i}{n}$$

$$= \frac{18.56 + 23.74 + 23.42 + 23.12 + 22.84 + 22.58}{6}$$

$$= \frac{134.26}{6} = 22.38 \text{ m}^2$$

A new profile for the temperature drop across the effects is then calculated

$\Delta T_1 = \Delta T_1 \, (A_1/A_m) = (8.7628)(18.59)/(22.38) = 7.28$ °C

$\Delta T_2 = \Delta T_2 \, (A_2/A_m) = (9.224)\,(23.74)/(\,22.38) = 9.78$ °C

$\Delta T_3 = \Delta T_3 \, (A_3/A_m) = (9.7095)\,(23.41)/(\,22.38) = 10.16$ °C

$\Delta T_4 = \Delta T_4 \, (A_4/A_m) = (10.2205)\,(23.12)/(\,22.38) = 10.56$ °C

$\Delta T_5 = \Delta T_5 \, (A_5/A_m) = (10.7584)\,(22.84)/(\,22.38) = 10.98$ °C

$\Delta T_6 = \Delta T_6 \, (A_6/A_m) = (11.3247)\,(22.58)/(\,22.38) = 11.43$ °C

A new iteration is then taken, which starts with temperature profiles and continues to the convergence criteria part. Since, the specified tolerance is small, a total of 8 iterations are executed. The error criterion after the last iteration is $5.7\text{x}10^{-5}$ m^2, i.e., the difference between the maximum and minimum areas is equal to this value. Summary of flow rates, concentrations, temperatures, and heat transfer areas in the last iteration are given in the following table

Effect	1	2	3	4	5	6
D (kg/s)	0.1708	0.1693	0.1677	0.1662	0.1646	0.1614
B (kg/s)	2.3292	2.16	1.9922	1.826	1.6614	1.5
X (ppm)	45078.9	48611.5	52704.4	57501.2	63198.6	70000
T (°C)	92.67	84.96	76.84	68.29	59.29	40
A (m^2)	22.1446	22.1445	22.1445	22.1446	22.1446	22.1446

Finally, the system performance parameters are calculated. To obtain the performance ratio it is necessary to determine the steam flow rate, where

$M_s = D_1 \, \lambda_{v_1}/\lambda_s = (0.1713)(2280.7)/(2256.04) = 0.1726$ kg/s

Since the total distillate flow rate is equal to 1 kg/s, then,

$PR = M_d/M_s = 1/0.1726 = 5.79$

This is an interesting result and is consistent with MEE practice, where the performance ratio is *approximately* equal to the total number of effects.

The condenser thermal load is calculated from

$Q_c = D_6\, \lambda_{v_6} = (0.1614)\ (2412.46) = 389.44$ kJ/s

The logarithmic mean temperature difference in the condenser is given by

$$\begin{aligned}(LMTD)_c &= (T_f - T_{cw})/Ln((T_6 - \Delta T_{loss} - T_{cw})/(T_6 - \Delta T_{loss} - T_f)) \\ &= (35\text{-}25)/Ln((40\text{-}2\text{-}25)/(40\text{-}2\text{-}35)) \\ &= 6.819\ ^{o}C\end{aligned}$$

The condenser heat transfer area in the condenser is then calculated from

$A_c = Q_c/(U_c\ (LMTD)_c) = 389.44/((1.75)(6.819)) = 32.628\ m^2$

The specific heat transfer area is calculated by the summing the heat transfer areas for the six evaporators and the condenser. This is

$$sA = \frac{\sum_{i=1}^{n} A_i + A_c}{M_d} = (132.86+32.628) = 165.49\ m^2$$

The cooling water flow rate is obtained from Eq. 39

$D_6\, \lambda_{v_6} = (M_f + M_{cw})\ Cp\ (T_f - T_{cw})$

$(0.1614)(2412.45) = (2.5+Mcw)\ (4.2)(35\text{-}25)$

which gives M_{cw} = 13.73 kg/s. The specific cooling water flow rate has the same value, since the total product flow rate is equal to 1 kg/s.

Detailed Mathematical Model of MEE

The steady-state MEE model includes a set of material and energy balances, heat transfer equations, and thermodynamic relations. The main features of the model include the following:

- It maintains constant heat transfer areas in the evaporators and the feed heaters. This is common industrial practice, which is necessary to reduce the cost of construction, spare parts stocking, and maintenance.
- It considers the effect of the vapor leak to the venting system.
- It takes into consideration variations in the thermodynamic losses within the system. This includes the boiling point elevation, the non-equilibrium allowance inside the evaporators and the flashing boxes, temperature depression corresponding to the pressure drop in the demister, vapor transmission lines, and during the condensation process.
- It includes the effect of boiling temperature, brine velocity inside the tubes of feed heaters, the tube material, and the tube bundle geometry on the required heat transfer area.
- It takes into consideration temperature and salinity effects on the water physical properties such as latent heat, heat capacity, density, thermal conductivity, and viscosity.
- It weights the effect of non-condensable gases on the heat transfer coefficient in the evaporators and the feed heaters.

Assumptions used in the model include:

- The vapor formed in the effects is salt free.
- Energy losses from the effects to the surroundings are negligible. This is because of operation at relatively low temperatures, between 100-40 ^{o}C, and the effects are well insulated.
- The heat transfer efficiency in the exchange units, which include evaporators, condensers, and preheaters, is assumed constant.
- The physical properties of various streams are calculated at the temperature average of influent and effluent streams.

The mathematical model is divided into three parts, which include material balances, energy balances, and the heat transfer rate equations. Also, the model includes equations for the heat transfer coefficient, thermodynamic losses, and the physical properties. Details for these equations are given in the appendices. The following section gives the equations used to determine flow rates of various streams, temperature profiles in the effects, preheaters, and flash boxes, and the heat transfer areas in the effects, preheaters, and the down condenser.

Material Balances

The overall material and salt balances are given by

$$M_f = M_d + M_b \tag{40}$$

$$M_b = M_f \, (X_f / X_b) \tag{41}$$

where M is the mass flow rate, X is the salt concentration, and the subscript b, d, and f denotes the brine, the distillate, and the feed seawater. The total distillate flow rate, M_d, is defined by

$$M_d = \sum_{k=1}^{n} D_k + \sum_{k=2}^{n} d_k \tag{42)}$$

where D and d are the amounts of vapor formed by boiling and flashing, respectively, and the subscripts k and n define the effect number and the total number of effects. The difference of the total seawater feed, M_f, and the amount of vapor formed in the first effect, D_1, gives the brine flow rate leaving the first effect

$$B_1 = M_f - D_1 \tag{43}$$

For effects 2 to n, the brine flow rate leaving effect j is given

$$B_j = M_f - \sum_{k=1}^{j} D_k - \sum_{k=2}^{j} d_k \tag{44}$$

In Eqs. 43 and 44 B is the brine flow rate. The salt balance in the brine stream leaving the first effect and effects 2 to n is

$$X_1 = \frac{M_f X_f}{M_f - D_1} \tag{45}$$

$$X_j = \frac{M_f X_f}{M_f - \sum_{k=1}^{j} D_k - \sum_{k=2}^{j} d_k} \tag{46}$$

Energy balances

In the first effect, the latent heat of the condensing steam is used to increase the temperature of feed seawater from t_2 to the boiling temperature T_1 and to provide the heat required to evaporate a controlled mass of vapor, D_1 at T_1. This gives

$$M_s \lambda_s = M_f C_p (T_1 - t_2) + D_1 \lambda_{v_1} \tag{47}$$

where Cp is the specific heat at constant pressure, λ is the latent heat, T is the effect temperature, t is the seawater temperature, and the subscripts 1, 2, v and s denotes the first effect, the preheater associated with the second effect, the vapor, and the heating steam. Correlations for the specific heat at constant pressure and the latent heat are given in Appendix A. In all effects, the boiling temperature, T_j, is higher than the vapor saturation temperature, T_{vj}, by the boiling point elevation, $(BPE)_j$, and the temperature rise caused by the hydrostatic pressure head, ΔT_{yj}. This is

$$T_j = T_{vj} + (BPE)_j + \Delta T_{yj} \tag{48}$$

The term, ΔT_{yj}, is negligible in horizontal falling films, because of the very small thickness of the boiling film.

The latent heat of the vapors formed by boiling in effect j-1 is used to boil off a smaller amount of vapor in the next effect, j. The decrease in the vapor amount is caused by the increase in the vapor latent heat upon the decrease in effect temperature, i.e., $T_{cj-1} > T_{vj}$ and $\lambda_{cj-1} < \lambda_{vj}$. This energy balance is

$$D_j = \frac{D_{j-1}\lambda_{cj-1}}{\lambda_{vj}} \tag{49}$$

In Eq. 49 the boiling process occurs on the outer surface of the evaporator tubes. The condensation temperature, T_{cj}, is lower than the effect temperature, T_j, by the boiling point elevation, $(BPE)_j$, and the saturation temperature depressions associated with pressure losses in the demister, $(\Delta P_p)_j$, transmission lines between the effects, $(\Delta P_t)_j$, and vapor condensation inside the tubes, $(\Delta P_c)_j$. The resulting condensation temperature is

$$T_{cj} = T_j - (BPE + \Delta T_p + \Delta T_t + \Delta T_c)_j \tag{50}$$

The pressure drop during condensation, ΔP_c, is defined as the algebraic sum of the decrease caused by friction, ΔP_r, and the increase caused by gravity (ΔP_g) and vapor deceleration (ΔP_a). This relation is given by

$$\Delta P_{cj} = (\Delta P_r - \Delta P_g - \Delta P_a)_j \tag{51}$$

Correlations for the pressure drop components, ΔP_p, ΔP_t, ΔP_r, ΔP_g, and ΔP_a are given in Appendix B.

As the brine enters the second effect, which is at a lower pressure, it flashes and consequently its temperature is reduced from T_1 to T'_2. The flashing process forms a small amount of vapor, d_2, which is used to preheat partially the feed seawater in the effect preheater. Similar, processes take place in effects 3 to n. The energy balance for this process in the second effect and effects 3 to n is given by

$$d_2 = (M_f - D_1)\, C_p \frac{T_1 - T_2'}{\lambda_{v_2}'} \tag{52}$$

$$d_j = \left(M_f - \sum_{k=1}^{j} D_k - \sum_{k=2}^{j} d_k \right) C_p \frac{T_{j-1} - T_j'}{\lambda_{v_j}'} \tag{53}$$

where λ'_{vj} is the latent heat of formed vapor at T'_j. In Eq. 53, the brine flowing into effects 3 to n is reduced by the amounts of boiled and flashed off vapors. In effects 2 to n, the boiling temperature within the effect, T_j, is lower than the temperature of flashing brine, T'_j, by the non-equilibrium allowance $(NEA')_j$;

$$T'_j = T_j + (NEA')_j \tag{54}$$

The correlation for the non-equilibrium allowance is given in Appendix B.

The formed vapor in the first effect, D_1, condenses as it releases its latent heat in the second effect. This condensate enters the flashing box associated with second effect. The flashing process reduces the temperature of condensed vapor from T_{c_1} to T''_2. The value of T''_2 is higher than the vaporization temperature within the flash box, T''_{v_2}, by the non-equilibrium allowance for the flash box $(NEA'')_2$. The same process takes place in the flashing boxes of effects 3 to n and the resulting relation between T''_{vj} and T''_j is given by

$$T''_j = T''_{vj} + (NEA'')_j \tag{55}$$

The energy balance in flash boxes in the second effect and effects 3 to n gives the flow rate of amount of formed vapor.

$$\bar{d}_2 = D_1 C_p \frac{(T_{c_1} - T_2'')}{\lambda_{v_2}''} \tag{56}$$

$$\bar{d}_j = \left(\sum_{k=1}^{j} D_k + \sum_{k=2}^{j-2} d_k \right) C_p \frac{\left(T_{c_{j-1}} - T_j''\right)}{\lambda_{v_j}''} \tag{57}$$

where λ''_{vj} is the latent heat of vaporization at T''_{vj}. As shown in Eqs. 56 and 57, the amount of condensing vapor entering the flash box in the second effect is equal to D_1. This amount increases in subsequent flash boxes by the amount of vapor formed by boiling and flashing within these effects.

At the other end of the flow diagram, in the down condenser, the temperature of the intake seawater, $M_{cw}+M_f$, is increased from T_{cw} to T_f. The heating energy is provided by condensation of the vapors formed by flashing and boiling in the last effect and by flashing in the associated flash box. This is given by

$$\eta_n \left(d_n \lambda_{c_n}' + \bar{d}_n \lambda_{c_n}'' + D_n \lambda_{c_n} \right) = \left(M_{cw} + M_f \right) C_p \left(T_f - T_{cw} \right) \tag{58}$$

where η is the heat exchange efficiency and the subscripts c, cw, and n denote the condensing vapors, the intake seawater, and the last effect. The energy source in the feed preheaters in effects 2 to n-1 is the latent heat of condensation for the vapors formed by flashing inside the effect and the flash boxes. This balance is

$$\eta_j \left(d_j \lambda_{c_j}' + \bar{d}_j \lambda_{c_j}'' \right) = M_f C_p \left(t_j - t_{j+1} \right) \tag{59}$$

In Eqs. 58 and 59 λ''_{cj} and λ''_{cj} are the latent heat of condensation of flashed vapors in the feed preheaters at T'_{cj} and T''_{cj}. These temperature are lower than the vapor temperatures, T'_{vj} and T''_{vj} by the depression in the saturation temperature caused by pressure loss in the demister and during condensation outside the preheater tubes. These relations are

$$T'_{cj} = T'_{vj} - \Delta T'_{pj} - \Delta T'_{cj} \tag{60}$$

$$T''_{cj} = T''_{vj} - \Delta T''_{pj} - \Delta T''_{cj} \tag{61}$$

The correlation for the pressure loss in the demister is given in Appendix B. As for the condensation pressure loss it is assumed negligible, sinco the friction losses are compensated by the hydrostatic deceleration gains, Muller, 1991.

Heat Transfer Design Equations

The design equations for the heat transfer area are developed for the evaporators, the preheaters, and the down condenser. For the evaporators, the heat transfer area, A_e, is

$$A_e = \frac{M_s \lambda_s}{U_{e1}(T_s - T_1)} = \frac{D_j \lambda_{v_j}}{U_{ej}(T_{c_{j-1}} - T_j)} \tag{62}$$

where U is the overall heat transfer coefficient, the subscript j defines effects 2 to n, and the subscript e refers to the evaporator. As discussed before and as shown in Eq. 47, the thermal load in the first effect differs from other effects by the energy consumed to increase the seawater temperature from t_2 to T_1. In other effects, the feed brine is at the saturation temperature and the effect thermal load is equivalent to the vaporization latent heat.

The following relation gives the heat transfer area in the preheaters of effects 2 to n-1

$$A_{h_j} = \frac{M_f C_p (t_j - t_{j+1})}{U_{h_j} (\mathrm{LMTD})_j} \tag{63}$$

$$(\mathrm{LMTD})_j = \frac{t_j - t_{j+1}}{\ln \dfrac{T'_{c_j} - t_{j+1}}{T'_{c_j} - t_j}} \tag{64}$$

Similarly, the heat transfer area of the down condenser is given by

$$A_c = \frac{(M_f + M_{cw}) C_p (T_f - T_{cw})}{U_c A_c (\mathrm{LMTD})_c} \tag{65}$$

$$(\mathrm{LMTD})_c = \frac{T_f - T_{cw}}{\ln \dfrac{T'_{c_n} - T_{cw}}{T'_{c_n} - T_f}} \tag{66}$$

The overall heat transfer coefficient in Eqs. 62, 63, and 65 is based on the outside surface area and is related to the individual thermal resistance by the following expression.

$$\frac{1}{U_o} = \frac{1}{h_i}\frac{r_o}{r_i} + R_{f_i}\frac{r_o}{r_i} + \frac{r_o \ln(r_o / r_i)}{k_w} + R_{f_o} + \frac{1}{h_o} \tag{67}$$

where h is the heat transfer coefficient, R_f is the fouling resistance, k_w is the thermal conductivity of tube material, and r is the tube radius. The subscript i and o refer to the inner and outer tube surface, respectively. Correlations for the individual heat transfer coefficient are given in Appendix C.

Solution Method of the Detailed MEE Model

The developed model contains a large number of highly non-linear algebraic equations. The equations are solved by a modified fixed point iteration technique developed by El-Dessouky and Bingulac, 1996. The method is simple, but yet powerful and has proved to have a rapid convergence rate. The solution process starts with setting values of system parameters, which include salinity of intake seawater and rejected brine, temperature of intake seawater, temperature of rejected cooling seawater, and boiling temperature in effect n, tube length and diameter, vapor and brine velocities inside the tubes, evaporator area (constant in all effects), and area of preheaters in effects 2 to n-1. Initial guess is made for the temperature profiles in the effects and the preheaters. Iterations are performed in two loops on the preheaters and the evaporators. Solution starts at the last effect and proceeds towards the first effect. Completion of the iterative procedure results in determination of the temperature profiles, salt concentration profile, and flow rates of brine and distillate. Results are used to determine other system parameters, which include the performance ratio, the specific heat transfer area, and the specific cooling seawater flow rate.

The system parameters used in generating the model results are:

- The seawater temperature, T_{cw}, and salinity, X_f, are 25°C and 42000 ppm.
- The salinity of rejected brine, X_b, is 70000 ppm
- The temperature of rejected cooling water, T_f, is 35°C.
- The boiling temperature in the last effect, T_n, is 40°C.
- The sum of the fouling heat transfer resistance inside and outside the tubes in the preheaters and the evaporators, $R_{f_i}+R_{f_o}$, is 1.75×10^{-4} m^2 °C/W.
- The thermal efficiency of the preheaters, η_i, is 90%.
- The tube outside and inside diameters, δ_o and δ_i, are 31.75 mm 19.75 mm.
- The brine velocity, V, inside the preheater tubes is 1.55 m/s.
- The range for the top brine temperature in the first effect is 60-110 °C.
- The range for the number of effects is 4-12.

4.2.3 System Performance

The developed model for MEE system is validated through analysis of the effect and preheater characteristics. This includes analysis of profiles for the temperature and the distillate flow rates across the effects. Further analysis includes variations in the system performance parameters as a function of the number of effects and the top brine temperature.

The temperature profiles in the effects and the preheaters are shown in Fig. 5. The nonlinear form of both profiles across the effects and the preheaters indicates higher temperature drop per effect close to the cold side of the effects, i.e., effect number n. Since the heat transfer area is constant in all effects and preheaters and the overall heat transfer coefficients are larger at higher temperatures, it is necessary to have larger temperature drop at the cold side of the effects in order to compensate the reduction in the coefficient value. Similar thermal loads in the effects and the preheaters dictate this behavior. This is shown in the relation given by Eq. 49, where the rate of the latent heat of condensation of formed vapor in effect j is equivalent to the rate of the latent heat of evaporation of formed vapor in effect j+1.

Figure 6 include profiles for the distillate flow rates generated in the flash box and in the effect by boiling and flashing. Results indicate that the major portion of the total product is formed by evaporation within the effect. In addition, evaporation rates are higher at the first effect and decreases in subsequent effects. The relation given by Eq. 49, where the latent heat of vaporization is smaller at higher temperatures, dictates this behavior. Figure 6, show that the amount of distillate formed by flashing inside each effect is negligible in comparison with that formed in the flash boxes. In each effect, the flow rate of flashing vapors is close to 10% of the amount formed by boiling. Irrespective of this, the small amount of flashing vapors posses sufficient heat to increase the temperature of the feed seawater from a low value of 25 °C to higher temperatures close to the top brine temperature.

Figure 7 illustrates the effect of the top brine temperature and the total number of effects on the performance ratio of the system. As is shown, the performance ratio is nearly independent of the top brine temperature and is strongly related to the number of effects. This behavior is explained in terms of the distillate flow rate profiles shown in Fig. 6 for a 12 effect system. As is shown, the amount of distillate formed at high temperature side is close to 1 kg/s. This rate decreases at the low temperature side of the effects to values close to 0.7 kg/s. Irrespective of this, the amount of distillate formed at the low temperature side a sizeable fraction of the total product flow rate. Therefore, increase of the number of effects allows for increase in the number of steam reuse and the formation of additional amounts of distillate.

Variations in the specific heat transfer area as a function of the top brine temperature and the number effects are shown in Fig. 8. As is shown, the required heat transfer area per unit mass of product water increases by using a larger number of effects and reducing the top brine temperature. The use of a larger number of effects decreases the temperature drop per effect or the driving force for heat transfer. Therefore, keeping the top brine temperature constant and increasing the number of effects results in the increase of the specific heat transfer area. On the other hand, keeping the number of effects constant and increasing the top brine temperature result in the increase of the overall heat transfer coefficient. This reduces the thermal resistance and gives smaller heat transfer areas. At the highest top brine, the specific heat transfer area is almost independent on the number of effects. As is show in Fig. 8, all profiles converges to lower value at the highest top brine temperature. This is caused by the increase in the temperature drop per effect, especially at a larger number of effects.

Effects of the top brine temperature and the number of effects on the specific cooling water flow rate are shown in Fig. 9. Variations in the specific cooling water flow rate of cooling are insensitive to the value of the top brine temperature. On the other hand, the specific cooling water flow rate decreases rapidly upon the increase of the number of effects. As previously shown in Fig. 8 the system performance ratio is independent on the top brine temperature, where the amounts of distillate generated and steam used vary slightly as the top brine temperature increases. As a result, the amount of vapor formed in the last effect, which is condensed by the cooling seawater, varies slightly as the top brine is increased. This results in negligible variations in the specific cooling water flow rate as the top brine temperature is increased. Increasing the number of effects increases the total amount of product fresh water and reduces the amount of distillate formed per effect. In turn, a smaller amount of cooling seawater is needed to operate the condenser. The net result is a rapid decline in the specific cooling water flow rate.

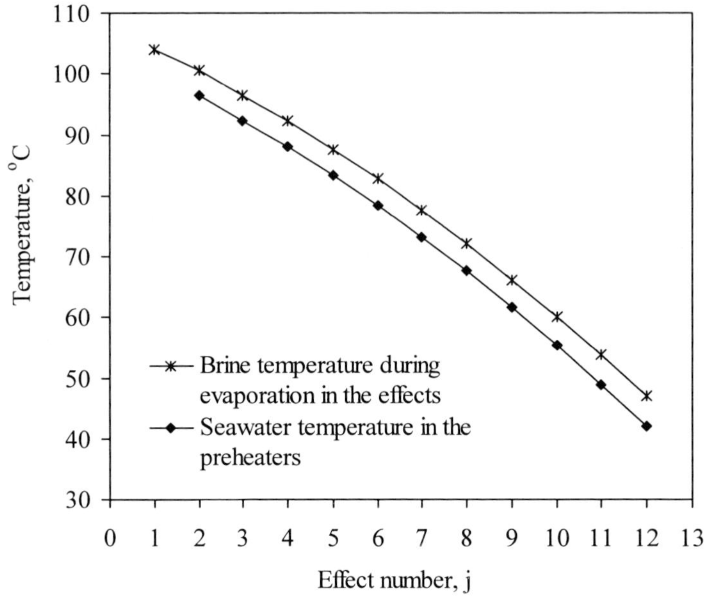

Fig. 5. Variation in temperature profiles in the evaporator and the preheater.

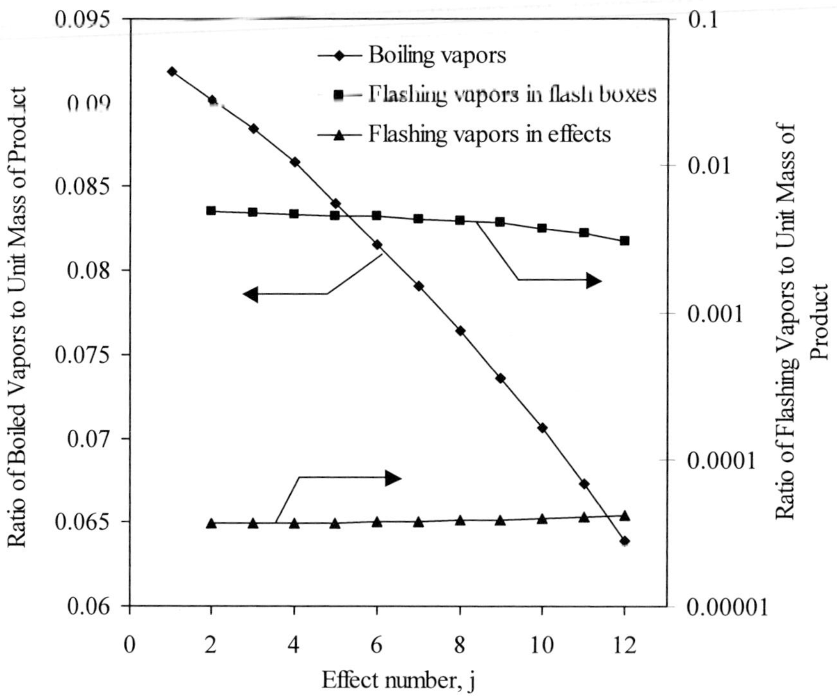

Fig. 6. Variation in distillate profiles as a function of effect number at T_s = 100 °C.

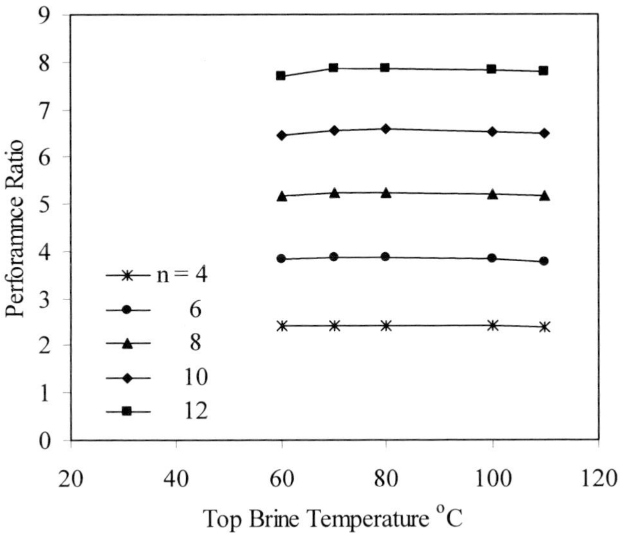

Fig. 7. Variation in the performance ratio as a function of top brine temperature and the number of effects.

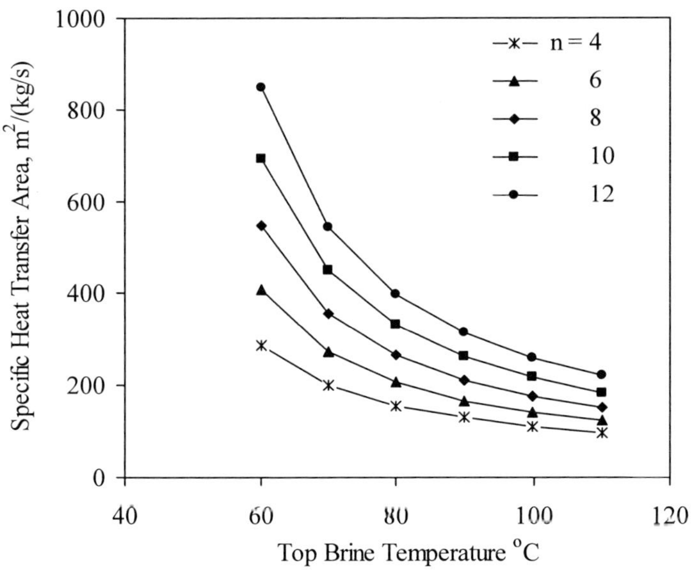

Fig. 8. Effect of top brine temperature and the number of effects on the specific heat transfer area.

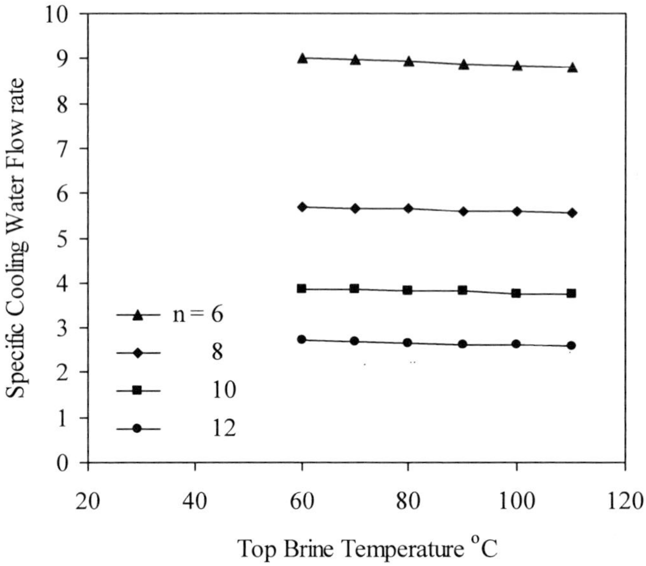

Fig. 9. Effect of top brine temperature and the number of effects on the specific cooling water flow rate.

4.2.4 Summary

The following conclusions are made in the light of the results and discussion given in the previous section:

- Modeling of the MEE system must take into consideration the nonlinear behavior of the governing equations. This is necessary to obtain complete descriptive model suitable for design, simulation, and analysis of existing and new systems. Simplified models with linear profiles have limited value and caution should be made in use of its predictions.
- Vapor formation by boiling and flashing is essential in modeling the effects, flash boxes and preheaters.
- The performance ratio of the MEE system is virtually independent of the top brine temperature and is strongly affected by the number of effects. A larger number of effects increase the number of vapor reuse and consequently the total amount of vapor formed.
- Operation of the MEE system at higher top brine temperature results in drastic decrease in the specific heat transfer area. This is because of the increase in the temperature driving force per effect and the heat transfer coefficient.

- The specific cooling water flow rate is independent of the top brine temperature. This is because the temperatures of the vapor in the last effect and the seawater leaving the down condenser are kept constant.
- The specific cooling water flow rate is reduced rapidly as the number of effects is increased, this is because of the reduction in the amount of vapor formed per effect, which reduces the thermal load in the down condenser.
- Comparison of values for the overall heat transfer coefficient predicted by the developed model show consistent behavior with literature data. The coefficient data with fouling are lower than literature data with clean surfaces. However, removal of the fouling effect gives values similar to literature correlations.

References

El-Dessouky, H.T., and Bingulac, S., Solving equations simulating the steady-state behavior of the multi-stage flash desalination process, Desalination, **107**(1996)171-193.

El-Dessouky, H.T., Alatiqi, I., Bingulac, S., and Ettouney, H.M., Steady-state analysis of the multiple effect evaporation desalination process, Chem. Eng. Tech., **21**(1998)437-451.

Muller, A.C., Condensers, Handbook of Heat Exchanger Design, Hewitt, G.F., Editor, Hemisphere, N.Y., 1991.

Rautenbach, R., and Schafer, S., Calculation of stagewise fouling factors from process data of large MSF distillers, Proceeding of the IDA World Congress on Desalination and Water Sciences, Madrid, Spain, October, 1997, Vol. I, pp 165-177.

Temstet, C., Laborie, J., and et al., Dual purpose desalination plant-high efficiency multiple-effect evaporator operating with a turbine for power production, Proceedings of the IDA World Congress on Desalination and Water Sciences, Abu Dhabi, November, 1995, Vol. III, 297-308.

Temstet, C., Canton, G., Laborie, J., and Durante, A., A large high-performance MED plant in Sicily, Desalination, **105**(1996)109-114.

Problems

1. A four effect MEE system operates at the following conditions:
   - Intake seawater temperature = 25 °C.
   - Intake seawater salinity = 35,000 ppm.
   - Rejected brine temperature = 35 °C.

- Top brine temperature = 95 °C.
- Flow rate of distillate product = 50 kg/s
- Heat transfer area of brine heater of the third effect = 80.7 m2.
- Heat transfer area of each effect = 723.3 m^2.
- Heat transfer coefficient in effect 1 = 5.2 kW/m^2 °C.
- Heat transfer coefficient in effect 2 = 3 kW/m^2 °C.
- Heat transfer coefficient in effect 3 = 2.1 kW/m^2 °C.
- Temperature of cooling water = 30 °C.
- Temperature of brine flow from third effect preheater = 50 °C.
- Boiling point elevation in each effect = 1 °C.
- The mass of vapor formed in each stage is constant.

Calculate the plant performance ratio, the specific heat transfer area, the mass flow rate of cooling water, and mass of vapor formed by flashing in the flashing box of the third effect.

2. A five effect MEE system operates at the following conditions:
   - Intake seawater temperature = 40 °C.
   - Intake seawater salinity = 42,000 ppm.
   - Heating steam temperature = 112 °C.
   - Temperature of vapor in last effect = 55 °C.
   - Thermodynamic losses other than BPE = 0.45 °C
   - Specific heat at constant pressure of seawater = 4.1 kJ/kg °C.
   - Flow rate of distillate product = 2000 kg/s
   - Temperature of cooling water = 45 °C.
   - Heat transfer coefficient in effect 1 = 5.25 kW/m^2 °C.
   - Heat transfer coefficient in effect 2 = 5.1 kW/m^2 °C.
   - Heat transfer coefficient in effect 3 = 4.85 kW/m^2 °C.
   - Heat transfer coefficient in effect 4 = 4.3 kW/m^2 °C.
   - Heat transfer coefficient in effect 5 = 3.7 kW/m^2 °C.

Calculate the plant performance ratio, the specific heat transfer area, the mass flow rate of cooling water, and mass of vapor formed by flashing in the flashing box of the third effect.

3. A three effect MEE system operates at the following conditions:
   - Plant capacity = 500 ton/day.
   - Steam temperature = 110 °C.
   - Intake seawater salinity = 42,000 ppm.
   - Temperature of vapor in last effect = 40 °C.
   - Intake seawater temperature = 20 °C.
   - Heat transfer coefficient in effect 1 = 3.123 kW/m^2 °C.
   - Heat transfer coefficient in effect 2 = 1.987 kW/m^2 °C.
   - Heat transfer coefficient in effect 3 = 1.136 kW/m^2 °C.

Calculate the plant performance ratio, the specific heat transfer area, and the mass flow rate of cooling water.

4. A three effect MEE system operates at the following conditions:
   - Plant capacity = 5 kg/s.
   - Steam temperature = 115 ^oC.
   - Intake seawater salinity = 42,000 ppm.
   - Specific heat at constant pressure of seawater = 4.18 kJ/kg ^oC.
   - Load of the third effect = 13.5 kN/m^2.
   - Intake seawater temperature = 27 ^oC.
   - Heat transfer coefficient in effect 1 = 4 kW/m^2 ^oC.
   - Heat transfer coefficient in effect 2 = 3 kW/m^2 ^oC.
   - Heat transfer coefficient in effect 3 = 2.5 kW/m^2 ^oC.

Calculate the plant performance ratio, the specific heat transfer area, and the mass flow rate of cooling water.

4.3 Parallel Feed Multiple Effect Evaporation

A large number of the parallel feed multiple effect evaporation is found in the desalination industry and it accounts for 3% of the total desalination market, IDA (2000). The process is found in the stand-alone mode or combined with thermal or mechanical vapor compression. The process has evolved from small production units with capacities less than 5000 m^3/d to larger units with capacities close to 20000 m^3/d, which are competitive to the MSF process.

Figures 2a and 2b show the operating lines for two possible configurations for the process as a function of the stream salinity and temperature. In both diagrams the horizontal line represent the feed stream to each effect. As is shown for all effects the feed has the same temperature and salinity. Inside the effect the feed temperature is increased to saturation conditions. This followed by evaporation and increase in salinity, which is represented by the vertical lines. Further discussion and details for this diagram are given in the following sections.

4.3.1 Process Description

Process schematics for the parallel-feed multiple-effect evaporation are shown in Figs. 10 and 11. The effects are numbered 1 to n from the left to right (the direction of the heat flow). Each effect constitutes a heat transfer area, vapor space, mist eliminator and other accessories. In the parallel feed system, the vapor flows from left to right, in the direction of falling pressure, while the feed seawater flows in a perpendicular direction. As for the parallel/cross flow system, Fig. 11, the brine stream leaving the first stage flows to the second, where it flashes and mixes with the feed seawater. Either system contains a number of evaporators, a train of flashing boxes, a down condenser, and a venting system. The parallel and the parallel/cross flow systems contain (n-1) flashing boxes for the distillate product. In the parallel/cross flow system, brine flashing takes place inside effects 2 to n. The two configurations utilize the horizontal falling film tubes, which are characterized by their ability to handle seawater scaling. This is because of the high wetting rates and efficient water distribution over the heat transfer surfaces by large spray nozzles. Thus, dry-patch formation or water mal-distribution is eliminated. This configuration offers the additional advantages of positive venting and disengagement of vapor products and/or non-condensable gases, high heat transfer coefficients, and monitoring of scaling or fouling materials.

The intake seawater is introduced into the down condenser, where it absorbs the latent heat of the condensing vapor from the last effect. As a result, intake seawater temperature increases to the feed temperature. Part of the heated intake seawater is rejected back to the sea, which is known as the cooling seawater. The function of cooling seawater is the removal of the excess heat added to the system in the first effect. The feed seawater stream is chemically treated, deaerated, and sprayed into the effects. The seawater spray falls in the form of thin film down the succeeding rows of tubes arranged horizontally. Within each effect, the brine temperature is increased to the boiling temperature corresponding to the pressure in the vapor space before a small portion of water vapor is formed. In the first effect, the heat required for preheating and evaporation is provided by condensing a controlled mass of saturated steam inside the tube bundle. The steam is supplied to the system from an external boiler. The high quality condensate from the first effect is returned back to the boiler.

The saturation temperature of the vapor formed in each effect is less than the brine boiling temperature inside the effect by the boiling point elevation. The vapor generated in each effect flows through a knitted wire mist separator known as wire mesh demister to remove the entrained brine droplets. The saturation temperature of the vapor departing the demister is less than that of the formed vapor due to the frictional pressure loss in the demister. The vapor flows from the demister has to be transported to the second effect. This transport inevitably involves a pressure drop and hence a corresponding decrease in the saturation temperature. Another pressure fall and consequent depression in the saturation temperature of the vapor is associated with vapor condensation inside the heat transfer tubes in the evaporators or over the heat transfer area in the preheaters. The latent heat of condensation of the vapor is exploited for further evaporation in the second effect.

In the parallel/cross system, the vapor formed in effects 2 to n is by boiling over the heat transfer surfaces and by flashing or free boiling within the liquid bulk. The temperature of the vapor formed by flashing is less than the effect boiling temperature by the boiling point elevation and the non-equilibrium allowance. Another small quantity of vapor is formed in the flashing box due to the flashing of distillate condensed in effect i. The flashed off vapor is produced at a temperature lower than the distillate condensation temperature by the non-equilibrium allowance. The flashing boxes offer a means for recovering heat from condensed fresh water and the brine stream. The boiling point elevation and temperature depression corresponding to pressure loss in the demister, transmission lines and during the condensation process reduces the available driving force for heat transfer in the evaporators and the preheaters. Thus, it is necessary to provide excess surface areas to compensate for these temperature degradations. In other words, the temperature losses present an extra resistance to the flow of heat between the condensing steam and the boiling seawater. Nonetheless, the temperature downgrading does not influence the plant thermal performance ratio or steam economy. The plant performance ratio depends on heat balance consideration and not on the rate of heat transfer. In the parallel/cross flow system the vapor formed by brine flashing from stage i are condensed inside the tube side of effect i+1. However, all vapors from the last effect are condensed on the shell side in the down condenser.

The amount of steam generated by evaporation in each effect is less than the amount generated in the previous effect. This is due to increase in the specific latent heat of vaporization with the decrease in the effect temperature. Consequently, the amount of vapor generated in an evaporator by boiling is less than the amount of condensing steam used for heating in the following evaporator. In either configuration, the salinity of the brine stream leaving each effect is close to solubility limit of $CaSO_4$, Figs. 2a. The brine stream leaving the last effect in the parallel or the parallel/cross systems is rejected back to the sea.

The down condenser is provided by good vents, first for purging during start-up and then for removing non-condensable gases, which may have been introduced with the feed or due to inleakage. The presence of the non-condensable gases not only impedes the heat transfer process but also reduces the temperature at which steam condenses at the given pressure. This occurs partially because of the reduced partial pressure of vapor in a film of poorly conducting gas at the interface. To help conserve steam economy venting is usually cascaded from the steam chest of one evaporator to another. The effects operate above atmospheric pressure are usually vented to the atmosphere. The non-condensable gases are always saturated with vapor. The vent for the bottom condenser must be connected to vacuum-producing equipment to compress the non-condensable gases to the atmosphere. This is usually a steam jet ejector if high-pressure steam is available. Steam jet ejectors are relatively inexpensive but also quite inefficient. Since the vacuum is maintained on the last effect, the unevaporated brine flows by itself from effect to effect and only a blow down pump is required in the last effect.

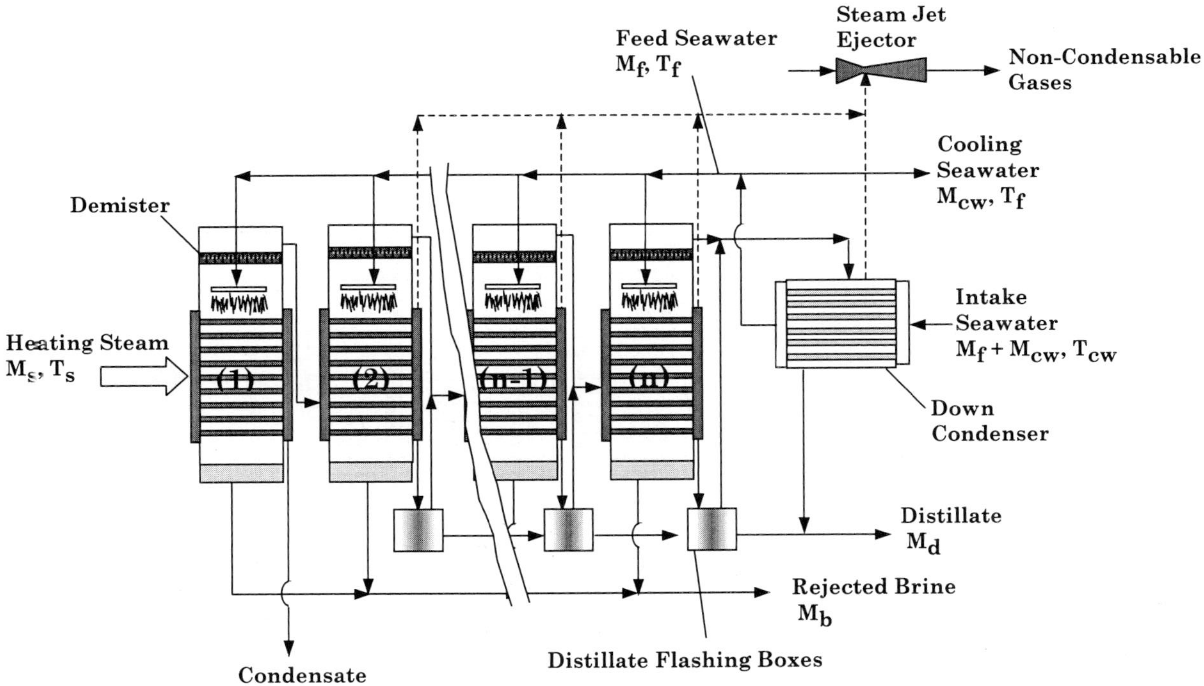

Fig. 10. Schematic of MEE parallel flow

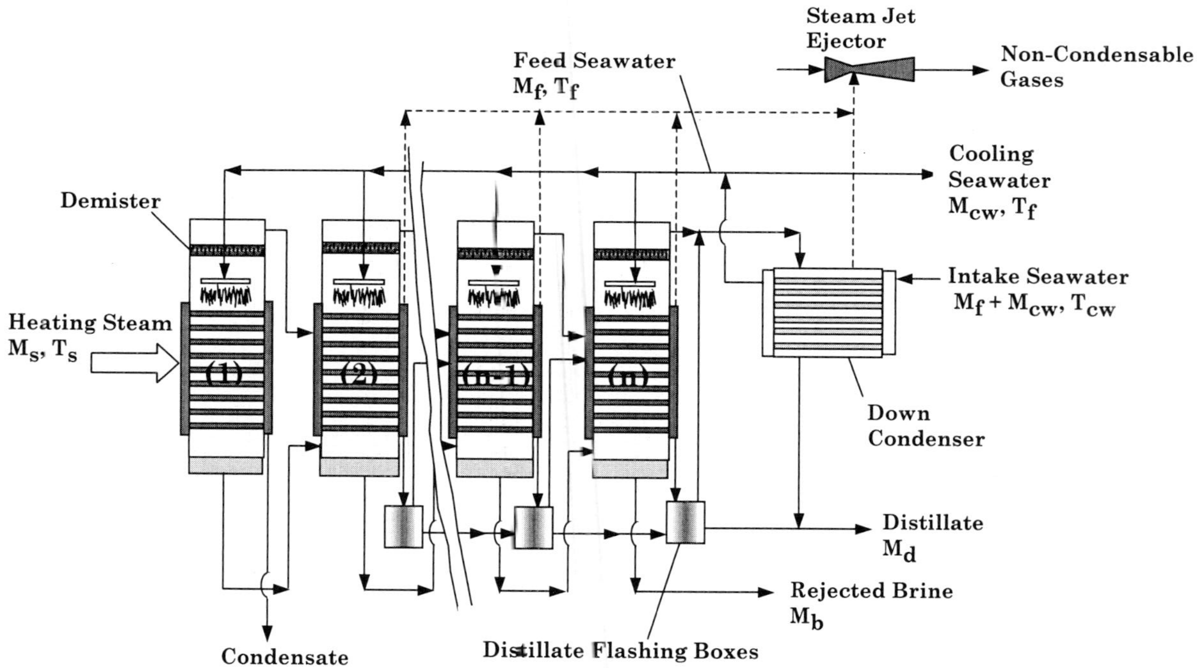

Fig. 11. Schematic of MEE parallel/cross flow

4.3.2 Process Modeling

The mathematical models of the parallel and parallel/cross flow MEE systems include basic material and energy balance equations as well as correlations for estimating the heat transfer coefficients, the thermodynamic losses, pressure drops, and physical properties. Results are reported in terms of the thermal performance ratio, the specific heat transfer area, the specific cooling water flow rate, and the conversion ratio. Other data include profiles of the effect temperature, pressure, flow rate, and salinity. The following two sections include model equations for the parallel and parallel/cross systems. Two assumptions are used in the analysis; the first assumes the system to be at steady state conditions and the second assumes that the distillate is salt free. The second assumption implies negligible entrainment of the brine droplets by the formed vapor.

Features of the developed mathematical models include the following:

- Constant and equal heat transfer areas in all effects, which is the standard practice in design of thermal desalination system.
- The heat transfer equations model the heat transfer area in each evaporator as the sum of the area for brine heating and the area for evaporation.
- Model variations in the thermodynamic losses (boiling point elevation, non-equilibrium allowance inside the evaporators and the flashing boxes, temperature depression corresponding to the pressure drop in the demister, vapor transmission lines, and during the condensation process) from one effect to another.
- Study the effect of boiling temperature, the velocity of brine flowing through the down condenser tubes, the tube material of construction, and the tube bundle geometry on the required specific heat transfer area.
- Variable physical properties of water.
- Weight the effect of the presence of non-condensable gases on the heat transfer coefficients in the evaporators and down condenser.

Mathematical Model of the MEE Parallel Flow

The mathematical model for the MEE parallel flow system includes the material and energy balance equations as well as the heat transfer equations for each effect, the flashing boxes, and the down condenser. The model includes the following equations:

- Total balance in effect i

$$F_i = D_i + B_i \tag{68}$$

- Salt balance in effect i

$$X_{F_i} F_i = X_{B_i} B_i \tag{69}$$

- Energy balance in effect i

$$D_{i-1} \lambda_{i-1} + d'_{i-1} \lambda'_{i-1} = F_i C_p (T_i - T_f) + D_i \lambda_i \tag{70}$$

In Eq. (70) the first term corresponds to the heat added to the effect by condensing the vapor generated in the previous effect. This only applies to effects 2 to n, since heating steam from an external source is used to drive the system and heat the first effect. The second term, which applies only to effects 3 to n, corresponds to the heat added to the effect by condensing the vapor generated in the distillate flashing box associated with the previous effect. The third term in Eq. 3 gives the amount of heat gained by the feed stream, where its temperature is increased inside the effect from the seawater temperature to the brine boiling temperature. The last term gives the amount of heat needed to generate the vapor inside the effect. In the above equation the specific heat at constant pressure depends on the brine salinity and temperature, while the latent heat depends on the vapor temperature. Correlations for the two properties are given in Appendix A.

- Vapor temperature in effect i

$$T_{v_i} = T_i - BPE_i \tag{71}$$

where T_v is the vapor temperature.

- The vapor condensation temperature

$$T_{c_i} = T_i - BPE_i - \Delta T_p - \Delta T_t - \Delta T_c \tag{72}$$

In Eq. 72, the condensation temperature, T_{c_i}, is lower than the brine boiling temperature, T_i, by the boiling point elevation and the losses caused by pressure depression in the demister (ΔT_p), friction in the transmission line (ΔT_t), and during condensation (ΔT_c).

- Flow rate of vapor flashed off in the distillate flashing boxes

$$d'_i = D_{i-1} C_p \frac{(T_{c_{i-1}} - T''_i)}{\lambda'_i} \tag{73}$$

with

$$T''_i = T_{v_i} + (NEA)_i \tag{74}$$

where $(NEA)_i$ is the non-equilibrium allowance and is equal to $(NEA)_i = 0.33 \frac{(T_{c_{i-1}} - T_i'')}{T_{v_i}}$, T_i'' is the temperature to which the accumulated distillate stream, formed in previous effects, cools down to as it enters the flashing box, Miyatake et al. (1973).

– Evaporator heat transfer area in effect i

$$D_{i-1}\lambda_{i-1} + d'_{i-1}\lambda'_{i-1} = F_i C_p (T_i - T_f) + D_i\lambda_i$$
$$= A_{1i} U_{1i} (LMTD)_i + A_{2i} U_{2i} (T_{v_i} - T_i) \quad (75)$$

$$\alpha(D_{i-1}\lambda_{i-1} + d'_{i-1}\lambda'_{i-1}) = D_i\lambda_i = A_{2i} U_{2i} (T_{v_i} - T_i) \quad (76)$$

$$(LMTD)_i = (T_i - T_f)/\ln((T_{v_i} - T_f)/(T_{v_i} - T_i)) \quad (77)$$

where A_{1i} is the heat transfer area for brine heating, A_{2i} is the heat transfer area for evaporation, U_{1i} and U_{2i} are the corresponding overall heat transfer coefficients, and α is the fraction of input heat consumed by vapor formation.

– Energy balance and heat transfer area of the down condenser

$$(d'_n + D_n)\lambda_n = (M_{cw} + M_f)\, Cp\, (T_f - T_{cw}) \quad (78)$$

$$(d'_n + D_n)\lambda_n = U_c A_c (LMTD)_c \quad (79)$$

$$(LMTD)_c = (T_f - T_{cw})/\ln((T_{v_n} - T_{cw})/(T_{v_n} - T_f)) \quad (80)$$

Mathematical Model of the MEE Parallel/Cross Flow

The mathematical model for the MEE parallel/cross flow system is developed in a similar manner to the parallel flow system and it includes the following equations:

– Total balance in effect i

$$F_i + B_{i-1} = D_i + B_i \quad (81)$$

– Salt balance in effect i

$$X_{F_i} F_i + X_{B_{i-1}} B_{i-1} = X_{B_i} B_i \quad (82)$$

– Energy balance for effect i

$$D_{i-1}\,\lambda_{i-1} + d_{i-1}\,\lambda_{i-1} + d'_{i-1}\,\lambda'_{i-1} = F_i\,C_p(T_i - T_f) + D_i\lambda_i \tag{83}$$

– Flow rate of vapor formed by brine flashing inside the effect

$$d_i = B_{i-1}\,Cp\,\frac{T_{i-1} - T_i'}{\lambda_i} \tag{84}$$

with

$$T_i' = T_i + NEA_i \tag{85}$$

where T_i' is the temperature to which the brine cools down to as it enters the effect. As given by Eq. 84 this temperature is lower than the effect brine temperature by the non equilibrium allowance.
Heat transfer area in effect i

$$D_{i-1}\,\lambda_{i-1} + d_{i-1}\,\lambda_{i-1} + d'_{i-1}\,\lambda'_{i-1} = F_i\,C_p(T_i - T_f) + D_i\lambda_i = A_{1i}\,U_{1i}\,(LMTD)_i + A_{2i}\,U_{2i}\,(T_{v_i} - T_i) \tag{86}$$

$$\alpha(D_{i-1}\,\lambda_{i-1} + d_{i-1}\,\lambda_{i-1} + d'_{i-1}\,\lambda'_{i-1}) = D_i\lambda_i = A_{2i}\,U_{2i}\,(T_{v_i} - T_i) \tag{87}$$

Energy balance and heat transfer area of the down condenser

$$(d_n + d'_n + D_n)\lambda_n = (M_{cw} + M_f)\,Cp\,(T_f - T_{cw}) \tag{88}$$

$$(d_n + d'_n + D_n)\lambda_n = U_c\,A_c\,(LMTD)_c \tag{89}$$

It should be noted that the model equations for the flow rate of vapor flashed off in the distillate flashing boxes and the logarithmic mean temperature differences in the effects and down condenser are identical to those given in the model of the MEE parallel flow system. Also, the symbols used in Eqs. 81-89 are the same as those for Eqs. 68-80. Models for the overall heat transfer coefficients in the evaporator and the down condenser are summarized in Appendix C.

Solution Algorithm

The model equations for either system are interlinked and highly nonlinear. Therefore, iterative solution is necessary to calculate the system characteristics. The solution algorithm starts with definition of the following parameters:
Number of effects are 4, 6, 8, or 12.
The heating steam temperature varies over a range of 60-100 °C.
The intake seawater temperature (T_{cw}) is 25°C.

The feed seawater temperature (T_f) is 35°C.
The boiling temperature in the last effect (T_n) is 40°C.
The seawater salinity has values of 34,000 ppm or 42,000 ppm.
The sum of the heat transfer resistances due to the tube material, fouling inside and outside the tube is 731×10^{-6} m^2 °C/W.
The tubes outside diameter (δ_o) is 31.75 mm and inside diameter (δ_i) is 19.75 mm.

The model equations for both systems are solved simultaneously by Newton's method to calculate the following:

- Flow rates of the feed, brine, and distillate in each effect.
- The steam flow rate.
- The brine temperature in effects 1 to n-1.
- The fraction of heat consumed by evaporation in each effects.
- The heat transfer areas for vapor formation and brine heating in each effect.

The iterative procedure is based on Newton's method with an iteration error of 1×10^{-4}. To facilitate the conversion procedure, each equation is scaled by the largest term found in the equation. Therefore, all equations are in the order of one. For example, the salt balance equation is rearranged into the following form

$$f(X_{F_i}, F_i, X_{B_i}, B_i) = 1 - (X_{F_i} F_i)/(X_{B_i} B_i)$$

Convergence of Newton's method is dependent on the initial guess, therefore, linear profiles are used for the flow rates, brine temperature, heat transfer area, and the ratio α. The guess for the steam flow rate is based on the approximate relation of the number of effects and the performance ratio.

4.3.3 System Performance

Performance of the two MEE systems is analyzed as a function of the intake seawater salinity, number of effects, and the top brine temperature. Performance parameters include the thermal performance ratio, the specific cooling water flow rate, conversion ratio, and the specific heat transfer area. Also, analysis is presented for the dependence of the heat transfer area for evaporation and brine heating on the system operating conditions. Finally, comparison is made between model predictions and the forward feed MEE and MSF systems.

Figure 12 shows the performance of the MEE parallel feed as function of the heating steam temperature and the seawater salinity. As is shown the decrease in thermal performance ratio decreases at higher heating steam temperature is caused by three factors, which includes:

- Increase in the amount of sensible heat required for increasing the temperature of the feed seawater to higher boiling temperatures, since the feed temperature (T_f) is kept constant at 35 °C.

- Increase in the amount of feed flow rate because of decrease in the conversion ratio.
- Decrease in the latent heat of the heating stream at higher temperatures.
- These factors result in the consumption of larger amount of steam and consequently reduction in the thermal performance ratio. Increase in the heating steam temperature reduces the specific heat transfer area due to the increase in the temperature drop per stage, which enlarges the driving force for heat transfer. Also, at higher temperatures the value of the overall heat transfer coefficient augments causing the decrease in the heat transfer area. Another effect is caused by the increase in the brine salinity at low temperatures, which results in an increase of the boiling point elevation. This lowers the vapor temperature and consequently the driving force for heat transfer. Therefore, at lower heating steam temperatures the area for heat transfer increases drastically. At higher temperatures, the decrease in the amount of the specific cooling water is associated with the increase in the amount of feed flow rate, which is caused by reduction in the conversion ratio. The decrease in the conversion ratio at higher top brine temperature is caused by the limitations imposed by the maximum salinity of the rejected brine.

Effects of the seawater salinity on the system performance are also shown in Fig. 12. As is shown larger differences in the performance ratio, the specific cooling seawater, and the conversion ratio are obtained at higher heating steam temperatures. This is caused by the decrease in the limit imposed on the salinity of the rejected brine, which results in large decrease of the conversion ratio and the subsequent increase in the feed flow rate. Combining Eqs. (1 and 2) can easily prove reduction in the amount of vapor formed per stage upon increase of the seawater salinity. The resulting relation, $D_i/F_i = (X_{B_i}-X_{F_i})/X_{B_i}$, show that increasing X_{F_i} at constant temperature (which implies constant X_{B_i}) would reduce the ratio on the right hand side of the equation and consequently the amount of vapor formed. As a result, the system thermal performance ratio, specific cooling water flow rate, and conversion ratio decreases at higher seawater salinity. As is shown, the specific heat transfer area is insensitive to changes in the seawater salinity, since it only depends on the thermal load, the heating steam temperature, the temperature drop per stage, and the overall heat transfer coefficient.

Results for increasing the number of effects for the MEE parallel feed are shown in Fig. 13. As is shown, increasing the number of effects gives higher thermal performance ratios and larger specific heat transfer areas. The increase in the specific heat transfer area is caused by reduction in the driving force for heat transfer, or the temperature drop per stage. This is because the heating steam temperature and the brine temperature in the last effect are kept constant. The increase in the system performance ratio for larger number of effects is a result of increasing the number of vapor reuse in the system. In the first effect, the latent heat of the heating steam is used to heat the feed seawater to the saturation temperature and to form a smaller amount of vapor. This process is repeated in subsequent effects, where the feed seawater is heated and an additional amount of vapor is formed. The decrease in the specific cooling water flow rate for larger number of effects is caused by the reduction in the amount of vapor formed per effect. The decrease in the conversion ratio is also caused by limits imposed on the maximum salinity of the rejected brine. As is shown in Fig. 13, operation of the 8-effect system in parallel mode is limited to a minimum heating steam temperature of 70 °C. At lower heating steam temperatures the temperature range for the brine in the first and last effects is small. Therefore, the combined effect of the boiling point elevation and the temperature drop per stage results in a heat transfer pinch, i.e., the vapor temperature in effect i is less than the brine temperature in effect i+1.

Analysis of variations in the heat transfer areas for evaporation and feed heating for the parallel flow system shows high sensitivity to the heating steam temperature and some dependence on the number of effects. For example, at a heating steam temperature of 100 °C and four effects the area for evaporation constitutes 78, 92, 96, and 98% of the total heat transfer area from the first to the last effects, respectively. For a lower heating steam temperature of 70 °C, the evaporation heat transfer area varies over a narrower range of 95 to 98% of the total heat transfer between the first and last effect. Increasing the number of effects increases the range over which the evaporation heat transfer area varies. For example, in the 8-effect system and at a heating steam temperature of 100 °C the evaporation heat transfer area varies over a wider range of 68 to 99% between the first and the eighth effects. From the above, it can be seen that the heat transfer area for evaporation is lower at higher heating steam temperatures. This is because of the increase in the amount of sensible heat required to increase the temperature of the feed seawater to the saturation temperature.

The performance of the MEE parallel/cross flow system differs from the MEE parallel flow system in the conversion ratio and the specific cooling water flow rate. On the other hand, variations in the thermal performance ratio and the specific heat transfer area for the two systems are similar. As is shown in Figs. 14 and 15, the system conversion ratio remains constant as the heating steam temperature increases. However, the conversion ratio increases at lower salinity for the feed seawater. For this system, the conversion ratio is independent of the heating steam temperature because the salinity of the brine leaving the last effect is defined at the same temperature, which is equal to 40 °C. Therefore, the total mass and salt balance of the system is defined by the relations ($M_F = M_B+M_D$) and ($M_F X_F = M_B X_B$), which combines to ($M_D/M_F = (X_B-X_F)/X_B$). Accordingly, the conversion ratio is independent of the heating steam temperature, since X_B and X_F are independent of the heating steam temperature. The same conclusion applies to variations in the conversion ratio as a function in the number of effects, Fig. 15. As is shown the conversion ratio is independent of the number of effects. This is cleared by inspection of the above relation, where the conversion ratio depends only on X_B and X_F. In this regard, X_B is a function of the brine temperature in the last effect, 40 °C, and X_F is an independent parameter. The small increase in the amount of cooling seawater at higher heating steam temperatures and seawater salinity is caused by the decrease in the system thermal performance ratio at higher heating steam temperatures, which implies increase in the specific thermal energy of the system.

Comparison of the parallel feed and the parallel/cross flow systems for n = 4, is shown in Fig. 16, which contains two sets of data for each system. The first set limits the maximum brine concentration to 95% of the $CaSO_4$ solubility limit and the second set has a maximum limit of 70,000 ppm. As is shown, the two systems have similar variations in the thermal performance ratio and the specific heat transfer area, where both parameters decrease at higher heating steam temperature. Differences among the two systems are found upon comparison of the specific cooling water flow rate and the conversion ratio. Selection among the four operating conditions show that the parallel/cross flow system with a salinity limit of 70000 ppm has the lowest specific flow rate for the cooling seawater, highest thermal performance ratio, and lowest specific heat transfer area. On the other hand, the highest conversion ratio is obtained for the parallel/cross flow system with a salinity limit set by the $CaSO_4$ solubility.

Comparison of the forward and parallel/cross feed systems is shown in Fig. 17. The data for the forward feed system is extracted from a previous study by El-Dessouky et al. (1998). The data for the forward feed MEE and the parallel/cross flow systems are obtained for 12 effects, feed salinity of 42000 ppm, rejected brine salinity of 70000 ppm, intake seawater temperature of 25 °C, feed seawater temperature of 35 °C, and rejected brine temperature of 40 °C. As is shown, the parallel/cross feed has higher specific heat transfer area than the forward feed system, especially at lower top brine temperatures. This is because of the lower driving force for heat transfer, which is manifested in the parallel/cross flow system due to heating of the feed seawater in each from the intake temperature to the saturation temperature. The performance ratio for both systems is almost independent of the heating steam temperature. Also, the performance ratio for the parallel/cross flow system is higher because it is not necessary to heat all the feed to the top brine temperature.

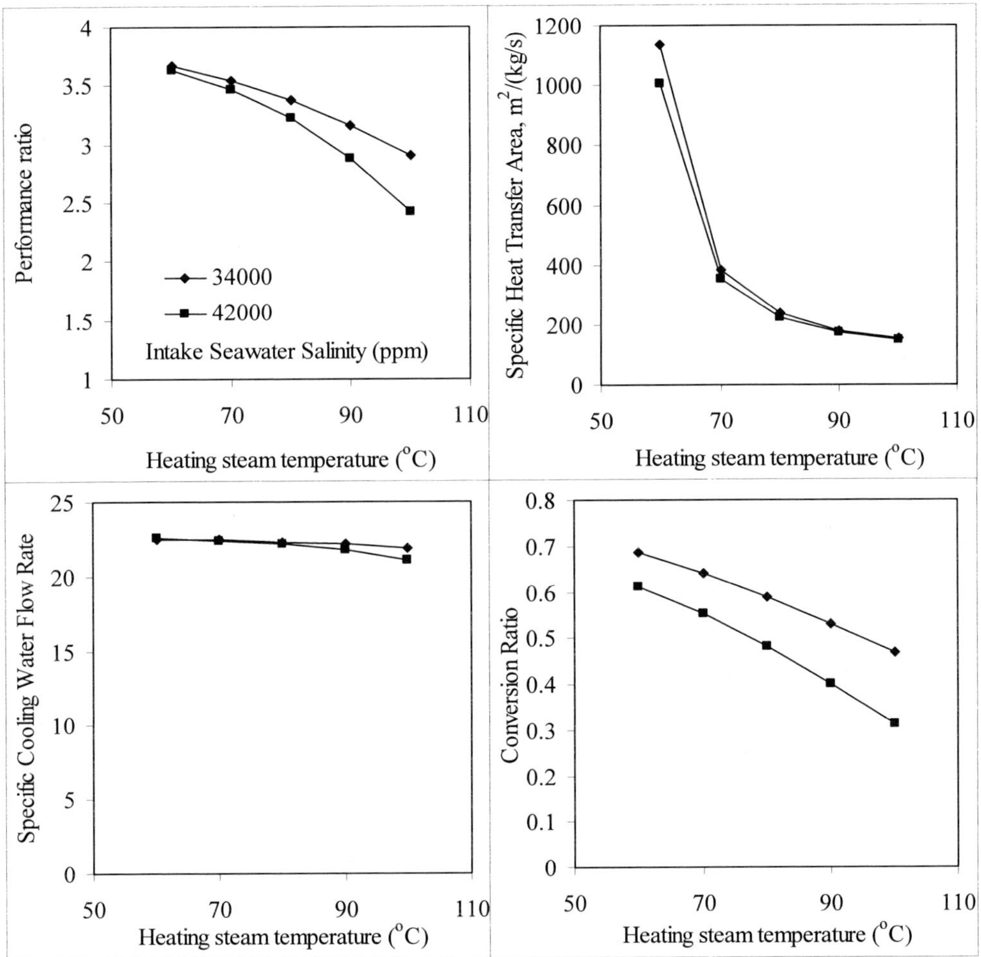

Fig. 12. Dependence of MEE parallel flow system on the heating steam temperature and seawater salinity for n = 4

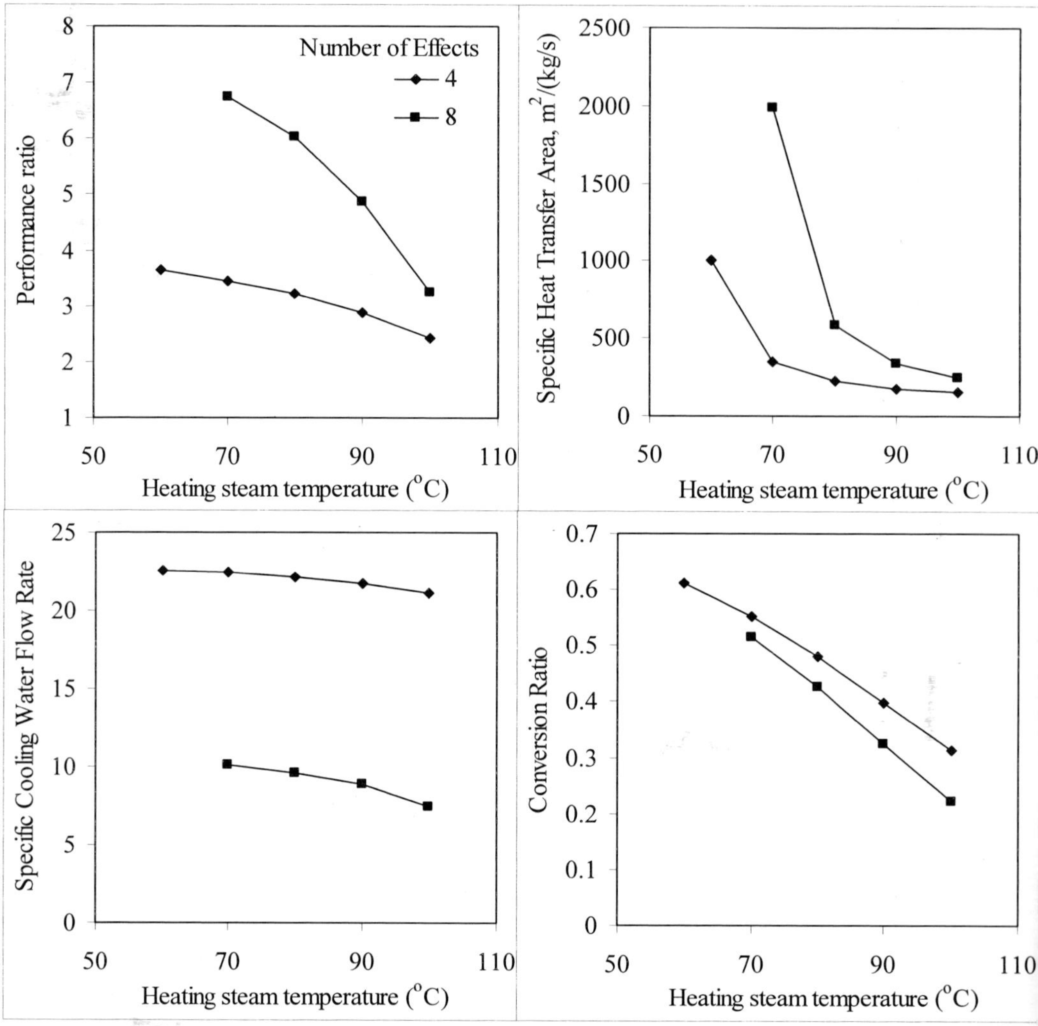

Fig. 13. Dependence of MEE parallel flow system on the heating steam temperature and number of effects for seawater salinity = 42000 ppm

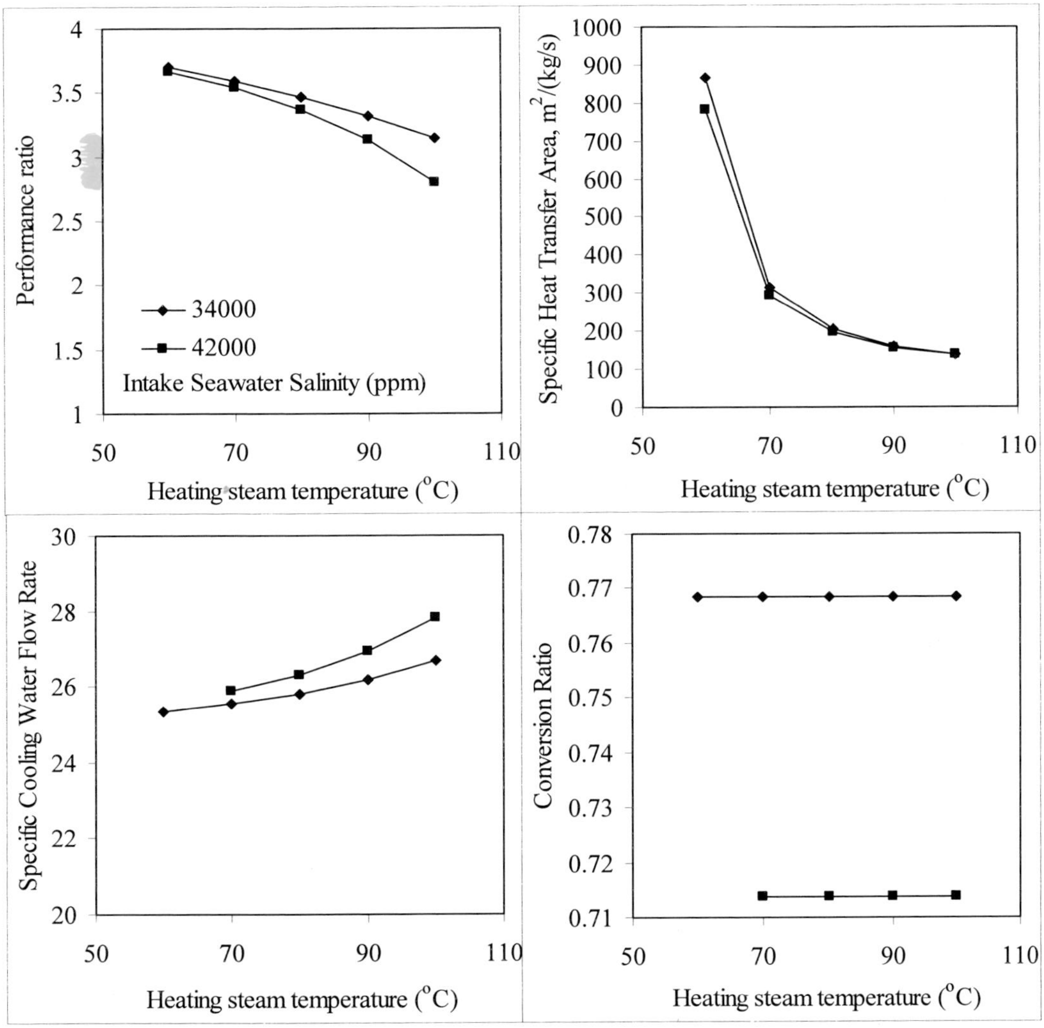

Fig. 14. Dependence of MEE parallel/cross flow system on the heating steam temperature and seawater salinity for n = 4

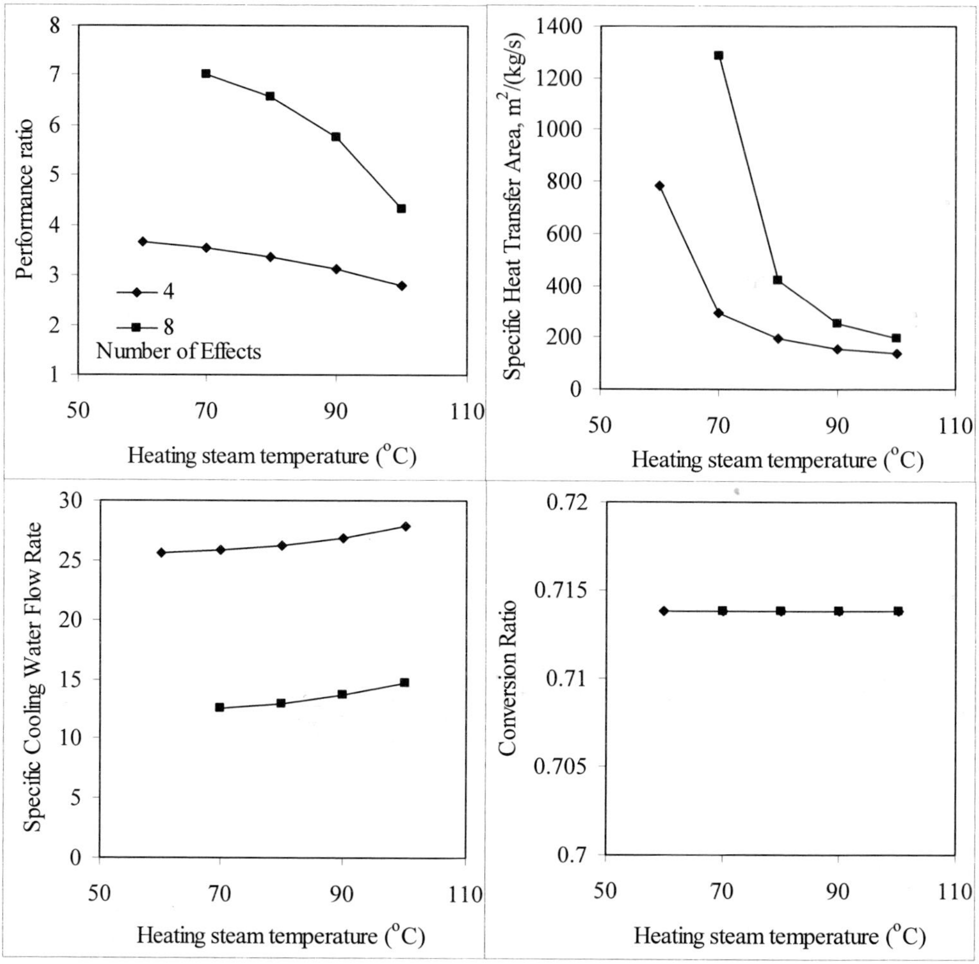

Fig. 15. Dependence of MEE parallel/cross flow system on the heating steam temperature and number of effects for seawater salinity = 42000 ppm

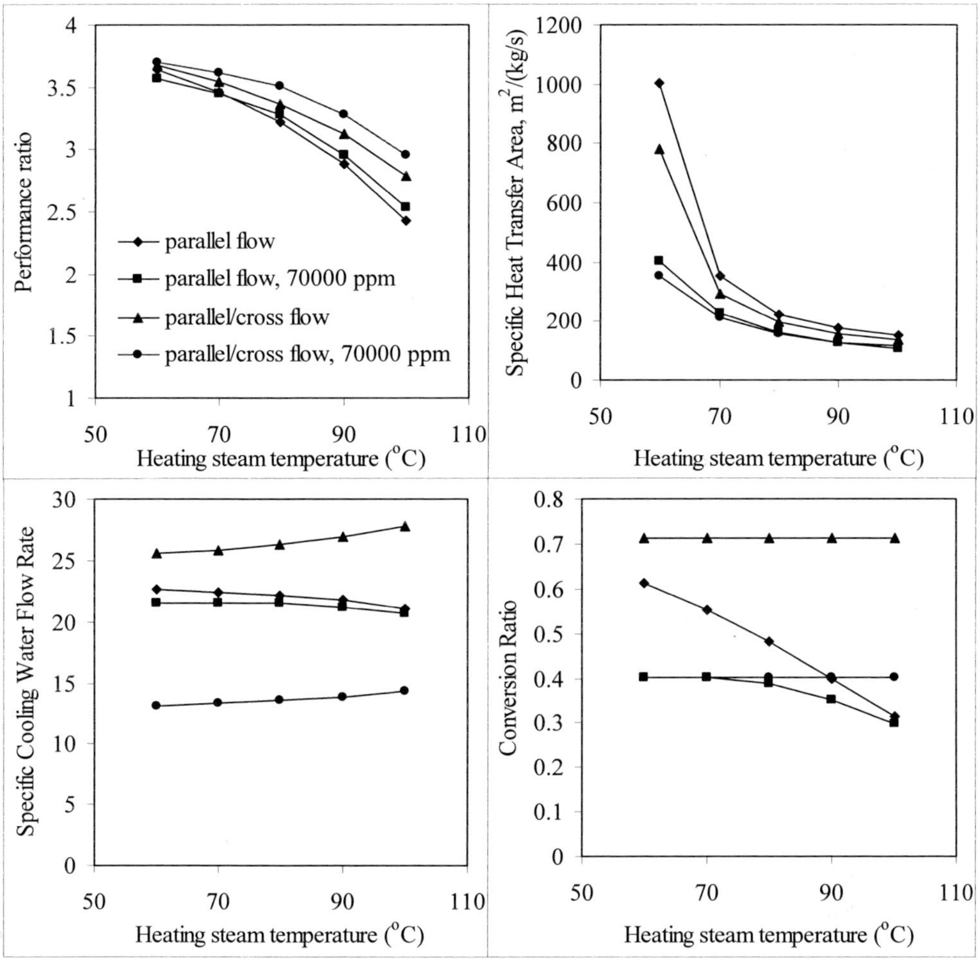

Fig. 16. Comparison of four MEE parallel configurations for n = 4 and seawater salinity = 42000 ppm

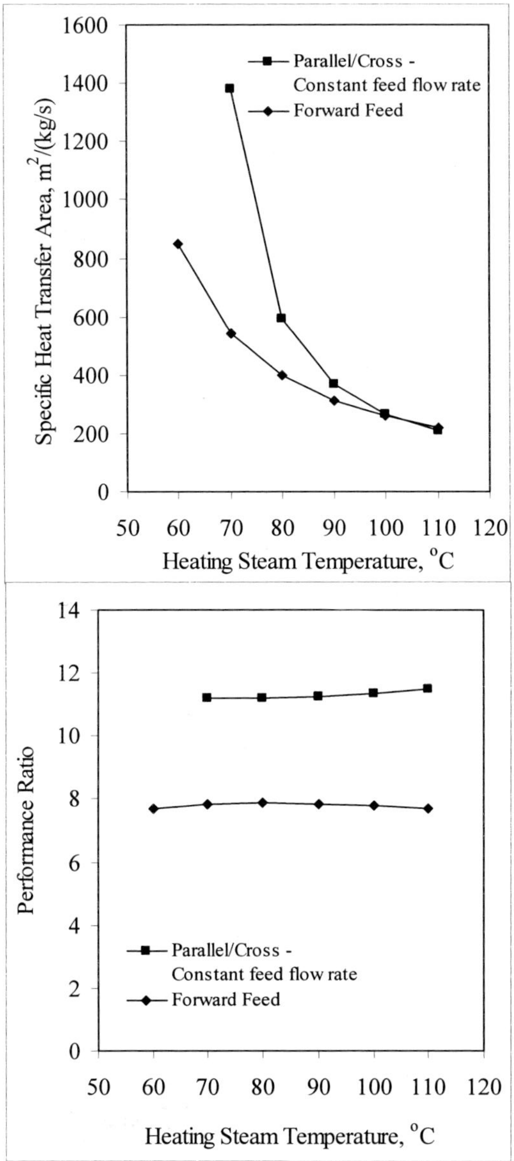

Fig. 17. Comparison of forward and parallel feed MEE for n = 12

4.3.4 Industrial Data and Practice

Comparing the performance of the parallel feed, the forward feed, and the conventional multistage flash system (MSF) is shown in Table 2. As is shown the performance ratio for the MSF system with 24 stages is 8, while the performance ratio for the MEE configurations with 8 effects varies from 4.9 to 5.2, and the 12 effect systems have an average value of 8. The specific heat transfer area for the MEE systems vary over a range of 200-500 $m^2/(kg/s)$ as the number of effects is increased from 8 to 12. As for the specific heat transfer area for the MSF system it has a value of 275 $m^2/(kg/s)$. It should be noted that the MEE forward feed system is not found on commercial scale and is limited to the conceptual design presented here.

Table 2: Comparison of MSF, forward feed MEE, parallel/cross flow MEE for intake seawater salinity of 42,000 ppm, heating steam temperature of 90 °C.

	MSF El-Dessouky et al. (1995)	MEE Forward feed El-Dessouky et al. (1998)	MEE Parallel	MEE Parallel Cross
Number of effects/stages	24	8	8	8
Performance Ratio	8	5.2	4.9	5.8
Specific heat transfer area	259	212	335	255
Conversion ratio	0.4	0.4	0.325	0.714
Salinity of rejected brine	70000	70000	62247	146776
Specific flow rate of cooling water	2.4	2.6	8.9	13.7
Specific pumping power	8.3	4.12	7.78	9.85

4.3.5 Summary

Performance analysis of various configurations shows that the best performance is obtained for the parallel/cross flow MEE. However, the parallel flow system has similar performance characteristics; moreover, its design, construction, and operation is simpler. Operation of both systems is favored at higher temperatures because of the drastic reduction in the specific heat transfer area. However, operation at lower temperatures gives higher thermal performance ratio and lower specific flow rate of the cooling water. Final selection of the most efficient and least expensive system and operating conditions necessitate full system optimization. The developed models should prove to be highly valuable in selecting and determining the characteristics of the optimum system.

Comparison of the MSF, forward feed, parallel, and parallel/cross flow MEE systems show several advantages of the forward feed MEE over the other systems. It is certain that the engineering design of the forward feed MEE is more energy efficient since it has the lowest specific power consumption, specific heat transfer area, and specific cooling water flow rate. Advantages of the forward feed MEE over the MSF system are found in the lower number of effects and specific power consumption. The forward feed and parallel flow MEE systems have similar or higher thermal performance ratio than the MSF system, however, the number of effects is only 12 for the MEE systems, while it is equal to 24 stages in the MSF system. Also, the MSF system has higher specific power consumption, which is required for pumping the brine circulation stream.

References

El-Dessouky, H.T., Shaban, H.I., and Al-Ramadan, H., Steady-state analysis of multi-stage flash desalination process, Desalination, **103**(1995)271-287.

El-Dessouky, H.T., Alatiqi, I., Bingulac, S., and Ettouney, Steady- H.M., state analysis of the multiple effect evaporation desalination process, Chem. Eng. Technol., **21**(1998)15-29.

Miyatake, O., Murakami, K., Kawata, Y., and Fujii, Fundamental Experiments with Flash Evaporation, Heat Transfer Jpn. Res., **2**(1973)89-100.

Chapter 5

Multiple Effect Evaporation Vapor Compression

Objectives

The objective of this chapter is to analyze and evaluate the performance of the multiple effect evaporation systems combined with various types of heat pumps. The analysis includes performance of the following systems:

- Parallel feed multiple effect evaporation with thermal or mechanical vapor compression heat pumps.
- Forward feed multiple effect evaporation with thermal, mechanical, absorption, or adsorption vapor compression heat pumps.

The performance of the parallel feed systems is compared against industrial data. However, the forward feed system presents only results of the system design, since there are no industrial units for these systems.

5.1 Parallel Feed Multiple Effect Evaporation with thermal and mechanical vapor compression

The parallel feed multiple effect evaporation is the industrial standard for seawater desalination using the multiple effect evaporation process. The parallel feed configuration has several attractive features including simple process layout, stable and wide operating range. The process model and performance has similar features to the forward feed configuration. The following sections include models and analysis for the thermal and mechanical vapor compression processes of the parallel and parallel/cross flow configurations.

As discussed in previous sections, the MEE-MVC system is thought to increase the system capacity. As will be shown later, use of this configuration has no effect on the specific power consumption. The market share of the MEE-MVC is less than 1%. On the other hand, the MEE-TVC has a higher share close to 5%. Both processes have attractive features that make them highly competitive against other well-established desalination processes that include the MSF and RO.

Limited number of field studies can be found on the MEE-TVC system, which include the following:

- Michels (1993) reported a number of outstanding features for the MEE process when combined with thermal vapor compression (MEE-TVC). These features include low corrosion and scaling, which is caused by low temperature operation (top brine temperature below 60°C). Other features include low energy consumption, short delivery time, easy operation and maintenance, proven reliability in the Gulf region. The cost of the plant erection, civil work, and the seawater intake is 35% cheaper than the MSF plants. Michels (1993) described three low capacity units of MEE with thermal vapor compression built in the remote western areas of the Emirate

of Abu Dhabi, UAE. The plants superseded the more classic multi stage flash (MSF) in the range of unit productions up to about $10x10^3$ ton/day.
- Temstet and Laborie (1996) outlined the main characteristic of a dual-purpose multi-effect desalination plant. The system is designed to switch automatically between two operating modes, which depends on the seasonal variations in power and water demand. The first mode combines the MEE system with a single-stage steam jet ejector, which compresses the vapor extracted from the last effect. The second mode of operation involves the use of low pressure heating steam. The plant operates over a low temperature ranges, includes 12 effects, and has a production capacity of 12000 m^3/day.

Other studies of the MEE-TVC system focus on modeling and performance evaluation. Examples for these studies include the following:
- Minnich et al. (1995) developed a simple model for the MEE-TVC system. The MEE system operates at low temperatures and in the parallel mode. The model is used to compare the performance and capital cost of the MEE-TVC versus the MSF and MEE systems. The capital cost for the three systems is based on the total heat transfer area. Several simplifying assumptions are used to develop the model and it includes:
  - Constant and equal temperature losses in all effects,
  - Constant and equal overall heat transfer coefficients in all effects,
  - Constant thermal load in all effects,
  - Negligible distillate flashing,
  - No feed preheaters,
  - Equal feed flow rates in all effects,
  - Negligible difference of latent heat and vapor enthalpy,
  - Constant specific heat and vapor enthalpy, and
  - Negligible pressure losses in the system components, demister and connecting tubes.

The model results show that operation of the MEE-TVC system at low top brine temperatures, 60 °C, gives higher heat transfer areas than the MSF system at performance ratios higher than 6. The capital cost the low temperature MEE-TVC system exceeds the MSF at performance ratios higher than 8. Merits of the MEE-TVC are only realized at higher top brine temperatures.
- Darwish and El-Dessouky (1995) developed a simple model for parallel feed MEE-TVC. The model includes balance equations for energy and mass in each effect and in the steam jet ejector. The ejector model is based on the graphical performance data for steam jet ejectors presented by Power (1994). The model assumes negligible pressure losses within the system components, constant and equal boiling point rise in all effects, and constant temperature drop per effect. In addition, the model did not include equations for the heat transfer areas and the distillate flashing boxes. The model is used to analyze a four-effect MEE-TVC system and results gave a performance ratio of 7.65 for a top brine temperature of 62 °C. The simplicity of the model imposes restrictions

on its use for system design or analysis. For example, a constant temperature drop per effect when used to calculations of the heat transfer area would result in varying area in the system effect. This result is the opposite of industrial practice, where constant heat transfer area is used in all effects to reduce construction and maintenance cost.

- El-Dessouky (1997) and El-Dessouky et al. (1998) developed extensive mathematical models for the single effect thermal vapor compression process (TVC) and the multiple effect systems (MEE). The model, results, and analysis for the single-effect TVC and the stand alone MEE form the basis for development of the more complex MEE-TVC model. Development of both models addressed the limitations found in previous literature studies. Discussion and details of the MEE system are presented in the previous chapter. As for the TVC model, it includes analysis of the evaporator/condenser and the steam jet ejector units. The model includes the energy and material balance equations for the evaporator/condenser, the ejector design equation, the heat transfer design equation for the evaporator/condenser, and correlations for the heat transfer coefficient, thermophysical properties, and thermodynamic losses. Predictions show that the performance ratio varies between 1 and 2 as the top brine temperature is increased from 60 to 100 ^{o}C. The performance ratio increases as the pressure of the motive steam is increased. This makes the motive steam capable of compressing larger amounts of the entrained vapor. As a result, the amount of motive steam is reduced causing the increase of the performance ratio. The system performance ratio is found to increase at lower compression ratios (pressure of compressed vapor/pressure of entrained vapor). At low compression ratios, the amount of motive steam required to compress the entrained vapor are smaller and as a result the system performance ratio increases. Lower heat transfer areas for the evaporator condenser are predicted at higher top brine temperatures, because of the increase in the overall heat transfer coefficient at higher temperatures. The specific flow rate of cooling water is found to decrease as the amount of entrained vapor to the steam ejector is increased. The behavior occurs at high top brine temperature, low motive steam pressures, and high compression ratios.
- El-Dessouky and Ettouney (1997) presented analysis of the MEE-TVC system. The developed MEE-TVC model is based on the two models developed by El-Dessouky (1997) for the single-effect TVC and the multiple effect MEE model developed by El-Dessouky et al. (1998). As a result, the MEE-TVC model is based on sound physical phenomena, which relates various processes occurring in the system. The model results show large increase in the system performance ratio over the stand alone MEE system, with increase varying from 20-50%. In addition, large reduction is obtained in the specific flow rate of cooling water.

5.1.1 Process Description

Figs. 1a and 1b show the MEE-P/TVC and MEE-PC/MVC processes. As is shown both systems include n effects and n-1 flashing boxes. Each effect includes a vapor space, demister, condenser/evaporator tubes, brine spray nozzles, and brine pool. In either system, the effects are numbered 1 to n from the left to right (the direction of the heat flow). Vapor flows from left to right, in the direction of falling pressure, while the feed seawater flows in a perpendicular direction. Compressed vapor is introduced into the tube side in the first effect; while, on the shell side feed seawater is sprayed on the tubes top rows. The brine spray forms a thin falling film on the succeeding rows within the evaporator. In the first effect, the brine falling film absorbs the latent heat of the compressed vapor. As a result, the brine temperature increases to saturation, where, evaporation commences and a smaller amount of vapor forms. This vapor is used to heat the second effect, where, it condenses on the tube side and releases its latent heat to the brine falling film. This process is repeated for all effects, until effect n.

In both systems, the condensed vapor in effects to 2 to n is introduced into the associated flashing box, where the temperature of the condensed vapor is reduced through flashing of a small amount of vapor. The flashed off vapor is routed into the tube side of the next effect together with the vapor formed by boiling or flashing within the previous effect.

In the MEE-P/TVC system, the vapor formed in the last effect is introduced into the down condenser. A controlled amount of intake seawater is routed into the tube side of the down condenser, where it condenses part of the vapor formed in the last effect. The steam jet ejector entrains the remaining part of the vapor, where it is compressed by the motive steam to the desired pressure and temperature. The warm intake seawater stream leaving the down condenser is divided into two parts; the first is the feed seawater stream, which is distributed among the evaporation effects, and the second is the cooling seawater stream, which is reject back to the sea. The cooling seawater stream removes the heat added to the system by the motive steam.

In the converging section of the steam jet ejector the kinetic energy of the motive steam increases drastically and its speed becomes supersonic near the contraction point. Consequently, its pressure drops to low values and allows for suction of the entrained vapor. Mixing of the motive steam and the entrained vapor takes place past the ejector contraction. In the diverging section, the mixture velocity is reduced, while, its pressure starts to increase. The compression process is controlled by the ejector geometry and the motive steam properties.

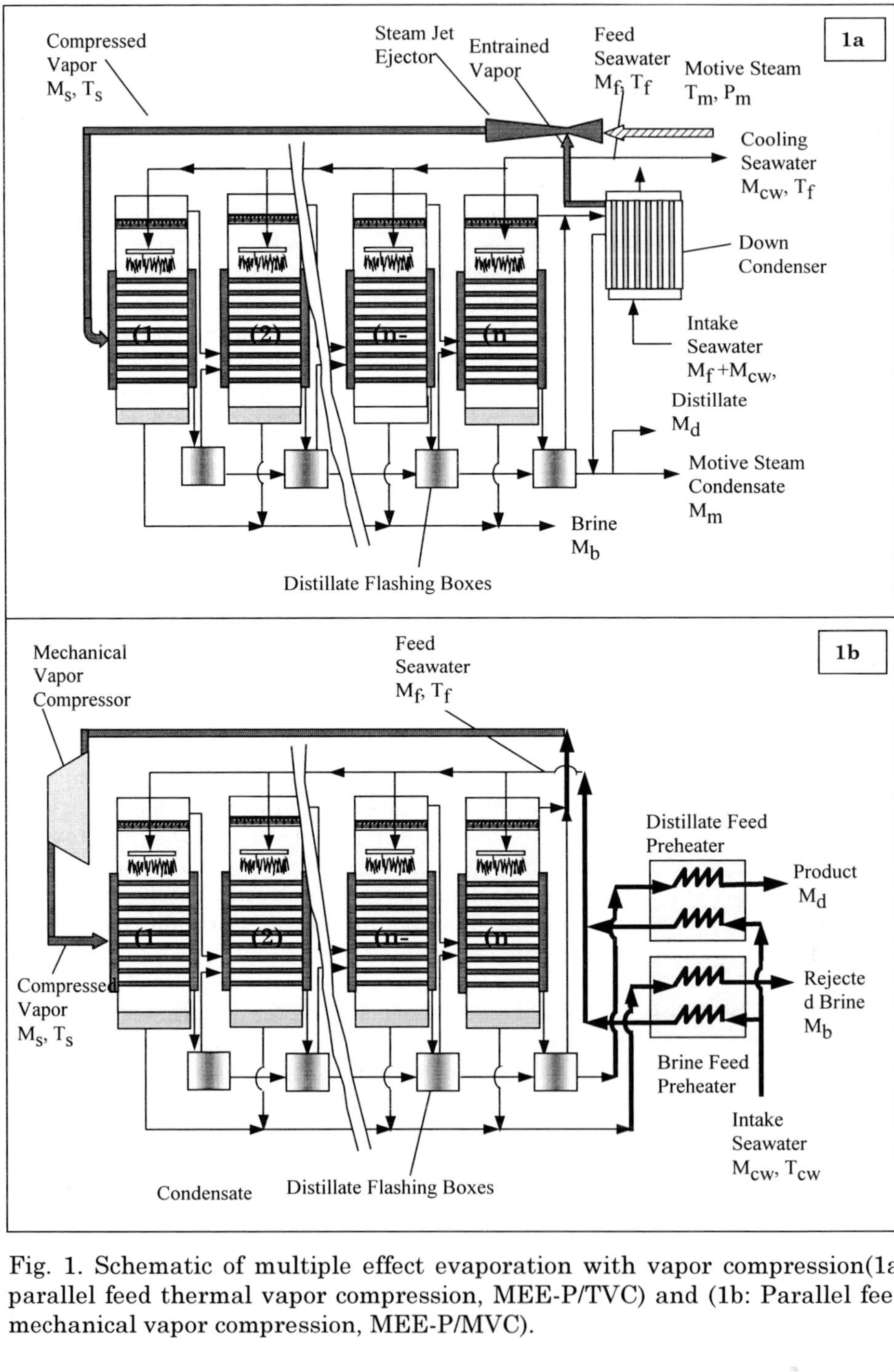

Fig. 1. Schematic of multiple effect evaporation with vapor compression(1a: parallel feed thermal vapor compression, MEE-P/TVC) and (1b: Parallel feed mechanical vapor compression, MEE-P/MVC).

The mechanical vapor compression system is distinguished by absence of the down condenser and use of the feed preheaters. Removal of the down condenser is a result of routing the entire vapor formed in the last effect to the mechanical vapor compressor, where the vapor is superheated to the desired temperature and pressure. At the other end, the feed preheaters recover part of the sensible heat found in the rejected brine and distillate product streams. This improves the system thermal efficiency and maintains production at the design levels, especially, during winter operation.

The main difference of the MEE-P and MEE-PC is that in the later system, the brine leaving effect (i) is introduced into the brine pool of effect (i+1). As a result of the positive temperature difference for the brine of effects (i) and (i+1), a small portion of the feed brine flashes off as it is introduced into effect (i+1). The flashed off vapors improves the system productivity and thermal efficiency. In effect (i+1), the flashed off vapors are added to the vapor formed by boiling within the same effect. As for the MEE-P process, the brine leaving each stage is directly rejected to the sea.

5.1.2 Process Modeling

Similarities among various systems considered in this analysis necessitate simultaneous development of the balance equations for various components within each system. Common assumptions among various models include steady state operation, constant heat transfer area in each effect, negligible heat losses to the surroundings, and salt free distillate product.

The following sections include discussion of the model equations for various components within the MEE-PC system. The model equations for the MEE-P system are not given, because of the similarity with the MEE-PC system. However, the discussion points to differences in balance equations of the MEE-P system. As for the correlations used to calculate the thermodynamic losses, pressure drops, and physical properties are given in the appendix. Fig. 2 shows a schematic for the system variables in the evaporator and the associated flash box in effect i. The figure includes flow rates, salinity, and temperatures of various streams as it enters and leaves the evaporator and the flashing box.

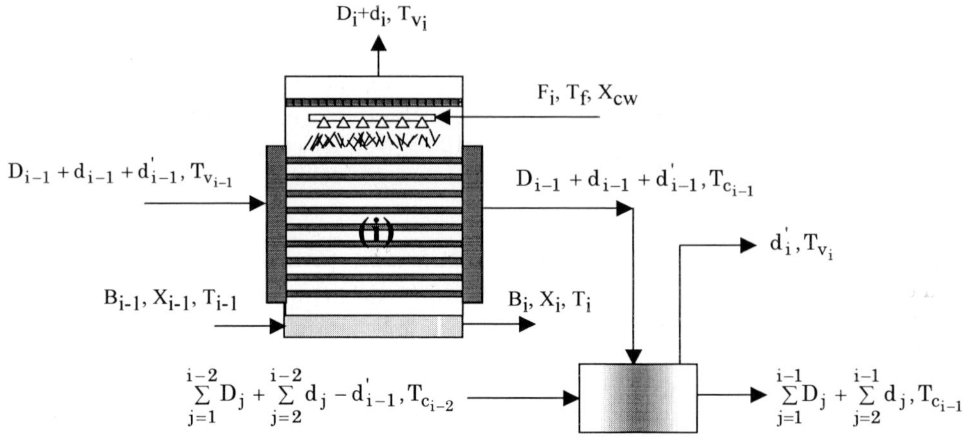

Fig. 2. Variables in evaporator and flash box of effect i.

Balance Equations for Each Effect

The mathematical model for each effect includes the material and energy balances as well as the heat transfer equation. The model includes the following equations:

– Total balance in effect i

$$F_i + B_{i-1} = D_i + B_i \tag{1}$$

– Salt balance in effect i

$$X_{F_i} F_i + X_{B_{i-1}} B_{i-1} = X_{B_i} B_i \tag{2}$$

In Eqs. 1 and 2, B, D, and F are the flow rates of brine, distillate, and feed, X is the salinity, and the subscripts B, F, and i designate the brine, feed, and the effect number.

– Rejected brine salinity

$$X_b = 0.9(457628.5 - 11304.11 T_b + 107.5781 T_b^2 - 0.360747 T_b^3) \tag{3}$$

This equation is used to calculate the reject brine salinity in each effect as a function of the brine temperature. This equation is obtained by curve fitting of the salinity/temperature relation for the solubility 90% of the solubility of $CaSO_4$. The upper limit on the rejected brine salinity is set at 70,000 ppm.

– Energy balance for effect i

$$D_{i-1}\,\lambda_{i-1} + d_{i-1}\,\lambda_{i-1} + d'_{i-1}\,\lambda'_{i-1} = F_i\,C_p(T_i - T_f) + D_i\lambda_i \tag{4}$$

In the above equation d is the amount of vapor formed by brine flashing in effect i-1, d' is the amount of vapor formed by flashing in the flashing boxes, λ is the latent, C_p is the specific heat at constant pressure, T_i is the brine boiling temperature, and T_f is the feed seawater temperature. In Eq. (4) the first term corresponds to the heat added to the effect by condensing the vapor generated in the previous effect. This only applies to effects 2 to n, since heating steam from an external source is used to drive the system and heat the first effect. In effect 3 to n, the second term in Eq. (4) defines the amount of heat associated with condensation of the vapor formed by brine flashing in the previous effect. The third term, which applies only to effects 3 to n, corresponds to the heat added to the effect by condensing the vapor generated in the distillate flashing box associated with the previous effect. The fourth term in Eq. 4 gives the amount of heat gained by the feed stream, where its temperature increased inside the effect from the seawater temperature to the brine boiling temperature. The last term gives the amount of heat consumed by the vapor generated inside the effect. In the above equation the specific heat at constant pressure depends on the brine salinity and temperature, while the latent heat depends on the vapor temperature. Correlations for the two properties are given in the appendix.

– Vapor temperature in effect i

$$T_{v_i} = T_i - BPE_i \tag{5}$$

where BPE is the boiling point elevation and T_v is the vapor temperature.

– The vapor condensation temperature

$$T_{c_i} = T_i - BPE_i - \Delta T_p - \Delta T_t - \Delta T_c \tag{6}$$

In Eq. 5, the condensation temperature, T_{c_i}, is lower than the brine boiling temperature, T_i, by the boiling point elevation and the losses caused by pressure depression in the demister (ΔT_p), friction in the transmission line (ΔT_t), and during condensation (ΔT_c).

– Amount of vapor formed by brine flashing inside the effect

$$d_i = B_{i-1}\,Cp\,\frac{T_{i-1} - T_i'}{\lambda_i} \tag{7}$$

with

$$T_i' = T_i + NEA_i \tag{8}$$

In Eq. 7, T_i' is the temperature to which the brine cools down as it enters the effect. Also, the latent heat λ_i is calculated at the effect vapor temperature, T_{v_i}. The term $(NEA)_i$ is the non-equilibrium allowance and is calculated from the correlation developed by Miyatake (1973):

$$(NEA)_i = \frac{33.0\,(T_{i-1} - T_i)^{0.55}}{T_{v_i}}$$

– Amount of vapor flashed off in the distillate flashing boxes

$$d_i' = D_{i-1} C_p \frac{(T_{c_{i-1}} - T_i'')}{\lambda_i'} \qquad (9)$$

with

$$T_i'' = T_{v_i} + (NEA)_i$$

where $(NEA)_i$ is the non-equilibrium allowance and is equal to $(NEA)_i = 0.33 \frac{(T_{c_{i-1}} - T_{v_i})}{T_{v_i}}$, T_i'' is the temperature to which the condensing vapor cools down to as it enters the flashing box.

– Heat transfer area in effect i

$$D_{i-1}\,\lambda_{i-1} + d_{i-1}\,\lambda_{i-1} + d_{i-1}'\,\lambda_{i-1}' = F_i\, C_p (T_i - T_f) + D_i \lambda_i$$
$$= A_{1i}\, U_{1i}\, (LMTD)_i + A_{2i}\, U_{2i}\, (T_{c_i} - T_i) \qquad (10)$$

$$\alpha(D_{i-1}\,\lambda_{i-1} + d_{i-1}\,\lambda_{i-1} + d_{i-1}'\,\lambda_{i-1}') = D_i \lambda_i = A_{2i}\, U_{2i}\, (T_{c_i} - T_i) \qquad (11)$$

$$(LMTD)_i = (T_i - T_f)/\ln((T_{c_i} - T_f)/(T_{c_i} - T_i)) \qquad (12)$$

where A_{1i} is the heat transfer area for sensible heating of the brine from the feed to the boiling temperature in each effect and A_{2i} is the heat transfer area for evaporation, U_{1i} and U_{2i} are the corresponding overall heat transfer coefficient, LMTD is the logarithmic heat transfer coefficient, and α is the fraction of input heat consumed by vapor formation.

Balance Equations for the Down Condenser

The down condenser balance equations include the energy balance and heat transfer rating equation.

– Energy balance of the down condenser

$$(d_n + d'_n + D_n)\lambda_n = (M_{cw} + M_f)\, Cp\, (T_f - T_{cw}) \tag{13}$$

– Rating of the down condenser

$$(d_n + d'_n + D_n)\lambda_n = U_c\, A_c\, (LMTD)_c \tag{14}$$

$$(LMTD)_c = (T_f - T_{cw})/\ln((T_{v_n} - T_{cw})/(T_{c_n} - T_f)) \tag{15}$$

where A_c, U_c, and $(LMTD)_c$ are the heat transfer area, overall heat transfer coefficient, and logarithmic mean temperature difference.

In presence of the steam jet ejector, the thermal load of the down condenser is lower since the part of the vapor formed in the last effect and the associated flashing box is entrained in the steam jet ejector. Therefore, the vapor formed in the last effect is defined by

$$M_{ev} + M_u = (d_n + d'_n + D_n) \tag{16}$$

where M_{ev} and M_u are the flow rates of the entrained and un-entrained vapor, respectively. In the following section, which includes the steam jet ejector model, the flow rate of the entrained vapor is obtained from the ejector entrainment ratio.

Model of the Steam Jet Ejector

The steam jet ejector is modeled by the semi-empirical model developed by El-Dessouky (1997). The model makes use of the field data collected over 35 years by Power (1994) for vapor entrainment and compression ratios of steam jet ejectors. The compression ratio, Cr, is the pressure ratio of the compressed and entrained vapors. The entrainment ratio is flow rate ratio of the motive steam and the entrained vapor. The entrainment ratio, Ra, is calculated from the following relation

$$Ra = 0.296 \frac{(P_s)^{1.19}}{(P_{ev})^{1.04}} \left(\frac{P_m}{P_{ev}}\right)^{0.015} \left(\frac{PCF}{TCF}\right) \tag{17}$$

where, P_m, P_s and P_{ev} are the pressures of the motive steam, compressed vapor, and entrained vapor respectively, PCF is the motive steam pressure correction factor and TCF is the entrained vapor temperature correction factor. The following two equations are used to calculate PCF and TCF

$$PCF = 3x10^{-7} (P_m)^2 - 0.0009 (P_m) + 1.6101 \quad (18)$$

$$TCF = 2x10^{-8} (T_{ev})^2 - 0.0006 (T_{ev}) + 1.0047 \quad (19)$$

where P_m is in kPa and T_{ev} is in °C. The previous equations are valid only for ejectors operating with steam as the motive fluid and the entrained gas is water vapor. These equations are valid in the following ranges: $Ra \leq 4$, $500 \geq T_{ev} > 10$ °C, $3500 \geq P_m \geq 100$ kPa, and $6 \geq Cr = \frac{P_s}{P_{ev}} \geq 1.81$.

The steam jet ejector must be designed and operated at critical conditions to allow normal and stable operation. This condition is associated with absence of violent fluctuations in the suction pressure. If the ejector is designed to operate with a full stable range, it will have a constant mass flow rate of the entrained vapor for different discharge pressures when the upstream conditions remain constant. The ejector is critical when the compression ratio is greater than or equal to the critical pressure ratio of the suction vapor. For water vapor this ratio is 1.81. That is, the suction pressure must be less than 0.55 times the discharge pressure to obtain critical or stable conditions in the steam jet ejector. The above limit on the compression ratio necessitates the use of two steam jet ejectors in series, Fig. 3, for a wide compression range. For example, in a single jet ejector that compresses a vapor to 80 °C and entrains vapor at 38 °C, the compression ratio in 7.14. This compression value requires the use of two ejectors in series, where the compression range is divided over the two ejectors. The corresponding balance equations for two ejectors in series include the following:

$$M_s = M_{s1} + M_{m2} \quad (20)$$

$$M_{s1} = M_{ev} + M_{m1} \quad (21)$$

$$Ra_1 = M_{m1}/M_{ev} \quad (22)$$

$$Ra_2 = M_{m2}/M_{s1} \quad (23)$$

$$Cr_1 = P_{s1}/P_{ev} \quad (24)$$

$$Cr_2 = P_s/P_{s1} \quad (25)$$

where M is the mass flow rate and the subscripts ev, m, s, 1, and 2 define the entrained vapor, the motive steam, the compressed, first and second ejector.

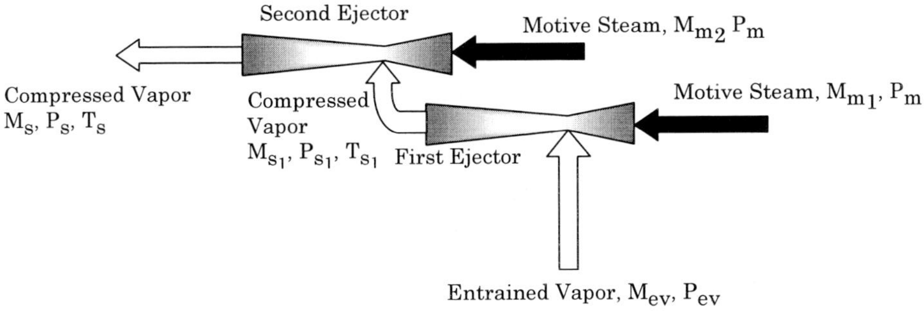

Fig. 3. Schematic of the two ejectors in series.

Model of the Mechanical Vapor Compressor

The specific power consumption of the compressor

$$Q_c = W\ \rho_d/3600 \tag{26}$$

where ρ_d is the density of distillate product, W is the actual specific work of the compressor, which is given by

$$W = H_s - H_v \tag{27}$$

The enthalpies H_s and H_v are calculated at the compressed vapor temperature, T_s, and the formed vapor temperature in the last effect, T_{vn}, which is lower than $\overline{T}_{vn}$ by the temperature depression caused by pressure drop in the demister. The compressor polytropic specific work is given by

$$\frac{W_m}{W_n} = \eta\left[\frac{(P_s/P_v)^{(\gamma-1/\gamma\eta)}-1}{(P_s/P_v)^{(\gamma-1/\gamma)}-1}\right] \tag{28}$$

In Eq. 28 the adiabatic compressibility factor is defined as

$$\gamma = \frac{1}{1-(1+X)^2(ZR/Cp_v)/Y} \tag{29}$$

where X = 0.1846 $(8.36)^{(1/Z)}$ – 1.539 and Y = 0.074 $(6.65)^{(1/Z)}$ + 0.509, ASHRAE (1997). In Eq. 29, the compressibility factor Z is set equal 1. The compressor adiabatic work, W_n, given in Eq. 28 is defined as the enthalpy difference of the in terms of the

$$W_n = H_n - H_v \tag{30}$$

In Eq. 30 H_n and H_v are calculated at T_n and T_{v_n}, respectively, where T_n is calculated from the relation

$$T_n = T_{v_n} \, (P_v/P_n)^{(\gamma-1)/\gamma} \tag{31}$$

The enthalpy and temperature of the superheated (or compressed vapor) are obtained from the following relations

$$\eta = \frac{W_m}{H_s - H_v} \tag{32}$$

$$H_s = H_d + Cp_v \, (T_s - T_d) \tag{33}$$

where H_d and T_d are the saturation enthalpy and temperature of the compressed vapor, and H_s and T_s are the superheat enthalpy and temperature of the compressed vapor.

Preheaters Models

Two preheaters are used to increase the intake seawater temperature in the MEE-P/MVC system. This temperature increase is an essential part in energy recovery within the system and it has a strong effect on the plant performance or the specific power consumption. Heating of the feed seawater is performed against the hot product and brine streams leaving the last effect. This process takes place in two plate type heat exchange units, where the intake seawater is divided into two portions, αM_f and $(1-\alpha)M_f$. In the first preheater, heat is exchanged between αM_f and the product water, and in the second preheater, heat is exchanged between $(1-\alpha)M_f$ and the rejected brine. The sum of the thermal load for the two heat exchangers is given in terms of the intake seawater temperature increase. This is

$$Q_h = M_f \, C_p \, (T_f - T_{cw}) \tag{34}$$

where Q_h is thermal load of the two preheaters, Cp is the specific heat at constant for the seawater, T_f is feed seawater temperature, and T_{cw} is the intake seawater temperature. Equation (34) can be also written in terms of the heat load of the product water and the rejected brine, which gives

$$Q_h = M_d\, C_p\, (T_{cn} - T_o) + M_b\, C_p\, (T_n - T_o) \tag{35}$$

Where T_{cn} and T_n are the temperatures of the product water and brine leaving the last effect and T_o is the temperature of both streams after leaving the preheaters. Equations 34 and 35 are equated and the result is used to determine the outlet temperature of the heating streams, T_o,

$$M_f\, C_p\, (T_f - T_{cw}) = M_d\, C_p\, (T_{cn} - T_o) + M_b\, C_p\, (T_{cn} - T_o) \tag{36}$$

The driving force for heat transfer in the preheaters is taken as the logarithmic mean of the temperature difference at both ends of the preheater. These equations are given by

$$A_d = \frac{M_d C_p (T_{c_n} - T_o)}{U_d (LMTD)_d} = \frac{\alpha M_f C_p (T_f - T_{cw})}{U_d (LMTD)_d} \tag{37}$$

$$\begin{aligned} A_b &= \frac{M_b C_p (T_n - T_o)}{U_b (LMTD)_b} \\ &= \frac{M_d (X_f/(X_b - X_f)) C_p (T_n - T_o)}{U_b (LMTD)_b} \\ &= \frac{(1-\alpha) M_f C_p (T_f - T_{cw})}{U_b (LMTD)_b} \end{aligned} \tag{38}$$

The $(LMTD)_d$ is defined as:

$$(LMTD)_d = \frac{(T_{c_n} - T_f) - (T_o - T_{cw})}{\ln \dfrac{T_{c_n} - T_f}{T_o - T_{cw}}} \tag{39}$$

The $(LMTD)_b$ is defined as:

$$(LMTD)_b = \frac{(T_n - T_f) - (T_o - T_{cw})}{\ln \dfrac{T_n - T_f}{T_o - T_{cw}}} \tag{40}$$

Solution Algorithm

The mathematical models for either system are interlinked and highly nonlinear. Therefore, iterative solution is necessary to calculate the system characteristics. The solution algorithm starts with definition of the following parameters:

- The number of effects varies over a range of 4-12.
- The heating steam temperature varies over a range of 60-100 °C.
- The seawater temperature (T_{cw}) is 25°C.
- The seawater salinity has values of 34,000 ppm or 42,000 ppm.
- The temperature of rejected cooling water or feed seawater (T_f) is less than condensing vapor temperature (T_{c_n}) by 5 °C.
- The boiling temperature in the last effect (T_n) is 40°C.
- The specific heat at constant pressure of the vapor, C_{p_v}, is 1.884 kJ/kg °C.
- The polytropic efficiency of the compressor, η, is 0.76 [24].

The solution algorithm for the thermal vapor compression system is shown in Fig. 4. As is shown, the model equations are solved simultaneously by Newton's method to calculate the following:

- The flow rates, salinity, and temperatures of the feed, brine, and distillate in each effect.
- The heat transfer area for evaporation and sensible heating in each effect.
- The fraction of heat consumed by evaporation in each effect.
- The above results are used to calculate the following:
- The heat transfer area in the condenser.
- The flow rate of cooling seawater.
- The entrainment ratio in the steam jet ejector.
- The amount of motive steam.

Figure 5 shows the solution algorithm for the mechanical vapor compression system. In this system, the amount of compressed vapor is known and is equal to the amount of vapor formed by boiling in the last effect as well as the amount of vapor formed by brine and distillate flashing. The energy and material balance model as well as the compressor model are solved simultaneously and iteratively by Newton's method. Simultaneous solution of the two models gives the following system variables:

- Temperature, salinity, and flow rate profiles of feed, distillate, and brine streams.
- The specific power consumption of the mechanical vapor compressor.
- The temperature of the compressed vapor.
- The heat transfer areas for vapor formation and brine heating in each effect.
- The heat transfer area of the feed preheaters.

The Newton's iterative procedure has an iteration error of $1x10^{-4}$. To facilitate the conversion procedure, each equation is scaled by the largest term found in the equation. Therefore, all equations are in the order of one. For example, the salt balance equation is rearranged into the following form

$$f(X_{cw}, F_i, X_{b_i}, B_i) = 1 - (X_{cw} F_i)/(X_{b_i} B_i)$$

Convergence of Newton's method is dependent on the initial guess, therefore, linear profiles are used for the flow rates, brine temperature, heat transfer areas, and the ratio α. The guess for the steam flow rate is based on the approximate relation of the number of effects and the performance ratio.

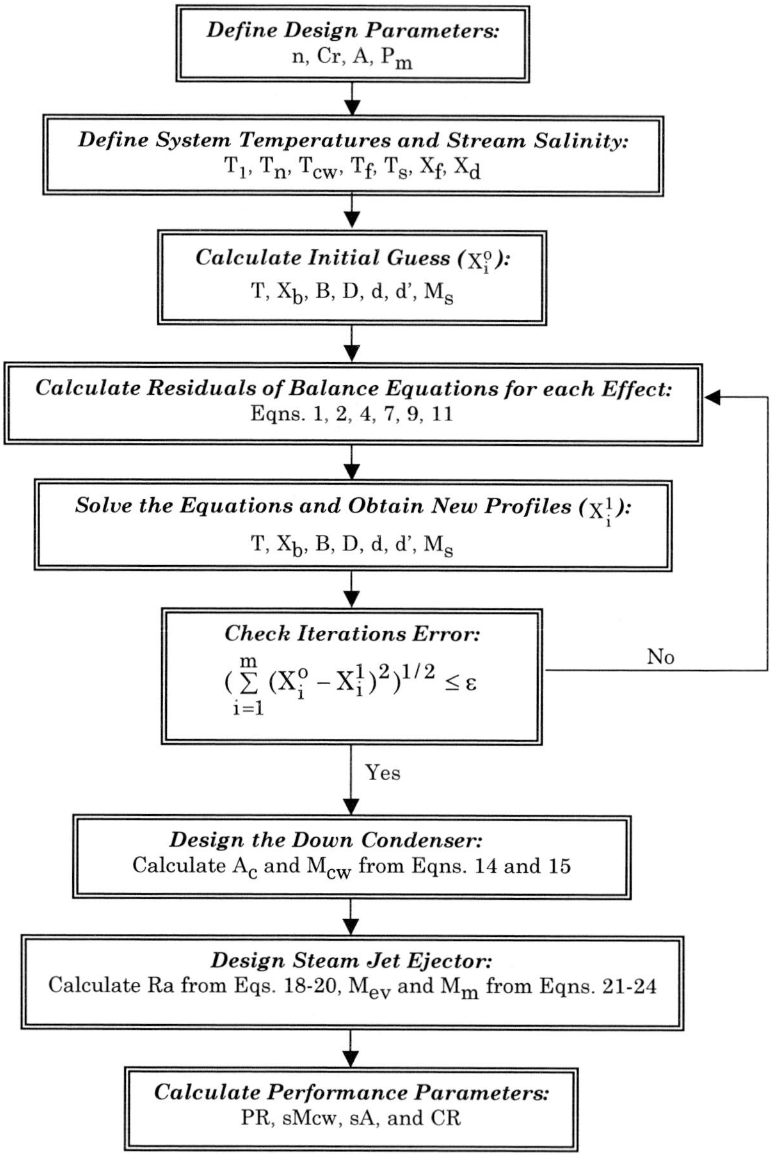

Fig. 4. Solution algorithm of the thermal vapor compression system.

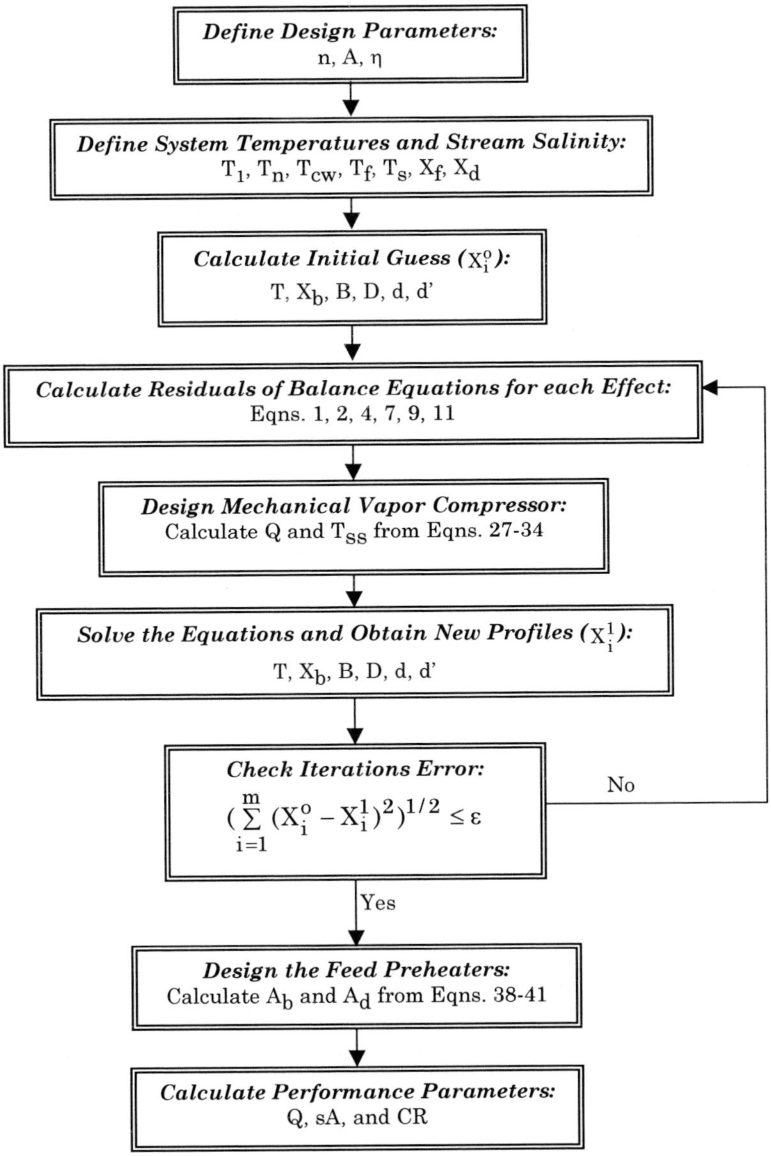

Fig. 5. Solution algorithm of the mechanical vapor compression system.

5.1.3 System Performance

Characteristics of the thermal vapor compression systems are obtained as a function of the heating steam temperature. Figure 6 shows variations in the thermal performance ratio for the MEE-P/TVC and MEE-PC/TVC for 8 effects, motive steam pressure of 1500 kPa, and a compression ratio of 4. As is shown, the performance ratio decreases with the increase in the top brine temperature. Also, at low top brine temperatures the thermal performance ratio for vapor compression units is close to 75-100% higher than the stand-alone systems. For example at a top brine temperature of 60 °C, the thermal performance ratio for the vapor compression units is 12.2 and is equal to 7.3 for the stand-alone units.

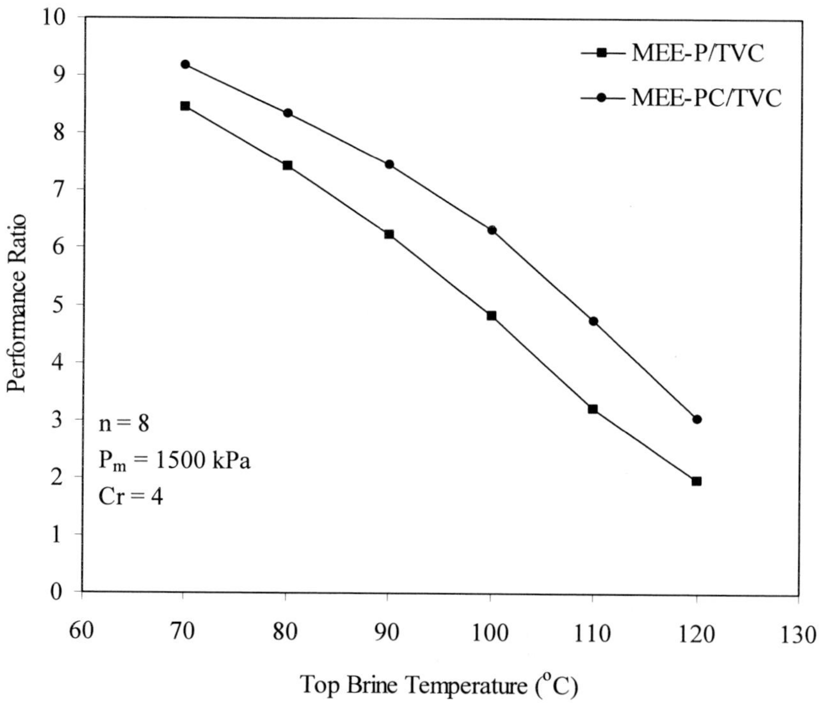

Fig. 6. Variation in the thermal performance ratio as a function of the top brine temperature

The reduction in the system thermal performance ratio at higher steam temperature is caused by the following factors:

- The reduction in compressed vapor latent heat, i.e., at 60 °C the latent heat is 2470 kJ/kg and at 110 °C it is equal to 2105 kJ/kg.

- The increase in the amount of feed sensible heating, since the feed temperature is kept constant at 35 °C.
- The increase in the amount of motive steam required for vapor compression at higher temperatures, since the entrained vapor is kept constant at a temperature below 40 °C.

Variations in the specific heat transfer area for both MEE-P/TVC and MEE-PC/TVC are shown in Fig. 7. As is shown the specific heat transfer area decreases rapidly as the heating steam temperature increases. The following effects cause this behavior:

- The increase in the overall heat transfer coefficient as a result of higher values for the physical properties of the brine and condensing vapor, which enhances the rate of heat transfer in either stream.
- The increase in the temperature driving force per effect, where at higher top brine temperatures and for the same number of effects causes the increase in the temperature drop per stage.

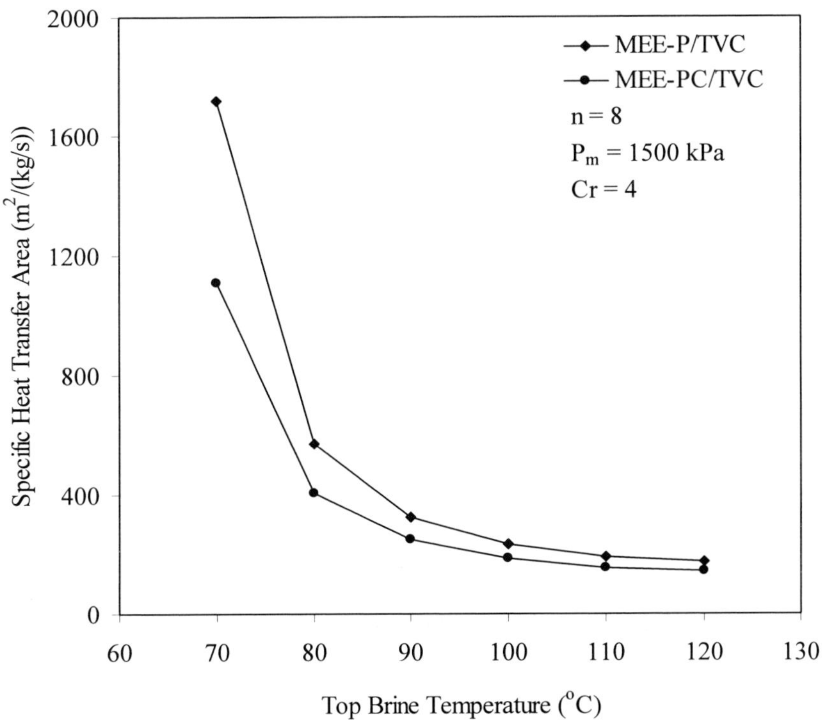

Fig. 7. Variation in the specific heat transfer area as a function of the top brine temperature

As is shown in Fig. 8, the conversion ratio for the MEE-PC/TVC is independent of the top brine temperature. On the other hand, the conversion ratio for the MEE-P/TVC system decreases with the increase of the top brine temperature. For the MEE-PC/TVC system the feed stream for all effects has a constant salinity of 42,000 ppm, and the salinity of the final brine stream is 70,000 ppm. Therefore, the balance equations for the system give a conversion ratio independent of the top brine temperature. As a result, the amount of feed seawater for the MEE-PC/TVC remains constant as the top brine temperature increases. As for the MEE-P/TVC system, the conversion ratio decreases with the increase in the heating stream temperature (Fig. 8). This is because of the reduction in the brine salinity at higher temperatures. Therefore, at higher temperatures the amount of feed seawater must be increased to account for the limits imposed on the brine salinity. This increase results in reduction in the amount of cooling seawater.

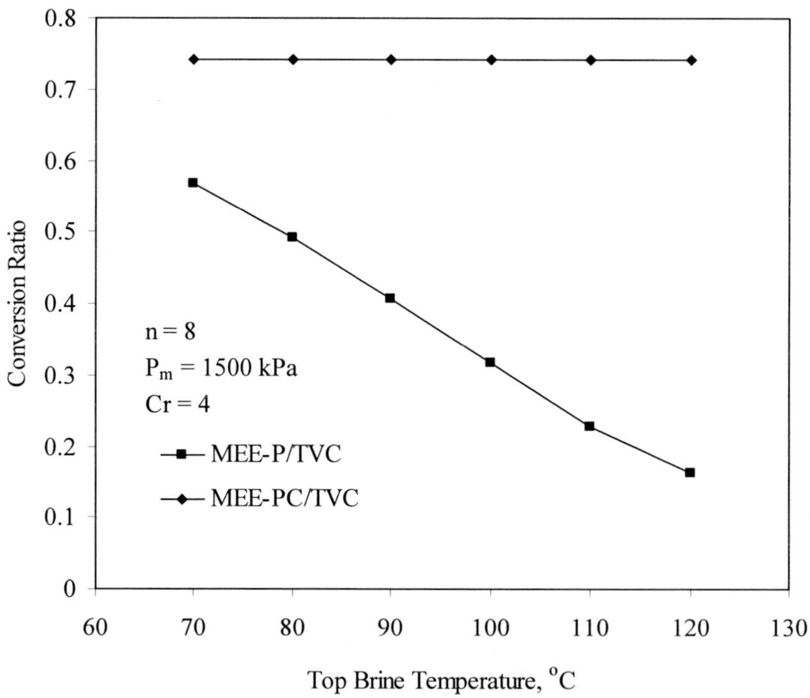

Fig. 8. Variations in the conversion ratio as a function of the top brine temperature.

Variations in the specific flow rate of cooling water for both systems are shown Fig. 9. As is shown, for MEE-PC/TVC system the specific flow rate of cooling water decreases with the increase in the top brine temperature. This is because of the increase in the specific thermal load, or the decrease in the thermal performance ratio, and the constant conversion ratio, or a constant feed flow rate. Both effects require the increase in the specific flow rate of cooling water. The decrease in the specific flow rate for the cooling water in the MEE-P/TVC system at higher top brine temperatures is also caused by the decrease in the system conversion ratio or the increase in the amount of feed seawater.

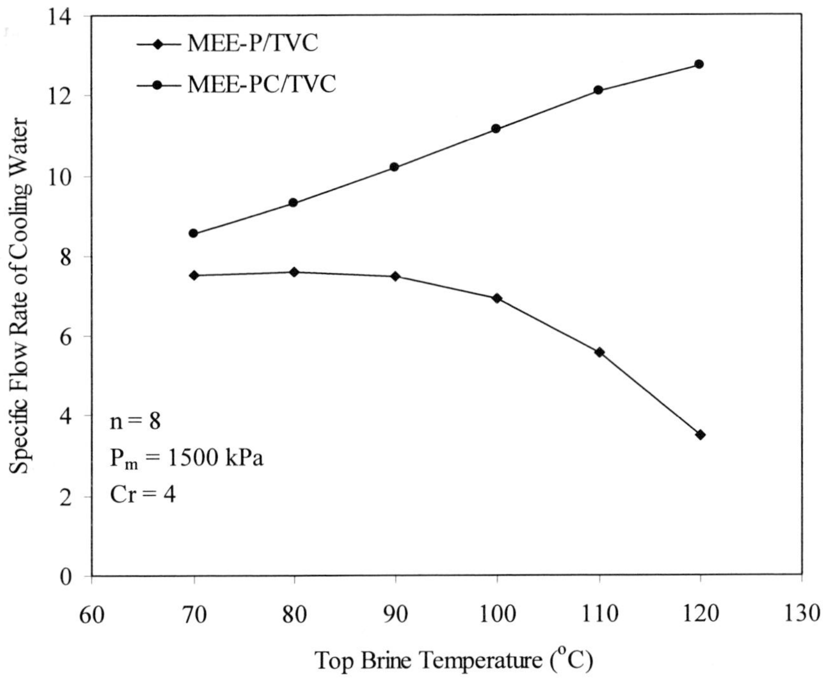

Fig. 9. Variation in the specific flow rate of cooling as a function of the top brine temperature

Analysis of the mechanical vapor compression systems shows high sensitivity to the range of operating parameters, especially, the temperature difference of the brine in the first and last effect and the temperature of the feed seawater. Calculations are performed for the following conditions:

- Top brine temperatures of 50, 60, 70, and 80 °C.
- Condensation temperatures of the compressed vapor are higher than the top brine temperature by 1, 2, 3, and 4 °C.

- Brine temperature in the last effect lower than the top brine temperature by 9 °C.
- Feed temperature lower than the brine temperature in the last effect by 2 °C.

The Results for four effects MEE-P/MVC system are shown in Figs. 10 and 11 for the specific heat transfer area and the specific power consumption, respectively. As is shown in Fig. 10, the specific heat transfer area decreases with the increase in the top brine temperature and the difference of the condensing vapor and top brine temperatures. On the other hand, the specific power consumption decreases with the increase in the top brine temperature and the decrease in the difference of the condensing vapor and the top brine temperatures, Fig. 11. The specific power consumption for the above set of parameters varies between low values close to 8 kWh/m^3 and higher values close to 12 kWh/m^3. Selection of the best design and operating conditions necessitates optimization among the specific heat transfer area and the specific power consumption.

In should be noted that the specific power consumption for the MEE-P/MVC and MEE-PC/VC have similar values at the same set of operating conditions (Fig. 11). This is consistent with the model of the compressor, since it depends on the amount of generated vapor in the last effect and flashing box, the compression range, and the temperatures of the intake and compressed vapor streams. For both systems the temperatures of the brine in the first effect, the intake vapor, and the compressed vapor are identical. However, in the MEE-PC/MVC system the amount of vapor generated in the last effect is slightly higher because of brine flashing.

As for the specific heat transfer area, values for the MEE-PC system are lower than the MEE-P system. This is because of direct rejection of the brine from each effect in the MEE-P system. On the other hand, the brine stream leaving each effect in the MEE-PC system is allowed to release part of its heat through flashing in subsequent effects.

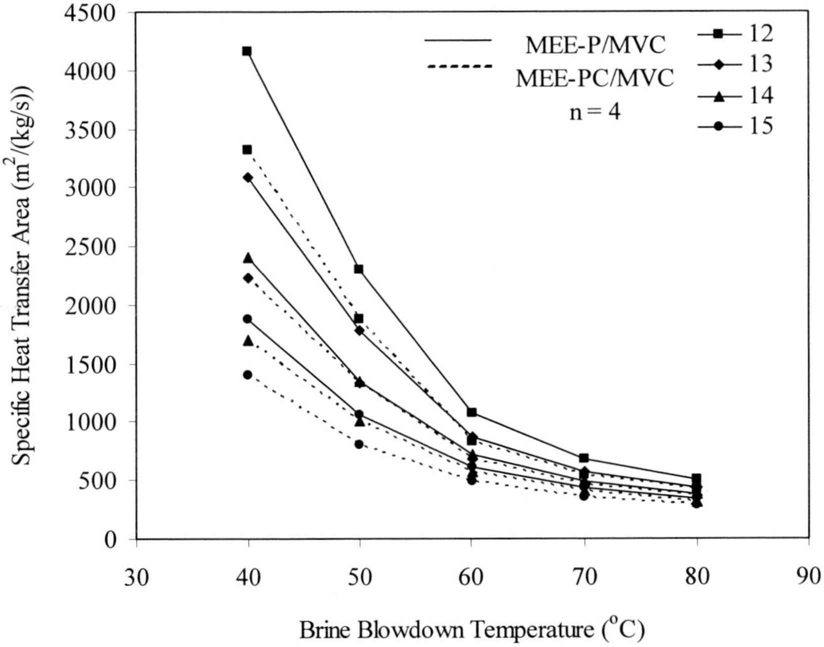

Fig. 10. Variation in the specific heat transfer area as a function of the brine blowdown temperature and the difference between condensing vapor and brine blowdown temperatures

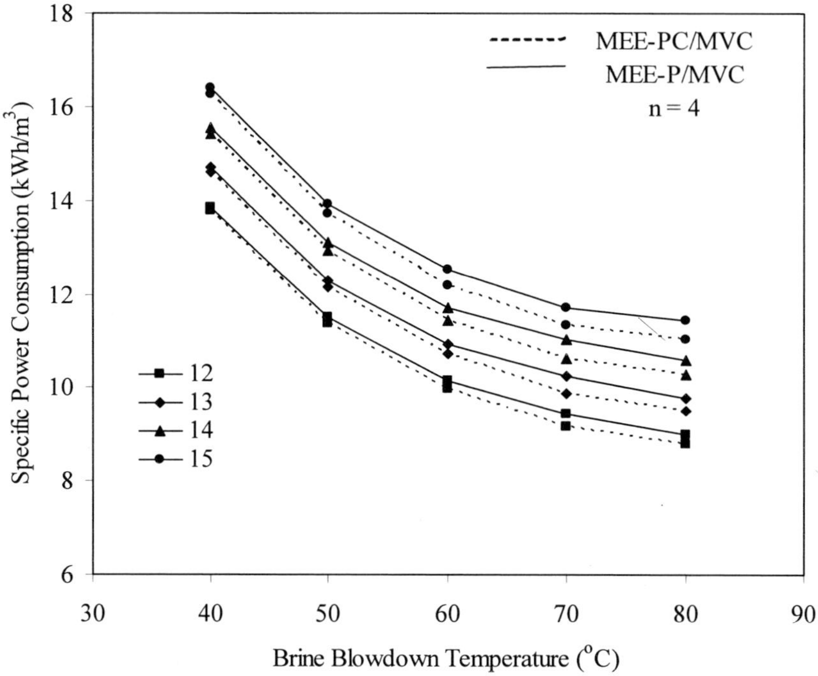

Fig. 11. Variation in the specific power consumption as a function of the brine blowdown temperature and the difference between condensing vapor and brine blowdown temperatures

5.1.4 Comparison with Industrial Data

Table 1 includes comparison of model predictions against two industrial MEE-PC/MVC systems. Literature review indicates that most of the existing MVC units are of the single effect type. It should be stressed that industrial use of the 3 and 4 effects systems is to increase the total system capacity rather than to decrease the specific power. Both systems operate in the MEE-PC mode, where the brine stream cascades across the effects. The results in Table 2 show good agreement between the predicted and actual specific power consumption. The relative error in the specific power consumption is below 9%. Comparison of the specific heat transfer area was not possible because no field data was available.

The data shown in Table 2 are obtained for multiple effect thermal vapor compression systems with 4, 6, and 12 effects. To obtain the model predictions, the system layout had to be arranged similar to the industrial configuration. Also, the temperatures of the heating steam, the last stage, the intake seawater, and feed seawater are all defined. Other system definitions include the salinity of the intake seawater and rejected brine. The model is used to calculate the specific heat transfer area, the specific flow rate of cooling water, and the performance ratio. The comparison includes only the performance ratio and the specific flow rate of cooling water. No comparison was made for the specific heat transfer area, because, the field data was not available. As is shown, the model predictions compares well with the industrial data. The relative percentage error of model predictions to the industrial data is limited to values below 15%.

Table 1
Comparison of model predictions against field data for MEE-MVC systems.

Reference	Lucas and Tabourier (1985)	Model	Ophir and Gendel (1999)	Model
N	4	4	3	3
M_d (m^3/d)	1500	1500	3000	3000
T_s (°C)	62.5	62.5	70	70
T_n (°C)	50.7	50.7		
T_{cw} (°C)	5	5		
T_f (°C)	49	49		
X_{cw} (ppm)	36000	36000	36000	36000
X_{bn} (ppm)	64800	64800		
CR	0.446	0.446		
sA_c (m^2/(kg/s))	–	2234		
Q (kWh/m^3)	11	10.7	6.9	6.3

Table 2
Comparison of model predictions against field data for MEE-TVC systems.

Process	Temstet et al. (1996)	Model	Weinberg and Ophir (1997)	Model	Michles (1993)	Model	Elovic and Willocks (1999)	Model
n	12	12	6	6	4	4	12	12
M_d (m^3/d)	$1.2x10^4$	$1.2x10^4$	$2.1x10^4$	$2.1x10^4$	$4.5x10^3$	$4.5x10^3$	$5.9x10^3$	$5.9x10^3$
T_s (°C)	70	70	62.9	62.9	62.7	62.7	71	71
T_n (°C)	38.5	38.5	36.3	36.3	48.4	48.4	40^+	40
T_{cw} (°C)	29.5	29.5	26	26	33	33	30^+	30
T_f (°C)	34.5	34.5	32	32	44	44	35^+	35
X_{cw} (ppm)	36000	36000	42000	42000	47000	47000	36000^+	36000
X_{b_n} (ppm)	51730	51730	52900	52900	71500	71500	52000^+	52000
CR	0.33	0.33	0.33	0.33	0.33	0.33	0.31^+	0.31
sM_{cw}	6.212	6.8	11.9	12.4	3.79	4.31	–	7.2
sA_c (m^2/(kg/s))	–	1385	–	734	–	523	–	1283
PR	13.4	14.1	5.7	6.2	8.6	9.3	11.5	11.9

+ Values assumed

5.1.5 Summary

System analysis is presented for two configuration of the parallel feed multiple effect evaporation. Each system is analyzed for the thermal and mechanical vapor compression modes. In the light of system analysis, the following conclusions are made:

- The thermal performance ratio for MEE-P/TVC and MEE-PC/TVC systems, especially at lower top brine temperatures, are more than 50-100% higher than the stand alone mode.
- The specific heat transfer area for all configurations, including thermal and mechanical vapor compression, decreases drastically at higher top brine temperatures because of the increase in the driving force for heat transfer.
- The specific power consumption for the mechanical vapor compression system have similar values for both systems since it depends on the temperature difference of the intake and compressed vapors as well as the top brine temperature, all of which were similar for both systems.
- The specific heat transfer area for the MEE-PC/MVC is lower than the MEE-P/MVC system. This is because of the increase in the total amount of product flow rate, which is caused by brine flashing within each effect.

References

ASHRAE Handbook – Fundamentals, ASHRAE, Atlanta, GA, USA, 1997.

Darwish M.A., and El-Dessouky, H., The heat recovery thermal vapour-compression desalting system: A Comparison with other thermal desalination Processes, Applied Thermal Engineering, **18**(1996)523-537.

El-Dessouky, H.T., Modelling and Simulation of Thermal Vapor Compression Desalination Process, Proceedings of International Atomic Energy Agency, Symposium on Desalination of Seawater with Nuclear Energy, Taujon, Korea, 26-30 May (1997).

El-Dessouky, H.T., and Ettouney, H.M., Hybrid multiple effect evaporation/heat pump water desalination systems, 1st IDA Int. Desalination Conference in Egypt, Cairo, Egypt, September, 1997.

El-Dessouky, H., Alatiqi, I., Bingulac, S., and Ettouney, H., Steady-state analysis of the multiple effect evaporation desalination process, Chem. Eng. Technol., **21**(1998)15-29.

Elovic, P., and Willocks, G., Case study of operating experience of 9 low-temperature MED plants in the US Virgin Islands, Proceedings of the IDA

World Conference on Desalination and Water Science, San Diego, USA, 1999, Vol. IV, pp. 143-152.

Lucas, M., and Tabourier, B., The Mechanical Vapour Compression Process Applied to Seawater Desalination: A 1500 ton/day Unit Installed in the Nuclear Power Plant of Flamanville, France, Desalination, **52**(1985)123-133.

Michles, T., Recent achievements of low temperature multiple effect desalination in the western areas of Abu Dhabi, UAE, Desalination, **93**(1993)111-118.

Minnich, K., Tonner, J., and Neu, D., A comparison of heat transfer requirement and evaporator cost for MED-TC and MSF, Proceedings of the IDA World Congress on Desalination and Water Science, Abu Dhabi, UAE, 1995, Vol. III, 233-257.

Miyatake, O., Murakami, K., Kawata, Y., and Fujii, Fundamental Experiments with Flash Evaporation, Heat Transfer Jpn. Res., **2**(1973)89-100.

Power, B.R., Steam Jet Ejectors for Process Industries, McGraw-Hill, New York, 1994.

Temstet, C., Canton, G., Laborie, J., and Durante, A., A large high-performance MED plant in Sicily, Desalination, **105**(1996) 109-114.

Weinberg, J., and Ophir, A., Ashdod experience and other dual purpose desalination plants based on multi effect desalination with aluminum tubes, Symposium of Desalination of Seawater with Nuclear Energy, Taejon, Republic of Korea, May, 1997.

P-1680 B
CONDENSATE
P-1680 A
CONDENSATE

5.2 Forward Feed Multiple Effect Evaporation Thermal Vapor Compression

The forward feed thermal vapor compression system is illustrated in Fig. 12. Process elements are similar to the forward feed system given in chapter 4. Also, combining the system with thermal vapor compression has identical features to the parallel feed system given in the previous section.

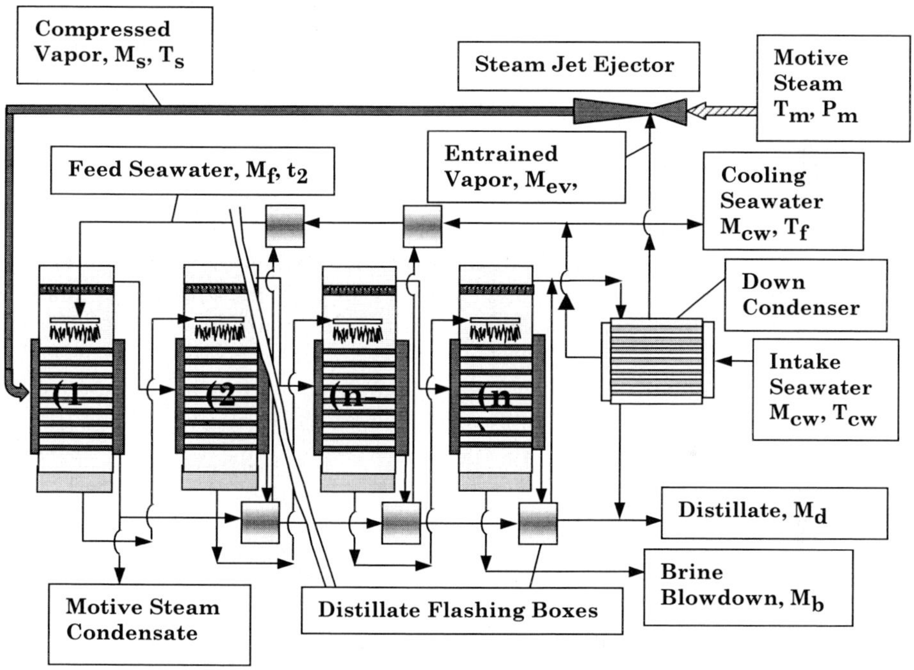

Fig. 12. Forward feed multiple effect thermal vapor compression.

5.2.1 Process Modeling

The mathematical model is divided into two parts; the first is for the MEE system and the second is for the steam jet ejector. Model equations and solution of the MEE system is given in the previous chapter. In addition, the steam jet ejector model is given in the previous section. Calculations of the MEE system variables are independent on the steam jet ejector equations. This includes temperature, flow rates, and concentration profiles as well as the heat transfer area in the effects. However, the performance parameters of the MEE system are dependent on the characteristics of the steam jet ejector. Also, the design of the steam jet ejector is affected by the vapor temperature in the last effect of MEE and the specification of the steam temperature (compressed vapor) required for operating the MEE system. The performance parameters in the MEE system, which are affected by the design of the steam jet ejector, are the performance ratio, the specific heat transfer area, and the specific cooling water flow rate. The following sections include brief listing of the model equations for the MEE and the steam jet ejector and are followed solution of different case studies.

MEE Model

The simplified mathematical model of the MEE (discussed in the previous chapter) is used to calculate the following:

- Brine and distillate flow rates.
- Brine concentration.
- Effect temperature.
- Evaporator heat transfer area.

The model equations exclude the flash boxes and preheaters. The model includes the governing equation for the down condenser and its solution is made upon completion of the effect iterations and design of the steam jet ejector.

- Total mass balance

$$M_f = M_d + B_n \quad (41)$$

- Total salt balance

$$X_f M_f = X_n B_n \quad (42)$$

where B is the brine mass flow rate, M is the mass flow rate, X is the salt concentration, and the subscripts d, f, and n define the product water, feed seawater, and last effect.

- Distillate flow rate in the first effect

$$D_1 = M_d /(1 + \lambda_{v_1}/\lambda_{v_2} + \ldots\ldots + \lambda_1/ \lambda_{v_{n-1}} + \lambda_1/ \lambda_{v_n}) \quad (43)$$

- Distillate flow rate in effects 2 to n

$$D_n = D_1 \lambda_{v_1}/\lambda_{v_n} \quad (44)$$

- Total temperature drop across the effects

$$\Delta T = T_s - T_n \tag{45}$$

where D is the distillate flow rate, λ_v is the latent heat of formed vapor, and T_n are the temperatures of the motive steam and the vapor formed in the last effect.

– Temperature drop in the first effect

$$\Delta T_1 = \frac{\Delta T_t}{U_1 \sum_{i=1}^{n} \frac{1}{U_i}} \tag{46}$$

– Temperature drop in effects 2 to n

$$\Delta T_i = \Delta T_1 \, U_1/U_i \tag{47}$$

– Temperature of the first effect

$$T_1 = T_s - \Delta T_1 \tag{48}$$

– Temperature of effects 2 to n

$$T_i = T_{i-1} - \Delta T_1 \, U_1/U_i \tag{49}$$

– Brine flow rate in the first effect

$$B_1 = M_f - D_1 \tag{50}$$

– Brine flow rate in effects 2 to n

$$B_i = B_{i-1} - D_i \tag{51}$$

– Brine salt concentration in the first effect

$$X_1 = X_f \, M_f/B_1 \tag{52}$$

– Brine salt concentration in effects 2 to n

$$X_i = X_{i-1} \, B_{i-1}/B_i \tag{53}$$

– Heat transfer area in the first effect

$$A_1 = M_s \, \lambda_s \, / \, (U_1 \, (T_s - T_1)) = (D_1 \, \lambda_1 + M_f \, C_p \, (t_{f_2} - T_1)) \, / \, (U_1 \, (T_s - T_1)) \tag{54}$$

– Heat transfer area in effects 2 to n

$$A_i = D_i \lambda_i / (U_i (\Delta T_i - \Delta T_{loss})) \qquad (55)$$

Performance Parameters

The performance ratio, PR, is defined as the flow rate ratio of distillate, M_d, and motive steam, M_m,

$$PR = M_d/M_m \qquad (56)$$

The motive steam flow rate, M_m, is defined as the difference of the flow rates for the compressed vapor and the entrained vapor

$$M_m = M_s - M_{ev}$$

From Eq. 17

$$M_{ev} = M_m/Ra$$

The above two equations are simplified and an expression for the motive steam flow rate is obtained as a function of the compressed vapor flow rate

$$M_m = M_s/(1+1/Ra) \qquad (57)$$

The compressed vapor flow rate is obtained from the thermal load for the first effect

$$M_s = (D_1 \lambda_{v_1} + M_f\, Cp\, (T_1 - t_{f_2}))/ \lambda_s \qquad (58)$$

where t_{f_2} is the seawater temperature leaving the last feed preheater.
The specific heat transfer area is

$$sA = \frac{\sum_{i=1}^{n} A_i + A_c}{M_d} \qquad (59)$$

where A_i is the heat transfer area in effect i and A_c is the down condenser heat transfer area, which is obtained from

$$A_c = \frac{Q_c}{U_c (LMTD)_c} \tag{60}$$

The $(LMTD)_c$ is defined as:

$$(LMTD)_c = \frac{t_f - t_{cw}}{\ln \dfrac{T_n - t_{cw}}{T_n - t_f}} \tag{61}$$

The condenser thermal load

$$Q_c = (D_n - M_{ev})\, \lambda_{v_n} \tag{62}$$

The specific cooling water flow rate

$$sM_{cw} = M_d/M_{cw} \tag{63}$$

The condenser energy balance,

$$(D_n - M_{ev})\, \lambda_{v_n} = (M_f + M_{cw})\, Cp\, (t_f - t_{cw}) \tag{64}$$

where M_{cw}, is the cooling water flow rate.

Solution Procedure

Solution of the MEE-TVC model proceeds as follows:

- Solution of the overall material and salt balances, Eqs. 41 and 42.
- Iterative solution of the MEE model, Eqs. 43-55.
- Solution of the steam jet ejector model (Eq. 16 in Section 5.1).
- Evaluation of the performance parameters, Eqs. 56-64.

The following set of specifications is used in the above solution procedure:

- The seawater temperature, $T_{cw} = 25$°C.
- The feed water temperature leaving the last preheater, $T_{f_2} = T_1 - 5$.
- The seawater salinity, $X_f = 42000$ ppm.
- The salinity of the rejected brine, $X_b = 70000$ ppm.
- The range for top brine temperature, 55-100 °C.
- The range for the motive steam pressure, 250-1750 kPa.
- The range for the number of effects in MEE, 4-12 effects.
- The vapor temperature in the last effect, $T_n = 40$ °C.
- The thermodynamic losses in each effect, $\Delta T_i = 2$ °C.
- The heat capacity of all liquid streams, $C_p = 4.2$ kJ/kg °C.

- The overall heat transfer coefficient in the condenser, U_c = 1.75 kW/m^2 °C.
- The overall heat transfer coefficient in the first effect, U_e = 2.4 kW/m^2 °C; this value decreases by 5% in each subsequent effect.

5.2.2 Case Study

A four-effect MEE-TVC system is designed using the model and solution procedure discussed above. Specifications of the system parameters are given in the previous section. However, calculations are made at the following conditions:
- Compressed vapor temperature, T_s, 60 °C.
- Pressure of motive steam, P_m, 250 kPa.

For a total distillate flow rate, M_d, of 1 kg/s, intake seawater salinity, X_f, 42000 ppm, and rejected brine salinity of 70000 ppm, the resulting feed flow rate, M_f, and rejected brine from the last effect, B_4, are

$$M_f = X_b/(X_b - X_f) = 70000/(70000 - 42000) = 2.5 \text{ kg/s}$$

$$M_b = M_f - M_d = 2.5 - 1 = 1.5 \text{ kg/s}$$

The temperature drop across the effects, T_s–T_4, is equal to 60 – 40 = 20 °C. The overall heat transfer coefficients in effects 1 to 4 are specified and are assumed to remain constant throughout the iterations. The overall heat transfer coefficient in the first effect, U_1, is set equal to 2.4 kW/m^2 °C. Values in subsequent effects are obtained from

$$U_{i+1} = 0.95\ U_i$$

Values of the overall heat transfer coefficient in all effects are summarized in the following table

U_1	U_2	U_3	U_4
2.4	2.28	2.16	2.0577

The summation of the inverse for the overall heat transfer coefficients is required to calculate the temperature drop per effect. This summation is

$$\frac{1}{\sum\limits_{i=1}^{6} U_i} = 1/U_1 + 1/U_2 + 1/U_3 + 1/U_4$$

$$= 1/2.4 + 1/2.28 + 1/2.16 + 1/2.0577 = 1.8029 \text{ m}^2\ {}^{o}\text{C/kW}$$

The temperature drop in the first effect is then calculated

$$\Delta T_1 = \frac{\Delta T_t}{U_1 \sum\limits_{i=1}^{n} \frac{1}{U_i}} = \frac{20}{(2.4)(2.8529)} = 4.6221\ {}^{o}\text{C}$$

The values of ΔT_i are calculated for effects 2 to 4

$\Delta T_2 = \Delta T_1\ (U_1/U_2) = (4.6221)(2.4)/(2.28) = 4.8654\ {}^{o}C$

$\Delta T_3 = \Delta T_1\ (U_1/U_3) = (4.8654)(2.4)/(2.166) = 5.1215\ {}^{o}C$

$\Delta T_4 = \Delta T_1\ (U_1/U_4) = (5.1215)(2.4)/(2.0577) = 5.391\ {}^{o}C$

The following table summarizes the above values

ΔT_1	ΔT_2	ΔT_3	ΔT_4
4.6221	4.8654	5.1215	5.391

The temperature profile in effects 1 to 4 is then calculated

$T_1 = T_s - \Delta T_1 = 60 - 4.6221 = 55.3779\ {}^{o}C$

$T_2 = T_1 - \Delta T_1\ (U_1/U_2) = 55.3779 - 4.6221\ (2.4/2.28) = 50.5\ {}^{o}C$

$T_3 = T_2 - \Delta T_1\ (U_1/U_3) = 50.5125 - 4.6221\ (2.4/2.166) = 45.4\ {}^{o}C$

To check the above values T_4 is calculated on

$T_4 = T_3 - \Delta T_1\ (U_1/U_4) = 45.391 - 4.6221\ (2.4/2.0577) = 40\ {}^{o}C$

This value checks with the initial specification of 40 °C. The following table includes summary of calculated temperatures as well as the temperature of the motive steam.

T_s	T_1	T_2	T_3	T_4
60	55.3779	50.5125	45.3910	40

The latent heat values in all effects are calculated using the correlation given in the appendix

$\lambda_s = 2499.5698 - 2.204864\ T_s - 2.304x10^{-3}\ T_s^2$

= 2499.5698 – 2.204864 (Chapter 5 Multiple Effect Evaporation – Vapor

= 2358.9 kJ/kg

$\lambda_{v_1} = 2499.5698 - 2.204864\ T_{v_1} - 2.304x10^{-3}\ T_{v_1}^2$
$= 2499.5698 - 2.204864\ (55.3779 - 2) - 2.304x10^{-3}\ (55.3779 - 2)^2$
$= 2375.3$ kJ/kg

$\lambda_{v_2} = 2499.5698 - 2.204864\ T_{v_2} - 2.304x10^{-3}\ T_{v_2}^2$
$= 2499.5698 - 2.204864\ (50.5125 - 2) - 2.304x10^{-3}\ (50.5125 - 2)^2$
$= 2387.1$ kJ/kg

$\lambda_{v_3} = 2499.5698 - 2.204864\ T_{v_3} - 2.304x10^{-3}\ T_{v_3}^2$
$= 2499.5698 - 2.204864\ (45.391 - 2) - 2.304x10^{-3}\ (45.391 - 2)^2$
$= 2399.5$ kJ/kg

$\lambda_{v_4} = 2499.5698 - 2.204864\ T_{v_4} - 2.304x10^{-3}\ T_{v_4}^2$
$= 2499.5698 - 2.204864\ (40 - 2) - 2.304x10^{-3}\ (40 - 2)^2$
$= 2412.4$ kJ/kg

Summary of the latent heat values is given in the following table, which includes the latent heat of motive steam.

λ_s	λ_{v_1}	λ_{v_2}	λ_{v_3}	λ_{v_4}
2358.9	2375.3	2387.1	2399.5	2412.4

The flow rate profiles of the distillate and brine as well as the brine concentrations are calculated from Eqs. 3,4 and 10-13. The distillate flow rate in the first effect is calculated from Eq. 3

$D_1 = M_d /(1 + \lambda_{v_1}/\lambda_{v_2} + \lambda_{v_1}/\lambda_{v_3} + \lambda_{v_1}/\lambda_{v_4})$
$= (1)/(1 + (2375.3/2387.1) + (2375.3/2399.5)$
$+ (2284.47/2412.4))$
$= 0.2519$ kg/s

Subsequently, the distillate flow rates in effects 2 to n are calculated

$D_2 = D_1\ \lambda_{v_1}/\lambda_{v_2} = 0.1712\ (2375.3/2387.1) = 0.2507$ kg/s

$D_3 = D_1\ \lambda_{v_1}/\lambda_{v_3} = 0.1712\ (2375.3/2399.5) = 0.2494$ kg/s

$D_4 = D_1\ \lambda_{v_1}/\lambda_{v_4} = 0.1712\ (2375.3/2399.5) = 0.248$ kg/s

The brine flow rates are obtained from Eqs. 10 and 11

$B_1 = M_f - D_1 = 2.5 - 0.2519 = 2.2481$ kg/s

$B_2 = B_1 - D_2 = 2.2481 - 0.2507 = 1.9974$ kg/s

$B_3 = B_2 - D_3 = 1.9974 - 0.2494 = 1.748$ kg/s

$B_4 = B_3 - D_4 = 1.748 - 0.248 = 1.5$ kg/s

This value of B_4 checks with the initial material balance calculations. The salt concentration profile is calculated from Eqs. 12 and 13.

$X_1 = X_f\, M_f / B_1 = 42000\ (2.5/2.2481) = 46706.5078$ ppm

$X_2 = X_1\ B_1/B_2 = 46706.5078\ (2.2481/1.9974) = 52567.9687$ ppm

$X_3 = X_2\ B_2/B_3 = 52567.9687\ (1.9974/1.748) = 60067.2617$ ppm

$X_4 = X_3\ B_3/B_4 = 60067.2617\ (1.748/1.5) = 70000$ ppm

The value of X_4 checks with the initial specification at 70,000 ppm. Summary for the values of distillate and brine flow rates and brine concentration are given in the following table.

Effect	1	2	3	4
D (kg/s)	0.2519	0.507	0.2494	0.248
B (kg/s)	2.2481	1.9974	1.748	1.5
X (ppm)	46706.5	52567.9	60067.2	70000

The heat transfer areas are calculated in effects 1 to 4. These values are calculated as follows:

$$A_1 = (D_1\ \lambda_{v_1} + M_f\ Cp\ (T_{f_2} - T_1))/(U_1(T_s - T_1))$$
$$= (0.2519)(2375.3) + (2.5)\ (4.2)\ (5))/(2.4(60 - 55.3779))$$

$= 58.67\ m^2$

$A_2 = D_2\ \lambda_{v_2}/(U_2(\Delta T_2 - \Delta T_{loss}))$
$= (0.2507)(2387.18)/(2.28(4.8654\text{-}2))$
$= 91.59\ m^2$

$A_3 = D_3\ \lambda_{v_3}/(U_3(\Delta T_3 - \Delta T_{loss}))$
$= (0.2494)(2399.56)/(2.166(5.1215\text{-}2))$
$= 88.5\ m^2$

$A_4 = D_4\ \lambda_{v_4}/(U_4(\Delta T_4 - \Delta T_{loss}))$
$= (0.248)(2412.45)/(2.0577(5.391\text{-}2))$
$= 85.757\ m^2$

The maximum difference in effect areas is equal to 32.9 m^2. Assuming an error criterion of less than 1 m^2 is required, therefore, a new iteration sequence has to be initiated. The second iteration starts with calculations of the new heat transfer area

$$A_m = \frac{\sum_{i=1}^{n} A_i}{n}$$
$= (58.675 + 91.5932 + 88.5045 + 85.7571)/4$
$= 324.529/4$
$= 81.13\ m^2$

A new profile for the temperature drop across the effects is then calculated

$\Delta T_1 = \Delta T_1\ (A_1/A_m) = (4.6221)(58.675)/(81.1324) = 3.3427\ ^oC$

$\Delta T_2 = \Delta T_2\ (A_2/A_m) = (4.8654)\ (91.5932)/(81.1324) = 5.4927\ ^oC$

$\Delta T_3 = \Delta T_3\ (A_3/A_m) = (5.1215)\ (88.5045)/(81.1324) = 5.5868\ ^oC$

$\Delta T_4 = \Delta T_4\ (A_4/A_m) = (5.391)\ (85.7571)/(81.1324) = 5.6983\ ^oC$

A new iteration is then taken, which starts with temperature profiles and continues to the convergence criteria part. The number of iterations executed to reach the above tolerance is 4. Summary of flow rates, concentrations, temperatures, and heat transfer areas in the last iteration are given in the following table

Effect	1	2	3	4
D (kg/s)	0.2518	0.2508	0.2498	0.2476
B (kg/s)	2.2482	1.9974	1.746	1.5
X (ppm)	46703	52567	60082	70000
T (°C)	56.54	52.9	49.07	40
A (m²)	78.3	77.5	78.2	78.7

Determination of the entrainment ratio (Ra) requires calculations of the correction factors, PCF and TCF, or

$$PCF = 3x10^{-7} (P_m)^2 - 0.0009 (P_m) + 1.6101$$
$$= 3x10^{-7} (250)^2 - 0.0009 (250) + 1.6101$$
$$= 1.40385$$

$$TCF = 2x10^{-8} (T_{v4})^2 - 0.0006 (T_{v4}) + 1.0047$$
$$= 2x10^{-8} (38)^2 - 0.0006 (38) + 1.0047$$
$$= 0.9819$$

$$Ra = 0.296 \frac{(P_6)^{1.19}}{(P_7)^{1.04}} \left(\frac{P_m}{P_7}\right)^{0.015} \left(\frac{PCF}{TCF}\right)$$

$$= 0.296 \frac{(19.87)^{1.19}}{(6.527)^{1.04}} \left(\frac{250}{6.527}\right)^{0.015} \left(\frac{1.4038}{0.981}\right)$$

$$= 2.228$$

The above results should be checked against permissible ranges specified in the steam ejector model, where $Ra \le 5$, $500 \ge T_m > 10$ °C, $3500 \ge P_m \ge 100$ kPa, and $(P_6/P_7) \ge 1.81$. The value of Ra is less than 5 and the ratio (P_6/P_7) is equal to 3.04 which is greater than 1.81. Also the values of T_m and P_m are within the specified range, where P_m is equal to 250 kPa and T_m is equal to 127.5 °C.

To obtain the performance ratio it is necessary to determine the flow rates of the motive steam, entrained vapor, and compressed vapor. The compressed vapor flow rate is given by

$$M_s = (D_1 \lambda_{v1}/+C_p M_f (T_1 - T_{f2}))/\lambda_s$$
$$= (0.2518)(2372.45) + 2.5) (4.2) (5))/(2358.98)$$
$$= 0.2754 \text{ kg/s}$$

The motive steam flow rate is obtained from

$$M_m = M_s/(1+1/Ra)$$
$$= 0.19 \text{ kg/s}$$

The entrained vapor flow rate is obtained from

$$M_{ev} = M_s - M_m$$
$$= 0.275 - 0.19$$
$$= 0.08532 \text{ kg/s}$$

Since the total distillate flow rate is equal to 1 kg/s, then,

$$PR = M_d/M_m = 1/0.19 = 5.26$$

The condenser thermal load is calculated from

$$Q_c = (D_4 - M_{ev})\,\lambda_{v_4} = (0.2476 - 0.085329)\,(2412.46) = 391.47 \text{ kW}$$

The logarithmic mean temperature difference in the condenser is given by

$$(LMTD)_c = (t_f - t_{cw})/Ln((T_4 - \Delta T_{loss} - t_{cw})/(T_4 - \Delta T_{loss} - t_f))$$
$$= (35\text{-}25)/Ln((40\text{-}2\text{-}25)/(40\text{-}2\text{-}35))$$
$$= 6.819\ ^oC$$

The condenser heat transfer area in the condenser is then calculated from

$$A_c = Q_c/(U_c\,(LMTD)_c) = 391.765/((1.75)(6.819)) = 32.79 \text{ m}^2$$

The specific heat transfer area is calculated by the summing the heat transfer areas for the six evaporators and the condenser. This is

$$sA = \frac{\sum_{i=1}^{n} A_i + A_c}{M_d} = (312.96+32.79) = 345.76 \text{ m}^2$$

The cooling water flow rate is obtained from Eq. 29

$$(D_4 - M_{ev})\,\lambda_{v_4} = (M_f + M_{cw})\,Cp\,(t_f - t_{cw})$$
$$(0.2476\text{-}0.08532)(2412.45) = (2.5+M_{cw})\,(4.2)(35\text{-}25)$$

which gives M_{cw} = 6.819 kg/s. The specific cooling water flow rate has the same value, since the total product flow rate is equal to 1 kg/s.

5.2.3 System Performance

Performance of the MEE-FF-TVC system together with the stand-alone MEE-FF system is shown in Figs. 13-15. The analysis is performed as a function of the number of effects and the heating steam temperature. As is shown performance ratio of the vapor compression system is higher, especially at low operating temperatures. The decrease in the performance ratio of the TVC system at higher temperatures is caused by the increase in the compression range. This is because the brine temperature in the last effect is kept constant in all calculations. Therefore, at higher temperature larger amount of motive steam is used to achieve the required compression range. This also affects the required amount of specific flow rate of cooling water. Results show the increase in the specific flow rate of cooling water for vapor compression system at higher operating temperatures and smaller number of effects. Increase in the system operating temperature increases the amount of motive steam, which increases the system thermal load and the required amount of cooling water per kg of distillate product.

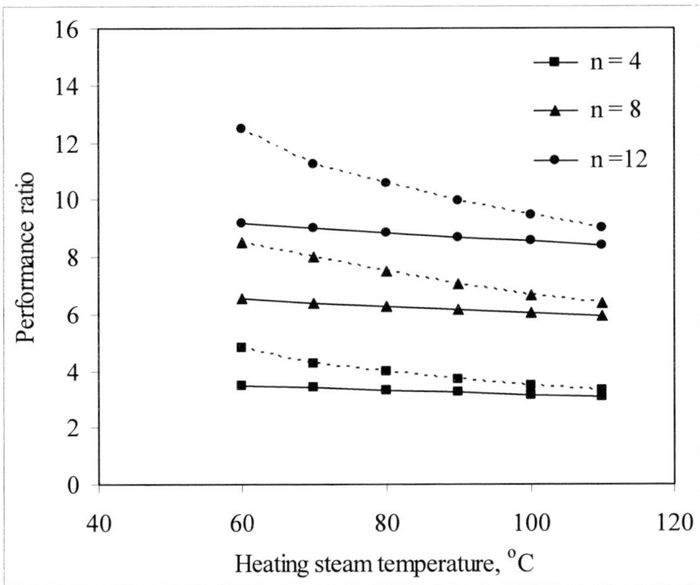

Fig. 13. Effect of heating steam temperature and the number of effects on the performance ratio of the MEE (—) and MEE-TVC (----) systems.

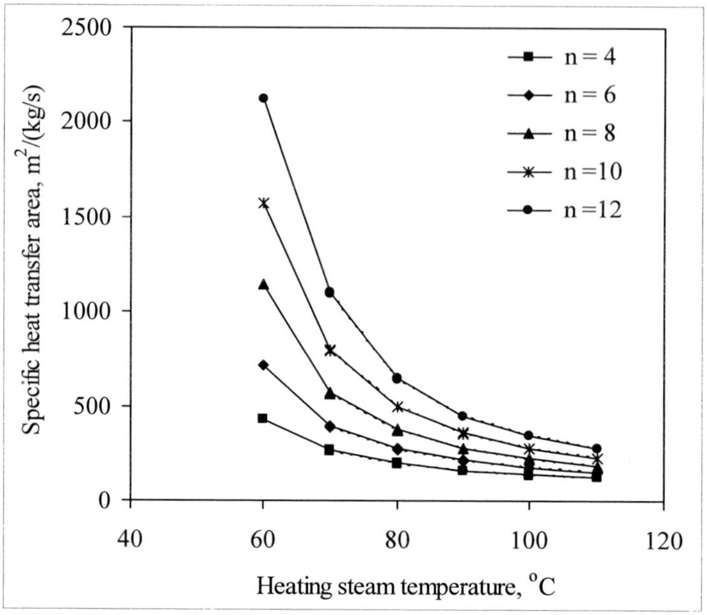

Fig. 14. Effect of heating steam temperature and the number of effects on the specific heat transfer area of the MEE (——) and MEE-TVC (----).

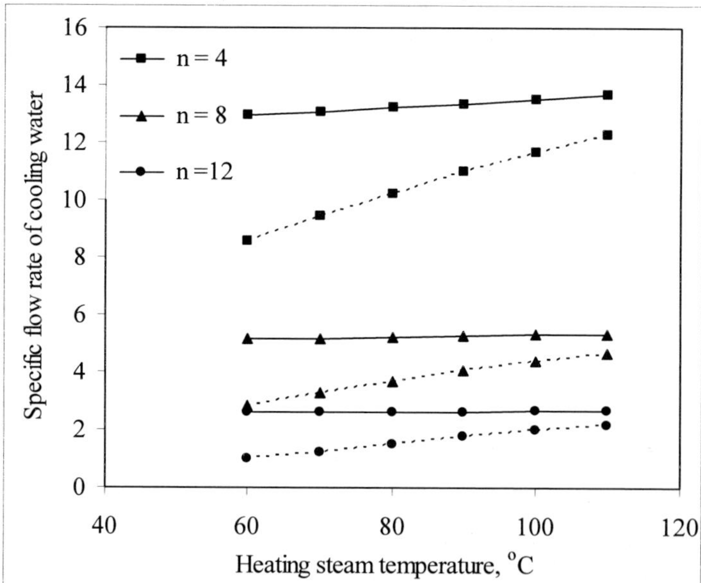

Fig. 15. Variation in specific flow rate of cooling water as a function of heating steam temperature and number of effects for MEE (——) and MEE-TVC (----).

Increase in the specific flow rate of cooling water at lower number of effects is caused by the increase in the amount of vapor generated per effect. Therefore, the amount of vapor generated in the last effect becomes larger and would require a larger amount of cooling water. It should be noted that the specific heat transfer area is slightly affected by the vapor compression process. This is because the effect temperatures remain the same for system operation with/without vapor compression. The main difference between the two systems comes in the heat transfer area for the down condenser, which is lower in the vapor compression mode due to vapor entrainment by the ejector.

5.2.4 Comparison of MEE and MEE-TVC

Performance characteristics of the MEE and MEE-TVC systems are compared for a four-effect system. Comparison includes performance ratio, specific heat transfer area, and specific cooling water flow rate. Results are summarized in the following Table. As is shown, the characteristics of the MEE-TVC out perform those for the MEE system; where

- The performance ratio is higher by 45%.
- The specific cooling water flow rate is lower by 41.8%.
- The condenser specific heat transfer area is lower by 34.4%.
- The total specific heat transfer area is lower by 4.75%.

Since, the characteristics of the MEE evaporators are identical in either configuration the same specific heat transfer area for the evaporators is obtained for both systems.

Process	MEE		MEE-TVC		MEE-TVC	
T_s	112.9	60.9	112.9	60.9	112.9	60.9
T_n	40	40	40	40	40	40
T_{cw_i}	25	25	25	25	25	25
T_{cw_o}	35	35	35	35	35	35
n	12	4	12	12	4	4
Power(1)	86.3	287.9	84.81	89.1	274.3	283.2
M_{cw}	4.47	14.2	4.11	2.2	8	12
A_c	16.37	39.5	15.34	15.8	36.2	37.5
sA	202.5	302.8	201.5	2119	280.5	428.7
Pr	8.67	2.52	9.2	12.5	3.3	4.8

(1) In kJ/kg and excluding the pumping power.

5.2.5 Summary

This section is focused on modeling and performance analysis of the forward feed multiple effect evaporation with thermal vapor compression. The analysis included a step-by-step calculation method for the MEE-TVC system, detailed mathematical model, performance as a function of the top brine temperature and the number of effects, and comparison against the stand-alone forward feed MEE system. Results show increase up to 50% in the thermal performance ratio and a similar decrease in the specific flow rate of cooling water. Operation of field units gives similar behavior.

Problems

Problem 1

A four effect parallel feed MEE-TVC system is shown in the attached figures. The system operates at the following conditions:

Temperature of the first effect = 58 °C
Temperature of the second effect = 54 °C,
Temperature of the third effect = 50 °C
Temperature of the fourth effect = 46 °C
Temperature of heating steam = 63 °C
Motive steam flow rate = 6.88 kg/s, Motive steam pressure = 15 bar
Entrained vapor flow rate = 8 kg/s
Amount of vapor formed in the first effect = 14 kg/s
Amount of vapor formed in the second effect = 13.2 kg/s
Amount of vapor formed in the third effect = 13.2 kg/s
Amount of vapor formed in the fourth effect = 14 kg/s
Flow rate of product water = 52.6 kg/s

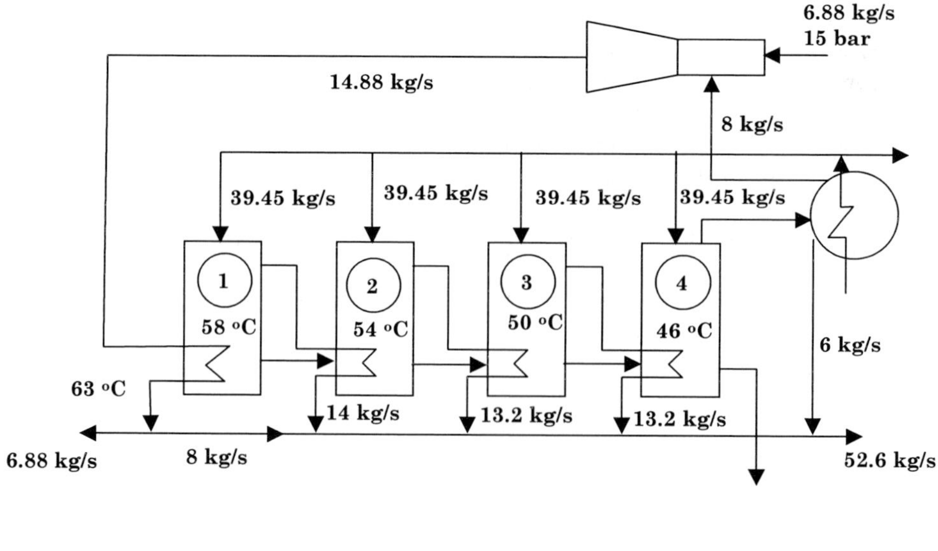

Calculate the following

- Plant performance ratio
- Specific heat transfer area
- Mass flow rate of rejected cooling water
- Pressure of the motive steam

Problem 2

A four effect forward feed MEE-TVC system operates at the following conditions:

Plant capacity = 1000 m^3/d
Motive steam pressure = 250 kPa
Ejector area ratio = 50
Temperature of vapor in the last effect = 65 °C
Temperature of feed seawater = 40 °C
Feed seawater salinity = 45000 ppm

The overall heat transfer coefficient is given by

$$U = 3.25 + 0.05\,(T - 60)$$

Where U is in kW/m^2 °C and T is in °C

Calculate the following

- Plant performance ratio
- Flow rates of cooling seawater and motive steam

Problem 3

A Three effect forward feed MEE-TVC system operates at the following conditions:

Plant capacity = 500 m^3/d
Temperature of feed seawater = 20 °C
Feed seawater salinity = 42000 ppm
Salinity of brine blow down = 70000 ppm
Temperature of vapor in the last effect = 45 °C
Motive steam pressure = 250 kPa
Ejector area ratio = 50
Temperature of compressed heating steam = 80 °C

The overall heat transfer coefficients in the three effects are equal to

$$U_1 = 3.123 \text{ kW/m}^2 \text{ °C}$$
$$U_2 = 1.987 \text{ kW/m}^2 \text{ °C}$$

$U_3 = 1.136$ kW/m^2 °C

Calculate the following

- Plant performance ratio
- The specific heat transfer area
- The specific flow rate of cooling water

5.3 Forward Feed Multiple Effect Evaporation with Mechanical Vapor Compression

A schematic diagram for the MEE-MVC system is shown in Fig. 16, where it has a similar set for the MEE configuration except for removal of the down condenser and addition of the feed preheaters, a flashing box for the first effect, and the mechanical compressor. The compressor unit operates on the entire vapor formed in the last effect, where it is compressed to the desired pressure and superheat temperature. This is necessary to take into consideration the lower amount of vapor formed in the last effect in comparison with that formed in the first effect. Routing the entire vapor formed in the last effect to compressor results in elimination of the down condenser. However, to maintain high thermal efficiency for the process plate preheaters are used to increase the temperature of the feed seawater from (T_{cw}) to (T_f). This is achieved by heat recovery from the brine blowdown and the distillate product streams in two separate feed preheaters.

5.3.1 System Model

Elements forming the model of the forward feed multiple effect evaporation are a combination of the mathematical model for the stand-alone forward feed system together with the mathematical model for the mechanical vapor compressor. Therefore, it is advisable to review elements of both models. The forward feed model is given in section 5.2 and the model on mechanical vapor compression is given in the chapter on single effect systems.

5.3.2 System Performance

Performance of the forward feed multiple effect system is shown in Figs. 17-18. Performance results are obtained for the specific power consumption and the specific heat transfer area. Results are presented as a function of the brine blowdown temperature, the temperature difference of the brine between the first and last effects, and the number of effects.

Fig. 17 shows variations in the specific power consumption for both systems, where it decreases at higher operating temperature and lower temperature differences of the brine. At higher operating temperatures, the specific volume of the vapor decreases, which reduces the power consumed for vapor compression. On the other hand, larger temperature differences of the saturation temperature of the compressed vapor and the brine blowdown result in increase in the compression range, which increases the power consumed for vapor compression. The specific power consumption for the both systems and the above set of parameters varies between low values close to 6 kWh/m^3 and higher values close to 14 kWh/m^3, which are consistent with literature data.

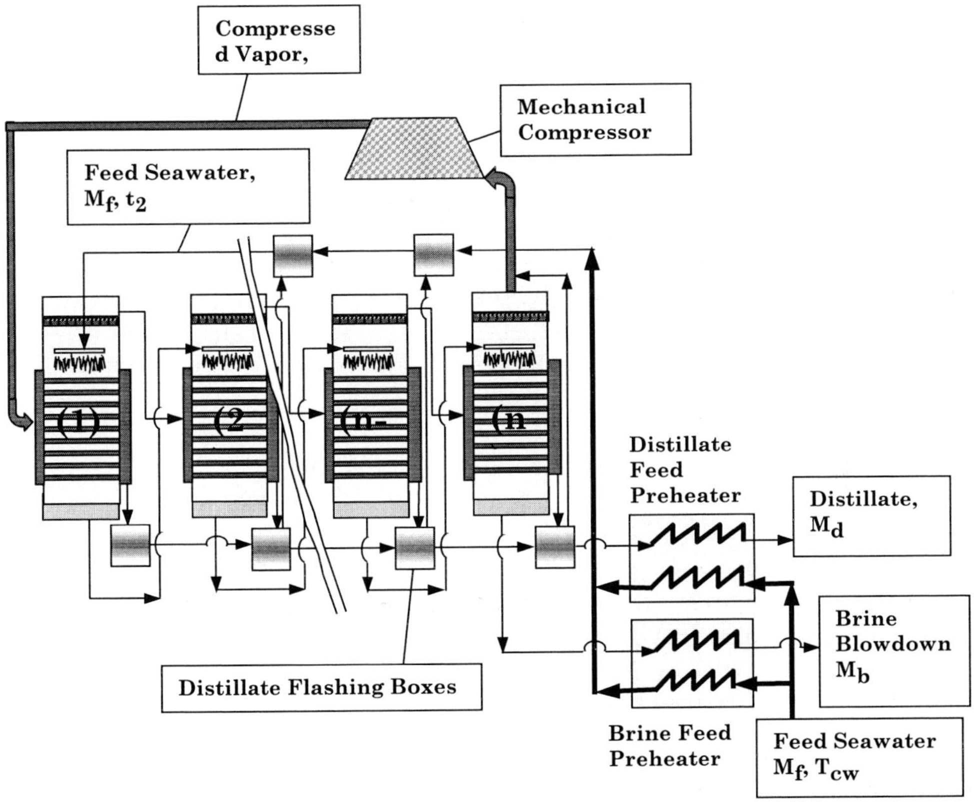

Fig. 16. Forward feed multiple effect evaporation with mechanical vapor compression

Figure 18 shows that the specific heat transfer has stronger dependence on the temperature drop per stage rather than the top brine temperature. The temperature drop per stage is affected by the number of effects and the temperature difference between the first and last effects. On the other hand, increase in the system temperature has smaller effect on the specific heat transfer area. Increase in the system temperature has a limited effect on the heat transfer coefficient and the heat transfer area.

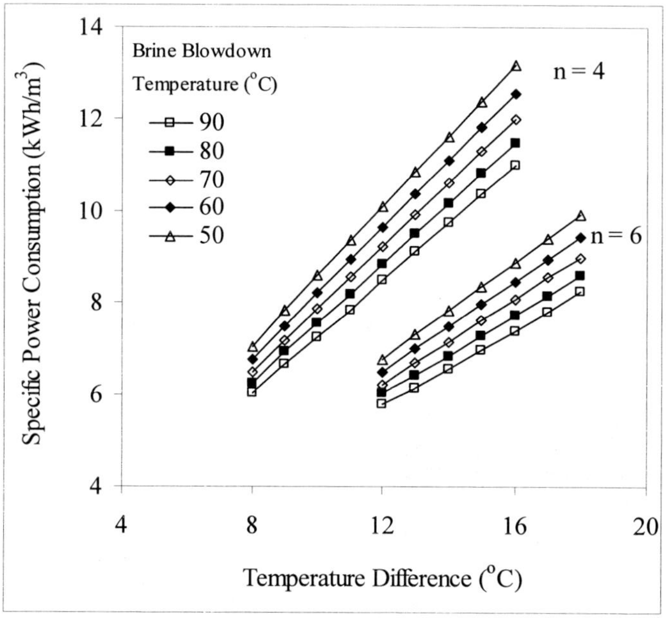

Fig. 17. Variation in specific power consumption for the forward feed multiple effect with mechanical vapor compression.

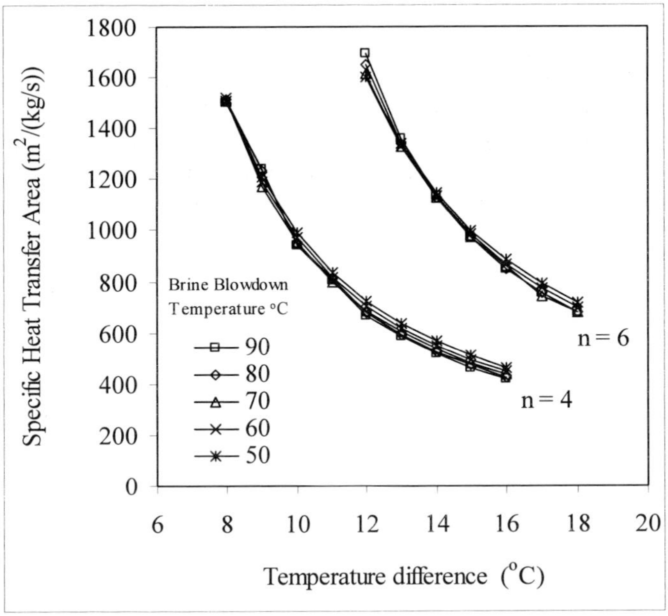

Fig. 18. Variation in specific heat transfer area for the forward feed multiple effect with mechanical vapor compression.

5.4 Forward Feed Multiple Effect Evaporation with Adsorption Vapor Compression

The MEE-ADS includes two beds for vapor adsorption and desorption, Fig. 19. Similar to MEE-MVC, the system does not include a down condenser. Therefore, the intake seawater is heated in two plate preheaters against the brine blowdown and the distillate product. Operation of the adsorption/desorption heat pump is transient and it involves simultaneous condensation/adsorption of vapor formed in the last effect/flash box into the adsorption bed and evaporation/desorption of heating steam from the desorption bed. In each bed zeolite solid is used as the adsorption/desorption medium. Adsorption proceeds at an equilibrium temperature corresponding to the vapor temperature in the last effect; while, while desorption proceeds at an equilibrium temperature corresponding to the heating steam temperature in the first effect. External heating/cooling sources are used to assist the adsorption/desorption processes. Cooling water is used to remove the latent heat of condensation from the adsorption bed, while, motive steam is used to add the latent heat of evaporation for the desorption bed. The desorption process reaches an equilibrium dry condition as most of the adsorbed water is released as vapor. Similarly, the wet equilibrium condition in the adsorption bed is reached as the bed solid phase becomes saturated with water. Upon completion of the adsorption/desorption processes, the external cooling/heating sources are disengaged. Subsequently, the thermal fluid circulates between the two beds to remove the sensible heat from the dry bed into the wet bed. This heat exchange process increases the temperature of the cold/wet bed to a higher value close to the required desorption temperature. Similarly, the temperature of dry/hot bed is reduced to a temperature that allows the start of the adsorption process. However, it should be stressed that reaching conditions required to start the adsorption/desorption process requires additional cooling/heating by the cooling water and the motive steam.

As discussed in the previous section, the mathematical model for this system is a combination of the forward feed multiple effect system, which is given in section 5.2, and the adsorption heat pump model, which is given in chapter 3.

Performance analysis of this system is limited to evaluation of the thermal performance ratio and comparison against the stand-alone system. This is because variations in other system parameters are similar to those of the stand-alone system. Variations in the thermal performance ratio as a function of the heating steam temperature and number of effects is shown in Fig. 20. As is shown, the thermal performance ratio increases by more than 100% over the stand-alone system, especially, at high temperatures. For example, the thermal performance ratio for the 12 effect system is equal to 24 at a heating steam temperature of 115 °C. On the other hand, the thermal performance ratio of the stand-alone system it is equal to a value of 8. It should be noted that the thermal performance ratio for the system increases with the increase of the system temperature. This is because of the decrease in the latent heat of desorption at higher temperatures.

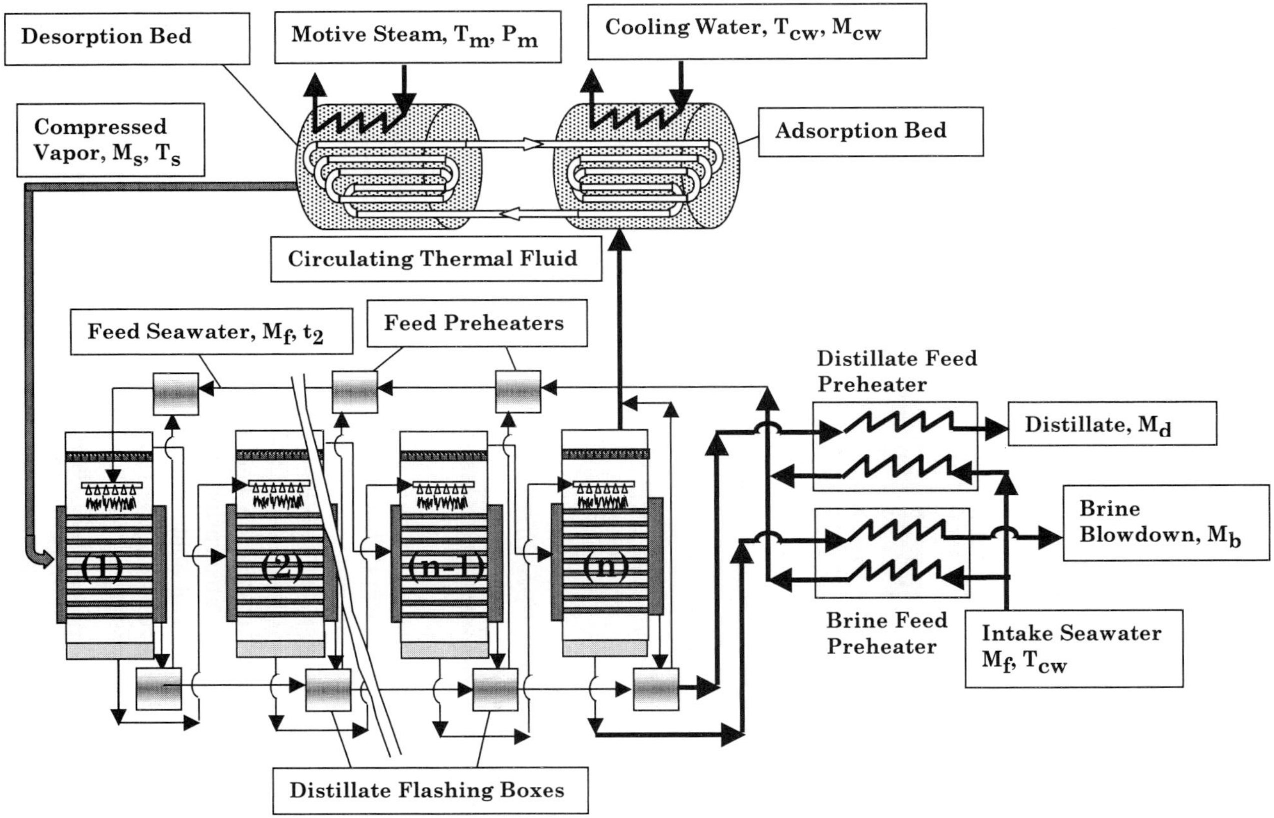

Fig. 19. Forward feed multiple effect evaporation with adsorption vapor compression

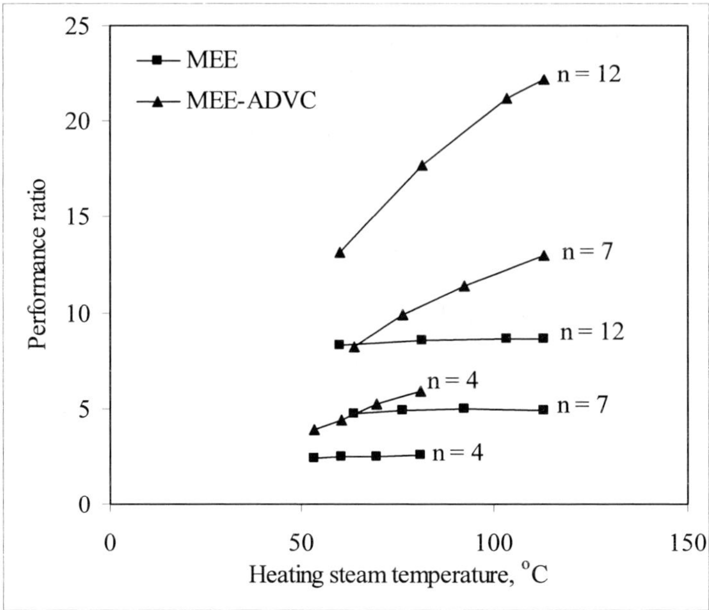

Fig. 20. Variation in the performance ratio for the forward feed multiple effect with adsorption vapor compression.

5.5 Forward Feed Multiple Effect Evaporation with Absorption Vapor Compression

The MEE-ABS system, shown in Fig. 21, includes evaporation effects, flashing boxes, feed preheaters, down condenser, stripping and absorption beds containing lithium bromide water solution (LiBr-H_2O). As is shown, part of the vapor formed in the last effect/flashing box is condensed in the down condenser, where it releases its latent heat to the feed seawater. The remaining part of the vapor flows through the absorption bed where it is absorbed the concentrated (LiBr-H_2O) solution. The absorption process is exothermic, where the feed seawater absorbs the released heat. This result in heating of the feed seawater to the saturation temperature and formation of a small amount of saturated vapor. The dilute LiBr-H2O solution is feed to the generator or stripper, where heat added by the motive steam results in water evaporation and increase in the solution concentration. The concentrated LiBr solution is pumped back to the absorption bed. The vapors formed in the generator and the absorber are combined together and are used to drive the evaporation effect number (1).

The mathematical model for this system is a combination of the forward feed multiple effect system, which is given in section 5.2, and the adsorption heat pump model, which is given in chapter 3. Similarly, the system performance, except for the thermal performance ratio, is similar to the stand-alone system. Variations in the thermal performance ratio as a function of the heating steam temperature and number of effects is shown in Fig. 22. As is shown, the thermal performance ratio increases by more than 100% over the stand-alone system, especially, at high temperatures. For example, the thermal performance ratio for the 12 effect system is equal to 27 at a heating steam temperature of 115 ^oC. On the other hand, the thermal performance ratio of the stand-alone system it is equal to a value of 8. It should be noted that the thermal performance ratio for the system increases with the increase of the system temperature. This is because of the decrease in the latent heat of desorption at higher temperatures.

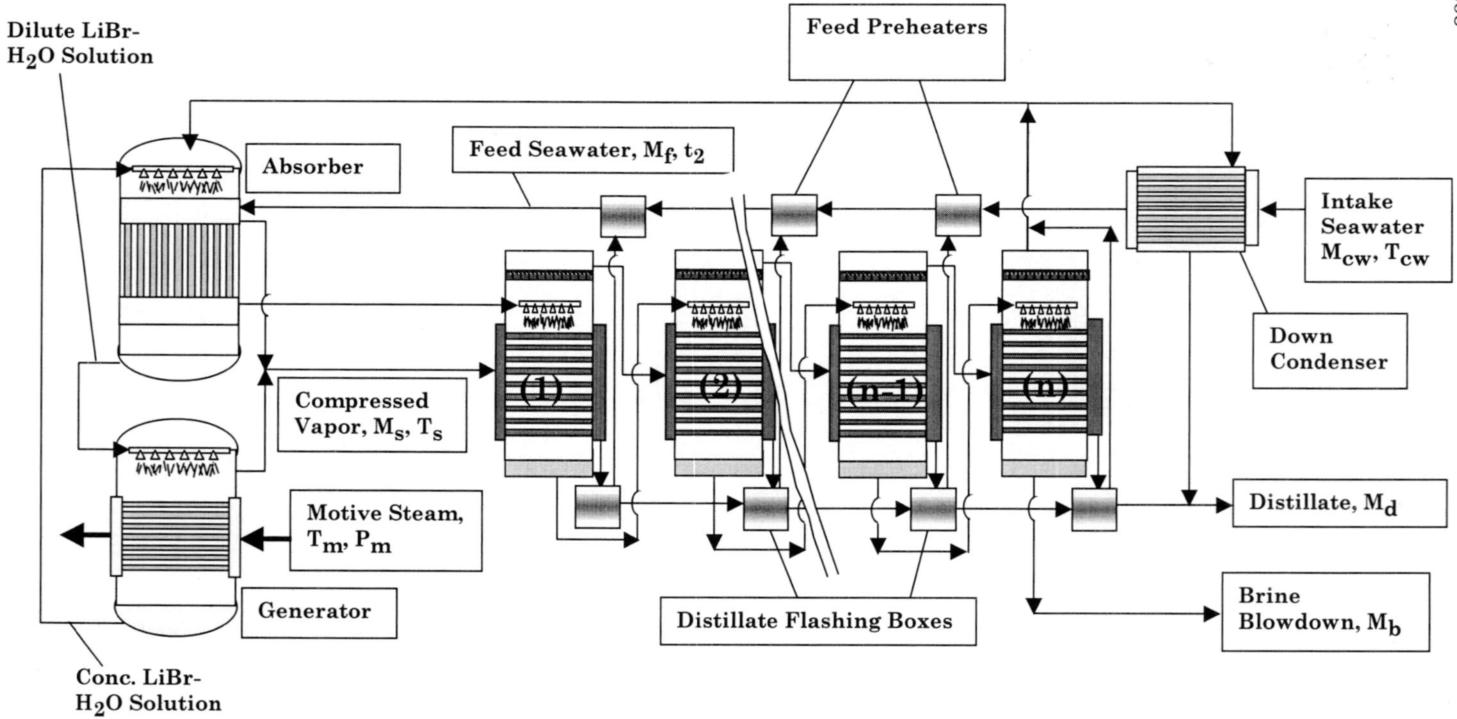

Fig. 21. Forward feed multiple effect evaporation with absorption vapor compression

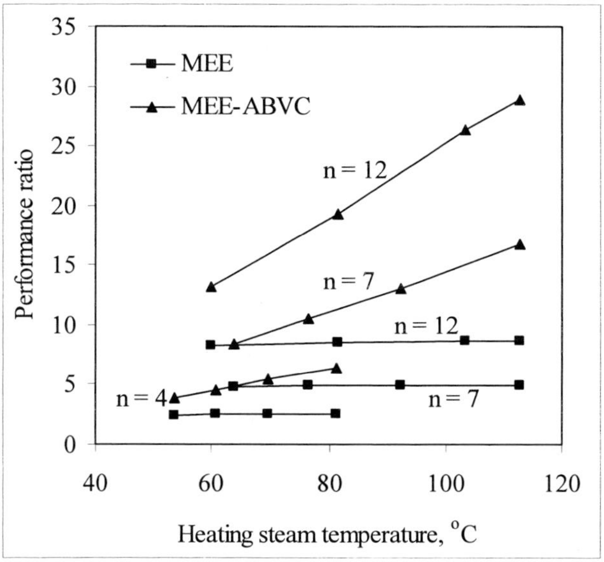

Fig. 22. Variation in the thermal performance ratio for the forward feed multiple effect evaporation with absorption vapor compression.

5.6 Summary

Four configurations are investigated for the forward feed MEE desalination system. The systems include combinations with thermal, mechanical, adsorption, and absorption vapor compression. The combined systems make efficient use of the characteristics of the MEE system and the combined heat pump. In the combined MEE-ADS and MEE-ABS systems, the first and last effects of the MEE system replace the condenser and evaporator units of the heat pump. This selection reduces the equipment cost of the heat pump. The combined system, also, allows for heating of utility water, which has a great value in industrial applications. These features are not found in the MEE, the MEE-MVC, and the MEE-TVC systems.

Mathematical models are developed for the proposed configurations. Results show large increase in the performance ratio of the hybrid MEE systems against that of the MEE configuration. In addition, results show the possibility of operating the hybrid systems at high steam temperatures. This was made possible by utilizing two steam ejectors in the MEE-TVC system and by using heat pumps in the MEE-ABS and MEE-ADS systems. In addition, results for the MEE-MVC system show the possibility of operation at high steam temperatures.

This is a major advantage for all systems, where at high steam temperatures the evaporator area in all effects is drastically reduced. This in turn will decrease considerably the construction cost of the MEE system.

In summary the following conclusions are made

- High increase in the performance ratio in the hybrid systems in comparison with conventional MEE.
- Increase of the top brine temperature reduces dramatically the required specific heat transfer area for all configurations.
- The MEE-MVC requires no cooling water, however, use of auxiliary heat is necessary to drive the first effect.
- The MEE-TVC requires less cooling water than conventional MEE.
- The MEE-ABS and MEE-ADS generates hot utility water, which can be used in other applications.
- Predictions of all models show very good agreement with industrial practice, i.e., performance ratio, power consumption, specific heat transfer area, and specific cooling water flow rate.
- Hybrid MEE-heat pump systems have great potential to replace conventional MSF (predominant in current desalination practice) in the near future.

Chapter 6

Multi-Stage Flash Desalination

Objectives

The objective of this chapter is to analyze and evaluate the performance of the multistage flash desalination system. This is made through discussion of the following:

- Process developments
- Standard features of the brine circulation MSF process, which is the most common process
- Modeling and analysis of the single stage flashing system, the once through multistage system, and the brine circulation multistage system.
- Features and performance of novel configurations, which include MSF with brine mixing and MSF with thermal vapor compression.

6.1 Developments in MSF

The MSF process is an innovative concept, where vapor formation takes place within the liquid bulk instead of the surface of hot tubes. The hot brine is allowed to flow freely and flash in a series of chambers; this feature keeps the hot and concentrated brine from the inside or outside surfaces of heating tubes. This is a major advantage over the original and simple concept of thermal evaporation, where submerged tubes of heating steam are used to perform fresh water evaporation. The performance of such configurations was far from satisfactory, where salt scale is formed progressively on the outside surface of the tubes. The formed scale has a low thermal conductivity and acts as an insulating layer between the heating steam and the boiling seawater. Consequently, the evaporation rate is drastically reduced and cleaning becomes essential to restore the process efficiency. Earlier designs were plagued with such problems, where operation lasted for less than two weeks and shutdown and cleaning lasted for more than four weeks.

The brine circulation MSF process is the industry standard. The process elements are illustrated in Fig. 1, where the flashing stages are divided among the heat recovery and heat rejection sections. The system is driven by heating steam, which increases the temperature of the brine recycle or feed seawater to the desired value in the brine heater. The hot brine flashes in the consecutive stages, where the brine recycle or the feed seawater flowing inside the condenser tubes recovers the latent heat of the formed vapor. In the heat rejection section of brine circulation system, the excess heat added to the system by the heating steam is rejected to the environment by the cooling seawater stream. In the MSF process the tubes are arrange in a long or cross tube configuration. The cross tube configuration is the original system design and its units have production capacities in the range of 27,276 – 32,731 m^3/d. In this configuration, the tubes are aligned along the width of the flashing chambers and are connected via external water boxes. The long tube arrangement is geared towards larger production volume with current unit capacities up to 57,734 m^3/d. In this system, a single bundle of tubes span the whole length of a limited number of flashing stages. This eliminates the water boxes found in the cross flow system and allows for the increase of the flow rate per chamber width, which reduces the required chamber width.

The brine circulation process has many attractive features, which makes it distinguishable among other desalination configurations. Since establishment in the late fifties, a huge field experience has been accumulating in process technology, design procedure, construction practice, and operation. This has resulted in development of simple and reliable operational procedures. In addition, the development addressed and solved various operational problems, which include scale formation, foaming, fouling, and corrosion. Gained experience in operation and design of the MSF plants has lead to use inexpensive construction material capable of standing harsh conditions at high salinity. The MSF system does not include moving parts, other than conventional pumps. Construction of the MSF plants is simple and contains a small number of connection tubes, which limits leakage problems and simplifies maintenance works. In the light of the above, we strongly believe that the MSF system will remain the main desalination process, especially in the Middle East. This is due to the following facts:

- The conservative nature of the desalination owner.
- The product is a strategic life-supporting element.
- Extensive experience in construction and operation.
- Process reliability.
- Limited experience, small database, and unknown risks with new technologies.

Since inception, several developments have been achieved in system design and operation. These developments include the following:

- Increasing the unit capacity from 454.6 m^3/d to a current conventional capacity of 27,276 – 32,731 m^3/d. Recently, larger units with a capacity of 57,734 m^3/d are commissioned. Each capacity doubling is associated with 24% reduction in unit product cost.
- Decreasing the specific power consumption from 25-70 kW/m^3 in 1955 to 4-10 kW/m^3.
- Plant operation is drastically improved with introduction of more efficient antiscalent and corrosion control chemicals and use of construction materials capable of withstanding the harsh operating conditions. Earlier, operation was plagued by excessive scaling and damaging corrosion resulting in limited production time followed by prolonged cleaning procedures, Temperley (1995). Currently, conventional MSF units operate continuously for more than 2

years, without the need for complete shut down. This is achieved in part by adoption of on-line acid or sponge ball cleaning.

- Other process developments include use of smaller specific equipment size, reduction of the stand-by units, and minimizing and simplifying control units. For example, more efficient interstage devices are developed to withstand erosion effects and increase the brine flow rate per chamber width.
- Development of more accurate and advanced models capable of various tasks, which includes process design, rating, evaluation of process economics, optimization, process dynamics, and system control, Helal et al. (1986), Darwish (1991), El-Dessouky et al. (1995), Darwish and El-Dessouky (1996), El-Dessouky and Bingulac (1996), El-Dessouky et al. (1998).

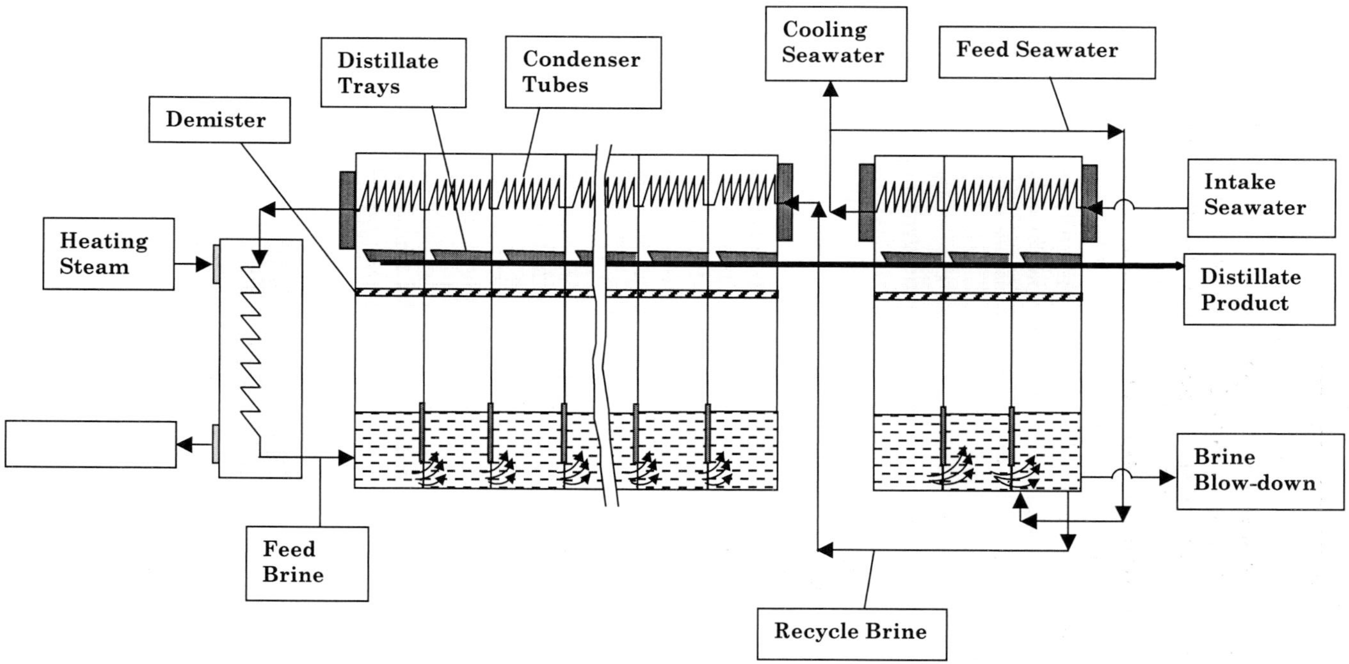

Fig. 1. Brine circulation multi-stage flash desalination process

6.2 MSF Flashing Stage

For conventional MSF systems with capacities of 27,000 up to 32,000 m^3/d, the flashing stage has dimensions of 18x4x3 m in width, height, and length. A schematic of the MSF flashing stage is shown in Fig. 2 and it includes the following:

- A large brine pool with similar width and length of the flashing stage and a depth of 0.2-0.5 m.
- A brine transfer device between the stages is designed to seal the vapor space between the stages and to enhance turbulence and mixing of the inlet brine stream. This device promotes flashing; controls formation of vapor bubbles, their growth, and subsequent release.
- The demister is formed of wire mesh layers and the supporting system. The demister function is to remove the entrained brine droplets from the flashed off vapor. This is essential to prevent increase in the salinity of product water or scale formation on the outer surface of the condenser tubes.
- The tube bundle of the condenser/preheater tubes, where the flashed off vapor condenses on the outer surface of the tubes. The released latent heat of condensation results in heating of the brine recycle stream flowing inside the tubes. This energy recovery is essential to maintain high system performance.
- Distillate tray, where the condensed distillate product is collected and cascade through the stages. The distillate product is withdrawn from the tray of the last stage.
- Water boxes at both ends of the tube bundle to transfer the brine recycle stream between adjacent stages.
- Connections for venting system, which removes non-condensable gases (O_2, N_2, and CO_2), which are dissolved in the feed seawater, even after deaeration. Also, CO_2 can be generated during decomposition of the bicarbonate compounds in the high temperature stages. Another important source for the non-condensable gases is air in-leakage from the ambient surroundings into the flashing stages operating at temperatures below 100 °C, which correspond to vacuum conditions.
- Instrumentation, which includes thermocouples, level sensor, and conductivity meter, are placed in the last and first flashing stages. The measured data from these stages are adopted by the control system of the process. Accordingly and subject to disturbances in the system parameters, i.e., feed seawater temperature, increase in fouling thermal resistance, available steam, etc., adjustments are made in the controllers to restore the desired operating conditions. The magnitude of these adjustments depends on the measurements made in the last and first stages.

The MSF process operates over a temperature range of 110-30 °C. This implies the majority of the flashing stages operate at a temperature below 100 °C or vacuum conditions. Therefore, all flashing stages are designed to withstand full vacuum. However, the bottom of the flashing stages is exposed to the hydrostatic pressure of the brine pool. Therefore, the system is designed to withstand a maximum pressure of 2 bar.

The walls, ceilings, and partitions of the flashing stages are constructed of carbon steel with stainless steel or epoxy cladding. Stainless steel cladding is used in locations where higher erosion or corrosion conditions can be found. All stages are reinforced with a stainless steel structure and heavily insulated to minimize heat losses.

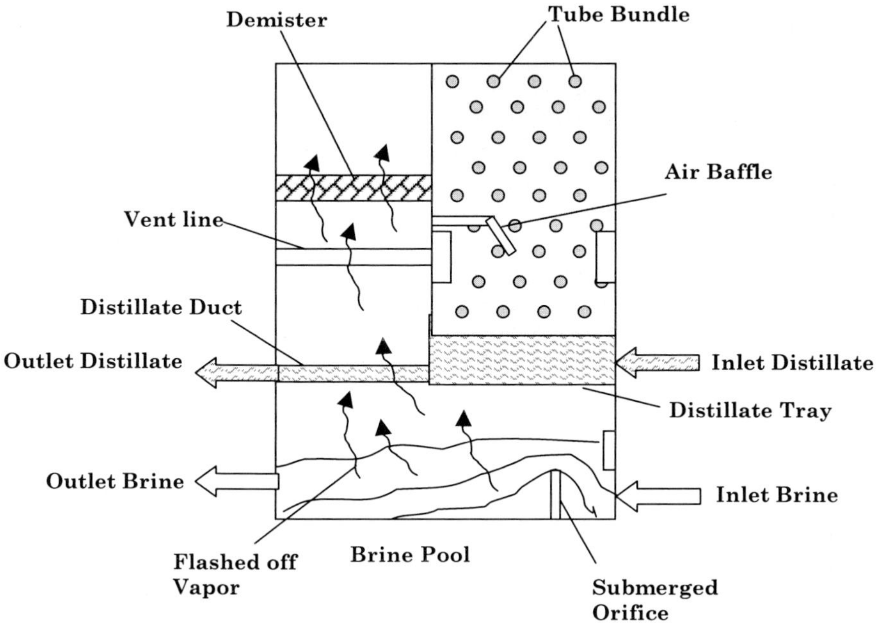

Fig. 2. MSF flashing stage

6.2.1 Condenser/Preheater Tubes

The condenser/preheater tubes are used to recover and reject heat in the MSF process. In the stages forming the heat recovery section, heat is recovered from the condensation of the flashed off vapor to heat the brine recycle stream flowing on the tube side. This heat recovery is essential in obtaining a high thermal performance ratio. In the heat rejection stages, the feed and cooling seawater are heated by absorbing the latent heat of the condensing flashed off vapor. Accordingly, the feed seawater is heated to a temperature equal to the temperature of the brine in the last flashing stage. This is necessary to prevent thermal shock upon mixing of the heated feed stream in the brine pool of the last stage. A thermal shock would result in decomposition of the calcium bicarbonate and calcium carbonate precipitation, which known as the soft scale. The decomposition process is also associated with release of carbon dioxide gas, which would promote corrosion reaction and increases the load on vacuum ejectors. The function of the cooling seawater stream in the heat rejection section is to remove the excess heat added to the system in the brine heater.

The heat transfer area of the condenser tube is a major design feature that controls the temperature of the brine recycle entering the brine heat. This parameter has a strong effect on the system performance ratio. If the heat transfer area is smaller than the thermal load of the condensing vapor, the stage pressure will increase due to accumulation of the non-condensed vapors. This pressure increase will reduce the amount of flashed of vapor. Eventually, the system will reach a new steady state with lower flashing rates and smaller flow rate of the distillate product. Also, the temperature of the brine recycle entering the brine heater will become smaller, which will result in increase of the required amount of heating steam and reduce of the system performance ratio. Although, initial system design provides sufficient heat transfer within various stages; however, poor operation and increase in fouling resistance will reduce the heat transfer coefficient and will create conditions, where the overall heat transfer rate is lower than the design thermal load. However, tube blockage may result in a similar result.

6.2.2 Tube Materials

Table 1 shows properties of the characteristics of preheater/condenser tubes used in the MSF process. As is shown, material selection depends on the stage temperature. In this regard, Cu/Ni 70/30 is used in stages with temperatures higher than 80 °C. On the other hand, in stages with lower temperatures a number of materials can be used, which includes Cu/Ni 90/10, aluminum brass, high steel alloys, and titanium. The highest thermal conductivity among these materials is the Cu/Ni 90/10 with a value of 44×10^{-3} kW/m °C. On the other hand, titanium tube provides the highest erosion resistance and the lowest wall thickness. Aluminum bronze provides a cheaper material, however, its copper content dissolves in the seawater and has an adverse impact on the receiving water bodies. The same problem is also found in other types of copper based tubes. In this regard, titanium, although more expensive than the copper alloys, it does not dissolve in the seawater.

6.2.3 Tube Configuration

The first tube configuration in the MSF system is the long tube arrangement, Fig. 3. In the configuration, the tubes are aligned in the same direction as the brine flow. Depending, on the available tube length, a single tube can span more than one stage. The limit on the tube length is imposed manufacturing companies and associated technical difficulties in transportation and handling. On-site tube welding may prove to be useful in constructing the required tube length. In practice, available long tube configurations are limited a maximum length of 28 m, which encompass on average 8-10 flashing stages. Features of the long tube configuration include the following:

- Fouling, blocking, and scaling of a single tube have a strong impact on the system performance, since; loss of a single tube implies reduction of the heat transfer in 8-10 stages for the lost tube.
- In maintenance and cleaning, tube removal is not a simple task. Also, conventional mechanical cleaning, blasting, would require specially designed equipment.
- Expansion consideration for long tubes requires special consideration in stage design.
- Vapor leakage between stages is a serious problem that needs special consideration in design, installation, and during maintenance and cleaning procedures.
- The main advantage for the long tube configuration is the reduction of the tube pressure drop by a factor of 25-30%. This reduces the associated pumping power.
- The long tube configuration can be thought as the optimum choice for plants with capacities higher than 50,000 m^3/d.
- Long tube configuration eliminates the water boxes on both sides of the flashing chamber, which is found in the cross tube configuration.

The second configuration is the cross tube arrangement, where the tubes are arranged in perpendicular direction to the brine flow, Fig. 4. This is a common configuration and is found in most of the MSF plants. Huge field experience is accumulated over the years for the cross tube configuration and it includes design, installation, operation, maintenance, and replacement. Also, less technical experience is required for construction, maintenance, and removal than the long tube system. Tube expansion in this system does not represent a problem in design and construction. The main disadvantage of this system is the need for installing water boxes on both ends of the tubes to transfer the brine recycle or feed seawater between the stages, which will increases the process capital, pressure drop, and pumping power.

Table 1
Properties of tube materials used in the brine heater and the condenser tubes

Material	Temperature	Thermal Conductivity kW/m °C	Brine Velocity m/s	Wall Thickness mm
Cu/Ni 70/30 (66% Cu, 30% Ni, 2% Fe, and 2% Mn)	Above 80 °C	29×10^{-3}	2-4	1.2
Aluminum Brass (76% Cu, 22% Zn, and 12% Al)	Below 80 °C	32×10^{-3}	1.5-2.5	1.2
Titanium	Below 80 °C	16.5×10^{-3}	3-20	0.5
Cu/Ni 90/10	Below 80 °C	44×10^{-3}	2-4	1.2
High Steel Alloy	Below 80 °C	19.9×10^{-3}	3-10	0.7

Table 2
Summary of fouling resistance in m^2 °C/W on the tube side

	Design value	Test data	Actual operation
Heat recovery section	1.5×10^{-4}	1.04×10^{-4}– 2.18×10^{-4}	0.68×10^{-4}– 2.23×10^{-4}
Heat rejection section	1.77×10^{-4}	2.42×10^{-4}	2.51×10^{-4}
Brine Heater	3.01×10^{-4}	1.49×10^{-4}	1.52×10^{-4}

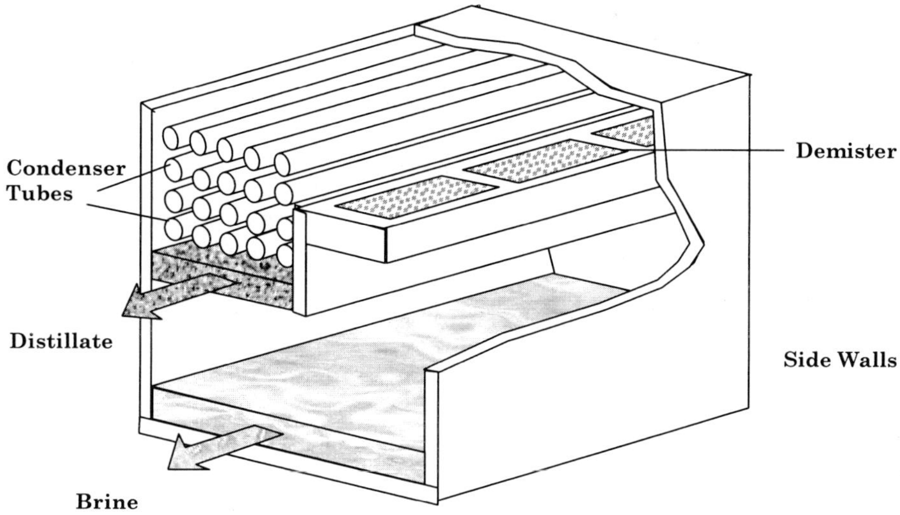

Fig. 3. Long Tube Configuration

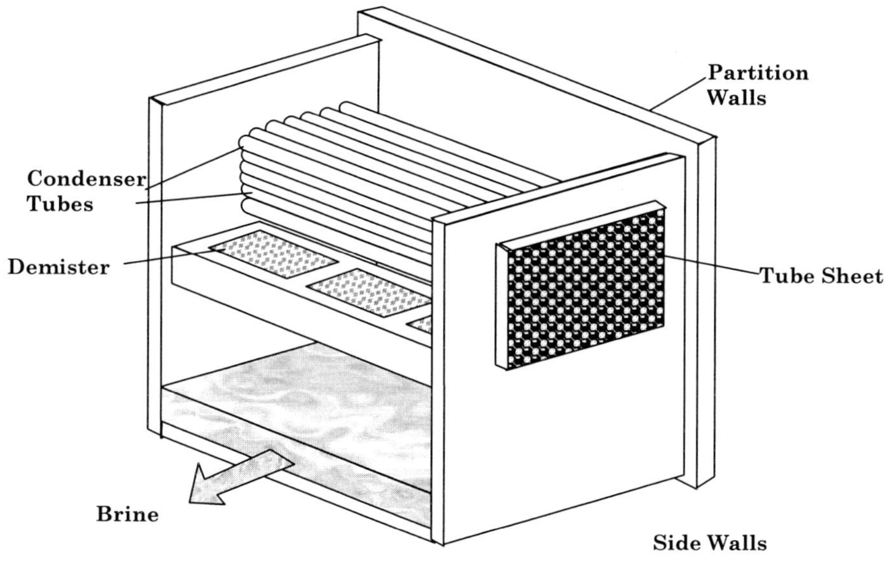

Fig. 4. Cross Tube Configuration

6.2.4 Features of MSF Brine Circulation Plants

The general features of the MSF plants include the following:

- Stable and reliable operation by insuring adequate heat transfer area, suitable materials, and proper corrosion allowance.
- The MSF desalination units operate with dual-purpose power generation plants. Design of the co-generation plants allows for flexible operation during peak loads for power or water. In Kuwait and the Gulf area peak loads for electricity and water occurs during the summer time due to the high ambient temperature, which is associated with massive use of indoor air-conditioning units and increase in domestic and industrial water consumption. The opposite is true for the winter season with mild temperatures and limited use of indoor heating units.
- The majority of the MSF plants are brine circulation, which are more superior to the once through design. Brine circulation result in higher conversion ratio, uses smaller amount of chemical additives, and gives good control on the temperature of the feed seawater.
- Cross tube design has simpler manufacturing and installation properties than long tube arrangement.
- All auxiliaries are motors driven have better operating characteristics than turbine drive units; even for the large brine circulation pumps.
- Additive treatment is superior to acid treatment, where acidic solutions may enhance corrosion rates of tubing, shells, and various metallic parts.
- Proper system design should allow for Load variations between 70-110% of the rated capacity.

6.3 MSF Process Synthesis

As discussed in Chapter 1 and the previous section, the MSF process accounts for more than 60% of the global desalination industry. In addition, it is the major source of fresh water in the Gulf countries. This section focuses on process fundamentals and developing better understanding for various elements forming the MSF process. The layout for the MSF process shown in Fig. 1 is quiet complex and understanding the functions and relations of various elements in the process is essential for successful system operation, analysis, optimization, and control. In addition, comprehensive analysis of the system flow sheet aids the development and design of novel and more efficient desalination processes. The following is a brief description of the plant flow diagram shown in Fig 1. The system includes three major sections: the brine heater, the heat recovery section, and the heat rejection section. The number of stages in the heat recovery section is larger than the heat rejection section. The brine heater drives the flashing process through heating the recycle brine stream to the top brine temperature. Flashing occurs in each stage, where a small amount of product water is generated and accumulated across the stages in the two sections. Vapor formation results because of the reduction of the brine saturation temperature; therefore, the stage temperature decreases from the hot to cold side of the plant. This allows for brine flow across the stages without the aid pumping power. The flashed off vapors condenses on the tubes of the preheater/condenser units. The released latent heat by the condensing vapor is used to preheat the brine recycle stream. On the cold side of the plant, the feed and the cooling seawater are introduced into the condenser/preheater tubes of the last stage in the heat rejection section. As this stream leaves the heat rejection section, the cooling seawater is rejected back to the sea and the feed seawater is mixed in the brine pool of the last stage in the heat rejection section. Also, two streams are extracted from the brine pool in this stage, which include the brine blow down and the brine recycle. The rejection of brine is necessary to control the salt concentration in the plant. As is shown, the brine reject is withdrawn prior to mixing of the feed seawater and the recycled brine is withdrawn from a location beyond the mixing point. The brine blow down is rejected to the sea and the brine recycle is introduced to the last stage in the heat recovery section. Additional units in the desalination plant include pretreatment of the feed and cooling seawater streams. Treatment of the intake seawater is limited to simple screening and filtration. On the other hand treatment of the feed seawater is more extensive and it includes dearation and addition of antiscalent and foaming inhibitors. Other basic units in the system include pumping units for the feed seawater and brine recycle. Also, gas-venting systems operate on flashing stages for removal of non-condensable gases.

From the above brief description, many questions arise regarding the specific arrangement of flashing stages and various streams. These questions include the following:

- Use of a large number of flashing stages.
- Upper limit on the top brine temperature.
- Need for two flashing sections (recovery and rejection).
- Minimum number of stages in the heat rejection section.
- The use of brine recycle.
- Function of the cooling seawater stream.

As mentioned before, complexity of the process makes it difficult for many people in the field to find the proper answers for the above questions. Therefore, finding the answers is pursued through simplifying the complicated MSF diagram to a number of simpler configurations. The simplest of these configurations is the single stage flashing system, Fig. 5. Results and analysis of this simple system are then used to modify the process diagram to a more detailed configuration, which solves the problems encountered in the simple system. As will be shown later, this process involved analysis of four simpler systems before reaching the conventional MSF system.

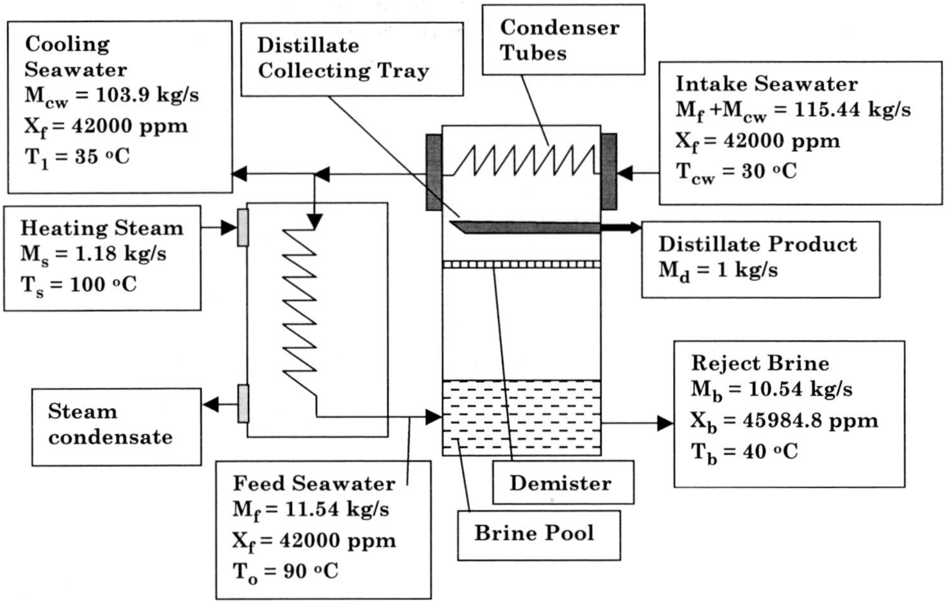

Fig. 5. Single stage flash desalination

The above task is achieved through mathematical modeling and analysis of various configurations. Performance of this type of analysis requires the use of analytical and non-iterative mathematical models rather than models based on numerical analysis. Examples for numerical models can be found in the studies performed by Omar, 1983, El-Dessouky and Assassa, 1985, Helal, et al. (1986), Darwish (1991), Montagna (1991), Hussain et al. (1993), Rosso et al. (1996), and El-Dessouky et al. (1995). This choice is necessary since analytical models generate closed form equations, which can be used to assess effect of various system parameters on the system performance. Of course use of analytical models should be made with caution to avoid simplifying assumptions that result in inaccurate predictions, which are not consistent with process characteristics.

Certainly, the following analysis will provide useful insights to the desalination community into the details of the MSF process. Results and analysis are also of great value to a number of the undergraduate and graduate engineering courses including plant design, process synthesis, modeling and simulation, energy conservation, flow sheet analysis, and of course water desalination. It is interesting to mention that the procedure outlined here can be used to analyze other complicated systems, i.e., multi-effect evaporation (MEE), distillation, etc.

6.3.1 Elements of Mathematical Analysis

To simplify the analysis procedures the following assumptions are used in development of various models:

- Distillate product is salt free. This assumption is valid since the boiling temperature of water is much lower than that of the salt.
- Specific heat at constant pressure, C_p, for all liquid streams, brine, distillate, and seawater is constant and equal to 4.18 kJ/kg °C.
- The overall heat transfer coefficient in the brine heater and preheaters is constant and equal to 2 kW/m^2 °C.
- Subcooling of condensate or superheating of heating steam has negligible effect on the system energy balance. This is because of the large difference of the vapor latent heat in comparison with the sensible heat value caused by liquid subcooling or vapor superheating of few degrees.
- The power requirements for pumps and auxiliaries are not included in the system analysis.
- The heat losses to the surroundings are negligible because the flashing stages and the brine heater are usually well insulated and operate at relatively low temperatures.

Performance analysis of various configurations is determined in terms of the following parameters:

- The thermal performance ratio, which is the ratio of distillate flow rate to the heating steam, $PR = M_d/M_s$.
- The specific heat transfer area is the ratio of the total heat transfer area to the distillate flow rate, $sA = A/M_d$.
- The specific feed flow rate is ratio of the feed to distillate flow rates, $sM_f = M_f/M_d$.
- The specific cooling water flow rate is the ratio of cooling water to distillate flow rates, $sM_{cw} = M_{cw}/M_d$

These variables are the most important factors controlling the cost of fresh water production.

The following data set is used to evaluate the performance of various configurations:

- Top brine temperature, T_o = 90 °C.
- Temperature of reject brine, T_b = 40 °C.
- Temperature of motive steam, $T_s = T_o$+10 °C.
- Temperature of intake seawater, T_{cw} = 30 °C.
- Thermodynamic loss, ΔT_{loss} = 2 °C.
- The condenser terminal temperature difference, TTD_c = 3 °C.

- The salinity of intake seawater, $X_f = 42,000$ ppm.
- The maximum attainable concentration of the rejected brine, $X_b = 70,000$ ppm. This value is imposed by scale formation limits of $CaSO_4$.

6.3.2 Single Stage Flashing

The basic elements of the SSF system include the brine heater and the flashing chamber, which contains the condenser/preheater tubes, the demister, the brine pool, and the collecting distillate tray. A schematic diagram for the unit is shown in Fig. 5. As is shown, saturated steam at a flow rate of M_s and a temperature of T_s drives the single unit. The heating steam condenses outside the tubes of the brine heater, where it releases its latent heat, λ_s. This energy increases the feed seawater temperature from T_1 to the top brine temperature, T_o. The type of chemical additive that used to control scale formation dictates the upper limit on T_o. For acid and modern chemical additives, the limit on T_o is 120 °C and for polyphosphate the limit is 90 °C. The hot brine enters the flashing chamber, which operates at a pressure lower than the saturation pressure corresponding to the temperature of the brine flowing into the stage, T_o. In other words, the temperature difference of T_o-T_1 gives the degree of superheating of the brine as it flows to the flashing stage. During the flashing process, a part of the sensible heat of the brine is changed to latent heat by evaporation of a small portion of the brine, M_d. Distillate formation also results in the increase of the brine salinity from X_f to X_b. The formed vapor flows through the demister pad and then releases its latent heat, λ_v, as it condenses on the seawater condenser/preheater tubes. The condensed vapor is collected on the distillate tray. The latent heat of condensation is transferred to the intake seawater, $M_{cw}+M_f$, and increases its temperature from T_{cw} to T_1. The cooling seawater, M_{cw}, is rejected and the feed seawater, M_f, is introduced into the brine heater. Recovery of the latent heat by the feed seawater improves the overall efficiency of the desalination process. This reduces the amount of heating steam required in the brine heater, since the feed seawater temperature is increased in the brine heater from T_1 to T_o instead of T_{cw} to T_o. From a thermodynamic point of view, the function of the feed seawater preheater is the recovery of the energy added to the system by the heating steam in the brine heater. Also, it controls the saturation pressure inside the flashing chamber. On the other hand, the function of the demister is the removal of any brine droplets entrained with the flashed off vapor. This is necessary to avoid product contamination and lowering of its quality. In addition, removal of entrained brine protects the preheater tubes from fouling.

The mathematical model for the single flash unit is simple and it includes total mass and salt balances, rate equations for the heat transfer units, as well as energy balances for the brine heater and the condenser. The total mass and salt balances are

$$M_f = M_b + M_d \tag{1}$$

$$X_f M_f = X_b M_b \tag{2}$$

Eq. 2 assumes that the salt concentration, X_d, in the formed vapor is zero. The brine heater and condenser energy balances are given respectively by

$$M_s \lambda_s = M_f C_p (T_o - T_1) \tag{3}$$

$$M_d \lambda_v = (M_{cw}+M_f) C_p (T_1 - T_{cw}) = M_f Cp (T_o - T_b) \tag{4}$$

The heat transfer rate equations for the brine heater is

$$M_s \lambda_s = U_h A_h (LMTD)_h \tag{5}$$

where

$$(LMTD)_h = (T_o - T_1)/\ln((T_s - T_1)/(T_s - T_o)) \tag{6}$$

The heat transfer rate equation for the condenser is

$$M_d \lambda_v = U_c A_c (LMTD)_c \tag{7}$$

where

$$(LMTD)_c = (T_1 - T_{cw})/\ln((T_v - T_{cw})/(T_v - T_1)) \tag{8}$$

In the above system of equations, A is the heat transfer area, Cp is the specific heat at constant pressure, M is the mass flow rate, T is the temperature, X is the salinity of seawater and brine, λ is the latent heat of evaporation. The subscripts 1, b, c, cw, d, h, s, and v refer to stage number, brine, condenser, intake seawater, distillate, brine heater, steam, and vapor, respectively.

The unit thermal performance ratio, defined as the mass ratio of fresh water produced per unit mass of heating steam, is obtained by dividing Eqs. 4 and 3, where

$$PR = \frac{M_d}{M_s} = \frac{M_f C_p (T_o - T_b)\lambda_s}{M_f C_p (T_o - T_1)\lambda_v}$$

which simplifies to

$$PR = \frac{M_d}{M_s} = \frac{(T_o - T_b)\lambda_s}{(T_o - T_1)\lambda_v} \tag{9}$$

The stage temperature drop (ΔT_{st}) is equal to the difference (T_o – T_b) and is known as the flashing range. On the other hand, the term (T_o – T_1), as is shown in Fig. 6, is equal to the sum of the stage temperature drop (ΔT_{st}) the stage terminal temperature difference, TTD_c, and the thermodynamic losses, ΔT_{loss}, or,

$(T_o - T_b) = \Delta T_{st}$, and

$(T_o - T_1) = \Delta T_{st} + \Delta T_{loss} + TTD_c$

The thermodynamic losses (ΔT_{loss}) are the temperature difference of the brine leaving the stage, T_b, and the condensation temperature of the vapor, T_v. In a single stage flashing unit, these losses are caused by the boiling point elevation, the non-equilibrium allowance, and the temperature drop corresponding to the pressure drop in the demister pad and during condensation. The terminal temperature difference of the condenser, TTD_c, is equal to temperature difference of the condensing vapor, T_v, and the seawater leaving the condenser, T_1. The value of TTD_c plays a very important role in the design of the MSF system and its value ranges between 3-5 °C. In the brine heater, the temperature difference of the condensing steam and the effluent brine gives the brine heater terminal temperature difference, TTD_h.

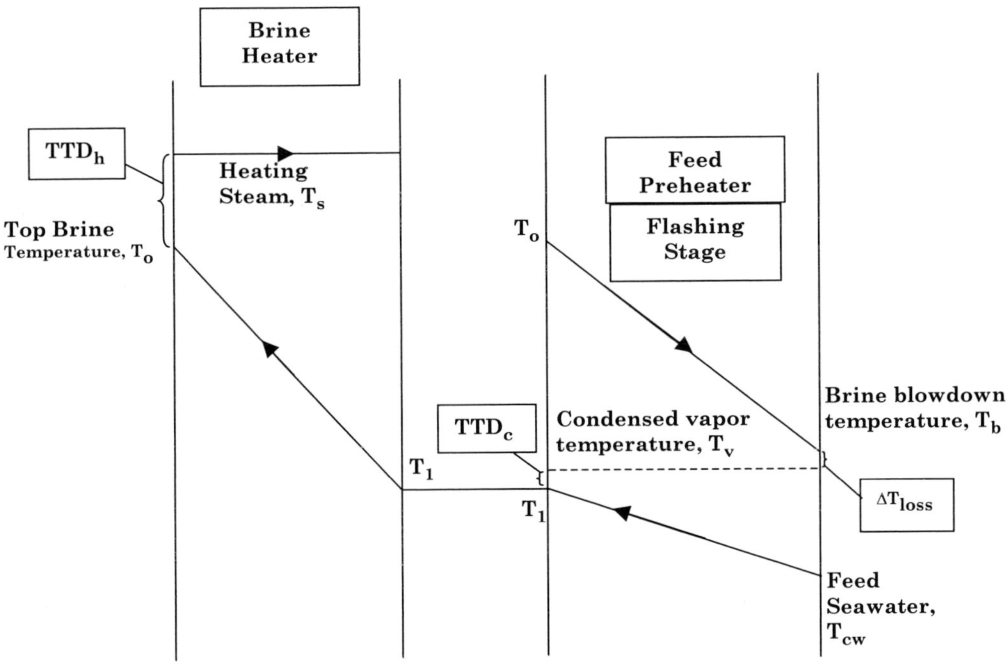

Fig. 6. Temperature profiles of heating steam, seawater, flashing brine, and condensate

The previous relations are substituted in the single stage model equations, Eqs. 3, 4, 6, 8, and 9. This results in the following equations, which are used to determine the single effect performance,

$$M_s \lambda_s = M_f C_p (\Delta T_{st}+\Delta T_{loss}+TTD_c) \tag{10}$$

$$M_d \lambda_v = M_f C_p \Delta T_{st} = (M_f+M_{cw}) C_p (T_o-\Delta T_{st}-\Delta T_{loss}-TTD_1-T_{cw}) \tag{11}$$

$$(LMTD)_h = (\Delta T_{st}+\Delta T_{loss}+TTD_c)/\ln((TTD_h+\Delta T_{st}+\Delta T_{loss}+TTD_c)/(TTD_h)) \tag{12}$$

$$(LMTD)_c = (\Delta T_{st})/\ln((\Delta T_{st}+TTD_c)/(TTD_c)) \tag{13}$$

$$PR = (\lambda_s)(\Delta T_{st})/((\Delta T_{st}+\Delta T_{loss}+TTD_c)(\lambda_v)) \tag{14}$$

The above data are used to calculate the stage temperature drop,

$\Delta T_{st} = T_o - T_b = 90 - 40 = 50\ ^oC$,

The terminal temperature difference for the brine heater,

$TTD_h = T_s - T_o = 100 - 90 = 10\ ^oC$,

The temperature of formed vapor,

$T_v = T_b - \Delta T_{loss} = 40 - 2 = 38\ ^oC$,

The latent heat of condensing vapor,

$\lambda_v = 2412.5$ kJ/kg (at 38 °C),

The latent heat of steam,

$\lambda_s = 2256$ kJ/kg (at 100 °C).

The performance ratio of the single stage flashing unit can be obtained from Eqs. 14, where

$$\begin{aligned} PR &= (\lambda_s)(\Delta T_{st})/((\Delta T_{st}+\Delta T_{loss}+TTD_c)(\lambda_v)) \\ &= ((2256)(50))/((50+2+3)(2412.5)) \\ &= 0.85 \text{ kg distillate/kg steam} \end{aligned}$$

The specific flow rates of the feed and cooling seawater are obtained from Eq. 11, where

$$M_d\ \lambda_v = M_f\ C_p\ \Delta T_{st}$$

$$\begin{aligned} M_f/M_d &= \lambda_v/(C_p\ (\Delta T_{st})) \\ &= 2412.5/((4.18)(50)) \\ &= 11.54 \text{ kg feed seawater/kg distillate} \end{aligned}$$

As for the specific flow rate of the cooling water it is given by

$$M_d\ \lambda_v = (M_f+M_{cw})\ C_p\ (T_o-\Delta T_{st}-\Delta T_{loss}-TTD_c-T_{cw})$$

which simplifies to

$$\begin{aligned} M_{cw}/M_d &= \lambda_v/(C_p\ (T_o-\Delta T_{st}-\Delta T_{loss}-TTD_c-T_{cw})) - (M_f/M_d) \\ &= (2412.5)/((4.18)(90-50-2-3-30)) - 11.54 \\ &= 103.9 \text{ kg cooling seawater/kg distillate} \end{aligned}$$

This result together with specification of the amount of distillate water can be used to calculate the brine flow rate and its salinity. Assuming that the distillate flow rate is equal to 1 kg/s gives

$M_b = 11.54 - 1 = 10.54$ kg brine/kg distillate

and

$$X_b = X_f\, M_f/M_b$$
$$= (42000)\,(11.54)/(10.54)$$
$$= 45984.8 \text{ ppm}$$

The specific heat transfer areas for the brine heater and the condenser are calculated from Eqs. 5, 7, 12, 13. Eqs. 12 and 13 are used to calculated the LMTD values in each unit, where,

$$(LMTD)_c = \Delta T_{st} / \ln((\Delta T_{st} + TTD_c)/(TTD_c))$$
$$= 50/\ln((50+3)/3)$$
$$= 17.4\ {}^{o}C$$

and

$$(LMTD)_h = \frac{\Delta T + \Delta T_{loss} + TTD_c}{\ln((TTD_h + \Delta T + \Delta T_{loss} + TTD_c)/(TTD_h))}$$
$$= (50+2+3)/\ln((10+50+3+2)/(10))$$
$$= 29.4\ {}^{o}C$$

The specific heat transfer area for the condenser and the brine heater are then calculated from Eqs. 5 and 7, where,

$$A_h/M_d = ((M_s)(\lambda_s))/((M_d)(U_h)(LMTD)_h)$$
$$= 2256/((0.85)(2)(29.4))$$
$$= 45.1 \text{ m}^2/(\text{kg/s})$$

and

$$A_c/M_d = \lambda_v/((U_c)(LMTD)_c)$$
$$= 2412.5/((2)(17.4))$$
$$= 69.3 \text{ m}^2/(\text{kg/s})$$

The above model and results shows that the main drawbacks of the single stage flash unit are:

- The performance ratio of the single stage flashing unit is always less than one. This result implies that the amount of distillate water produced is less than the amount of heating steam. Thus, the single stage flash unit can not be used on industrial scale
- The flow rate of feed seawater is much larger than the amount of distillate generated. This ratio is above ten. Therefore, a high rate of chemical additives and treatment for the feed/unit product is high.
- The specific flow rate of cooling water is very high. This would increase first cost of intake seawater pumping unit and power consumption.

On the other hand, the single stage flash unit is characterized by:

- The salinity of reject brine is much smaller than allowable design value set by $CaSO_4$ solubility limits.
- The heat transfer area for the brine heater and condenser are small, because of the large temperature driving force.
- The specific heat transfer area for the brine heater is inversely proportional to the performance ratio.
- The thermodynamic losses affect the area of the brine heater, however, it has no effect on the condenser area.

It is important to emphasis that most of the heat added to the system is rejected with the cooling seawater. In other words, the flashing stage does not consume most of the energy provided by the heating steam, but simply it degrades its quality.

6.3.3 Once Through MSF

The objective of the once through MSF system is to overcome the main drawback of the single stage flash unit that is to improve the system performance ratio. This is achieved by dividing the flashing range over a larger number of stages and as a result reducing the stage temperature drop.

A schematic diagram for the system is shown in Fig. 7. As is shown, the system includes a number of stages, n, and the brine heater. The stage elements are identical to those of the single stage unit. In the once through system, the temperature of intake seawater is increased as it flows through the preheater tubes of each stage from T_{cw} to t_1. The intake seawater flows from stage n to stage 1, i.e., from the low temperature to the high temperature side of the plant. The seawater leaving the last condenser enters the brine heater, where its temperature is increased from t_1 to T_o. The heated brine flashes off as it flows through the successive stages, where its temperature decreases from T_o to T_n. Simultaneously, the flashing vapor condenses around the condenser tubes in each stage, where it heats the seawater flowing through the tubes. The collected distillate in the distillate-collecting tray flows across the stages, where it leaves the plant from stage n. The flashing process reduces the brine temperature and increases its salinity from X_f to X_b. The brine leaving the last stage is rejected to the sea. This designates the process as the once through system, since no recycle of any portion of the unevaporated brine is made into the system. It is worth mentioning that the MSF-OT system does not contain a cooling water stream. This is because the reject brine stream, which has a low temperature and a large flow rate, contains the energy that must be removed from the system.

The performance of the once through system is developed in a similar manner to that of the singe stage unit. For this system, the energy balance of the brine heater is identical to Eq. 10. The amount of distillate formed in stage i is determined approximately from the energy balance on unit. This is

$$D_i \, \overline{\lambda}_v = M_f \, Cp \, (t_{i+1} - t_i) = B_{i-1} \, Cp \, (T_{i-1} - T_i)$$

where $\overline{\lambda}_v$ is the average latent heat of vapor condensation, evaluated at $T_{av} = (T_o+T_n)/2$. It should be noted that in the previous equation the brine flow rate in the first stage is equal to feed seawater flow rate, M_f. Summation of the above equation for all stages gives the total amount of distillated

$$M_d = \sum_{i=1}^{n} D_i = M_f \, C_p \, (T_o - T_n)/\overline{\lambda}_v$$

The flashing range term $(T_o - T_n)$ in the above equation is replaced by the stage temperature drop, which gives

$$M_d = \sum_{i=1}^{n} D_i = M_f \, C_p \, (n\Delta T_{st})/\overline{\lambda}_v \qquad (15)$$

Division of Eqs. 15 and 10 gives the performance ratio for the once through system, where,

$$PR = \frac{M_d}{M_s} = \frac{M_f C_p (n\Delta T_{st}) \lambda_s}{M_f C_p (\Delta T_{st} + \Delta T_{loss} + TTD_c) \overline{\lambda}_v}$$

Simplifying the above equation gives

$$PR = \frac{M_d}{M_s} = \frac{(n\Delta T_{st}) \lambda_s}{(\Delta T_{st} + \Delta T_{loss} + TTD_c) \overline{\lambda}_v} \qquad (16)$$

Eq. 7 is modified to calculate the condenser heat transfer area in each flashing stage, where

$$M_f\, C_p\, \Delta T_{st} = U_c\, A_c\, (LMTD)_c \tag{17}$$

The remainder of the model equations is similar to those of the single stage unit, where, X_b and M_b are obtained from Eqs. 1 and 2, A_h is calculated by Eq. 5, and the LMTD values are determined from Eqs. 12 and 13.

For the same set of system specifications, given for the single stage unit, the performance characteristics of the once through system are calculated for a 23-stage plant. The temperature drop per stage is first determined, where

$$\Delta T_{st} = (90\text{-}40)/23 = 2.174\ ^oC$$

The latent heat of condensation is calculated at the average stage temperature, $T = (T_o + T_n)/2$, which equal to 65 °C. The latent heat value at this temperature is 2346.5 kJ/kg. The performance ratio for the system is calculated from Eq. 16,

$$\begin{aligned} PR &= \frac{n\Delta T_{st}\ \lambda_s}{\left(\Delta T_{st} + \Delta T_{loss} + TTD_c\right)\overline{\lambda}_v} \\ &= (23)(2.174)(2256)/((2.174+2+3)(2346.5)) \\ &= 6.7 \text{ kg distillate water/kg steam} \end{aligned}$$

The specific feed flow rate is obtained from Eq. 17, where

$$\begin{aligned} M_f/M_d &= \overline{\lambda}_v /(C_p\ (T_o - T_n) \\ &= 2346.5/((4.18)(90\text{-}40)) \\ &= 11.22 \text{ kg intake seawater/kg distillate} \end{aligned}$$

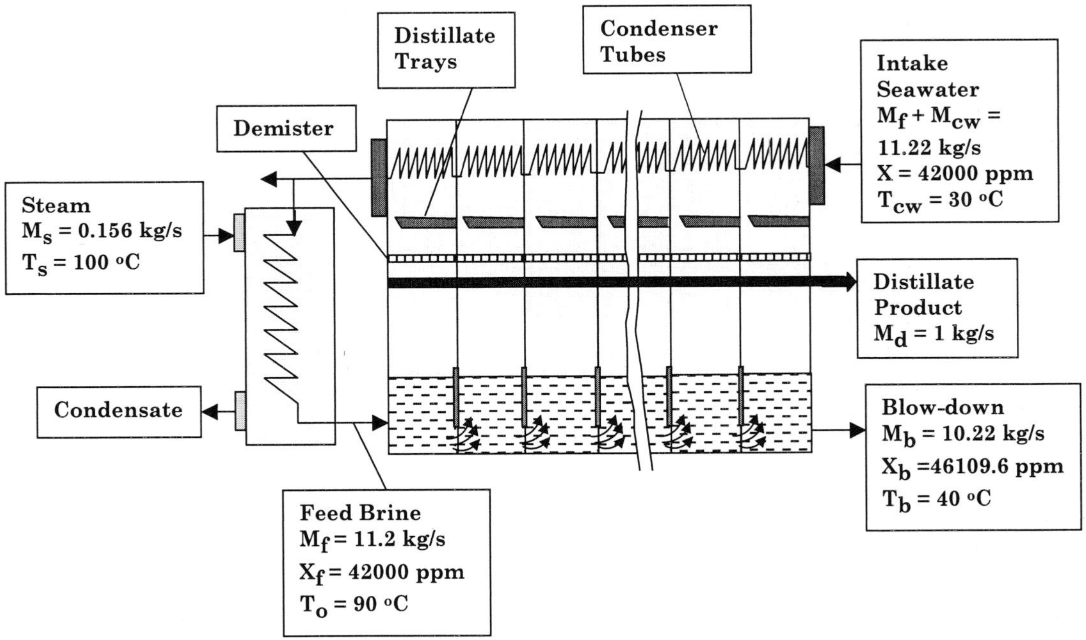

Fig. 7. Once-through multi-stage flash desalination process

This value together with specification of the amount of distillate water is used to calculate the rejected brine flow rate and its salinity. Assuming that the distillate flow rate is equal to 1 kg/s gives

$M_b = 11.22 - 1 = 10.22$ kg/s

and

$$X_b = X_f M_f/M_b$$
$$= (42000)(11.22)/(10.22)$$
$$= 46106.6 \text{ ppm}$$

The LMTD values are calculated from Eqs. 12 and 13, where,

$$(LMTD)_c = \Delta T_{st}/\ln((\Delta T_{st}+TTD_c)/(TTD_c))$$
$$= 2.174/\ln((2.174+3)/3)$$
$$= 3.98\ {}^{o}C$$

and

$$(LMTD)_h = \frac{\Delta T_{st} + \Delta T_{loss} + TTD_c}{\ln((TTD_h + \Delta T_{st} + \Delta T_{loss} + TTD_c)/(TTD_h))}$$
$$= (2.174+2+3)/\ln((10+2.174+2+3)/(10))$$
$$= 13.27\ {}^{o}C$$

The specific heat transfer area for the condenser and the brine heater are then calculated from Eqs. 5 and 17, where,

$$A_h/M_d = M_s \lambda_s/((M_d)(U_h)(LMTD)_h)$$
$$= 2256/((6.7)(2)(13.27))$$
$$= 12.7 \text{ m}^2/(\text{kg/s})$$

and

$$A_c/M_d = M_f C_p \Delta T_{st}/((U_c)(LMTD)_c)$$
$$= (11.22)(4.18)(2.174)/((2)(3.98))$$
$$= 12.8 \text{ m}^2/(\text{kg/s})$$

The above heat transfer areas are used to calculate the total specific heat transfer area, where

$$sA = A_h + n A_c$$
$$= 12.7 + (23)(12.8)$$

$= 307.1\ m^2/(kg/s)$

The above results show that the MSF-OT system is distinguished from the single stage unit by the drastic increase in the thermal performance ratio, which increases from a value below one for the single stage flash unit to a value above six for the once through system. Also, the MSF-OT system does not use cooling water for removal of excess heat added in the brine heater. Irrespective of this improvement, the MSF-OT still has a high flow rate ratio for the intake seawater to the distillate product. Also, the total specific heat transfer area for the condenser increases from 69.3 $m^2/(kg/s)$ for the single stage unit to 294.4 $m^2/(kg/s)$ in the once through MSF. This is caused by the decrease in the temperature driving force from 17.4 °C in the single unit to 3.98 °C in the once through system.

Other features of the MSF-OT system include the following:

- Operation at low salinity for the feed and flashing brine streams, with values of 42000 and 46106.6 ppm, respectively. This reduces fouling and scaling problems in the condenser tubes and the brine heater and the boiling elevation, which reduces the value of the thermodynamic losses.
- The specific heat transfer area for the brine heater is inversely proportional to the performance ratio. This is because the heat added to the system in the brine heater is divided over a larger amount of distillate water. Therefore, the increase in the performance ratio for the once through system reduces the specific heat transfer area from 45.1 $m^2/(kg/s)$ found in the single unit to 12.7 $m^2/(kg/s)$ for the once through system.

6.3.4 Brine Mixing MSF

The purpose of brine recirculation is to decrease the flow rate of the feed seawater. As a result, this lowers the chemical additive consumption rate and the size of the pretreatment facilities for the feed stream. Also, since the recycled brine contains higher energy than the feed seawater, the process thermal efficiency will improve. The simplest brine circulation system is made through mixing part of the blow-down brine with the feed stream. This simple configuration is shown in Fig. 8. In this system, a portion of the blow-down brine, M_r-M_f, is mixed with the intake seawater stream, M_f. The resulting mixture, M_r, has a higher salinity and temperature than the intake seawater. The remaining elements of the system are similar to those of the once through MSF.

The brine recycle system allows for achieving the maximum limit on the salinity of the blow-down brine, which is equal to 70000 ppm. Assuming that the distillate is salt free, the salinity of the feed seawater is 42000, and the distillate flow rate is equal to 1 kg/s, gives

$70000 (M_b) = 42000 (M_b + 1)$

This results in M_b = 1.5 kg/s and M_f = 2.5 kg/s. Thus, the brine recirculation reduces the specific feed flow rate from a high value above 11 in the once through system to a lower value of 2.5 in the new system.

The thermal performance ratio of the system is identical to Eq. 16. The system variables, which include T_n, t_1, T_r, and M_r, are determined by performing the following:

- Overall energy balance,

$$M_s \lambda_s = M_b C_p (T_n - T_{cw}) + M_d Cp (T_d - T_{cw}) \quad (18)$$

- Energy balance on the brine heater,

$$M_s \lambda_s = M_r C_p (T_o - t_1) \quad (19)$$

- Overall balance on the flashing brine,

$$M_r C_p (T_o - T_n) = M_d \overline{\lambda}_v \quad (20)$$

- Energy balance on the mixing point,

$$(M_r - M_f) C_p (T_n - T_{cw}) = M_r C_p (T_r - T_{cw}) \quad (21)$$

Eqs. 18 and 19 are combined to express M_r in terms of M_d, where,

$$M_r C_p (T_o - t_1) = (X_f/(X_b\text{-}X_f)) M_d C_p (T_n - T_{cw}) + M_d Cp (T_d - T_{cw})$$

Neglecting the effect of the thermodynamic losses in the last stage, i.e., setting $T_n = T_d$, simplifies the above equation to

$$M_r C_p (T_o - t_1) = (1+X_f/(X_b\text{-}X_f)) M_d C_p (T_n - T_{cw})$$

Or

$$M_r C_p (T_o - t_1) = (X_b/(X_b\text{-}X_f)) M_d C_p (T_n - T_{cw}) \quad (22)$$

Eliminating M_r from Eqs. 20 and 22 gives

$(M_d \overline{\lambda}_v)(T_o - t_1)/(T_o - T_n) = (X_b/(X_b\text{-}X_f))\ M_d\ C_p\ (T_n - T_{cw})$

which reduces to

$(\overline{\lambda}_v)(T_o - t_1)/(T_o - T_n) = C_p\ (X_b/(X_b\text{-}X_f))\ (T_n - T_{cw})$

Recalling that the ratio of $(T_o - t_1)/(T_o - T_n)$ is approximately equal to 1/PR, changes the above equation to the following form

$$PR = \overline{\lambda}_v /(C_p\ (X_b/(X_b\text{-}X_f))\ (T_n - T_{cw})) \qquad (23)$$

The same procedure is applied to Eq. 21, where M_r and M_f are expressed in terms of M_d. The resulting relation is

$$(\overline{\lambda}_v)(T_n - T_f)/(T_o - T_n) = C_p\ (X_b/(X_b\text{-}X_f))\ (T_n - T_{cw}) \qquad (24)$$

Division of Eqs. 23 and 24, show that

$(T_n - T_f) = (T_o - t_1)$

or

$$(t_1 - T_f) = (T_o - T_n) \qquad (25)$$

This relation implies that the total temperature drop of the flashing brine is equal to the total temperature increase of the brine recycle flowing inside the preheater/condenser tubes.

Iterative solution of the above system of equations is dictated by the dependence of $\overline{\lambda}_v$ on the value of T_n. However, a simple solution procedure can provide quick estimate for the system variables. Starting with Eq. 23 and by assuming that the system performance ratio PR is equal to 8, and the average latent heat of condensing vapor is assumed equal to 2346.5 kJ/kg (at 65 °C), gives

$(\overline{\lambda}_v) = (PR)\ C_p\ (X_b/(X_b\text{-}X_f))\ (T_n - T_{cw})$
$(2346.5) = (8)\ (4.18)\ (70000/(70000\text{-}42000))\ (T_n - 30)$

Solution of the above equation gives

$T_n = 58\ ^oC$.

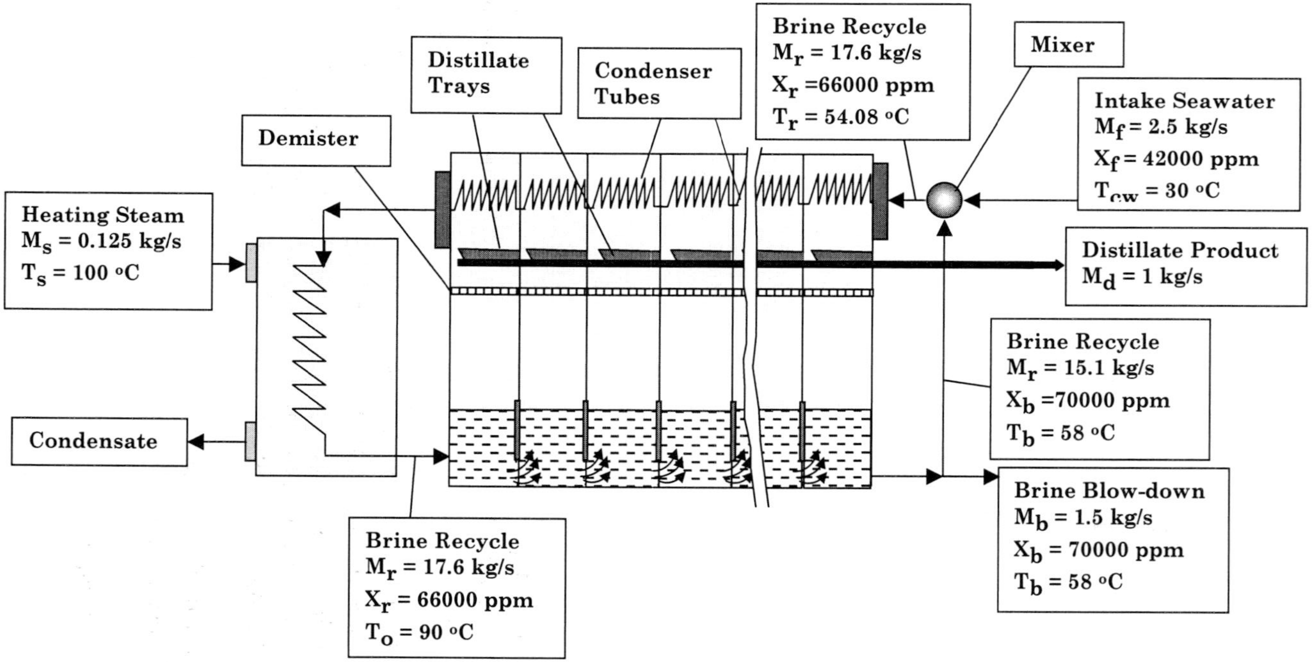

Fig. 8. Simple mixer brine recycle MSF desalination process

This value results in a stage temperature drop of (90-58)/23 or 1.39 °C. The flow rate of recycled brine, M_r, is obtained from Eq. 21, where

$$M_r C_p (n\Delta T_{st}) = M_d \overline{\lambda}_v$$

$$M_r (4.18) (23) (1.39) = (1) (2346.5)$$

This gives

$$M_r = 17.6 \text{ kg/s}$$

The flow rate of the recycle brine can be used to calculate the salinity of the feed seawater, X_r. A salt balance at the mixer is given by

$$M_r X_r = X_f M_f + X_n (M_r - M_f)$$

$$X_r (17.6) = (42000) (2.5) + (70000) (17.6 - 2.5)$$

Therefore, X_r is equal to 66018 ppm. The feed seawater temperature, T_r, is calculated from the mixer energy balance, Eq. 22, where

$$(M_r - M_f) C_p (T_n - T_{cw}) = M_r C_p (T_r - T_{cw})$$

$$(17.6 - 2.5) (4.18) (58 - 30) = (17.6) (4.18) (T_r - 30)$$

which gives

$$T_r = 54.08 \text{ °C}$$

The temperature of feed seawater leaving the first stage is determined from Eq. 25, where,

$$(t_1 - T_f) = (T_o - T_n)$$

$$(t_1 - 54.08) = (90 - 58.07)$$

This gives

$$t_1 = 86.01 \text{ °C}$$

The LMTD values are calculated from Eqs. 12 and 13, where,

$$(LMTD)_c = \Delta T_{st}/\ln((\Delta T_{st}+TTD_c)/(TTD_c))$$
$$= 1.39/\ln((1.39+3)/(3))$$
$$= 3.65\ ^oC$$

and

$$(LMTD)_h = \frac{\Delta T_{st} + \Delta T_{loss} + TTD_c}{\ln((TTD_h + \Delta T_{st} + \Delta T_{loss} + TTD_c)/(TTD_h))}$$
$$= (1.39+2+3)/\ln((10+1.39+2+3)/(10))$$
$$= 12.93\ ^oC$$

The specific heat transfer area for the condenser and the brine heater are then calculated from Eqs. 5 and 17, where,

$$A_h/M_d = \lambda_s/((PR)(U_h)(LMTD)_h)$$
$$= 2256/((8)(2)(12.93))$$
$$= 10.9\ m^2/(kg/s)$$

and

$$A_c/M_d = M_r\ C_p\ \Delta T_{st}/((U_c)(LMTD)_c)$$
$$= (17.58)(4.18)(1.39)/((2)(3.65))$$
$$= 13.97\ m^2/(kg/s)$$

This value gives a total specific heat transfer area of

$$sA = A_h + n\ A_c$$
$$= 10.9 + (23)\ (13.97)$$
$$= 332.3\ m^2/(kg/s)$$

The main result of the above analysis is the high temperature of the rejected brine flowing from the last flashing stage, which is larger than the practical limit of 40 °C. This reduces the flashing range of the system and in turn results in the following:

- Increase of the recycle brine flow rate per unit of distillate product. This is necessary to account for the reduction in the temperature flashing range, $T_o - T_n$, across the stages.
- Increase of the specific pumping power for the recycle brine.
- Increase of the salt concentration in the first stage. The feed seawater salinity is quite high, 66018 ppm. This value would result in severe operational problems, because of enhanced formation rates for scale and fouling in the preheater/condenser tubes in the flashing stage close to the brine heater. Also,

the high salinity increases the thermodynamic losses, which is caused by the boiling point elevation.

- The temperature drop per stage is low, 1.39 °C, in comparison with 2.174 °C, found in the once through MSF unit. This reduces the temperature driving force for heat transfer and results in the increase of the increase of the heat transfer area of preheater/condenser tubes.

6.3.5 MSF with Brine Recirculation and a Heat Rejection Section

The main aim of adding heat rejection section is the removal of the excess heat added to the system in the brine heater. The heat rejection section is used to control the temperature of the recycled brine. This is achieved through recovery of a controlled amount of energy from the flashing brine into the brine recycle and rejection of the remaining energy into the cooling seawater stream, Fig. 9. A major advantage of the heat rejection section is the reduced pretreatment applied to the large stream of intake seawater, which requires inexpensive screening and filtration. Therefore, extensive pretreatment, which includes deareation, anti-foam, and anti-scalent additions, is applied only to the feed stream, which is less than the feed flow rate in the once through process.

In this analysis, the heat rejection section is assumed to contain a single stage. Figure 9 does not show the elements forming the brine heater and the heat recovery section, which are similar to those of the once through system. As for the heat rejection section, which includes a single stage, it receives the brine leaving the heat recovery section, where it flashes and generates a small amount of vapor. The vapor condenses around the preheater/condenser tubes, where its latent heat is absorbed by the intake seawater, M_f+M_{cw}. As a result, the intake seawater temperature is increased from T_{cw} to T_n. The cooling seawater stream, M_{cw}, is rejected, while the feed seawater stream is mixed in the brine pool within the stage. A portion of the brine is recycled and enters the tubes of the condenser tubes of the last stage in the heat recovery section. The remaining brine is returned back to the sea.

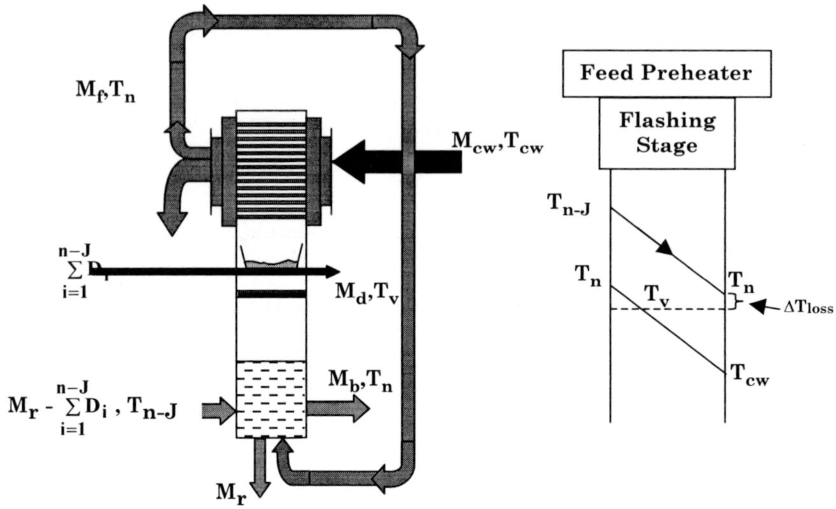

Fig. 9. Single stage heat rejection section and temperature profiles in the condenser

First, analysis is made for an MSF system containing a single stage heat rejection. The stage temperature profile, shown in Fig. 9, is used to determine applicability of this configuration. As is shown, the temperature of the intake seawater is increased from T_{cw} to T_n, where T_n is equal to flashing brine temperature in the stage. This temperature increase is necessary to avoid thermal shock upon mixing of the feed seawater and the flashing brine inside the stage. Since, the condensing vapor temperature, T_v, is less than the flashing brine temperature, T_n, by the thermodynamic losses. Therefore, the hot stream temperature profile, T_v, intersects the cold stream profile, T_{cw}-T_n, and as a result the required heat transfer area will be infinite. This simple analysis indicates that a single stage heat rejection section can not be implemented industrially.

The layout of the two-stage heat rejection and brine recycle MSF system is similar to conventional MSF, shown in Fig. 1, except for the number of stages in the heat rejection section. To determine feasibility of the two-stage system its heat transfer characteristics are compared against those of conventional MSF. Comparison of the two systems is based on performance of the condenser/preheater units in each stage and the value of the terminal temperature difference for these units. Common practice for similar heat exchange units puts a minimum value of 2 °C on the terminal temperature difference of heat exchange units. Operation at values lower than this limit do not yield the desired design values, which may include heating, cooling, condensation, or evaporation.

Results for the analysis are shown in Fig. 10. As is shown for the two-stage system, the values of TTD_{21} and TTD_{22} are 0.273 and 3 °C. On the other hand, the values of TTD_{21}, TTD_{22}, and TTD_{23} for conventional MSF are 2.35, 3.5, and 4.67 °C, respectively. In the light of the above, design and operation of the MSF system with a two-stage heat rejection is feasible, nevertheless, the low terminal temperature difference will increase dramatically the required heat transfer area. Addition of the third heat rejection unit in conventional MSF solves the limit imposed on the terminal temperature difference for the two stage system and consequently the required heat transfer area.

6.3.6 Conventional MSF

The model and analysis performed for the MSF system is made for a three-stage heat rejection section and twenty stages heat recovery section. Most of the model equations are similar to those previously given for other configurations. The overall material and salt balance equations are identical to Eqs. 1 and 2 given in the model of the single stage flash unit. The equations for the amounts of heating steam and recycle brine are identical to Eqs. 19 and 20, respectively. In addition, the heat transfer areas for the brine heater and the condensers in the heat recovery section are given by Eqs. 5 and 17. As for the LMTD values for both units it is calculated from Eqs. 12 and 13.

The salt concentration in the recycle stream, X_r, is obtained by performing salt balance on the heat rejection section. This balance is

$$X_r\, M_r + M_b\, X_b = X_f\, M_f + (M_r - M_d)\, X_{n-j}$$

The above balance is arranged to

$$X_r = (X_f\, M_f + (M_r - M_d)\, X_{n-j} - M_b\, X_b)/M_r$$

Assuming that $X_{n-j} = X_b$, simplifies the above equation to

$$X_r = (X_f\, M_f + (M_r - M_d)\, X_b - M_b\, X_b)/M_r$$

Since $M_f = M_b + M_d$, then,

$$X_r = ((X_f - X_b)\, M_f + M_r\, X_b)/M_r \qquad (26)$$

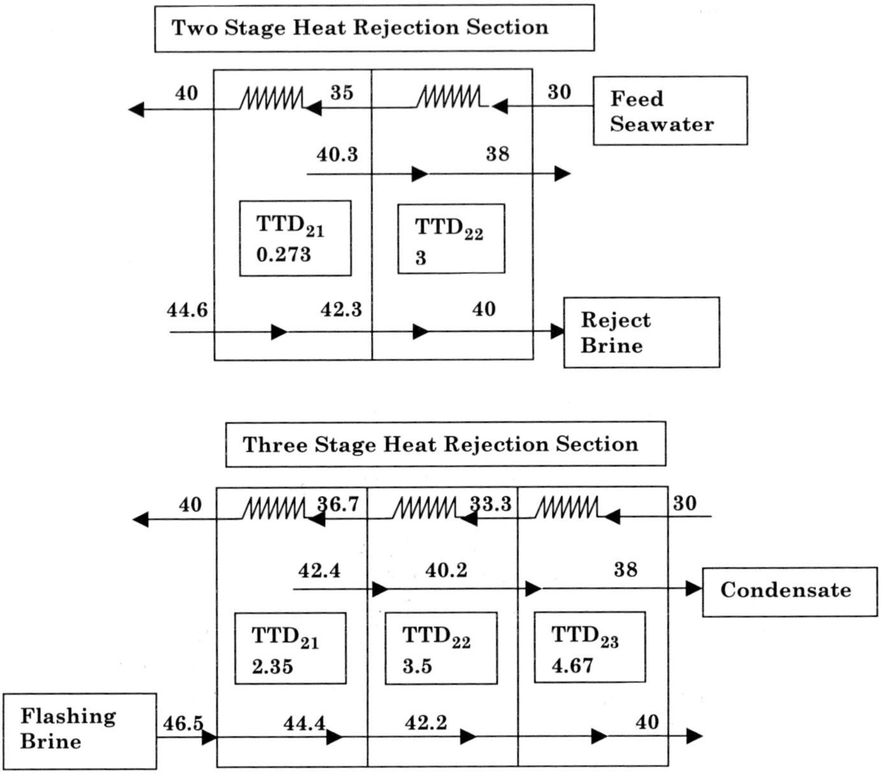

Fig. 10. Variation in stage terminal temperature difference and temperatures of flashing brine, condensate, and feed seawater.

The cooling water flow rate, M_{cw}, is obtained from an overall energy balance around the desalination plant, Fig. 1. The intake seawater temperature, T_{cw}, is used as the reference temperature in the energy balance. This gives

$$M_s \lambda_s = M_{cw} C_p (T_n - T_{cw}) + M_b C_p(T_n - T_{cw}) + M_d C_p (T_n - T_{cw})$$

The above equation is arranged to obtain an expression for M_{cw}

$$M_{cw} = (M_s \lambda_s - M_f C_p(T_n - T_{cw}))/ (C_p (T_n - T_{cw})) \quad (27)$$

The heat transfer area for the condensers in the heat rejection section, A_j, are obtained from the following equation

$$A_j = (M_f+M_{cw})\ Cp\ (t_i - t_{i+1})/(U\ (LMTD)_j)$$
$$= (M_f+M_{cw})\ Cp\ (\Delta t_j)/(U\ (LMTD)_j) \qquad (28)$$

$$(LMTD)_j = (t_i - t_{i+1})_j/\ln((T_{v_i} - t_{i+1})/(T_{v_i} - t_i))$$
$$= \Delta t_j/\ln((\Delta t_j + TTD_j)/(TTD_j)) \qquad (29)$$

The value of TTD_j is set equal to 3 °C and Δt_j is obtained from

$$\Delta t_j = (T_n - T_{cw})/j$$

The total heat transfer area for all condensers in the heat recovery and rejection sections as well as the brine heater is then obtained from

$$sA = A_h + (n\text{-}j)\ A_c + (j)\ A_j \qquad (30)$$

As discussed before, a product flow rate of 1 kg/s and salinity of 42000 ppm and 70000 ppm for the feed and blow down, respectively, gives a feed flow rate of 2.5 kg/s and brine blow down flow rate of 1.5 kg/s. The temperature drop per stage is obtained from

$$\Delta T_{st} = (T_o - T_n)/n = (90\text{-}40)/23 = 2.174\ °C$$

The recycle flow, M_r, is then calculated from Eq. 20, where

$$M_d = M_r\ C_p\ (n\Delta T_{st})/\bar{\lambda}_v$$
$$1.0 = M_r\ (4.18)\ (23)(2.174)/2346.5$$
$$M_r = 11.22\ kg/s$$

The salinity of the recycle brine is calculated from Eq. 26,

$$X_r = ((X_f - X_b)\ M_f + M_r\ X_b)/M_r$$
$$X_r = ((42000 - 70000)\ (2.5) + (11.22)\ (70000))/(11.22)$$

which reduces to

$$X_r = 63765\ ppm$$

Equation 19 is used to determine the flow rate of the heating steam. In this equation, the steam latent heat, λ_s, is equal to 2256 kJ/kg (at 100 °C), thus,

$$M_s = M_r\ C_p\ (\Delta T+\Delta T_{loss}+TTD_c)/\lambda_s$$
$$= (11.22)\ (4.18)\ (2.174+2+3)/2256$$
$$= 0.149\ kg/s$$

Since the total flow rate of the distillate product, M_d, is equal to 1 kg/s, the system thermal performance ratio can be readily obtained

$$PR = M_d/M_s = 1/0.149 = 6.7$$

The cooling water flow rate is then calculated from Eq. 27, where

$$M_{cw} = (M_s\ \lambda_s - M_f\ C_p(T_n - T_{cw}))/(C_p\ (T_n - T_{cw}))$$
$$= ((0.149)(2256) - (2.5)(4.18)(40\text{-}30))/((4.18)(40\text{-}30))$$
$$= 5.55\ kg/s$$

The heat transfer areas are obtained from Eqs. 5, 17, and 28. The heat transfer area for the brine heater is calculated from Eq. 5,

$$Ms\ \lambda s = U\ Ah\ (LMTD)h$$
$$(0.149)\ (2256) = (2)\ (Ah)(13.26)$$

This gives, $A_h = 12.69\ m^2$. The $(LMTD)_h$ is calculated from the Eq. 12, where

$$(LMTD)_h = \frac{\Delta T_{st} + \Delta T_{loss} + TTD_c}{\ln((TTD_h + \Delta T_{st} + \Delta T_{loss} + TTD_c)/(TTD_h))}$$
$$= (2.174+2+3)/\ln((10+2.174+2+3)/(10))$$
$$= 13.26\ ^oC$$

The heat transfer area for the preheater/condenser in each stage in the heat recovery section is calculated from Eq. 17

$$A_c = (M_r)\ Cp\ (\Delta T_{st})/(U\ (LMTD)_c)$$
$$= (11.22)\ (4.18)\ (2.174)/((2)(3.315))$$
$$= 12.78\ m^2$$

The value of $(LMTD)_c$ is calculated using Eq. 13, thus,

$$(LMTD)_c = \Delta T_{st}\ /\ln((\Delta T_{st} + TTD_c)/(TTD_c))$$
$$= (2.174)/\ln((2.174+3)/(3))$$

$= 3.98\ ^{o}C$

The same procedure is applied to the stages in the heat rejection section, where the condenser area in rejection stages is given by

$$A_j = (M_f+M_{cw})\ Cp\ (\Delta t_j)/(U\ (LMTD)_j)$$
$$= (2.5+5.55)\ (4.18)\ (3.33)/((2)(4.46))$$
$$= 12.57\ m^2$$

where

$$(LMTD)_j = \Delta t_j/\ln((\Delta t_j + TTD_j)/(TTD_j))$$
$$= (3.33)/\ln((3.33+3)/(3)$$
$$= 4.46\ ^{o}C$$

The total specific heat transfer area is obtained from Eq. 30, where,

$$sA = A_h + (n\text{-}j)\ A_c + (j)\ A_j$$
$$= 12.69 + (20)\ (12.78) + (3)\ (12.57) = 306.2\ m^2$$

Examining the above results for the MSF system show the following characteristics:

- Reduction in the feed flow rate in comparison with the MSF-OT.
- Low total specific heat transfer area.
- High Performance ratio.
- Low salinity for the recycle brine in comparison with MSF-R.
- Low temperature for the reject brine in comparison with MSF-R.

6.3.7 Effects of Operating Variables

Sensitivity of conventional MSF, the once through (MSF-OT), and the single stage flashing (SSF) are analyzed as a function of the top brine temperature, the number of flashing stages, and the thermodynamic losses. Analysis includes effects of system parameters on the thermal performance ratio, the total specific heat transfer area, the salinity of recycle and blow down brine, and the specific flow rates of feed, cooling water, recycle brine, and blow down brine. Analysis is performed over a temperature range of 90-110 °C for the top brine temperature, a total number of stages of 20 to 29, and a thermodynamic loss range of 0.5-2 °C.

Figure 11 shows variations in the thermal performance ratio for the three systems as a function of the top brine temperature for different values of the thermodynamic losses. As is shown, the performance ratio for the SSF system is less than 1 and is insensitive to variations in the top brine temperature as well as the thermodynamic losses. This behavior is cleared by inspection of Eq. 11, which is dominated by the flashing range. This parameter is much larger than the thermodynamic losses as well as the terminal temperature difference. As for the MSF and MSF-OT, their performance ratio is identical and is much larger than that for the single stage flashing system. This is because the flashing range is divided over n stages. As is shown in Fig. 11, the thermal performance ratio for the MSF and MSF-OT systems increases by the decrease of the thermodynamic losses and the increase of the top brine temperature. As given in Eq. 16, which applies to either system, the thermal performance ratio is inversely proportional to the thermodynamic losses. In addition, increase of the top brine temperature increases the flashing range and results in the increase of the amount of distillate product per unit mass of recirculated brine.

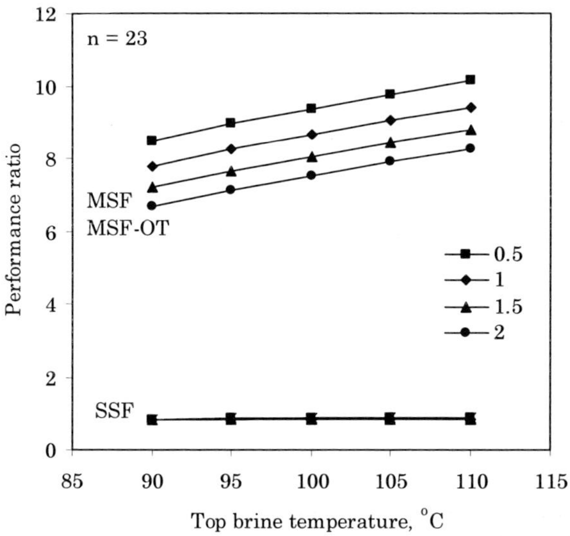

Fig. 11. Variation in system performance ratio as a function of top brine temperature and the thermodynamic losses.

Variations in the specific heat transfer area for the three systems are shown in Fig. 12. As is shown, the specific heat transfer area decreases with the increase of the top brine temperature and the decrease of the thermodynamic losses. Increase of the top brine temperature increases the temperature driving force, which enhances the rate of heat transfer, and as a result reduces the heat transfer area. The same effect is found upon lowering of the thermodynamic losses, which implies increase of the temperature of the condensing vapor. Thus, the temperature driving force increases and results in lowering of the heat transfer area.

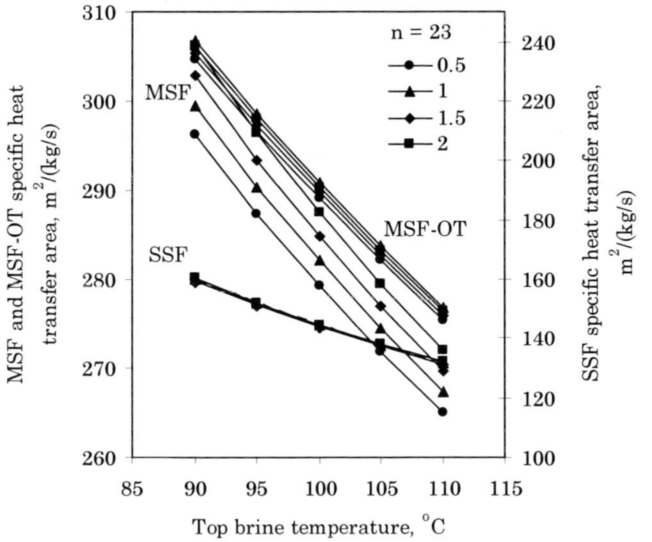

Fig. 12. Variation in specific heat transfer area for MSF, MSF-OT, and SSF as function of the top brine temperature and thermodynamic losses

Figure 13 shows variations of specific system parameters for the MSF configuration, which include the specific flow rate and salinity of the brine recycle as well as the specific cooling water flow rate. Variations are given as a function of the top brine temperature and at different values of the thermodynamic losses. As is shown, the flow rate and the salinity of the brine recycle are independent of the thermodynamic losses. Examining the balance equations for either parameter, Eqs. 20 and 26, respectively, show no dependence on the thermodynamic losses. However, both parameters are inversely proportional to the flashing range or the top brine temperature. The decrease of the specific flow rate of the cooling water with the decrease of the thermodynamic losses and the increase of the top brine temperature is associated with simultaneous increase of the system thermal performance ratio. At higher performance ratios, lower amounts of the heating steam are used, thus, the rate of heat removal in the heat rejection section is reduced, i.e., the specific flow rate of the cooling water.

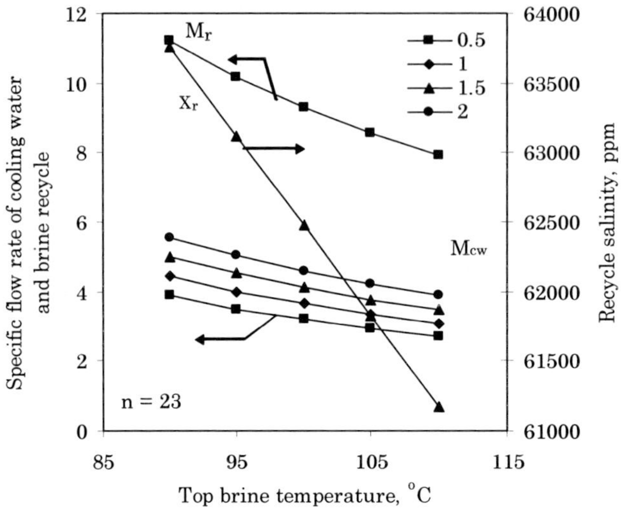

Fig. 13. Variation in specific cooling water flow rate, specific recycle flow rate, and salinity of brine blow down for MSF as a function of the top brine temperature and thermodynamic losses.

Dependence of the system parameters for the MSF and MSF-OT on the number of stage is shown in Figs. 14-16. Identical variations in the system thermal performance ratio and specific total heat transfer area are obtained for the MSF and MSF-OT as a function of the number of stage, Figs. 14 and 15. In either system, the thermal performance ratio is proportional with the number of stages. Increase of the number of stages reduces the stage temperature drop and as a result increases the system thermal performance. Increase in the total specific heat transfer area in both systems with the increase of the number of stages is caused by the reduction in the stage temperature drop and consequently the temperature driving force for heat transfer.

Variations in the MSF parameters as a function of the number of stage, which include the specific flow rates of brine recycle and cooling seawater as well as the salinity of the brine recycle, are shown in Fig. 16. As is shown the salinity and specific flow rate of brine recycle are independent of the number of stages. This is because the specific flow rate of the brine recycle depends only the total flashing range, which is independent of the number of stages. The decrease of specific flow rate of the cooling water with the increase of the number of stages is caused by the increase in the system thermal performance ratio. Increase of the performance ratio reduces the heat load in the brine heater and consequently the amount of heat removal by the cooling seawater.

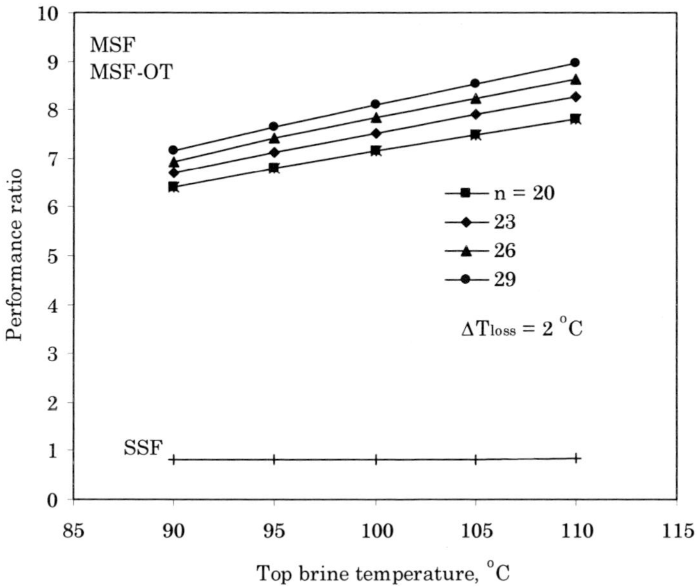

Fig. 14. Variation in performance ratio as a function of the top brine temperature and the number of stages.

Other system parameters, which include the specific flow rates of feed and brine blow down as well as the salinity of blow down brine, are shown in Figs. 17-19. As is shown these parameters are independent of the thermodynamic losses as well as the number of stages. In addition, these parameters are also independent of the top brine temperature for the MSF system. This is because the thermal performance of the MSF system is dependent on the properties of the brine recycle stream rather than the feed seawater stream. For the MSF system, the salinity and specific flow rate of the reject brine are kept constant at 70,000 ppm and 1.5 kg/s, respectively. In addition, the specific feed flow rate in the MSF system is also kept constant at 2.5 kg/s. The decrease in the specific feed flow rate for the MSF-OT and SSF system upon the increase in the top brine temperature is caused by the increase the flashing range and consequently the specific amount of product per unit mass of feed seawater. This decrease is also associated with simultaneous decrease in the specific flow rate of brine blow down for the MSF-OT and SSF, which is less than the specific feed flow rate by 1 kg/s or the distillate flow rate. Decrease in the specific flow rate of the brine blow down causes the increase of the stream salinity, Fig. 19.

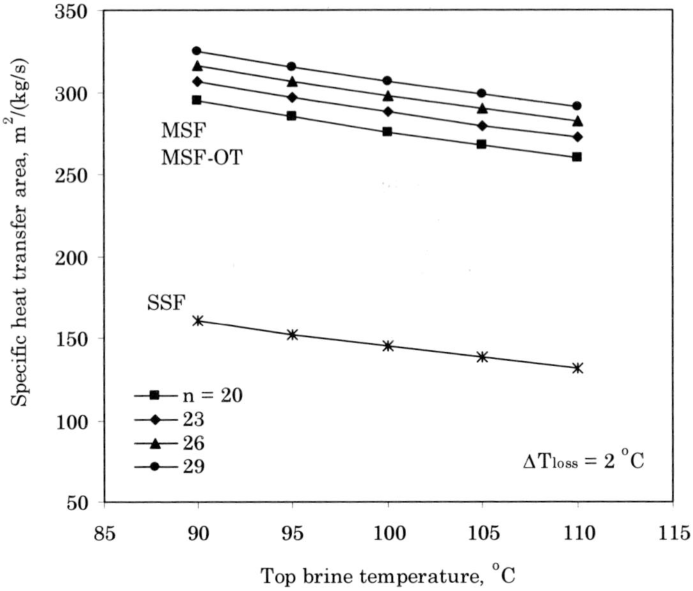

Fig. 15. Variation in specific heat transfer area for MSF, MSF-OT, and SSF as function of the top brine temperature and the number of stages

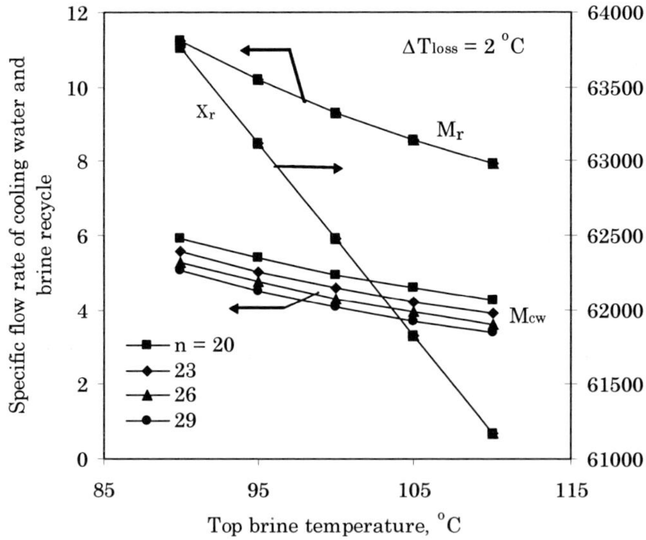

Fig. 16. Variation in specific cooling water flow rate, specific recycle flow rate, and salinity of brine blow down for MSF as a function of number of stages and the top brine temperature.

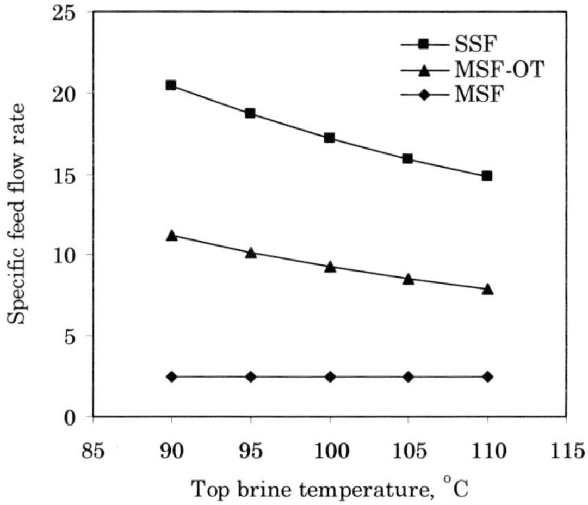

Fig. 17. Variation in specific feed flow rate for MSF, MSF-OT, and SSF as a function of the top brine temperature.

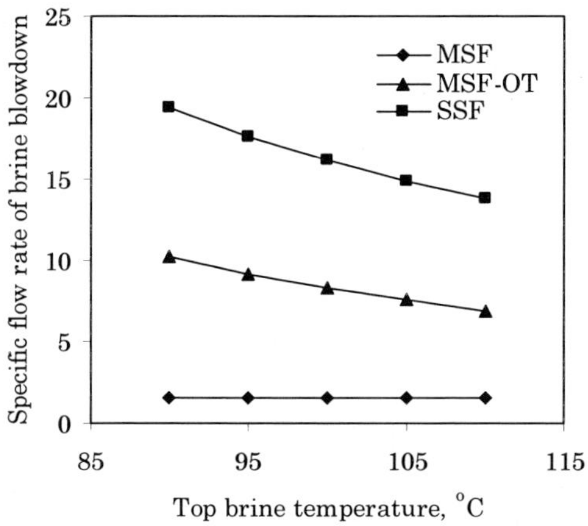

Fig. 18. Variation in specific flow rate of brine blow down for for MSF, MSF-OT, SSF as a function of the top brine temperature.

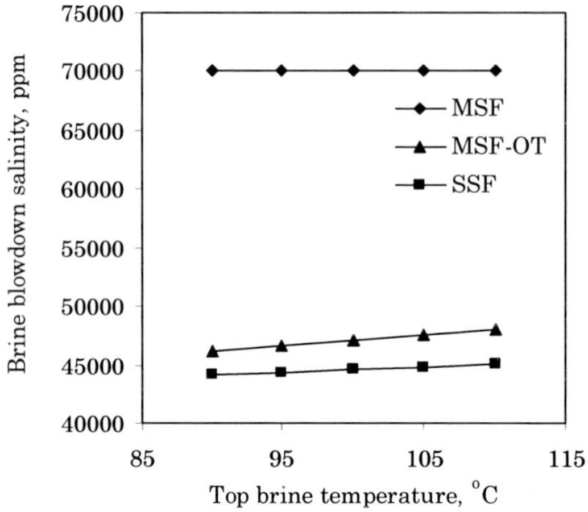

Fig. 19. Variation in salinity of brine blow down for MSF, MSF-OT, SSF as a function of the top brine temperature.

6.3.8 Summary

The following summary is made in the light of results and conclusions made for each flashing system. Figure 20 shows a schematic for the proposed configurations and their drawbacks and merits.

- A single effect flashing unit can not be used for water desalination. Specific power consumption for such system is very high. This is because the amount of steam used is larger than the amount of distillate water. In addition, the pumping power for intake seawater is very high, considering its large amount in comparison with the generated amount of distillate water. These two problems are addressed in the once through and the brine recirculation type desalination systems.
- The once through MSF system solves the performance ratio problem found in the single unit configuration, where the thermal performance ratio is increased from below one to values above 6. However, the problem of the large seawater intake is not solved. The apparent solution of this problem is to recycle part of the blow-down brine and to mix the recycle stream with the intake seawater.
- The simple mixer brine-recycle MSF system improves the thermal performance ratio, where it increases to a higher value of 8. However, several problems are found in the simple mixer configuration, i.e., high brine recirculation rates, and high salinity of the feed seawater.
- In the heat rejection stages, the intake seawater must be heated to the same temperature as that of the brine of the last stage. This is essential to prevent

thermal shock upon mixing of the two streams in the last stage. The thermal shock causes decomposition of the bicarbonate salts and formation of carbonate precipitates and carbon dioxide gas. The formed gas reduces the heat transfer efficiency around the condenser tubes and has harmful effect the steam jet ejector.

- Use of the single-stage heat rejection section is not possible, because of the intersection temperature profiles of the hot and cold streams in the preheater/condenser tubes.
- Use of the two-stage heat rejection section is not practical because of the low terminal temperature difference found in the first flashing stage. This value is well below limiting values that allow for stable and steady operation.
- The MSF system, which includes three stages in the heat rejection gives the desired system performance ratio, salinity for the recycled and blow-down brine, and practical values for the specific cooling water flow rate and heat transfer area.
- Performance of the MSF and MSF-OT systems improve at larger number of stages and higher top brine temperatures. This improvement is reflected in the increase of thermal performance ratio and the specific heat transfer area, and the reduction of the specific flow rate of cooling water. Increase of the thermal performance ratio and reduction of the specific flow rate of cooling water reduces the specific power consumption as well as the capital cost of the cooling water pumping unit. However, a higher capital cost is incurred at larger number of stages. In addition, the specific heat transfer area increases at higher number of stages. This increase has a direct effect on the capital and maintenance cost.

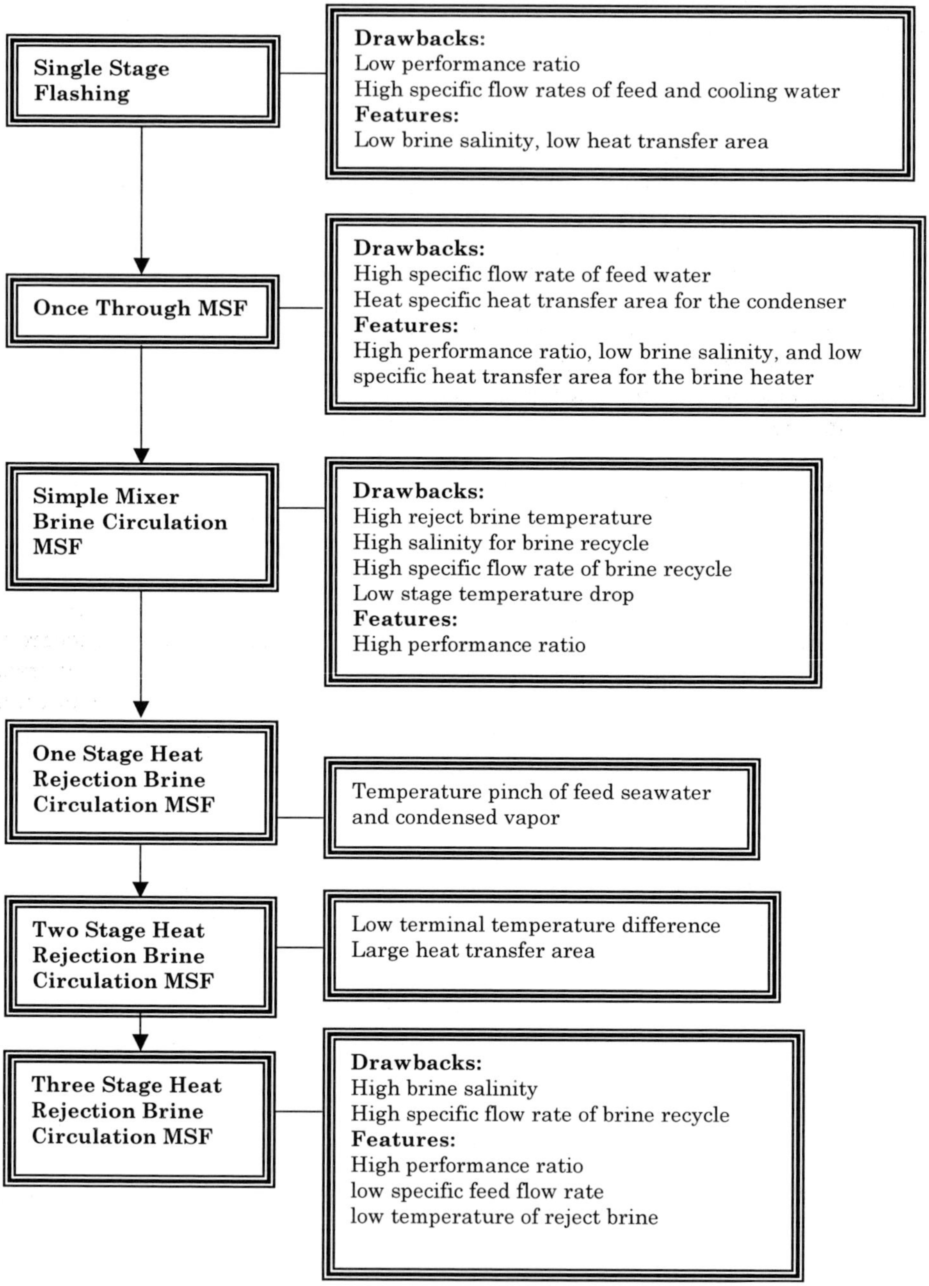

Fig. 20. Performance summary of various MSF configurations.

References

Darwish, M.A., Thermal analyis of multi stage flash desalination systems, Desalination **85**(1991)59-79.

Darwish, M.A., and El-Dessouky, H.T., The heat recovery thermal vapour compression desalting system: A Comparison with other thermal desalination process, Applied Thermal Engineering, **16**(1996)523-537.

El-Dessouky, H.T., Assassa, G., Losses in MSF desalination plants, Proc. Of the 3rd Joint Conf. On Mech Eng. Technology, Cairo, April, 1985.

El-Dessouky, H.T., Shaban, H.I., and Al-Ramadan, H., Steady-state analysis of multi-stage flash desalination process, Desalination, **103**(1995)271-287.

El-Dessouky, H.T., and Bingulac, S., Solving equations simulating the steady-state behavior of the multi-stage flash desalination process, Desalination, **107**(1996)171-193.

El-Dessouky, H.T., Alatiqi, I., Bingulac, S., and Ettouney, H.M., Steady-state analysis of the multiple effect evaporation desalination process, Chem. Eng. Tech., **21**(1998)15-29.

Helal, A.M., Medani, M.S., Soliman, M.A., and Flow, J.R., A tridiagonal matrix model for multistage flash desalination plants, Compu. Chem. Engrg. **10**(1986)327-324.

Hussain, A., Hassan, A., Al-Gobaisi, D.M., Al-Radif, A., Woldai, A., and Sommariva, C., Modelling, simulation, optimization, and control of multi stage flashing (MSF) desalination plants, Part I: Modelling and simulation, Desalination **92**(1993)21-41.

Montagna, J.M., Scennar, N.J., and Melli, T., Some theoretical aspects in simulation of multi stage desalination systems, Proc. Of the 12th Int. Symp. on Desalination and Water Re-Use, Malta (1991)437-448.

Omar, A.M., Simulation of MSF desalination plants, Desalination **45**(1983)65-67.

Rosso, M., Beltramini, A., Mazzotti, M., and Morbidelli, M., Modeling multistage flash desalination plants, Desalination, **108**(1996)365-374.

Temperley, T.G., The coming age of desalination, Proceeding of the IDA World Congress on Desalination and Water Sciences, Abu-Dhabi, UAE, November, 1995, Vol. I, pp 219-228.

Problems

Problem 1

Calculate the performance ratio, specific heat transfer area, specific flow rate of cooling water, conversion ratio, and salinity of brine blowdown for a single stage flash desalination unit operating at the following conditions:

- Feed salinity = 45000 ppm
- Feed temperature = 25 °C
- Heating steam temperature = 90 °C
- Production capacity = 1 kg/s
- Brine blowdown temperature = 35 °C
- Top brine temperature = 80 °C
- Terminal temperature difference in the condenser = 3 °C
- Thermodynamic losses = 2 °C.

Problem 2

An MSF-OT system operates at the following conditions

- Feed salinity = 45000 ppm
- Heating steam temperature = 100 °C
- Feed temperature = 25 °C
- Production capacity = 1 kg/s
- Brine blowdown temperature = 35 °C
- Top brine temperature = 90 °C
- Terminal temperature difference in the condenser = 3 °C
- Thermodynamic losses = 2 °C.

Calculate the system performance parameters if the number of stages is equal to 30.

Problem 3

Compare the performance of MSF-OT and MSF-M systems at the following conditions:

- Feed salinity = 45000 ppm
- Heating steam temperature = 112 °C
- Feed temperature = 28 °C
- Production capacity = 1 kg/s
- Brine blowdown temperature = 38 °C
- Top brine temperature = 105 °C
- Terminal temperature difference in the condenser = 3 °C

- Thermodynamic losses = 2 °C.
- Number of stages = 40.

Problem 4

A brine circulation MSF system has the following operating data
- Feed salinity = 57000 ppm
- Brine blowdown salinity = 70000 ppm
- Heating steam temperature = 116 °C
- Production capacity = 1 kg/s
- Brine blowdown temperature = 40 °C
- Feed temperature = 30 °C
- Top brine temperature = 106 °C
- Terminal temperature difference in the condenser = 3 °C
- Number of stages = 24 (with 3 stages in the heat rejection section).

Compare the system performance if the thermodynamic losses are equal to 1.5 °C.

Problem 5

A brine circulation MSF system operates at the following conditions:
- Feed salinity = 34000 ppm
- Feed temperature = 25 °C
- Heating steam temperature = 100 °C
- Production capacity = 1 kg/s
- Brine blowdown temperature = 35 °C
- Top brine temperature = 90 °C
- Terminal temperature difference in the condenser = 3 °C
- Thermodynamic losses = 1.5 °C

Calculate the system performance parameters for the following conditions:

a) Number of stages 40 with 3 stages in the heat rejection section.
b) Number of stages 19 with 3 stages in the heat rejection section.

6.4. Once Through MSF

The flow diagram of the MSF-OT is shown in Fig. 21. As is shown process is primarily formed of the flashing stages and the brine heater. In this system, the flashing stages forming the heat rejection section are eliminated. This simplifies the process layout and system operation. Irrespective, the MSF-OT is found on a limited industrial scale because of the following drawbacks:

- Absence of control system on the temperature of the feed seawater. This limits the use of the MSF-OT process in regions with large seasonal temperature changes. For example, the seawater temperature in the Gulf area drops to low values of 15 oC during the winter. Operation at these conditions would result in drastic reduction in the system performance ratio or the production capacity during wintertime. A thermal performance ratio of 3 is reported for the MSF-OT units during winter operation.
- Reduction in the thermal performance ratio at low intake seawater temperatures can be met by increasing the temperature range or reduction of the last stage temperature. At low temperatures the specific volume of the flashing vapor increases, which would require increasing the flash chamber size. This is necessary to keep the vapor velocity to values below 4 m/s, which limits entrainment of brine droplets.

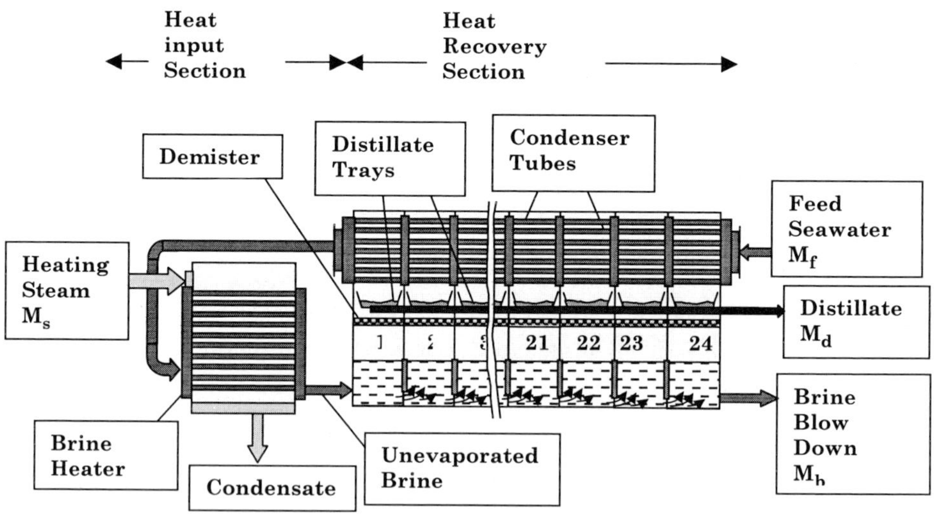

Fig. 21. Multistage flash desalination once through process.

- The flow rate ratio of the feed to product in the MSF-OT process is approximately 10:1. On the other hand, this ratio in the brine circulation MSF is 2.5:1. As a result, the MSF-OT system would consume a larger amount of chemicals for scale, foaming, and corrosion control.

Irrespective of the drawbacks of the MSF-OT process its use can be prove useful especially in equatorial regions, where the seawater temperature remains nearly constant throughout the year.

6.4.1 Process Description

Schematic of the brine circulation MSF process is shown in Fig. 21 and process variables in the brine heater and the flashing stages are shown in Fig. 22. Details of the MSF process are described below:

- The feed seawater (M_f) is deaerated and chemically before being last flashing stage in the heat rejection section, where it flows from stage (n) to stage.
- The seawater temperature increases due to absorption of the latent heat of the condensing fresh water vapor.

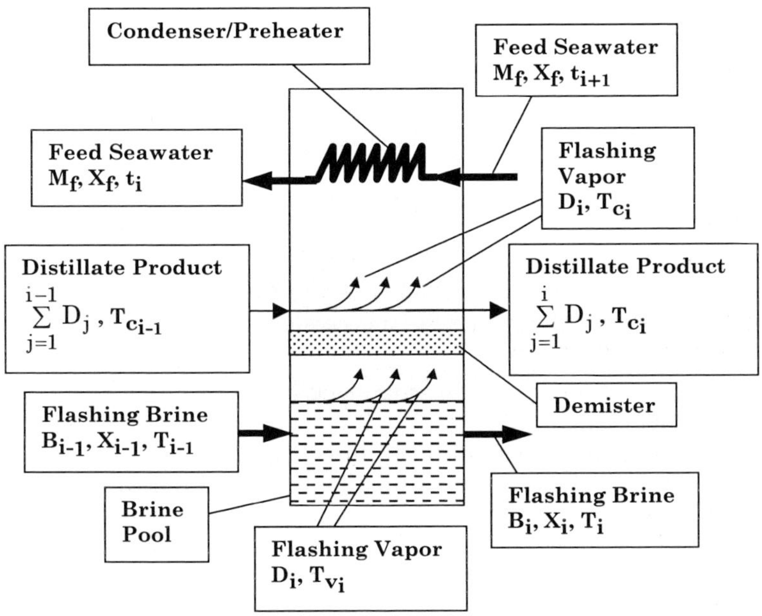

Fig. 22a. Schematics of model variables in flashing stage

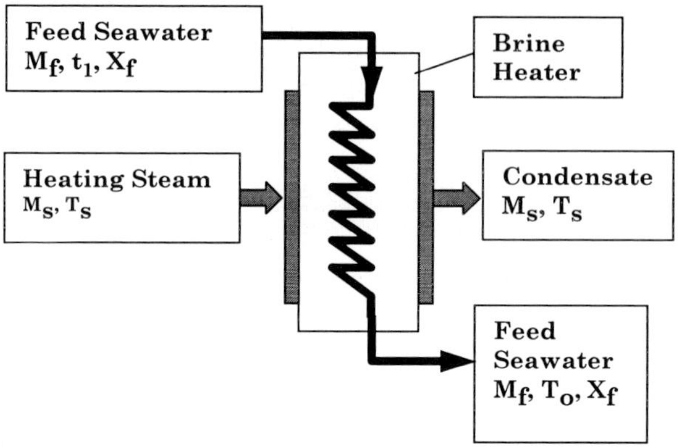

Fig. 22b. Schematics of model variables in brine heater

- The feed seawater (M_f) enters the brine heater tubes, where the heating steam (M_s) is condensed on the outside surface of the tubes. The feed seawater (M_f) absorbs the latent heat of condensing steam and its temperature increases to its maximum design value known as the top brine temperature (T_o). This value depends on the nature of chemicals used to control the scale formation.
- The feed seawater (M_f) enters the flashing stages, where a small amount of fresh water vapor is formed by brine flashing in each stage. The flashing process takes place due to decrease in the stage saturation temperature and causes the reduction in the stage pressure.
- In each stage, the flashed off vapors condenses on the outside surface of the condenser tubes, where the feed seawater (M_f) flows inside the tubes from the cold to the hot side of the plant. This heat recovery improves the process efficiency because of the increase in the feed seawater temperature.
- The condensed fresh water vapor outside the condenser tubes accumulates across the stages and forms the distillate product stream (M_d). This stream cascades in the same direction of the flashing brine from stage to stage and is withdrawn from the last stage.
- The flashing process and vapor formation is limited by increase in the specific vapor volume at lower temperatures and difficulties encountered for operation at low pressures. Common practice limits the temperature of the last stage to range of 30 to 40 °C, for winter and summer operation, respectively. Further reduction in these temperatures results in drastic increase of the stage volume and its dimensions.
- In MSF, most of flashing stages operating at temperatures below 100 °C have vacuum pressure. This increases the possibilities of in-leakage of the outside air. Also, trace amounts of dissolved gases in the flashing brine, which are not

removed in the deaerator or formed by decomposition of $CaHCO_3$. At such conditions, air and other gases are non-condensable and its presence in the system may result in severe reduction in the heat transfer rates within the chamber, increase of the tendency for corrosion, and reduction of the flashing rates. This condition necessitates proper venting of the flashing stages to enhance the flashing process and to improve the system efficiency.

- Treatment of the feed seawater (M_f) is limited simple screening and filtration. On the other hand, treatment of the feed seawater stream is more extensive and it includes dearation and addition of chemicals to control scaling, foaming, and corrosion.

6.4.2 Mathematical Model

The MSF-OT simplified model is a very useful tool, which can be used to obtain quick design data, evaluate system performance, and develop a good initial guess for more detailed mathematical models. The simplified model does not need iterative solution and requires minimal computational effort. Development of the simplified model is based on the following assumptions:

- Constant and equal specific heat for all liquid streams, Cp.
- Equal temperature drop per stage for the flashing brine.
- Equal temperature drop per stage for the feed seawater.
- The latent heat of vaporization in each stage is assumed equal to the average value for the process.
- Effects of the non-condensable gases have negligible effect on the heat transfer process.
- Effects of the boiling point rise and non-equilibrium losses on the stage energy balance are negligible, however, their effects are included in the design of the condenser heat transfer area.

The simplified model includes the following elements:

- Overall material balance.
- Stages and condensers temperature profiles.
- Stage material and salt balance.
- Condensers and brine heater heat transfer area.
- Stage dimensions.
- Performance parameters.

The following sections include the model equations for each of the above items:

Overall Material Balance

The MSF-OT flow diagram shows one input stream, the feed seawater, M_f, and two output streams, the distillate product, M_d, and the rejected brine, M_b.

Therefore, the excess energy added to the system in the brine heater is removed in brine blowdown and distillate product streams.

The overall material balance equations is given by

$$M_f = M_d + M_b \tag{31}$$

Where M is the mass flow rate and the subscript b, d, and defines the brine, distillate, and feed. The overall salt balance is given by

$$X_f\, M_f = X_b\, M_b \tag{32}$$

Where X is the salt concentration. Equation (32) assumes that the distillate is salt free.

Stages and Condensers Temperature Profiles

The temperature distribution in the MSF-OT system is defined in terms of four temperatures; these are the temperatures of the steam, T_s, the brine leaving the preheater (top brine temperature), T_o, the brine leaving the last stage, T_n, and the feed seawater, T_f. A linear profile for the temperature is assumed for the flashing brine and the seawater flowing inside the condenser tubes. The temperature drop per stage, ΔT, is obtained from the relation

$$\Delta T = (T_o - T_n)/n \tag{33}$$

where n is the number of stages. Therefore, the temperature in the first stage is given by

$$T_1 = T_o - \Delta T$$

As for the second stage temperature it is equal to

$$T_2 = T_1 - \Delta T$$

Substituting for T_1 in the above equation gives

$$T_2 = T_o - \Delta T - \Delta T = T_o - 2\,\Delta T$$

The same procedure is repeated for subsequent stages and a general expression is developed for the temperature of stage i

$$T_i = T_o - i\,\Delta T \tag{34}$$

The temperature of the feed seawater, M_f, which flows inside the condenser tubes, increases by Δt in the condenser of each stage. This temperature increase, Δt, is equal to the decrease in the brine temperature in each stage, ΔT. This result is arrived at by performing an energy balance on stage i, which gives

$$D_i \; Cp \; T_{v_i} + B_i \; Cp \; T_i - D_{i+1} \; Cp \; T_{v_{i+1}} - B_{i+1} \; Cp \; T_{i+1} = M_f \; C_p \; (t_i - t_{i+1})$$

Assuming the temperature difference, $T_i - T_{v_i}$, is small and has a negligible effect on the stage energy balance. Thus, the above equation reduces to

$$(D_i + B_i) \; Cp \; T_i - (D_{i+1} + B_{i+1}) \; Cp \; T_{i+1} = M_f \; C_p \; (t_i - t_{i+1})$$

Recalling that the sum $(D_i + B_i)$ in each stage is equal to M_f, would simplify the above equation to

$$M_f \; Cp \; T_i - M_f \; Cp \; T_{i+1} = M_f \; C_p \; (t_i - t_{i+1})$$

Elimination of the like terms on both sides of the equation gives the pursued relation, thus,

$$T_i - T_{i+1} = t_i - t_{i+1}$$

or

$$\Delta T_i = \Delta t_i$$

The seawater temperature, which leaves the condenser of the first stage, is then defined by

$$t_1 = T_f + n \; \Delta t$$

The seawater temperature leaving the condenser of the second stage, T_2, is less than T_1 by Δt, where

$$t_2 = t_1 - \Delta t$$

Substituting for T_1 in the above equation gives

$$t_2 = T_f + (n - 1) \; \Delta t$$

Similar to equation (34), a general equation is obtained for the condenser temperature in stage i

$$t_i = T_f + (n - (i - 1))\ \Delta t \tag{35}$$

Stage Material and Salt Balance

The amount of flashing vapor formed in each stage obtained by conservation of energy within the stage, where the latent consumed by the flashing vapor is set equal to the decrease in the brine sensible heat. This is

$$D_1 = y\ M_f$$

where D_1 is the amount of flashing vapor formed in the first stage, M_f is the feed seawater flow rate, and y is the specific ratio of sensible heat and latent heat and is equal to

$$y = Cp\ \Delta T/\lambda_{av} \tag{36}$$

Where Cp is the specific heat capacity and λ_{av} is the average latent heat calculated at the average temperature

$$T_{av} = (T_o + T_n)/2 \tag{37}$$

The amount of distillate formed in the second stage is equal to

$$D_2 = y\ (M_f - D_1)$$

Substituting the value of D_1 in the above equation gives

$$D_2 = y\ (M_f - y\ M_f)$$

This simplifies to

$$D_2 = M_f\ y\ (1 - y)$$

The balance equations for D_2 and D_3 will reveal the general form for the formula of D_i. The D_3 balance is

$$D_3 = y\ (M_f - D_1 - D_2)$$

Substituting for the values of D_1 and D_2 in the above equation gives

$D_3 = y\ (M_f - M_f\, y - M_f\, y\ (1 - y))$

Taking (M_r) as a common factor in the above equation gives

$D_3 = M_f\, y\ (1 - y - y + y^2)$

This simplifies to

$D_3 = M_f\, y\ (1 - y)^2$

Accordingly, the resulting general formula for D_i is

$$D_i = M_f\, y\ (1 - y)^{(i-1)} \qquad (38)$$

The total distillate flow rate is obtained by summing the values of D_i for all stages. The summation is performed in steps in order to obtain a closed form equation. Therefore, the summation of D_1 and D_2 gives

$$\begin{aligned} D_1 + D_2 &= M_f\ (y + y\ (1 - y)) \\ &= M_f\ (2\ y - y^2) \\ &= M_f\ (1 - (1 - y)^2) \end{aligned}$$

Addition of D_3 to the above gives

$D_1 + D_2 + D_3 = M_f\ ((2\ y - y^2) + y(1 - y)^2)$

This simplifies to

$$\begin{aligned} D_1 + D_2 + D_3 &= M_f\ (2\ y - y^2 + y - 2\ y^2 + y^3) \\ &= M_f\ (3\ y - 3\ y^2 + y^3) \\ &= M_f\ (3\ y - 3\ y^2 + y^3) \\ &= M_f\ (1 - (1 - y)^3) \end{aligned}$$

Comparison of the summations of D_1+D_2 and D_1+D_2+D_3 gives the general form for the total summation of the distillate formed in all stages, M_d, which is given by

$$M_d = M_f\ (1 - (1 - y)^n) \qquad (39)$$

Equation (39) is used to calculate the feed flow rate, since the distillate flow is always specified in a design problem.

The flow rate of the brine stream leaving stage (i) is given by

$$B_i = M_f - \sum_{k=1}^{i} D_k \tag{40}$$

The salt concentration in the brine stream leaving stage i is given by

$$X_i = M_f X_f / B_i \tag{41}$$

The flow rate of the heating steam, Ms, is obtained the energy balance equation for the brine heater, where

$$M_s \lambda_s = M_f \, Cp \, (T_o - t_1)$$

The above equation is arranged to calculate M_s

$$M_s = M_f \, Cp \, (T_o - t_1) / \lambda_s \tag{42}$$

Brine Heater and Condensers Heat Transfer Area

The brine heater area is given by

$$A_b = M_s \lambda_s / (U_b \, (LMTD)_b) \tag{43}$$

Where LMTD is given by

$$(LMTD)_b = ((T_s - T_o) - (T_s - t_1)) / \ln((T_s - T_o)/(T_s - t_1)) \tag{44}$$

and U_b is given by

$$U_b = 1.7194 + 3.2063\text{x}10^{-3} \, T_s + 1.5971\text{x}10^{-5} \, (T_s)^2 - 1.9918\text{x}10^{-7} \, (T_s)^3 \tag{45}$$

The heat transfer area for the condenser in each stage is assumed equal. Therefore, the calculated heat transfer area for the first stage is used to obtain the total heat transfer area in the plant. The condenser heat transfer area in the first stage is obtained from

$$A_c = M_f \, Cp \, (t_1 - t_2) / (U_c \, (LMTD)_c) \tag{46}$$

where

$$U_c = 1.7194 + 3.2063x10^{-3} T_{v_1} + 1.5971x10^{-5} T_{v_1}^{2} - 1.9918x10^{-7} T_{v_1}^{3} \quad (47)$$

$$T_{v_1} = T_1 - BPE_1 - NEA_1 - \Delta T_{d_1} \quad (49)$$

and

$$(LMTD)_c = ((T_{v_1} - t_1) - (T_{v_1} - t_2))/\ln((T_{v_1} - t_1)/(T_{v_1} - t_2)) \quad (50)$$

In the above equations (BPE) is the boiling point elevation, (NEA) is the non-equilibrium allowance, (T_v) is the condensing vapor temperature, (ΔT_d) is the temperature drop in the demister, and (U_c) is the condenser overall heat transfer coefficient. Expressions for correlations used calculate the BPE, NEA, and ΔT_d, are given in the appendix.

The total heat transfer in the plant is obtained by summing the heat transfer area for all condensers and the brine heater

$$A = A_b + n\,A_c \quad (51)$$

Stage Dimensions

Calculations of the stage dimensions include the gate height, the height of the brine pool, the stage width, and the stage length. The length of all stages is set equal to the length of the last stage and the width of all stages is set equal to the width of the first stage. The height of the brine pool must be higher than the gate height, this is necessary to prevent bypass of the vapors between stages (vapor blow through). The gate height (GH) is obtained in terms of the stage pressure drop (ΔP), the brine density (ρ_b), the weir friction coefficient (C_d), the stage width (W), and the feed flow rate (M_f). For stage i the gate height is

$$GH_i = (M_f - \sum_{j=1}^{i-1} D_j)\,(2\,\rho_{b_i}\,\Delta P_i)^{(-0.5)}/(C_d\,W) \quad (52)$$

The brine pool height is set higher than the gate height by 0.2 m.

$$H_i = 0.2 + GH_i \quad (53)$$

Where

$$\Delta P_i = P_i - P_{i+1}$$

$$W = M_f / V_b \tag{54}$$

where P_i and P_{i+1} are the pressures in stages i and i+1, and V_b is the brine mass velocity per chamber width. The length of the last stage is determined as a function of the vapor flow rate, D_n, the vapor density, ρ_{v_n}, the vapor allowable velocity, V_{v_n}, and the stage width, W. This is

$$L = D_n / (\rho_{v_n} V_{v_n} W) \tag{55}$$

The cross section area for each stage, A_s, is then calculated

$$A_s = L\,W \tag{56}$$

Performance Parameters

The system performance parameters are defined by the thermal performance ratio, PR and the specific heat transfer area, sA. The performance ratio is the defined as the amount of distillate product produced per unit mass of the heating steam. This is

$$PR = M_d / M_s \tag{57}$$

The specific heat transfer area is defined by

$$sA = (A_b + n\,A_c) / M_d \tag{58}$$

Solution Method

Solution of the MSF-OT simplified model is non-iterative. The solution proceeds as follows:

- The average stage temperature, T_{av}, is calculated from Eq. (37) and the corresponding latent heat value is obtained from the correlation given in the appendix.
- The temperature drop per stage, ΔT, is calculated from Eq. (33).
- The ratio of the stage sensible and latent heat, y, is calculated from Eq. (36).
- The feed flow rate is calculated from Eq. (39).
- The flow rate of the brine blow down, M_b, as well as its salinity, X_b, are obtained from Eqs. (31) and (32).

- The brine temperatures leaving stages 1 and 2, T_1 and T_2, are calculated from Eq. (34).
- The seawater temperatures leaving the condensers of the first and second stages, t_1 and t_2, are determined from Eq. (35). Also, the seawater temperature leaving the condenser of the last stage, t_n, is calculated from Eq. (35).
- The heat transfer area for the brine preheater (A_b), the first stage condenser (A_c) are calculated from Eqs. (43) and (46), respectively.
- The stage length, width, and cross section areas are determined from Eqs. (54-56). Also, the height of the brine pool, H, and the gate height for stage i are obtained from Eqs. (52) and (53).
- The system performance parameters, PR and sA, are calculated from Eqs. (57) and (58).

The above solution procedure requires specification of the following variables:

- Total distillate flow rate, M_d.
- Total number of stages.
- Feed seawater temperature, T_f.
- Top brine temperature, T_o.
- Steam temperature, T_s.
- Temperature of brine blowdown, T_n.
- Intake seawater salt concentration, X_f.
- Heat capacity of liquid streams, Cp.
- Weir friction coefficient, C_d.
- Vapor velocity in the last stage, V_{vn}.
- Brine mass flow rate per stage width, V_b.

6.4.3 Case Study

An MSF system with 24 stages is used to produce 7.2 MGD of product water. The following specifications are made to obtain the system design parameters and performance characteristics:

- Feed seawater temperature, $T_f = 25$ °C.
- Steam temperature, $T_s = 116$ °C.
- Top brine temperature, $T_o = 106$ °C.
- Brine temperature in the last stage, $T_n = 40$ °C.
- Heat capacity of liquid streams, $C_p = 4.18$ kJ/kg °C.
- Salinity of feed seawater, $X_f = 42000$ ppm.

- Vapor velocity in the last stage, $V_{v24} = 6$ m/s
- Brine mass flow rate per stage width, $V_b = 180$ kg/ms.
- Weir friction coefficient, $C_d = 0.5$.

Before proceeding, the product volume flow rate is converted to mass rate in SI units; this is necessary for solution of the energy balance equations. The distillate flow rate in kg/s is

$$M_d = \frac{\left(7.2\left(\frac{MG}{d}\right)\right)\left(10^6\left(\frac{G}{MG}\right)\right)\left(\frac{1}{24x3600}\left(\frac{d}{s}\right)\right)}{\left(\frac{1}{219.96}\left(\frac{m^3}{G}\right)\right)10^3\left(\frac{kg}{m^3}\right)}$$

$M_d = 378.8$ kg/s

Calculation of the y ratio from Eq. (36) is preceded by evaluation of T_{av} and λ_{av}. The T_{av} value is

$$\begin{aligned} T_{av} &= (T_o + T_n)/2 \\ &= (106+40)/2 \\ &= 73\ ^\circ C \end{aligned}$$

At this temperature λ_{av} is equal to 2330.1 kJ/kg from the correlation given in the appendix. The value of y is calculated

$$\begin{aligned} y &= Cp\ \Delta T/\lambda_{av} \\ &= (4.18)\ (2.75)/2330.1 \\ &= 4.933x10^{-3} \end{aligned}$$

The feed flow rate is obtained from equation (39)

$$\begin{aligned} M_f &= M_d/(1-(1-y)^n) \\ &= 378.8/(1-(1-4.933x10^{-3})^{24}) \\ &= 3384.8 \text{ kg/s} \end{aligned}$$

Arranging equations (31) and (32) gives the following expression, which is used to calculate the salinity of the brine blow down

$$\begin{aligned} X_b &= M_f\, X_f/(M_f - M_d) \\ &= (3384.8)(42000)/(3384.8-378.8) \end{aligned}$$

$= 47292.6$ ppm

The flow rate of the blow-down brine is then obtained from the overall balance given in equation (31)

$$
\begin{aligned}
Mb &= Mf - Md \\
&= 3384.8 - 378.8 \\
&= 3006 \text{ kg/s}
\end{aligned}
$$

The temperature drop in each effect is obtained from Eq. (33), where

$$
\begin{aligned}
\Delta T &= (T_o - T_n)/n \\
&= (106 - 40)/24 \\
&= 2.75\ ^oC
\end{aligned}
$$

This value is used to calculate the temperatures of the first and second stages, T_1 and T_2, where

$$
\begin{aligned}
T_1 &= T_o - \Delta T \\
&= 106 - 2.75 \\
&= 103.25\ ^oC
\end{aligned}
$$

As for the second stage temperature it is equal to

$$
\begin{aligned}
T_2 &= T_1 - \Delta T \\
&= 103.25 - 2.75 \\
&= 100.5\ ^oC
\end{aligned}
$$

Also the temperatures of the seawater leaving the condensers in the first and second stages are calculated. This is

$$
\begin{aligned}
t_1 &= T_f + n\ \Delta t \\
&= 25 + (24)\ (2.75) \\
&= 91\ ^oC
\end{aligned}
$$

and

$$
\begin{aligned}
t_2 &= t_1 - \Delta t \\
&= 91 - 2.75 \\
&= 88.25\ ^oC
\end{aligned}
$$

The steam flow rate is calculated from Eq. (42)

$$M_s = M_f\, Cp\, (T_o - t_1)/\lambda_s$$
$$= (3384.8)\,(4.18)\,(106 - 91)/(2222.33)$$
$$= 95.49 \text{ kg/s}$$

The heat transfer area for the brine heater, A_b, is calculated from Eq. (43). This requires calculations of the logarithmic mean temperature difference $(LMTD)_b$ and the overall heat transfer coefficient. The value of $(LMTD)_b$ is obtained from Eq. (44), which gives

$$(LMTD)_b = (t_1 - T_o)/\ln((T_s - T_o)/(T_s - t_1))$$
$$= (91 - 106)/\ln((116 - 106)/(116 - 91))$$
$$= 16.37\ {}^{o}C$$

The overall heat transfer coefficient is obtained from Eq. (45) where

$$U_b = 1.7194 + 3.2063x10^{-3}\, T_s + 1.5971x10^{-5}\, (T_s)^2 - 1.9918x10^{-7} (T_s)^3$$
$$= 1.7194 + 3.2063x10^{-3}\, (116) + 1.5971x10^{-5}\, (116)^2 - 1.9918x10^{-7}\, (116)^3$$
$$= 2 \text{ kW/m}^2\ {}^{o}C$$

The brine heater area, A_b, is then calculated from Eq. (43)

$$A_b = M_s\, \lambda_s/(U_b\, (LMTD)_b)$$
$$= (95.49)(2222.33)/((2)(16.37))$$
$$= 6481.68 \text{ m}^2$$

The condenser area, A_c, is determined for the first stage. This requires calculations of the vapor condensation temperature, T_{v_1}, the logarithmic mean temperature difference $(LMTD)_c$, and the overall heat transfer coefficient, U_c. The vapor temperature is given by Eq. (49), where

$$T_{v_1} = T_1 - BPE_1 - NEA_1 - \Delta T_{d_1}$$

The boiling point elevation is calculated from the correlation given in the appendix. This requires calculations of X_1, which is obtained in the following sequence

$$D_1 = y\, M_f$$
$$= (4.933x10^{-3})(3384.8)$$
$$= 16.697 \text{ kg/s}$$

$$B_1 = M_f - D_1$$

$= 3384.8 - 16.697$
$= 3368.1$ kg/s

$X_1 = M_f X_f / B_1$
$= (3384.8)(42000)/3368.1$
$= 42208$ ppm

The values of B and C in the correlation for the boiling point elevation are

$$B = (6.71 + 6.34x10^{-2}(T_1) + 9.74x10^{-5}(T_1)^2)10^{-3}$$
$$= (6.71 + 6.34x10^{-2}(103.25) + 9.74x10^{-5}(103.25)^2)10^{-3}$$
$$= 0.0143$$

$$C = (22.238 + 9.59x10^{-3}(T_1) + 9.42x10^{-5}(T_1)^2)10^{-8}$$
$$= (22.238 + 9.59x10^{-3}(103.25) + 9.42x10^{-5}(103.25)^2)10^{-8}$$
$$= 2.423x10^{-7}$$

Substituting the values of B and C in the BPE correlation gives

$$BPE_1 = X_1\ (B + (X_1)(C))10^{-3}$$
$$= 42208\ (0.0143 + (42208)(2.423x10^{-7}))10^{-3}$$
$$= 1.035\ ^{o}C$$

The non-equilibrium allowance, NEA_1, in the first stage is calculated from the correlation given in the appendix for the MSF system. This involves calculations of the gate height, GH_1, the height of the brine pool, H_1, the stage width, W, the stage pressure drop, P_1-P_2, and the brine density. The width of the first stage is given by

$W = M_f / V_b$
$= 3384.8/180$
$= 18.8$ m

The stage length is calculated for the last stage, where

$D_{24} = 14.9$ kg/s

$\rho_{v_n} = 0.0512$ kg/m^3,

$L = D_n/(\rho_{v_n} V_{v_n} W)$
$= 14.9/((0.0512)(6)(18.8))$
$= 2.58$ m

The brine density in the first stage is 1002.413 kg/m^3, which is obtained from the correlation given the appendix at a salinity of 42208 ppm and a temperature of 103.25 °C. The pressures of the first and second stage are obtained from the saturation pressure correlations, where, at T_1 = 103.25 °C, P_1 = 113.72 kPa, and at T_2 = 100.5 °C, P_2 = 103.23 kPa. The resulting gate height in the first stage, GH_1, is calculated from Eq. (52), where

$GH_1 = M_f (2 \rho_{b_1} \Delta P_1)^{(-0.5)}/(C_d W)$
$=(3384.8)((2)(1002.41)(113.72-103.23)x10^3)^{(-0.5)}/((0.5)(18.8)$
$= 0.078$ m

It should be noted that the pressure drop in the above equation is in Pa and not kPa. The corresponding brine pool height is obtained by simply adding 0.2 m to the value of GH_1, or

$H_1 = 0.278$ m

The non-equilibrium allowance is then calculated using the correlation given in the appendix

$NEA_1 = (0.9784)^{T_o} (15.7378)^{H_1} (1.3777)^{V_b x 10^{-6}}$
$= (0.9784)^{(106)} (15.7378)^{(0.278)} (1.3777)^{(180x10^{-6})}$
$= 0.213$ °C

The temperature drop in the demister is assumed negligible in comparison with the values of BPE_1 and NEA_1. Therefore, the vapor temperature in the first is

$T_{v_1} = T_1 - BPE_1 - NEA_1 - \Delta T_{d_1}$
$= 103.25 - 1.035 - 0.213 - 0$
$= 102.002$ °C

The vapor temperature, T_{v_1}, is used to calculate U_c and $(LMTD)_c$, where

$(LMTD)_c = (t_2\text{-}t_1)/\ln((T_v-t_1)/(T_v-t_2))$
$= (2.75)/\ln((102.002-91)/(102.002-88.25))$
$= 12.32$ °C

$U_c = 1.7194 + 3.2063x10^{-3}\ T_{v_1} + 1.5971x10^{-5}\ (T_{v_1})^2 - 1.9918x10^{-7}(T_{v_1})^3$
$= 1.7194 + 3.2063x10^{-3}\ (101.207) + 1.5971x10^{-5}\ (101.207)^2$
$- 1.9918x10^{-7}\ (101.207)^3$
$= 2\ kW/m^2\ {}^oC$

$A_c = M_f\ Cp\ (t_2\text{-}t_1)/(U_c\ (LMTD)_c)$
$= ((3384.8)\ (4.18)\ (97.75\text{-}95))/((2)\ (12.32))$
$= 1579\ m^2$

The total heat transfer area is obtained from

$A = A_b + n\ A_c$
$= 6481.68 + (24)\ (1579)$
$= 44377.7\ m^2$

Performance Parameters

Performance Ratio

$PR = M_d/M_s$
$= 378.8/95.49$
$= 3.96$

The specific heat transfer area

$sA = A/M_d$
$= (44377.7)/378.8$
$= 117.2\ m^2/(kg/s)$

Solution Summary

Flow Rates

$M_d = 378.8$ kg/s

$M_b = 3006$ kg/s

$M_f = 3384.8$ kg/s

$M_s = 52.52$ kg/s

Heat Transfer Areas

$A_b = 6481.68 \text{ m}^2$

$A_c = 1579 \text{ m}^2$

$A = 44377.7 \text{ m}^2$

Stage Dimensions

$W = 18.8 \text{ m}$

$L = 2.56 \text{ m}$

$GH_1 = 0.078 \text{ m}$

$H_1 = 0.278 \text{ m}$

Performance Parameters

$PR = 3.96$

$sA = 117.2 \text{ m}^2/(\text{kg/s})$

Stage Profiles

The MSF-OT simplified model is used to calculate the temperature and concentration profiles for the 24 stages. The calculations also include the brine and distillate flow rate, the gate height and the brine level in each stage. In the following table, the flow rates are in kg/s, the temperature in °C, the salinity in ppm, and the height in m.

Stage	D	$\sum D$	B	X	T	T_f	GH	H
1	16.70	16.70	3368.1	42208.2	103.25	91	0.078	0.278
2	16.61	33.31	3351.5	42417.5	100.5	88.25	0.081	0.281
3	16.53	49.84	3335.0	42627.7	97.75	85.5	0.084	0.284
4	16.45	66.30	3318.5	42839.1	95	82.75	0.087	0.287
5	16.37	82.67	3302.1	43051.4	92.25	80	0.09	0.29
6	16.29	98.96	3285.8	43264.9	89.5	77.25	0.094	0.294
7	16.21	115.16	3269.6	43479.3	86.75	74.5	0.097	0.297
8	16.13	131.29	3253.5	43694.9	84	71.75	0.101	0.301
9	16.05	147.34	3237.5	43911.5	81.25	69	0.105	0.305
10	15.97	163.31	3221.5	44129.2	78.5	66.25	0.11	0.31
11	15.89	179.21	3205.6	44348.0	75.75	63.5	0.114	0.314
12	15.81	195.02	3189.8	44567.8	73	60.75	0.119	0.319
13	15.74	210.75	3174.0	44788.8	70.25	58	0.124	0.324
14	15.66	226.41	3158.4	45010.8	67.5	55.25	0.13	0.33
15	15.58	241.99	3142.8	45233.9	64.75	52.5	0.136	0.336
16	15.50	257.50	3127.3	45458.2	62	49.75	0.143	0.343
17	15.43	272.92	3111.9	45683.5	59.25	47	0.15	0.35
18	15.35	288.27	3096.5	45910.0	56.5	44.25	0.157	0.357
19	15.28	303.55	3081.3	46137.6	53.75	41.5	0.165	0.365
20	15.20	318.75	3066.1	46366.3	51	38.75	0.175	0.375
21	15.12	333.87	3050.9	46596.2	48.25	36	0.185	0.385
22	15.05	348.92	3035.9	46827.2	45.5	33.25	0.197	0.397
23	14.98	363.90	3020.9	47059.3	42.75	30.5	0.211	0.411
24	14.90	378.80	3006.0	47292.6	40	27.75	0.211	0.411

Problems

Problem 1

An MSF-OT circulation plant has the following design data:

Plant capacity: Unknown
Seawater temperature: 32 °C
Seawater salinity: 49400 ppm
Top brine temperature: 100 °C
Performance ratio: 8
Number of stages: 22
Overall heat transfer coefficient in brine heater: 2 kW/m^2 °C
Overall heat transfer coefficient in heat rejection section: 1.9 kW/m^2 °C
Overall heat transfer coefficient

in heat recovery section:	2.4 kW/m^2 °C

Calculate the following:

- The heating steam flow rate.
- The plant capacity
- The specific heat transfer area

Problem 2

An MSF-OT plant has the following design data:

Plant capacity:	Unknown
Top brine temperature:	Unknown
Brine flow rate per chamber width:	Unknown
Number of stages:	20
Boiling temperature in last stage:	40 °C
Heat transfer area in the brine heater:	1000 m^2
Overall heat transfer coefficient in all sections:	2.527 kW/m^2 °C
Mass flow rate of heating steam:	16.782 kg/s
Heating steam temperature:	120 °C
Number of tubes in the brine heater:	1000 tube
Specific flow rate of feed water:	8.422
Diameter of tubes used in brine heater:	31.8 mm

Calculate the following:

- The plant performance ratio
- The specific heat transfer area
- The dimensions of stage 7.

Problem 3

An MSF-OT plant has the following design data:

Seawater temperature:	34 °C
Seawater salinity:	42000 ppm
Top brine temperature:	100 °C
Temperature in the last stage:	40 °C
Temperature of heating steam:	110 °C
Specific flow rate of brine circulation:	8.478
Heat transfer area in the brine heater:	80 m^2
Overall heat transfer coefficient	

in brine heater: 1.5 kW/m^2 °C

Calculate the following

- The plant performance ratio
- The terminal temperature difference of the first stage.
- The specific heat transfer area.
- The dimensions of stage 7.

Problem 4

An MSF brine circulation plant has the following design data:

Distillate flow rate:	5000 m^3/d
Seawater temperature:	30 °C
Seawater salinity:	44000 ppm
Top brine temperature:	112 °C
Temperature in the last stage:	40 °C
Temperature of heating steam:	116.7 °C
Terminal temperature difference:	3 °C
Number of stages:	20

The overall heat transfer coefficient in the brine heater or the flashing stages is given by the relation

$$U = 6.5 - 0.03(115\text{-}T)$$

With U in kW/m^2 °C and T in °C.

Calculate the following

- Thermodynamic losses in the first stage
- Plant performance ratio
- Brine heater surface area
- Flow rate of feed water.

Also, calculate the following parameters for stage number 5

- Boiling point elevation.
- Gate height
- Liquid level
- Demister temperature loss
- Stage height
- Stage length
- Preheater surface area.

Problem 5

An MSF-OT plant has the following design data:

Distillate flow rate:	22750 m^3/d
Seawater temperature:	28 ^oC
Seawater salinity:	45000 ppm
Top brine temperature:	90 ^oC
Temperature in the last stage:	40 ^oC
Temperature of heating steam:	100 ^oC
Terminal temperature difference:	3 ^oC
Number of heat rejection stages:	3
Number of heat recovery stages:	25
Width of stage 10:	16 m
Length of stage 10:	3.5 m

Calculate the following

- The temperature profile of flashing brine
- The temperature profile of seawater flowing in the preheaters
- The flow rates of heating steam, feed water, and brine blowdown.
- The pressure in stages 9,10, and 11.
- The distillate product in stage 10.
- The vapor velocity in stage 10.

6.5. Brine Circulation MSF

The MSF process with brine circulation is one of the major processes of the desalination industry. As mentioned before, the process is developed in the late fifties and since then it has gone several modifications to optimize its performance. At present, this process is considered the most suitable for large scale production capacity, where the conventional capacity of a single unit amounts to 25,000 m^3/d. Regardless, further developments remains to be necessary to improve performance and reduce product cost. Features and performance of the brine circulation process is discussed in this section, while possible process developments are proposed in the next sections.

6.5.1 Process Description

Schematic of the brine circulation MSF process is shown in Fig. 23 and process variables in the brine heater and the flashing stages are shown in Fig. 24. Details of the MSF process are described below:

- The intake seawater stream (M_f+M_{cw}) is introduced into the condenser tubes of the heat reject section, where its temperature is increased to a higher temperature by absorption of the latent heat of the condensing fresh water vapor.
- The warm stream of intake seawater is divided into two parts: the first is the cooling seawater (M_{cw}), which is rejected back to the sea and the second is the feed seawater (M_f), which is deaerated, chemically treated and then mixed in the brine pool of the last flashing stage in the heat rejection section.
- The brine recycle stream (M_r) is extracted from the brine pool of the last stage in the heat rejection section and is introduced into the condenser tubes of the last stage in the heat recovery section. As the stream flows in the condenser tubes across the stages it absorbs the latent heat of condensation from the flashing vapor in each stage.
- The brine recycle stream (M_r) enters the brine heater tubes, where the heating steam (M_s) is condensed on the outside surface of the tubes. The brine stream absorbs the latent heat of condensing steam and its temperature increases to its maximum design value known as the top brine temperature (T_o). Its value depends on the nature of chemicals used to control the scale formation.
- The hot brine enters the flashing stages in the heat recovery section and then in the heat rejection section, where a small amount of fresh water vapor is formed by brine flashing in each stage. The flashing process takes place due to decrease in the stage saturation temperature and causes the reduction in the stage pressure.

- In each stage of the heat recovery section, the flashed off vapors condenses on the outside surface of the condenser tubes, where the brine recycle stream (M_r) flows inside the tube from the cold to the hot side of the plant. This heat recovery improves the process efficiency because of the increase in the feed seawater temperature.
- The condensed fresh water vapor outside the condenser tubes accumulates across the stages and forms the distillate product stream (M_d). This stream cascades in the same direction of the flashing brine from stage to stage and is withdrawn from the last stage in the heat rejection section.
- The flashing process and vapor formation is limited by increase in the specific vapor volume at lower temperatures and difficulties encountered for operation at low pressures. Common practice limits the temperature of the last stage to range of 30 to 40 ^{o}C, for winter and summer operation, respectively. Further reduction in these temperatures results in drastic increase of the stage volume and its dimensions.
- In MSF, most of flashing stages operating at temperatures below 100 ^{o}C have vacuum pressure. This increases the possibilities of in-leakage of the outside air. Also, trace amounts of dissolved gases in the flashing brine, which are not removed in the deaerator or formed by decomposition of $CaHCO_3$. At such conditions, air and other gases are non-condensable and its presence in the system may result in severe reduction in the heat transfer rates within the chamber, increase of the tendency for corrosion, and reduction of the flashing rates. This condition necessitates proper venting of the flashing stages to enhance the flashing process and to improve the system efficiency.
- Treatment of the intake seawater (M_f+M_{cw}) is limited simple screening and filtration. On the other hand, treatment of the feed seawater stream is more extensive and it includes dearation and addition of chemicals to control scaling, foaming, and corrosion.

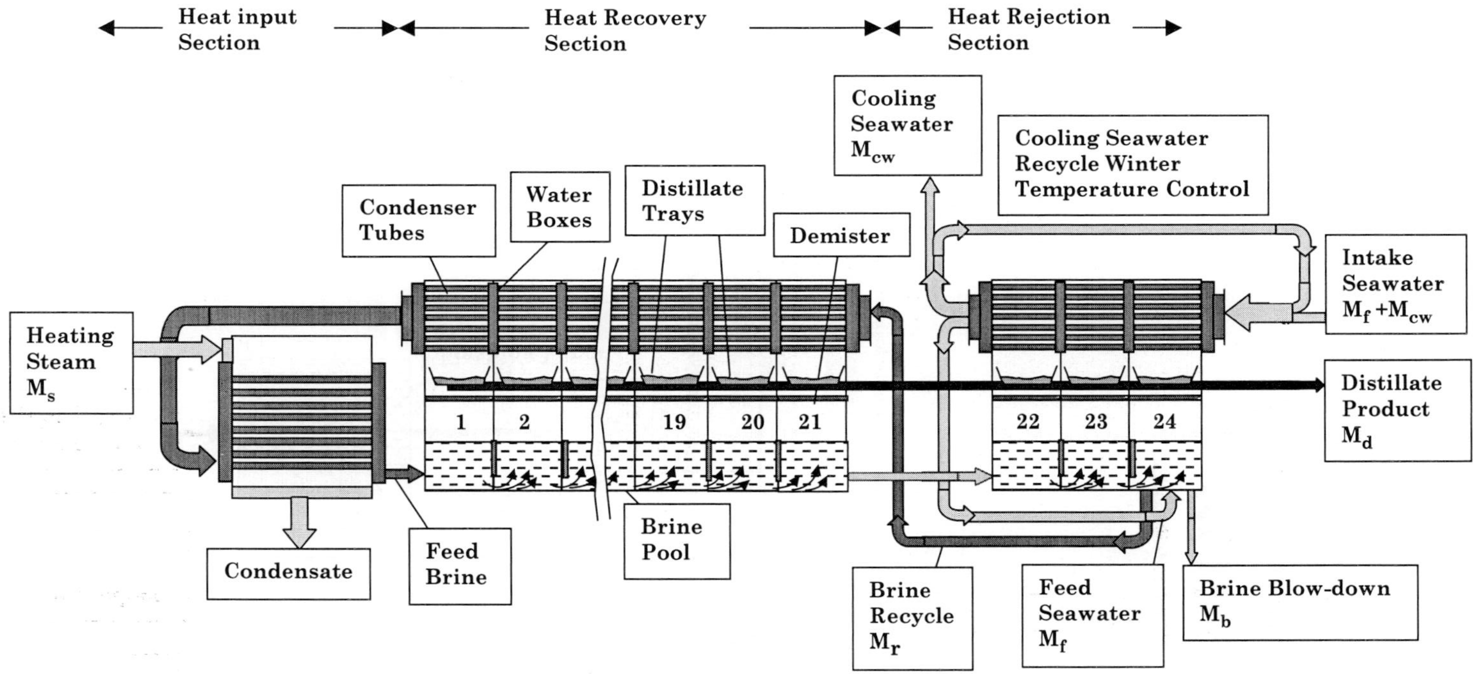

Fig. 23. Multistage flash desalination with brine circulation (MSF)

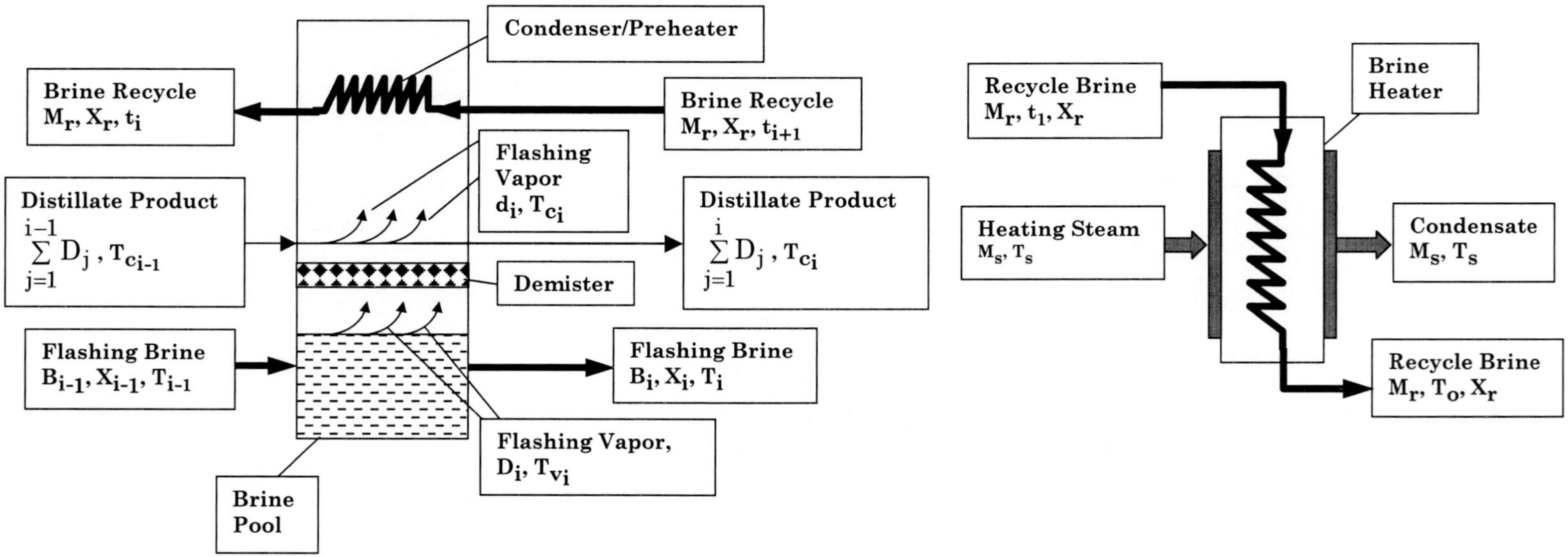

Fig. 24. Schematics of model variables in brine heater and flashing stage.

6.5.2 Mathematical Model

The MSF simplified model is a very useful tool for obtaining quick design data, evaluating system performance, and developing a good initial guess for more detailed mathematical models. This model does not need iterative solution and requires minimal computational effort. Model assumptions include the following:

- Constant and equal specific heat for all liquid streams.
- Equal temperature drop per stage for the flashing brine.
- Equal temperature drop per stage for the feed seawater.
- The latent heat of vaporization in each stage is assumed equal to the average value for the process.
- The non-condensable gases have negligible effect on the heat transfer process.
- Effects of the boiling point rise and non-equilibrium losses on the stage energy balance are negligible; however, their effects are included in the design of the condenser heat transfer area.
- The temperature of the feed seawater leaving the rejection section is equal to the brine temperature in the last stage.

Overall Material Balance

The overall material balance equations is given by

$$M_f = M_d + M_b \tag{59}$$

where M is the mass flow rate and the subscript b, d, and defines the brine, distillate, and feed. The overall salt balance is given by

$$X_f M_f = X_b M_b \tag{60}$$

where X is the salt concentration. Equation (60) assumes that the distillate is salt free. Equations (59) and (60) can be rearranged to obtain the expression for the total feed flow rate in terms of the distillate flow rate; this is

$$M_f = X_b/(X_b - X_f) M_d \tag{61}$$

Equation (61) is used to calculate M_f, since the values of X_b, X_f, and M_d are known.

Stages and Condensers Temperature Profiles

The temperature distribution in the MSF system is defined in terms of four temperatures; these are the temperatures of the steam, T_s, the brine leaving the preheater (top brine temperature), T_o, the brine leaving the last stage, T_n, and the intake seawater, T_{cw}. A linear profile for the temperature is assumed in the stages and the condensers. First, the temperature drop per stage, ΔT, is obtained from the relation

$$\Delta T = (T_o - T_n)/n \tag{62}$$

where n is the number of recovery and rejection stages. Therefore, the temperature in the first stage is given by

$$T_1 = T_o - \Delta T$$

As for the second stage temperature it is equal to

$$T_2 = T_1 - \Delta T$$

Substituting for T_1 in the above equation gives

$$T_2 = T_o - \Delta T - \Delta T = T_o - 2\,\Delta T$$

The same procedure is repeated for subsequent stages and a general expression is developed for the temperature of stage i

$$T_i = T_o - i\,\Delta T \tag{63}$$

The recycle seawater, which flows into the condensers of the heat recovery section, has a temperature equal to T_n. This temperature is assumed to increase by ΔT_r in the condenser of each unit. This temperature increase, ΔT_r, is equal to the decrease in the brine temperature in each stage, ΔT. This result is arrived at by performing an energy balance on stage i, which gives

$$D_i\ Cp\ T_{v_i} + B_i\ Cp\ T_i - D_{i+1}\ Cp\ T_{v_{i+1}} - B_{i+1}\ Cp\ T_{i+1} = M_r\ C_p\ (T_{r_i} - T_{r_{i+1}})$$

Assuming the temperature difference, $T_i - T_{v_i}$, is small and has a negligible effect on the stage energy balance. Thus, the above equation reduces to

$$(D_i + B_i)\ Cp\ T_i - (D_{i+1} + B_{i+1})\ Cp\ T_{i+1} = M_r\ C_p\ (T_{r_i} - T_{r_{i+1}})$$

Recalling that the sum $(D_i + B_i)$ in each stage is equal to M_r, would simplify the above equation to

$$M_r\ Cp\ T_i - M_r\ Cp\ T_{i+1} = M_r\ C_p\ (T_{r_i} - T_{r_{i+1}})$$

Elimination of the like terms on both sides of the equation gives the pursued relation, thus,

$$T_i - T_{i+1} = T_{r_i} - T_{r_{i+1}}$$

or

$$\Delta T_i = \Delta T_{r_i}$$

The seawater temperature, which leaves the condenser of the first stage, is then defined by

$$T_{r_1} = T_n + (n - j)\ \Delta T$$

The seawater temperature leaving the condenser of the second stage, T_{r_2}, is less than T_{r_1} by ΔT, where

$$T_{r_2} = T_{r_1} - \Delta T$$

Substituting for T_{r_1} in the above equation gives

$$T_{r_2} = T_n + (n - j)\ \Delta T - \Delta T$$

Similar to Eq. (63), a general equation is obtained for the condenser temperature in stage i

$$T_{r_i} = T_n + (n - j)\ \Delta T - (i - 1)\ \Delta T \quad (64)$$

As for the temperature drop of the seawater in the condensers of the heat rejection section, it is obtained from the stage energy balance. This is

$$D_i\ Cp\ T_{v_i} + B_i\ Cp\ T_i - D_{i+1}\ Cp\ T_{v_{i+1}} - B_{i+1}\ Cp\ T_{i+1} = (M_f + M_{cw})\ C_p\ (T_{f_j} - T_{f_{j+1}})$$

Assuming the small temperature difference, $T_i - T_{v_i}$, has a negligible effect on the stage energy balance. Thus, the above equation reduces to

$$(D_i + B_i)\ Cp\ T_i - (D_{i+1} + B_{i+1})\ Cp\ T_{i+1} = (M_f + M_{cw})\ C_p\ (T_{f_i} - T_{f_{i+1}})$$

Recalling that the sum (D_i + B_i) in each stage is equal to M_r, would simplify the above equation to

$$M_r\ Cp\ T_i - M_r\ Cp\ T_{i+1} = (M_f + M_{cw})\ C_p\ (T_{j_i} - T_{j_{i+1}})$$

Elimination of the like terms on both sides of the equation gives the pursued relation, thus,

$$(T_{j_i} - T_{j_{i+1}}) = (T_i - T_{i+1})\ (M_r/(M_f + M_{cw}))$$

or

$$(\Delta T_{j_i}) = \ \Delta T_i\ (M_r\ /(M_f + M_{cw}))$$

Since the temperature profile is assumed linear, the above relation can also be obtained from the following simple relation

$$(\Delta T_{j_i}) = (T_n - T_{cw})/j$$

The seawater temperature, which leaves the condenser of the last stage, is then defined by

$$T_{j_n} = T_{cw} + (\Delta T_{j_i})$$

This gives the general relation for the seawater temperature in the rejection section

$$T_{j_i} = T_{cw} + (n\text{-}i+1)(\Delta T_{j_i}) \qquad (65)$$

Stage Material and Salt Balance

The amount of flashing vapor formed in each stage obtained by conservation of energy within the stage, where the latent consumed by the flashing vapor is set equal to the decrease in the brine sensible heat. This is

$$D_1 = y\ M_r$$

where D_1 is the amount of flashing vapor formed in the first stage, M_r is the recycle brine flow rate, and y is the specific ratio of sensible heat and latent heat and is equal to

$$y = Cp\ \Delta T/\lambda_{av} \qquad (66)$$

where Cp is the specific heat capacity and λ_{av} is the average latent heat calculated at the average temperature

$$T_{av} = (T_o + T_n)/2 \tag{67}$$

The amount of distillate formed in the second stage is equal to

$$D_2 = y\ (M_r - D_1)$$

Substituting the value of D_1 in the above equation gives

$$D_2 = y\ (M_r - y\ M_r)$$

Which simplifies to

$$D_2 = M_r\ y\ (1 - y)$$

The balance equations for D_2 and D_3 will reveal the general form for the formula of D_i. The D_3 balance is

$$D_3 = y\ (M_r - D_1 - D_2)$$

Substituting for the values of D_1 and D_2 in the above equation gives

$$D_3 = y\ (M_r - M_r\ y - M_r\ y\ (1 - y))$$

Taking M_r as a common factor in the above equation gives

$$D_3 = M_r\ y\ (1 - y - y + y^2)$$

This simplifies to

$$D_3 = M_r\ y\ (1 - y)^2$$

Accordingly, the resulting general formula for D_i is

$$D_i = M_r\ y\ (1 - y)^{(i-1)} \tag{68}$$

The total distillate flow rate is obtained by summing the values of D_i for all stages. The summation is performed in steps in order to obtain a closed form equation. Therefore, the summation of D_1 and D_2 gives

$$D_1 + D_2 = M_r\,(y + y\,(1 - y))$$
$$= M_r\,(2\,y - y^2)$$
$$= M_r\,(1 - (1 - y)^2)$$

Addition of D_3 to the above gives

$$D_1 + D_2 + D_3 = M_r\,((2\,y - y^2) + y(1 - y)^2)$$

This simplifies to

$$D_1 + D_2 + D_3 = M_r\,(2\,y - y^2 + y - 2\,y^2 + y^3)$$
$$= M_r\,(3\,y - 3\,y^2 + y^3)$$
$$= M_r\,(3\,y - 3\,y^2 + y^3)$$
$$= M_r\,(1 - (1 - y)^3)$$

Comparison of the summations of D_1+D_2 and D_1+D_2+D_3 gives the general form for the total summation of the distillate formed in all stages, M_d, which is given by

$$M_d = M_r\,(1 - (1 - y)^n) \quad (69)$$

Equation 69 is used to calculate the brine recycle flow rate, since the distillate flow is always specified in design problems.

The salt concentration in the recycle stream, X_r, is obtained by performing salt balance on the loop shown in Fig. 25. This balance is

$$X_r\,M_r + M_b\,X_b = X_f\,M_f + (M_r - M_d)\,X_n$$

The above balance is arranged to

$$X_r = (X_f\,M_f + (M_r - M_d)\,X_n - M_b\,X_b)/M_r$$

Assuming that $X_n = X_b$, simplifies the above equation to

$$X_r = (X_f\,M_f + (M_r - M_d)\,X_b - M_b\,X_b)/M_r$$

Since $M_f = M_b + M_d$, then,

$$X_r = ((X_f - X_b)\, M_f + M_r\, X_b)/M_r \tag{70}$$

The flow rate of the brine stream leaving stage i is given by

$$B_i = M_r - \sum_{k=1}^{i} D_k \tag{71}$$

The salt concentration in the brine stream leaving stage i is given by

$$X_i = \left(M_r - \sum_{k=1}^{i} D_k \right) \Big/ B_i \tag{72}$$

The last item in this section is to determine the cooling water flow rate, M_{cw}. This is required to obtain the specific cooling water flow rate, sM_{cw}, which affects the process economics. This flow rate is obtained from an overall energy balance around the desalination plant, Fig. 26. The intake seawater temperature, T_{cw}, is used as the reference temperature in the energy balance. This gives

$$M_s\, \lambda_s = M_{cw}\, C_p\, (T_n - T_{cw}) + M_b\, C_p (T_n - T_{cw}) + M_d\, C_p\, (T_n - T_{cw})$$

The above equation is arranged to obtain an expression for M_{cw}

$$M_{cw} = (M_s\, \lambda_s - M_f\, C_p (T_n - T_{cw}))/\, (C_p\, (T_n - T_{cw})) \tag{73}$$

Brine Heater and Condensers Heat Transfer Area

The motive steam provides the brine heater with the necessary energy to increase the feed seawater temperature from T_{f_1} to the top brine temperature, T_o. This requires calculation of the motive steam flow, which is obtained from the brine heater energy balance

$$M_s\, \lambda_s = M_r\, Cp\, (T_o - T_{f_1})$$

The above equation is arranged to calculate M_s

$$M_s = M_r\, Cp\, (T_o - T_{f_1})/\, \lambda_s \tag{74}$$

The brine heater area is given by

$$A_b = M_s \lambda_s/(U_b (LMTD)_b) \tag{75}$$

where LMTD is given by

$$(LMTD)_b=((T_s-T_o)-(T_s-T_{f_1}))/\ln((T_s-T_o)/(T_s-T_{f_1})) \tag{76}$$

and U_b is given by

$$U_b=1.7194+3.2063x10^{-3}\ T_s+1.5971x10^{-5}\ T_s^2 -1.9918x10^{-7}T_s^3 \tag{77}$$

The heat transfer area for the condenser in each stage in the heat recovery section is assumed equal. The same assumption is made for the condenser heat transfer area in the heat rejection section. Therefore, the calculated heat transfer area for the first stage is used to obtain the total heat transfer area in the heat recovery section. The condenser heat transfer area in the first stage is obtained from

$$A_r = M_r\ Cp\ (T_{r_1} - T_{r_2})/(U_r\ (LMTD)_r) \tag{78}$$

where

$$U_r= 1.7194+3.2063x10^{-3}\ T_{v_1}+1.5971x10^{-5}T_{v_1}^2 -1.9918x10^{-7}T_{v_1}^3 \tag{79}$$

$$T_{v_1} = T_1 - BPE_1 - NEA_1 - \Delta T_{d_1} \tag{80}$$

and

$$(LMTD)_r=((T_{v_1}-T_{r_1})-(T_{v_1}-T_{r_2}))/\ln((T_{v_1}-T_{r_1})/(T_{v_1}-T_{r_2})) \tag{81}$$

In the above equations BPE is the boiling point elevation in the first stage, NEA, is the non-equilibrium allowance, T_v, is condensing vapor temperature, ΔT_d, is the temperature drop in the demister, and U_r is the condenser overall heat transfer coefficient. Expressions for correlations used calculate the BPE, NEA, and ΔT_d, are given in the appendix.

The same procedure is applied to the stages in the heat rejection section, where the condenser area in rejection stages is given by

$$A_j = (M_f + M_r)\ Cp\ (T_{j_n} - T_{cw})/(U_j\ (LMTD)_j) \tag{82}$$

where

$$U_j = 1.7194+3.2063x10^{-3}T_{v_n}+1.5971x10^{-5}T_{v_n}{}^2- 1.9918x10^{-7}T_{v_n}{}^3 \quad (83)$$

$$T_{v_n} = T_n - BPE_n - NEA_n - \Delta T_{d_n} \quad (84)$$

and

$$(LMTD)_j=((T_{v_n}-T_{cw})-(T_{v_n}-T_{j_n}))/\ln((T_{v_n}-T_{cw})/(T_{v_n}-T_{j_n})) \quad (85)$$

The total heat transfer area for all condensers in the heat recovery and rejection sections is then obtained from

$$A_c = (n-j)\ A_r + (j)\ A_j \quad (86)$$

Stage Dimensions

Calculations of the stage dimensions include the gate height, the height of the brine pool, the stage width, and the stage length. The length of all stages is set equal to the length of the last stage and the width of all stages is set equal to the width of the first stage. The height of the brine pool must be higher than the gate height, this is necessary to prevent bypass of the vapors between stages (vapor blow through). The gate height, GH, is obtained in terms of the stage pressure drop, ΔP, the brine density, ρ_b, the weir friction coefficient, C_d, the stage width, W, and the brine recycle flow rate, M_r. For stage i the gate height is

$$GH_i = (M_r - \sum_{j=1}^{i-1} D_j)\ (2\ \rho_{b_i}\ \Delta P_i)^{(-0.5)}/(C_d\ W) \quad (87)$$

The brine pool height is set higher than the gate height by 0.2 m.

$$H_i = 0.2 + GH_i \quad (88)$$

where

$$\Delta P_i = P_i - P_{i+1}$$

$$W = M_r / V_b \quad (89)$$

where P_i and P_{i+1} are the pressures in stages i and i+1, and V_b is the brine mass velocity per chamber width. The length of the last stage is determined as a function of the vapor flow rate, D_n, the vapor density, ρ_{vn}, the vapor allowable velocity, V_{vn}, and the stage width, W. This is

$$L = D_n/(\rho_{vn} V_{vn} W) \tag{90}$$

The cross section area for each stage, A_s, is then calculated

$$A_s = L W \tag{91}$$

Performance Parameters

The system performance parameters are defined by the thermal performance ratio, PR, the specific heat transfer area, sA, and the specific cooling water flow rate, sM_{cw}. The performance ratio is the ratio of product to steam flow rates. This is

$$PR = M_d/M_s \tag{92}$$

The specific heat transfer area is defined by

$$sA = (A_b + A_c)/M_d \tag{93}$$

The specific cooling water flow rate is given by

$$sM_{cw} = M_{cw}/M_d \tag{94}$$

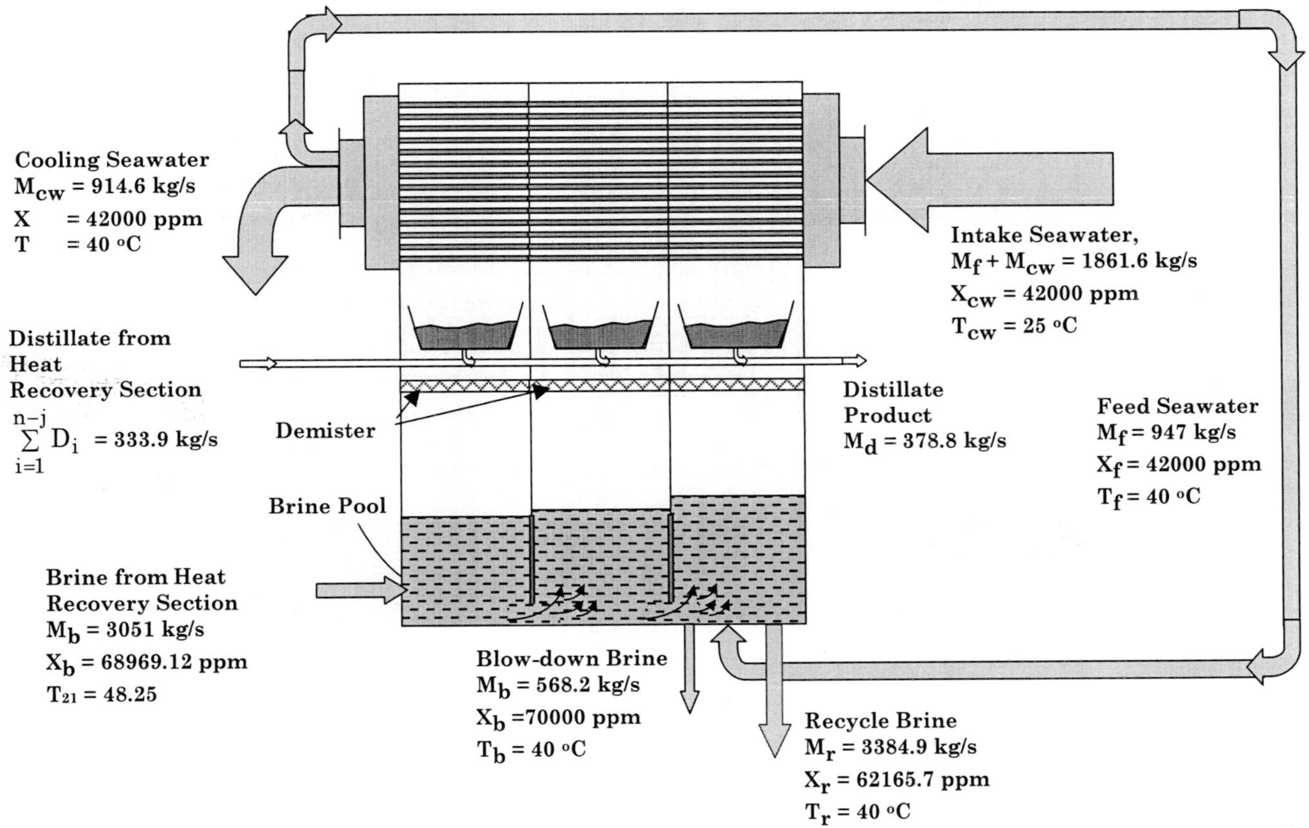

Fig. 25. Material and salt balance on the heat rejection section.

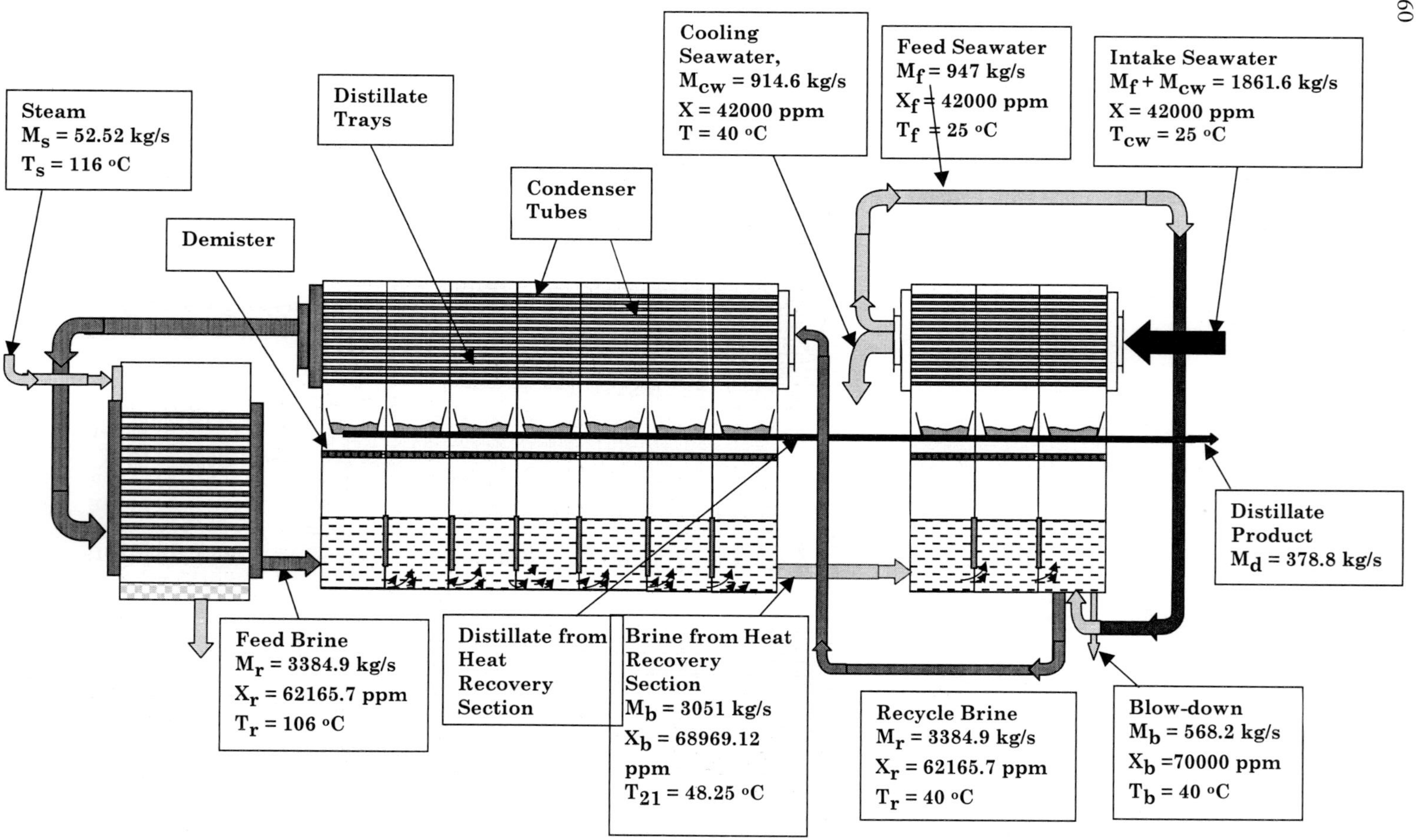

Fig. 26. Stream flow rate, salinity, and temperature in MSF

Solution Method

Solution of the MSF simplified model is non-iterative and can be limited to calculations of the system performance parameters, which include the performance ratio, the specific cooling water flow rate, and the specific heat transfer area. This reduces calculations of the temperature, flow rate, and concentration profiles to those of the first stage and the recycle brine. The solution proceeds as follows:

- The flow rates of the feed seawater, M_f, and blow down brine, M_b, are obtained from Eqs. 60 and 58, respectively.
- The temperature drop per stage, ΔT, is calculated from Eq. 62.
- The temperatures of stages 1 and 2, T_1 and T_2, are calculated from Eq. 63.
- The seawater temperatures leaving the condensers of the first and second stages, T_{r_1} and T_{r_2}, are determined from Eq. 64. Also, the seawater temperature leaving the condenser of the last stage, T_{j_n}, is calculated from Eq. 65.
- The average stage temperature, T_{av}, is calculated from Eq. 67 and the corresponding latent heat value is obtained from the correlation given in the appendix.
- The ratio of the stage sensible and latent heat, y, is calculated from equation 66.
- The flow rates of recycle brine, M_r, cooling seawater, M_{cw}, and steam, M_s, are calculated from Eqs. 69, 73, and 74, respectively.
- The heat transfer area for the brine preheater (A_b) the first stage condenser in the recovery section (A_r) and the last condenser in the rejection section (A_j) are calculated from Eqs. 75, 78, and 82, respectively.
- The stage length, width, and cross section areas are determined from Eqs. 89-91. Also, the height of the brine pool, H, and the gate height for stage i are obtained from Eqs. 87-88.
- The system performance parameters, PR, sA, and sM_{cw}, are calculated from Eqs. 92-94.

The above solution process requires specification of the following variables:

- Total distillate flow rate, M_d.
- Total number of stages (recovery and rejection).
- Number of heat rejection stages, j.
- Intake seawater temperature, T_{cw}.
- Top brine temperature, T_o.
- Steam temperature, T_s.
- Brine temperature of stage n.
- Brine salt concentration, X_b.

- Intake seawater salt concentration, X_f.
- Heat capacity of liquid streams, Cp.
- Weir friction coefficient, C_d.
- Vapor velocity in the last stage, V_{vn}.
- Brine mass flow rate per stage width, V_b.

Case Study

An MSF system with 24 stages is used to produce 7.2 MGD of product water. The system contains 21 stages in the heat recovery section and 3 stages in the heat rejection section. The following specifications are made to obtain the system design parameters and performance characteristics. The specifications include the following:

- Intake seawater temperature, T_{cw} = 25 °C.
- Steam temperature, T_s = 116 °C.
- Top brine temperature, T_o = 106 °C.
- Brine temperature in the last stage, T_n = 40 °C.
- Heat capacity of liquid streams, C_p = 4.18 kJ/kg °C.
- Salinity of intake seawater, X_f = 42000 ppm.
- Salinity of the brine blow-down, X_b = 70000 ppm.
- Vapor velocity in the last stage, V_v = 6 m/s
- Brine mass flow rate per stage width, V_b = 180 kg/ms.
- Weir friction coefficient, C_d = 0.5.

Before proceeding, the product volume flow rate is converted to mass rate in SI units; this is necessary for solution of the energy balance equations. The kg/s value of M_d is

$$M_d = \frac{\left(7.2\left(\frac{MG}{D}\right)\right)\left(10^6\left(\frac{G}{MG}\right)\right)\left(\frac{1}{24x3600}\left(\frac{D}{s}\right)\right)}{\left(\frac{1}{219.96}\left(\frac{m^3}{G}\right)\right)10^3\left(\frac{kg}{m^3}\right)}$$

$$= 378.8 \text{ kg/s}$$

The seawater feed flow rate is obtained from Eq. 60, where

$$M_f = X_b/(X_b - X_f)\, M_d$$
$$= 70000/(70000 - 42000)\,(378.8)$$

$= 947$ kg/s

The flow rate of the blow-down brine is then obtained from the overall balance given in Eq. 58,

$$
\begin{aligned}
Mb &= Mf - Md \\
&= 947 - 378.8 \\
&= 568.2 \text{ kg/s}
\end{aligned}
$$

The temperature drop in each effect is obtained from Eq. 62, where

$$
\begin{aligned}
\Delta T &= (T_o - T_n)/n \\
&= (106 - 40)/24 \\
&= 2.75\ ^{o}C
\end{aligned}
$$

This value is used to calculate the temperatures of the first and second stages, T_1 and T_2, where

$$
\begin{aligned}
T_1 &= T_o - \Delta T \\
&= 106 - 2.75 \\
&= 103.25\ ^{o}C
\end{aligned}
$$

As for the second stage temperature it is equal to

$$
\begin{aligned}
T_2 &= T_1 - \Delta T \\
&= 103.25 - 2.75 \\
&= 100.5\ ^{o}C
\end{aligned}
$$

Also the temperatures of the seawater leaving the condensers in the first and second stages are calculated. This is

$$
\begin{aligned}
T_{r_1} &= T_n + (n - j)\ \Delta T \\
&= 40 + (24 - 3)\ (2.75) \\
&= 97.75\ ^{o}C
\end{aligned}
$$

and

$$
\begin{aligned}
T_{r_2} &= T_{r_1} - \Delta T \\
&= 97.75 - 2.75 \\
&= 95\ ^{o}C
\end{aligned}
$$

Calculation of the y ratio from Eq. (64) is preceded by evaluation of T_{av} and λ_{av}. The T_{av} value is

$$T_{av} = (T_o + T_n)/2$$
$$= (106+40)/2$$
$$= 73\ ^oC$$

At this temperature λ_{av} is equal to 2330.1 kJ/kg from the correlation given in the appendix. The value of y is calculated

$$y = Cp\ \Delta T/\lambda_{av}$$
$$= (4.18)\ (2.75)/2330.1$$
$$= 4.933x10^{-3}$$

The brine recycle flow rate is obtained from Eq. 59

$$M_r = M_d/(1 - (1 - y)^n)$$
$$= 378.8 = M_r\ (1 - (1 - 4.933x10^{-3})^{24})$$
$$= 3384.8\ kg/s$$

The steam flow rate is calculated from Eq. 74

$$M_s = M_r\ Cp\ (T_o - T_{r_1})/\ \lambda_s$$
$$= (3384.8)\ (4.18)\ (106\text{-}97.75)/(2222.33)$$
$$= 52.52\ kg/s$$

The heat transfer area for the brine heater, A_b, is calculated from Eq. 73. This requires calculations of the logarithmic mean temperature difference $(LMTD)_b$ and the overall heat transfer coefficient. The value of $(LMTD)_b$ is obtained from Eq. 76, which gives

$$(LMTD)_b = ((T_s - T_o) - (T_s - T_{r_1}))/ln((T_s - T_o)/(T_s - T_{r_1}))$$
$$= ((116 - 106) - (116 - 97.75))/$$
$$ln((116 - 106)/(116 - 97.75))$$
$$= 13.71\ ^oC$$

The overall heat transfer coefficient is obtained from Eq. 77. This gives

$$U_b = 1.7194\text{+}3.2063x10^{-3}\ T_s\text{+}1.5971x10^{-5}\ (Ts)^2 - 1.9918x10^{-7}(Ts)^3$$
$$= 1.7194 + 3.2063x10^{-3}\ (116) + 1.5971x10^{-5}\ (116)^2 - 1.9918x10^{-7}\ (116)^3$$
$$= 2\ kW/m^2\ ^oC$$

The brine heater area, A_b, is then calculated from Eq. 75

$$A_b = M_s \lambda_s/(U_b\ (LMTD)_b)$$
$$= (52.52)(2222.33)/((2)(13.71))$$
$$= 4256.6\ m^2$$

The condenser area in the heat recovery section, A_r, is determined for the first stage. Determination of this value requires calculations of the vapor condensation temperature, T_{v_1}, the logarithmic mean temperature difference $(LMTD)_r$, and the overall heat transfer coefficient, U_r. The vapor temperature is given by Eq. 80, where

$$T_{v_1} = T_1 - BPE_1 - NEA_1 - \Delta T_{d_1}$$

The boiling point elevation is calculated from the correlation given in the appendix. Solution of this equation requires calculation of the recycle brine concentration, X_r, which is defined by Eq. 70. This gives

$$X_r = (X_f\, M_f + (M_r - M_d)\, X_n - M_b\, X_b)/M_r$$
$$= ((42000)(947)+(3384.8 - 378.8)(70000) - (568)(70000))/(3384.8)$$
$$= 62170\ ppm$$

The values of B and C in the correlation for the boiling point elevation are

$$B = (6.71 + 6.34x10^{-2}(T_1) + 9.74x10^{-5}(T_1)^2)10^{-3}$$
$$= (6.71 + 6.34x10^{-2}(103.25) + 9.74x10^{-5}(103.25)^2)10^{-3}$$
$$= 0.0143$$

$$C = (22.238 + 9.59x10^{-3}(T_1) + 9.42x10^{-5}(T_1)^2)10^{-8}$$
$$= (22.238 + 9.59x10^{-3}(103.25) + 9.42x10^{-5}(103.25)^2)10^{-8}$$
$$= 2.423x10^{-7}$$

Substituting the values of B and C in the BPE correlation gives

$$BPE_1 = X_r\ (B + (X_r)(C))10^{-3}$$
$$= 62165\left(0.0143 + (62165)\left(2.423x10^{-7}\right)\right)10^{-3}$$
$$= 1.83\ ^oC$$

The non-equilibrium allowance, NEA_1, in the first stage is calculated from the correlation given in the appendix for the MSF system. This involves calculations of the gate height, GH_1, the height of the brine pool, H_1, the stage width, W, the stage pressure drop, P_1-P_2, and the brine density. The width of the first stage is given by

$$W = M_r / V_b$$
$$= 3384.8/180$$
$$= 18.8 \text{ m}$$

The stage length is calculated for the last stage, where

$D_{24} = 14.904$ kg/s and

$\rho_{v_n} = 0.0512$ kg/m3,

$$L = D_n/(\rho_{v_n} V_{v_n} W) = 14.904/((0.0512)(6)(18.8))$$
$$= 2.58 \text{ m}$$

The brine density in the first stage is 1002.413 kg/m^3, which is obtained from the correlation given the appendix at a salinity of 62473.9 ppm and a temperature of 103.25 °C. The pressures of the first and second stage are obtained from the saturation pressure correlations, where, at $T_1 = 103.25$ °C, $P_1 =$ 113.72 kPa, and at $T_2 = 100.5$ °C, $P_2 = 103.23$ kPa. The resulting gate height in the first stage, GH_1, is calculated from Eq. 82. This result in

$$GH_1 = M_r\,(2\,\rho_{b_1}\,\Delta P_1)^{(-0.5)}/(C_d\,W)$$
$$= (3384.8)((2)(1002.41)(113.72-103.23)\text{x}10^3)^{(-0.5)}/((0.5)(18.8)$$
$$= 0.078 \text{ m}$$

It should be noted that the pressure drop in the above equation is in Pa and not kPa. The corresponding brine pool height is obtained by simply adding 0.2 m to the value of GH_1, or

$$H_1 = 0.278 \text{ m}$$

The non-equilibrium allowance is then calculated using the correlation given in the appendix

$$NEA_1 = (0.9784)^{T_o}\,(15.7378)^{H_1}\,(1.3777)^{V_b \text{x} 10^{-6}}$$

$= (0.9784)^{(106)}\ (15.7378)^{(0.278)}\ (1.3777)^{(180\text{x}10^{-6})}$
$= 0.213\ ^oC$

The temperature drop in the demister is assumed negligible in comparison with the values of BPE_1 and NEA_1. Therefore, the vapor temperature in the first is

$$
\begin{aligned}
T_{v_1} &= T_1 - BPE_1 - NEA_1 - \Delta T_{d_1} \\
&= 103.25 - 1.83 - 0.213 - 0 \\
&= 101.207\ ^oC
\end{aligned}
$$

The vapor temperature, T_{v_1}, is used to calculate U_r and $(LMTD)_r$, where

$$
\begin{aligned}
(LMTD)_r &= ((T_v-T_{r_1})-(T_v-T_{r_2}))/\ln((T_v-T_{r_1})/(T_v-T_{r_2})) \\
&= ((101.207-97.75)-(101.207-95))/ \\
&\qquad \ln((101.207-97.75)/(101.207-95)) \\
&= 4.69\ ^oC
\end{aligned}
$$

$$
\begin{aligned}
U_r &= 1.7194+3.2063\text{x}10^{-3}\ T_{v_1}+1.5971\text{x}10^{-5}\ (T_{v_1})^2 - 1.9918\text{x}10^{-7}(T_{v_1})^3 \\
&= 1.7194 + 3.2063\text{x}10^{-3}\ (101.207) + 1.5971\text{x}10^{-5}\ (101.207)^2 \\
&\qquad - 1.9918\text{x}10^{-7}\ (101.207)^3 \\
&= 2\ \text{kW/m}^2\ ^oC
\end{aligned}
$$

$$
\begin{aligned}
A_r &= M_r\ Cp\ (T_{r_1} - T_{r_2})/(U_r\ (LMTD)_r) \\
&= ((3384.8)\ (4.18)\ (97.75\text{-}95))/((2)\ (4.68)) \\
&= 4156.9\ \text{m}^2
\end{aligned}
$$

The condenser area in the heat rejection section, A_j, is determined for the last stage. Determination of this value requires calculations of the vapor condensation temperature, T_{vn}, the logarithmic mean temperature difference $(LMTD)_j$, and the overall heat transfer coefficient, U_j. The vapor temperature is given by Eq. 84, where

$$
T_{vn} = T_n - BPE_n - NEA_n - \Delta T_{dn}
$$

The values of B and C in the correlation for the boiling point elevation are

$$
\begin{aligned}
B &= \left(6.71 + 6.34\text{x}10^{-2}(T_n) + 9.74\text{x}10^{-5}(T_n)^2\right)10^{-3} \\
&= \left(6.71 + 6.34\text{x}10^{-2}(40) + 9.74\text{x}10^{-5}(40)^2\right)10^{-3} \\
&= 0.0094018
\end{aligned}
$$

$$
\begin{aligned}
C &= \left(22.238 + 9.59\text{x}10^{-3}(T_n) + 9.42\text{x}10^{-5}(T_n)^2\right)10^{-8} \\
&= \left(22.238 + 9.59\text{x}10^{-3}(40) + 9.42\text{x}10^{-5}(40)^2\right)10^{-8} \\
&= 2.277\text{x}10^{-7}
\end{aligned}
$$

Substituting the values of B and C in the BPE correlation gives

$$
\begin{aligned}
BPE_n &= X_n\ (B + (X_n)(C))10^{-3} \\
&= 70000\left(0.0094018 + (70000)\left(2.277\text{x}10^{-7}\right)\right)10^{-3} \\
&= 1.77\ ^{o}C
\end{aligned}
$$

The gate height and the height of the brine pool in the last stage are assumed equal to those in the previous stage. The brine density in stage n-1 is 1042.4 kg/m^3 , which is calculated at a salinity of 69654.7 ppm and a temperature of 35 ^{o}C. The pressures of stages n-1 and n are P_{n-1} = 8.35 kPa and P_n = 7.23 kPa, which are calculated at T_{n-1} = 35 ^{o}C and T_n = 40 ^{o}C. The resulting gate height in the stage n-1, GH_{n-1}, is calculated from Eq. 87. This result is

$$
\begin{aligned}
GH_{n-1} &= B_{n-2}\ (2\ \rho_{b_{n-1}}\ \Delta P_{n-1})^{(-0.5)}/(C_d\ W) \\
&= (3036)((2)(1043)(8.35\text{-}7.23)\text{x}10^3)^{(-0.5)}/((0.5)(18.8) \\
&= 0.21\ \text{m}
\end{aligned}
$$

The corresponding brine pool height is obtained by simply adding 0.2 m to the value of GH, or

$$H_{n-1} = 0.41\ \text{m}$$

The non-equilibrium allowance is then calculated using the correlation given in the appendix

$$
\begin{aligned}
NEA_n &= (0.9784)^{T_{n-1}}\ (15.7378)^{H_{n-1}}\ (1.3777)^{V_b\text{x}10^{-6}} \\
&= (0.9784)^{(42.75)}\ (15.7378)^{(0.41)}\ (1.3777)^{(180\text{x}10^{-6})} \\
&= 1.217\ ^{o}C
\end{aligned}
$$

The temperature drop in the demister is assumed negligible in comparison with the values of BPE_n and NEA_n. Therefore, the vapor temperature in the last stage is

$$T_{v_n} = T_n - BPE_n - NEA_n - \Delta T_{d_n}$$

$= 40 - 1.77 - 1.217 - 0$
$= 37\ ^{o}C$

The vapor temperature, T_{v_n}, is used to calculate U_j and $(LMTD)_j$, where

$$(LMTD)_j = ((T_{vn}-T_{jn})-(T_{vn}-T_{cw}))/\ln((T_{vn}-T_{j_n})/(T_{vn}-T_{cw}))$$
$$= ((37-30) - (37\text{-}25))/\ln((37-30)/(37-25))$$
$$= 9.28\ ^{o}C$$

$$U_j = 1.7194+3.2063\text{x}10^{-3}\ T_{v_n}+1.5971\text{x}10^{-5}\ (T_{v_n})^2 - 1.9918\text{x}10^{-7}(T_{v_n})^3$$
$$= 1.7194 + 3.2063\text{x}10^{-3}\ (37) + 1.5971\text{x}10^{-5}\ (37)^2 - 1.9918\text{x}10^{-7}\ (37)^3$$
$$= 1.85\ kW/m^2\ ^{o}C$$

$$A_j = (M_f+M_{cw})\ Cp\ (T_{j_n} - Tcw)/(U_j\ (LMTD)_j)$$
$$= ((1861.7)\ (4.18)\ (30-25))/((1.85)\ (9.28))$$
$$= 2266.4\ m^2$$

The total condenser area is then calculated

$$A_c = n\text{-}j\ A_r + j\ A_j$$
$$= (21)\ (4156.9) + (3)\ (2266.4)$$
$$= 94094.1\ m^2$$

The cooling water flow rate is calculate from Eq. 73, where

$$M_{cw} = (M_s\ \lambda_s - M_f\ C_p(T_n - T_{cw}))/\ (C_p\ (T_n - T_{cw}))$$
$$= ((52.52)(2222.33) - (947)(4.18)(40\text{-}25))/((4.18)(40\text{-}25))$$
$$= 914.64\ kg/s$$

Performance Parameters

Performance Ratio

$$PR = M_d/M_s$$
$$= 378.8/52.52$$
$$= 7.21$$

The specific heat transfer area

$$sA = (A_b + A_c\ )/M_d$$
$$= (4256.6 + 94094.1)/378.8$$
$$= 259.6\ m^2/(kg/s)$$

The specific cooling water flow rate

$sM_{cw} = M_{cw}/M_d$
$= 914.6/378.8$
$= 2.41$

Solution Summary

Flow Rates

$M_d = 378.8$ kg/s

$M_f = 947$ kg/s

$M_b = 568.2$ kg/s

$M_r = 3384.8$ kg/s

$M_s = 52.52$ kg/s

$M_{cw} = 914.64$ kg/s

Heat Transfer Areas

$A_b = 4256.6\ m^2$

$A_r = 4156.9\ m^2$

$A_j = 2266.4\ m^2$

$A_c = 94094.1\ m^2$

Stage Dimensions

$W = 18.8$ m

$L = 2.56$ m

$GH_1 = 0.078$ m

$H_1 = 0.278$ m

Performance Parameters

PR = 7.21

sA = 259.6 $m^2/(kg/s)$

sM_{cw} = 2.41

Stage Profiles

The MSF simplified model is used to calculate the temperature and concentration profiles for the 24 stages. The calculations also include the brine and distillate flow rate, the gate height and the brine level in each stage. In the following table, the flow rates are in kg/s, the temperature in °C, the salinity in ppm, and the height in m.

Stage	D	$\sum D$	B	X	T	T_f	GH	H
1	16.699	16.699	3368.284	62473.96	103.25	97.75	0.078	0.278
2	16.617	33.316	3351.667	62783.69	100.5	95	0.081	0.281
3	16.535	49.851	3335.133	63094.95	97.75	92.25	0.084	0.284
4	16.453	66.304	3318.68	63407.75	95	89.5	0.087	0.287
5	16.372	82.676	3302.308	63722.11	92.25	86.75	0.09	0.29
6	16.291	98.967	3286.017	64038.03	89.5	84	0.094	0.294
7	16.211	115.178	3269.806	64355.51	86.75	81.25	0.097	0.297
8	16.131	131.309	3253.675	64674.56	84	78.5	0.101	0.301
9	16.051	147.36	3237.624	64995.2	81.25	75.75	0.105	0.305
10	15.972	163.332	3221.652	65317.43	78.5	73	0.11	0.31
11	15.893	179.225	3205.759	65641.25	75.75	70.25	0.114	0.314
12	15.815	195.04	3189.944	65966.68	73	67.5	0.119	0.319
13	15.737	210.777	3174.208	66293.73	70.25	64.75	0.124	0.324
14	15.659	226.436	3158.548	66622.39	67.5	62	0.13	0.33
15	15.582	242.018	3142.966	66952.68	64.75	59.25	0.136	0.336
16	15.505	257.523	3127.461	67284.62	62	56.5	0.143	0.343
17	15.429	272.952	3112.033	67618.19	59.25	53.75	0.15	0.35
18	15.352	288.304	3096.68	67953.42	56.5	51	0.157	0.357
19	15.277	303.581	3081.404	68290.31	53.75	48.25	0.165	0.365
20	15.201	318.782	3066.202	68628.88	51	45.5	0.175	0.375
21	15.126	333.908	3051.076	68969.12	48.25	42.75	0.185	0.385
22	15.052	348.96	3036.024	69311.05	45.5	40	0.197	0.397
23	14.977	363.937	3021.047	69654.67	42.75	35	0.211	0.411
24	14.904	378.841	3006.143	70000	40	30	0.211	0.411

Approximate Form for the Performance Ratio

The approximate form for the performance ratio is based on the assumption that the temperature profile is linear throughout the stages and the condensers, Fig. 27. As is shown the two right-angle triangles, abc and $\overline{a}\overline{b}T_c$, are conforming. Also, the two reciprocal triangles, aT_oT_1 and $\overline{a}\overline{b}T_c$, are identical, so that

$$T_o - T_1 = T_n - T_{cw}$$

The following identity is then written for the two conforming triangles, abc and $\overline{a}\overline{b}T_c$,

$$\frac{ab}{\overline{a}\overline{b}} = \frac{bT_c}{\overline{b}T_c}$$

This relation is expressed as a function of temperatures and number of stages,

$$\frac{T_o - T_{cw}}{T_n - T_{cw}} = \frac{n+j}{j}$$

The numerator on the left-hand side in the above equation is arranged into the following form

$$\frac{T_o - T_n + T_n - T_{cw}}{T_n - T_{cw}} = \frac{n}{j} + 1$$

Simplifying the above equation gives

$$\frac{T_o - T_n}{T_n - T_{cw}} + 1 = \frac{n}{j} + 1$$

The dominator in the above equation is replaced with the relation obtained for the two identical-reciprocal triangles, aT_ot_1 and $\overline{a}\overline{b}T_c$,

$$\frac{T_o - T_n}{T_n - T_1} = \frac{n}{j} \tag{95}$$

The temperature differences in equation (95) are then expressed in terms of the process thermal loads. The total amount of energy gain by the recycle seawater in heat recovery section is defined in terms of the temperature difference, T_o-T_n, which is approximated by

$$D\ \lambda_{av} = M_r\ Cp\ (T_o - T_n) \tag{96}$$

where D is total amount of distillate product (λ_{av}) is the average latent heat of condensing vapor. The steam thermal load is then defined in terms of the temperature increase of the recycle seawater in the brine heater, T_o-T_1, where

$$M_s\ \lambda_s = M_r\ Cp\ (T_o - T_1) \tag{97}$$

Dividing equations (96) and (97) gives

$$\frac{D}{M_s} = \frac{T_o - T_n}{T_o - T_1}$$

The previous equation assumes that the ratio of the latent heat of steam and distillate product is approximately one. The previous equation gives the process thermal performance ratio, PR = D/M_s. Also, the left hand side of the previous equation is defined in Eq. (95) as n/j. Making these simplifications gives the desired approximate relation for the system thermal performance ratio, where

$$PR = n/j \tag{98}$$

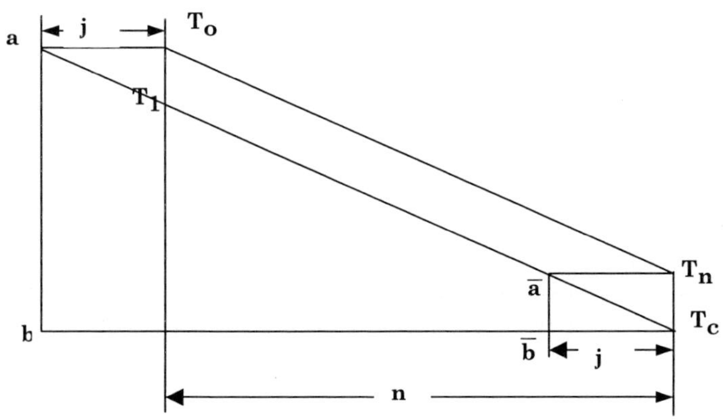

Fig. 27. Schematic of triangles for approximate PR calculations

6.5.3 System Performance

Relations among various system variables of the MSF system include the following:
- Temperature profiles of the flashing brine and seawater flowing into the condenser tubes.
- Thermal performance ratio, defined as the flow rate ratio of the product fresh water to the heating steam.
- Specific heat transfer area, defined as the ratio of the total heat transfer area of the condenser tubes and the brine heater to flow rate of product fresh water.
- Specific flow rate of cooling water, which is defined as the flow rate ratio of cooling seawater to the product fresh water.

The above system variables have a strong effect on the product unit cost. The thermal performance ratio is a measure of the process efficiency, where higher ratios imply larger production rate per unit flow rate of heating steam. As discussed before, a thermal performance ratio of 8 is common among various MSF designs. However, system operation at clean conditions normally gives thermal performance ratios close to 10. As operation proceeds, where fouling and scaling starts to reduce the overall heat transfer coefficient, the thermal performance ratio drops gradually and it is normal to allow the system to operate at thermal performance ratios below 7, where full shutdown takes place and acid cleaning and in some events mechanical cleaning might be necessary to restore the unit characteristics to the original design conditions, Al-Falah and Al-Shuaib, 2000.

As for the specific heat transfer area it is gives a simultaneous measure for the capital cost contribution on the product unit cost as well as efficiency of system operation. Design values vary between 200 to 300 $m^2/(kg/s)$. For clean systems, this parameter has low values close 200 m2/(kg/s) and as fouling and scaling takes place the parameter value reaches the maximum range. Fouling and scaling reduces the overall heat transfer coefficient as well the seawater temperature entering the brine heater. Compensation for this effect and to maintain constant production rate requires increase of the heating steam flow rate. Eventually, the system reaches limiting operating conditions, where fouling effects can no longer be met with the increase in the heating steam flow rate. At these conditions the specific heat transfer area starts to increase and reaches values close to the maximum limit.

The specific flow rate of cooling seawater is also a simultaneous measure for the contribution of the system capital that includes cost of the pumping units and the intake system as well as the operation efficiency on the product unit cost. The cooling seawater is used to remove excess heat added to the system by the heating steam in the brine heater. Proper system design aims at minimizing the amount of cooling seawater. Therefore, the size of the pumping units is reduced as well as the intake system. As discussed above, larger amounts of heating steam are used as fouling and scaling builds inside the condenser and the brine heater tubes. Considering the process thermodynamics and energy conservation implies the increase in the flow rate of the cooling seawater at such conditions. This is necessary to remove the excess heat added to the system.

Figure 28 shows a plot of the temperature distribution for both the brine flowing insider the condenser tubes and in the flashing chambers. Examination of the figure indicates that the temperature distribution for the brine flow inside the tubes of the preheaters deviates more from the straight line when compared to that for the flashing brine. Also, the rate of increase in temperature of the brine flowing in the preheaters at the heat rejection section is higher than the rate of increase at the heat recovery section. For any stage the vertical distance between these two nearly straight line curves is the summation of the stage flash down, non-equilibrium allowance, boiling point elevation, temperature decrease corresponding to the pressure drop in the demister, and the terminal temperature difference.

The effects of number of stages on the plant performance ratio, specific heat transfer area, specific cooling water flow rate and specific flow rate of brine circulation at different values of top brine temperature are shown in Figs. 29-32, respectively. Figure 29 shows the increase in the performance ratio upon the increase in the number of flashing stages and the top brine temperature. Increase of the number stages decreases the temperature drop per stage, which is inversely proportional to the performance ratio. Also, operation at higher top brine temperatures increases the flashing range, which results in the increase of the amount of product water per unit mass of heating steam.

Fig. 30 shows increase in the specific heat transfer area at larger number of stages and lower top brine temperatures. Since all calculations are made at constant total production rate, increasing the number of stages increases the total heat transfer area, which results in the increase of the specific heat transfer area. Similarly, operation at lower top brine temperature reduces the values of the heat transfer coefficient, which in turn necessitates use of larger heat transfer area.

Figure 31 illustrates that the top brine temperature has very little influence on the specific cooling water flow rate. On the other hand, use of larger number of stages reduces the specific flow rate of the cooling seawater. This is explained in terms of the results obtained in Fig. 29, where the performance ratio increases at larger number of flashing stages, which implies use of smaller amount of heating steam at constant production rate. Reduction in the amount of the heating steam is associated with decrease in the amount of energy added to the system as well as decrease in the amount of energy removed by the cooling seawater.

Figure 32 shows the independence of specific flow rate of circulated brine on the number of stages. This value decreases as the brine temperature is increased, which reflects increase in the system efficiency. Figures 33-36 demonstrate the effects of top brine temperature on the same variables, namely, the plant performance ratio, specific heat transfer area, specific cooling water flow and specific recirculated brine flow rate at different values of number of stages.

Figures 29-36 also show the data of six different plants in operation for a long time in Arabian Gulf countries. Plants from the C_3, C_4 and C_5 groups are in Saudi Arabia, Al-Mudaiheem et al. (1993). Umm Al-Nar and Taweelah B plants are in the United Arab Emirates, Hornburg and Watson (1993) and Hornburg et al. (1993). In the Doha-West plants are located in Kuwait, Darwish (1991). It is worthwhile mentioning that design and operation data for these plants are unfortunately scarce. The comparison of the data of operating plants with the developed model predictions shows adequate quantitative agreement, and no qualitative divergence has been observed.

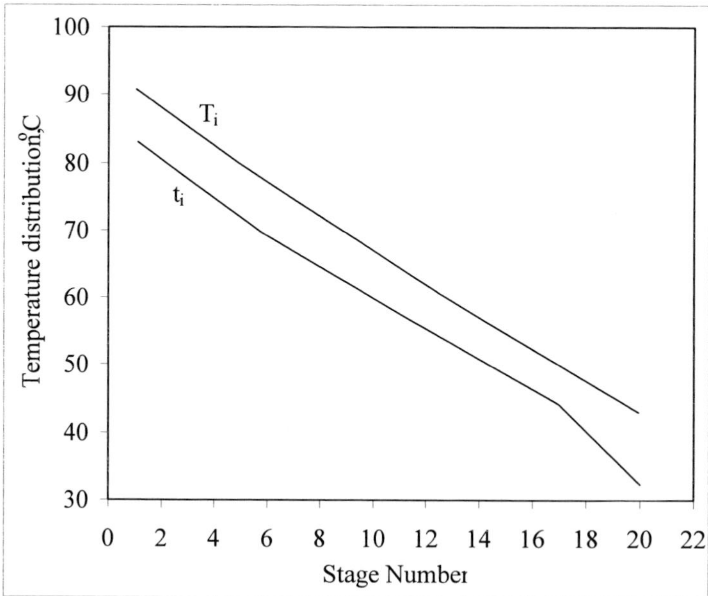

Fig. 28. Temperature distribution for brines flowing in the preheaters and through the flashing chambers

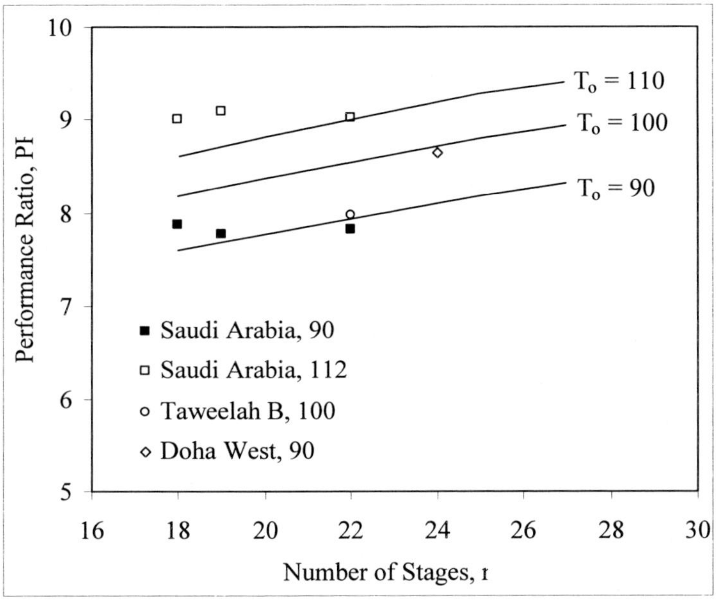

Fig. 29. Effect of number of stages on the plant performance ratio

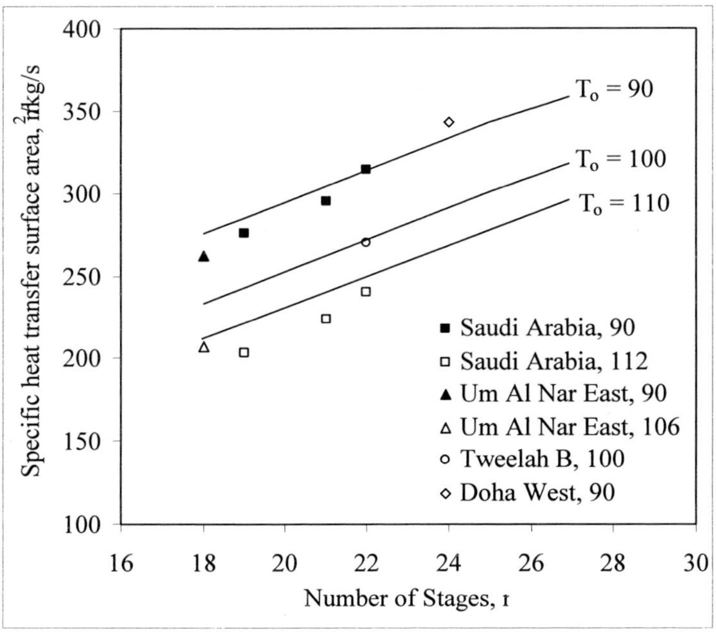

Fig. 30. Effect of number of stages on the specific heat transfer area

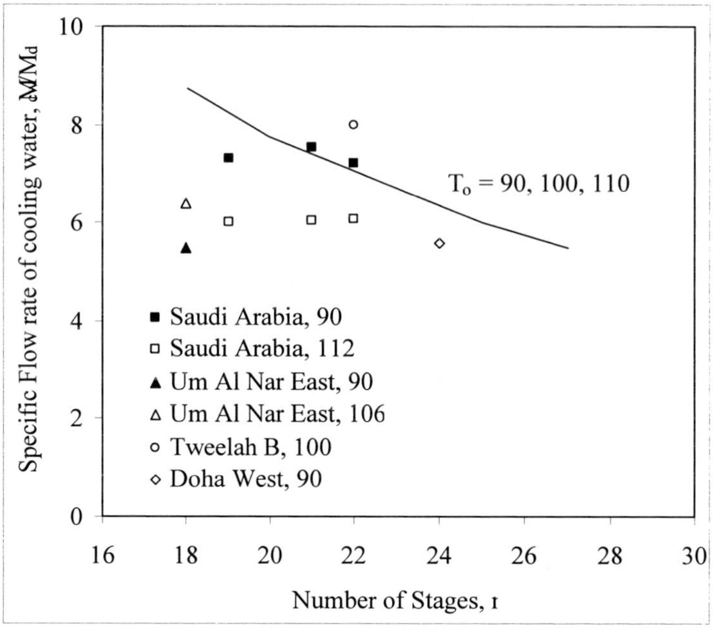

Fig. 31. Effect of number of stages on specific cooling water flow rate

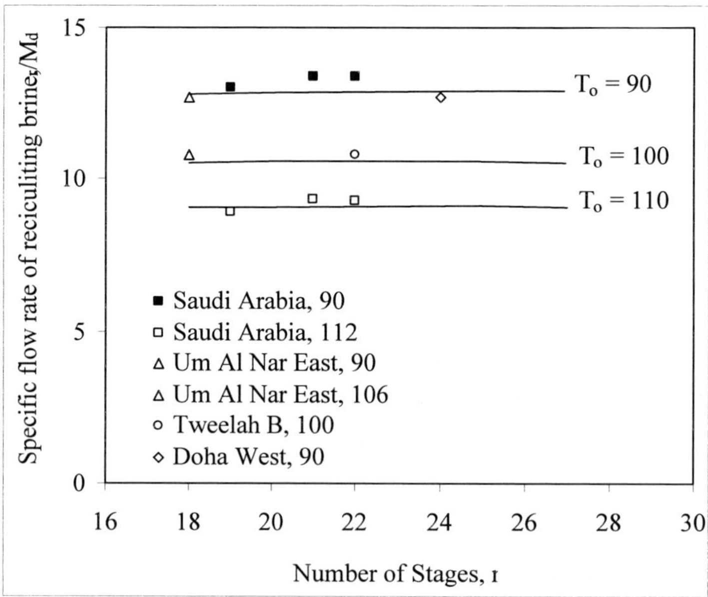

Fig. 32. Effect of number of stages on the specific of recirculating brine flow rate

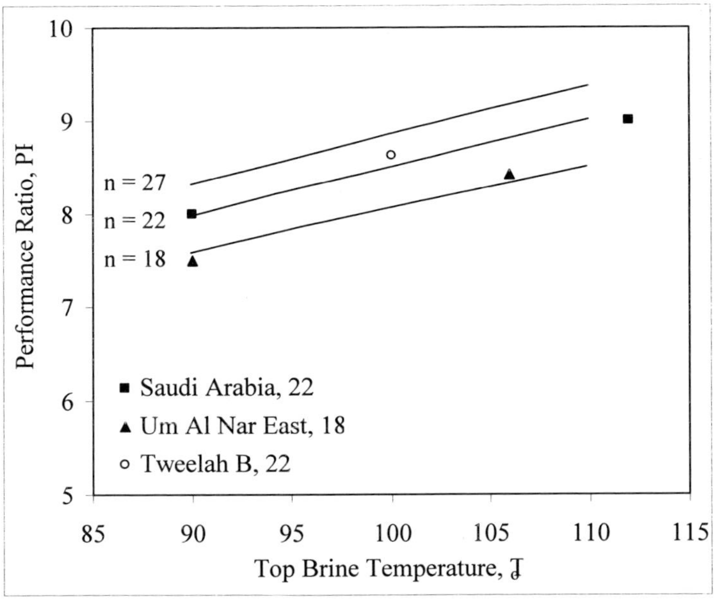

Fig. 33. Effect of top brine temperature on the plant performance ratio

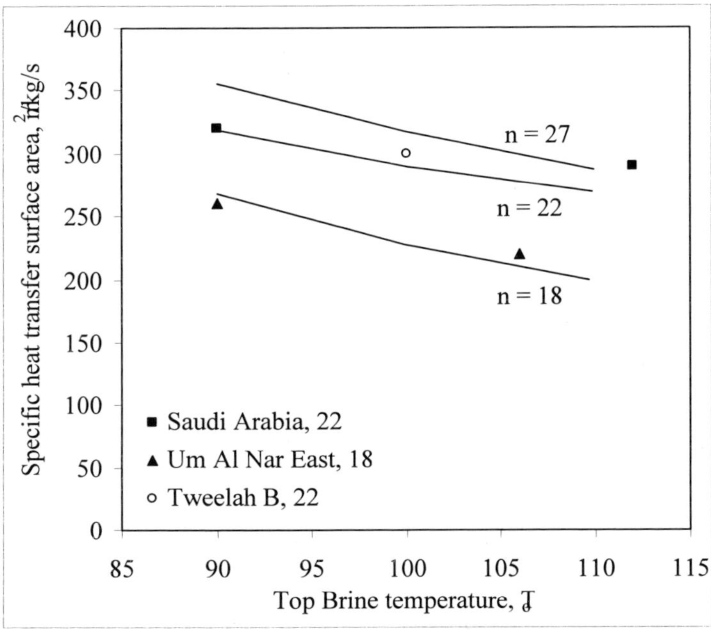

Fig. 34. Effect of number of stages on the specific heat transfer area

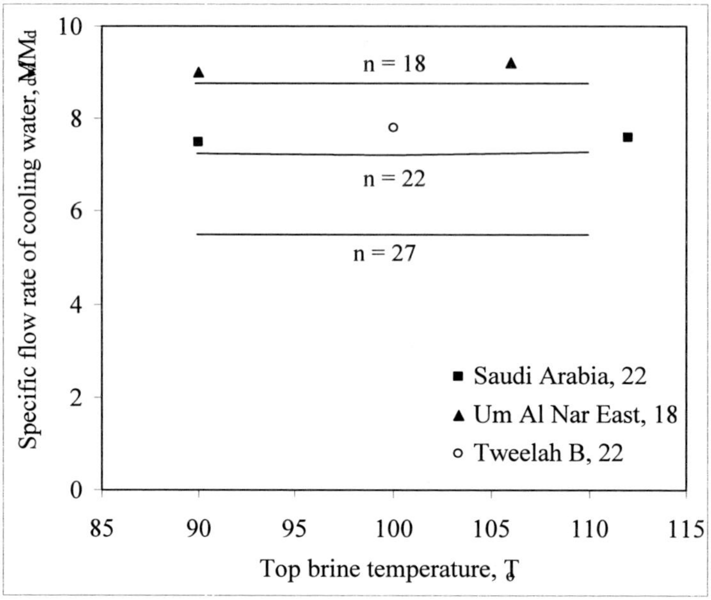

Fig. 35. Effect of number of stages on the specific flow rate of cooling water

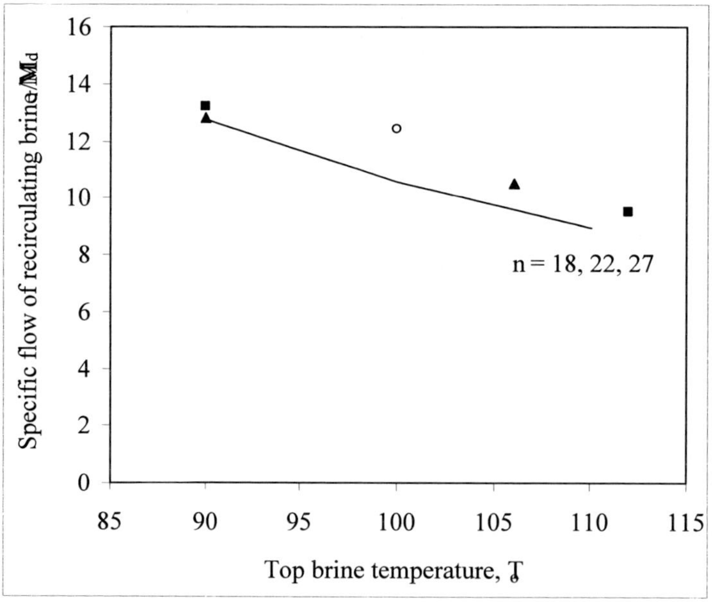

Fig. 36. Effect of top brine temperature on specific recirculating brine flow rate

References

Al-Mudaiheem, A.M., Al-Sofi, M.A.K., Al-Omran, A.A., and Al-Jordan, A.A., Desalination, **92**(1993)1.

Darwish, M.A., Thermal analysis of multi-stage flash desalting system, Desalination, **85**(1991)59-79.

Hornburg, C.D., and Watson, B.M., Desalination, **92**(1993)333.

Hornburg, C.D., Todd, B., and Tuthill, A.H., Heat transfer tubing selection for MSF desalination plants, Proceedings of the IDA World Congress on Desalination and Water Sciences, Abu Dhabi, November, 1995, Vol. III, pp. 131-148.

Problems

Problem 1

An MSF brine circulation plant has the following design data:

Plant capacity:	Unknown
Seawater temperature:	32 °C
Seawater salinity:	49400 ppm
Top brine temperature:	100 °C
Performance ratio:	8
Number of heat recovery stages:	19
Number of heat rejection stages:	3
Maximum brine concentration:	68600 ppm
Specific flow rate of cooling water:	1
Ratio of demister cross Sectional area to the chamber area:	0.75
Maximum vapor velocity in demister:	4 m/s
Maximum allowable flow rate Of brine per chamber width:	600×10^3 kg/m hr
Overall heat transfer coefficient in brine heater:	2 kW/m^2 °C
Overall heat transfer coefficient in heat rejection section:	1.9 kW/m^2 °C
Overall heat transfer coefficient in heat recovery section:	2.4 kW/m^2 °C

Calculate the following:

- The specific heat transfer area
- The vapor velocity in flashing chamber number 6.
- The liquid depth in flashing chamber number 10.

Problem 2

An MSF brine circulation plant has the following design data:

Plant capacity:	Unknown
Top brine temperature:	Unknown
Brine flow rate per chamber width:	Unknown
Number of stages:	20
Number of heat rejection stages:	3
Boiling temperature in last stage:	40 °C
Heat transfer area in the brine heater:	1000 m^2
Overall heat transfer coefficient in all sections:	2.527 kW/m^2 °C
Mass flow rate of heating steam:	16.782 kg/s
Heating steam temperature:	120 °C
Number of tubes in the brine heater:	1000 tube
Specific flow rate of brine circulation:	8.422
Diameter of tubes used in brine heater:	31.8 mm

Calculate the following:

- The plant performance ratio
- The specific heat transfer area
- The specific flow rate of cooling water
- The dimensions of chamber 7.

Problem 3

An MSF brine circulation plant has the following design data:

Seawater temperature:	34 °C
Seawater salinity:	42000 ppm
Top brine temperature:	100 °C
Temperature in the last stage:	40 °C
Temperature of heating steam:	110 °C
Specific flow rate of brine circulation:	8.478
Heat transfer area in the brine heater:	80 m^2
Number of heat rejection stages:	2

Overall heat transfer coefficient
in brine heater: 1.5 kW/m^2 °C

Calculate the following

- The plant performance ratio
- The specific flow rate of cooling water
- The terminal temperature difference of the first stage.
- The specific heat transfer area.
- The dimensions of stage 7.

Problem 4

An MSF brine circulation plant has the following design data:

Distillate flow rate:	5000 m^3/d
Seawater temperature:	30 °C
Seawater salinity:	44000 ppm
Top brine temperature:	112 °C
Temperature in the last stage:	40 °C
Temperature of heating steam:	116.7 °C
Terminal temperature difference:	3 °C
Number of heat rejection stages:	2
Number of heat recovery stages:	18

The overall heat transfer coefficient in the brine heater or the flashing stages is given by the relation

$$U = 6.5 - 0.03(115\text{-}T)$$

With U in kW/m^2 °C and T in °C.

Calculate the following

- Thermodynamic losses in the first stage
- Plant performance ratio
- Cooling water flow rate
- Brine heater surface area
- Flow rate of make up water.

Also, calculate the following parameters for stage number 5

- Boiling point elevation.
- Gate height
- Liquid level

- Demister temperature loss
- Stage height
- Stage length
- Preheater surface area.

<u>Problem 5</u>

An MSF brine circulation plant has the following design data:

Distillate flow rate:	22750 m^3/d
Seawater temperature:	28 °C
Seawater salinity:	45000 ppm
Top brine temperature:	90 °C
Temperature in the last stage:	40 °C
Temperature of heating steam:	100 °C
Terminal temperature difference:	3 °C
Number of heat rejection stages:	3
Number of heat recovery stages:	25
Width of stage 10:	16 m
Length of stage 10:	3.5 m

Calculate the following

- The temperature profile of flashing brine
- The temperature profile of seawater flowing in the preheaters
- The flow rates of heating steam, makeup water, brine circulation, cooling water, and brine blowdown.
- The pressure in stages 9, 10, and 11.
- The distillate product in stage 10.
- The vapor velocity in stage 10.

6.6. MSF with Thermal Vapor Compression

Attempts to improve the overall performance of the MSF system include design of hybrid configurations with the MVC or RO systems. The study by Genthner and El-Allawy (1983) focused on analysis of hybrid parallel configuration of MSF and MVC. As a result, several of the MVC devices are eliminated since their functions are provided from the hybrid MSF system. These functions include feed preheating, venting of non-condensable gases, chemical dosing, and pumps for distillate and brine blowdown. Results show lower chemical and specific fuel consumption with values of 50% and 65-75%, respectively. As for the RO-MSF hybrid system, it has similar advantages since the RO feed is extracted from the cooling seawater stream. This combination eliminates part of the pumping power requirements and treatment of the feed water for the RO system, Darwish et al. (1989) and Glueckstern (1995). Reported economics of the hybrid system are better than those for either system in a stand-alone mode. Regardless, actual use of the RO-MSF or the MVC-MSF systems are not found on industrial scale.

Development of the MSF brine circulation with thermal vapor compression (MSF-TVC) is motivated by the drastic increase in the thermal performance ratio of MEE upon combination with vapor compression. The field study by Temstet and Laborie (1995) and the performance studies by El-Dessouky et al. (1998) show that the thermal performance ratio of the stand-alone MEE system is approximately equal to the number of effects raised to the power 0.9. This gives thermal performance ratios of 5 and 9.4 for systems with 6 and 12 effects, respectively. The field study by de Gunzbourg and Larger (1998) and the performance evaluation of El-Dessouky and Ettouney (1997) show that combining the forward feed MEE system with lithium bromide vapor compression heat pumps increases the thermal performance ratio to a value of 21 for a 14-effect system. Similar findings are found in the studies by El-Dessouky et al. (2000) for the parallel feed MEE combined with thermal and mechanical vapor compression.

This section outlines features, model and analysis of the multistage flashing desalination combined with thermal vapor compression (MSF-TVC). The analysis considers several modes for vapor compression, which differs in the stage from which vapor is extracted. System evaluation focuses on calculations of the parameters that affect the unit product cost, which includes the thermal performance ratio, the specific heat transfer area, and the specific flow rate of cooling water. This report is the first, to the authors' knowledge, to propose and evaluate the performance of a combined MSF and thermal vapor compression. The next sections include brief description of the brine circulation MSF, the vapor compression modes, system model, results, and conclusions.

6.6.1 Process Description

Schematic diagrams for the proposed MSF-TVC processes are shown in Figs. 37 and 38. The systems constitute the brine heater, the flashing stages, and the steam jet ejector. The flashing stages are divided among the heat recovery and rejection sections. Each flashing stage includes brine pool, submerged orifice, vapor space, demister, distillate collection tray, and condenser tubes. The system also includes a cascade of venting units, which removes and prevents accumulation of non-condensable gases within the vapor space. The process is described in the following steps:

- The brine recycle stream (M_r) enters the brine heater tubes, where the compressed vapor (M_s) is condensed on the outside surface of the tubes. The brine stream absorbs the latent heat of condensing steam and its temperature increases to its maximum design value known as the top brine temperature (T_o). Its value depends on the nature of chemicals used to control the scale formation.
- The hot brine enters the flashing stages in the heat recovery section and then in the heat rejection section, where a small amount of fresh water vapor is formed by brine flashing in each stage. The flashing process takes place due to decrease in the stage saturation temperature and causes the reduction in the stage pressure.
- In each stage of the heat recovery section, the flashed off vapors condenses on the outside surface of the condenser tubes, where the brine recycle stream (M_r) flows inside the tube from the cold to the hot side of the plant. This heat recovery improves the process efficiency because of the increase in the feed seawater temperature.
- The condensed fresh water vapor outside the condenser tubes accumulates across the stages and forms the distillate product stream (M_d). This stream cascades in the same direction of the flashing brine from stage to stage and is withdrawn from the last stage in the heat rejection section.
- The intake seawater stream (M_f+M_{cw}) is introduced into the condenser tubes of the heat reject section, where its temperature is increased to a higher temperature by absorption of the latent heat of the condensing fresh water vapor.
- The warm stream of intake seawater is divided into two parts: the first is the cooling seawater (M_{cw}), which is rejected back to the sea and the second is the feed seawater (M_f), which is deaerated, chemically treated and then mixed in the brine pool of the last flashing stage in the heat rejection section.
- The brine recycle stream (M_r) is extracted from the brine pool of the last stage in the heat rejection section and is introduced into the condenser tubes of the last stage in the heat recovery section. As the stream flows in the condenser

tubes across the stages it absorbs the latent heat of condensation from the flashing vapor in each stage.

- The remaining brine in the last stage of the heat rejection section, known as the brine blowdown (M_b), is rejected to the sea.
- The steam jet ejector entrains a specified portion of the vapor formed in the flashing stages in the heat rejection (Fig. 37) or the heat recovery (Fig. 38). The motive steam compresses the entrained vapor to the desired temperature and pressure. The compressed vapor is then used to heat the brine recycle stream in the brine heater.
- Treatment of the intake seawater is limited simple screening. On the other hand, treatment of the feed seawater stream is more extensive and it includes dearation and addition of chemicals to control scaling, foaming, and corrosion.
- Another steam jet ejector is used to remove the non-condensable gases, which are released during flashing from each stage. Presence and accumulation of the gases result in reduction of the overall heat transfer coefficient for the condenser tubes and in turn reduces the process efficiency and the system productivity.
- The steam jet ejector is formed of a converging-diverging nozzle, a converging-diverging diffuser, a throat, a mixing zone, and a suction or entrainment chamber, Fig. 39. The motive steam, M_m, expands in the nozzle from state 1 to state 3, where its static pressure energy is converted to kinetic energy. The motive steam velocity becomes larger than the speed of sound (supersonic) as it leaves the nozzle. The suction chamber keeps the nozzle properly positioned with respect to the diffuser and directs the entrained vapor. The entrained vapor M_{ev} enters the suction chamber at pressure P_3 where it mixes violently and rapidly with the motive steam at point 4. The two streams mix together as they pass through the converging section of the venturi diffuser (from point 4 to 5). The mixture enters the throat section of the diffuser, completely mixed, at the sonic velocity of the mixture. The combined mixed streams is self compressed through the diverging section of the venture diffuser, where the cross sectional area increases and the velocity decreases, converting the kinetic energy of the mixture to static pressure energy. The mixture leaves the ejector at a pressure P_s that is intermediate to the motive (P_m) and suction (P_v) pressures. The steam jet ejector is designed to operate at the critical condition, where the supersonic shock wave is located at the nozzle exit. This condition occurs as the pressure compression ratio (P_m/P_v) is greater than 1.8. During operation, the shock wave may move downstream upon reduction in the condenser pressure. This has a negligible effect on the ejector performance, however, if conditions preventing formation of the shock wave are prevailed, the ejector performance deteriorates and no entrainment or compression takes place. In part stable operation is associated with absence of violent fluctuations in the suction pressure. If the ejector is designed to operate with a full stable range, it will have a constant mass flow rate of the

entrained vapor for different discharge pressures when the upstream conditions remain constant.

Two configurations are considered in this study that differs in the location of vapor entrainment, Figs. 37 and 38. In the first configuration (Fig. 37) vapor of equal thermal load and not flow rate is entrained from the stages forming the heat rejection section. This is necessary because of the equality of the heat transfer area and the thermal load of the flashed off vapor in these stages. Therefore, an equal rise occurs in the temperature of the feed seawater stream flowing on the inside of the condenser tubes of the heat rejection section. This scheme has some similarity to vapor compression in the MEE system, where vapor entrained by the steam jet ejector is at a low temperature of 40 ^{o}C, El-Dessouky and Ettouney (1997). However, the difference between MEE and MSF is the requirement to compress the vapor to a higher temperature of 90-110 ^{o}C in the MSF, while in MEE the compressed vapor temperature can be as low as 60 ^{o}C. In MSF constraints imposed by the flashing range and the temperature drop per stage dictates vapor compression to high temperatures. As is shown, mode (a) requires the use of two steam jet ejectors in series. This is because vapor compression from a temperature of 40 ^{o}C to 110 ^{o}C gives a pressure compression ratio of 19.4 between the compressed and the entrained vapor. Design and operation of steam jet ejectors puts a maximum limit of 5 on the compression ratio, Power (1994). As a result, operating the first ejector at the maximum limit gives an outlet pressure of 36.92 kPa for the compressed vapor corresponding to a saturation temperature of 73.95 ^{o}C. The remaining compression range between 73.95 and 110 ^{o}C is 3.88, which is achieved in the second ejector.

In the second configuration (Fig. 38) vapor of equal thermal load is equally entrained from 2 to 4 stages in the heat recovery section. Again the equal heat transfer area and the flashing thermal load in the heat recovery section dictate the above equality constraint. However, the higher vapor temperature in the heat recovery section would require one steam jet ejector only. For example, vapor compression from 70 ^{o}C to 110 ^{o}C gives a compression ratio of 4.6, which is lower than the maximum limit. As is shown in Figs. 37 and 38 separate ejectors are used to entrain and compress vapor from each stage. Use of a single ejector is not possible, because of differences in the pressure of entrained vapor.

In summary, differences in system design of the conventional and thermal vapor compression MSF include the following:

- As will be discussed later the MSF-TVC system has a higher thermal performance ratio than conventional MSF. This implies reduction in the amount of input energy to the system per unit mass of product distillate water. As a result, lower specific heat transfer area for the condenser tubing and smaller pump capacity for the cooling seawater are required for the MSF-TVC system.

- Elimination of the control loop and associated devices on the temperature and pressure of the low pressure steam used in conventional MSF. This loop reduces the steam temperature and pressure from 155 °C and 5.5 bar to the saturation conditions, which may vary from 100-120 °C. In the proposed MSF-TVC system the motive steam is directly extracted from the medium pressure steam line at 15 bar.
- Increase in the steam pressure for the MSF-TVC would result in reducing the size of the steam piping system. This is because of reduction in the specific volume of the vapor at higher pressures. In conventional MSF use of lower pressure steam, which has higher specific volume, would require use of larger pipe diameter.

6.6.2 Mathematical Model and Solution Procedure

Mathematical model and solution procedure for MSF with thermal vapor compression are similar to the model of the stand-alone brine circulation system. The model elements include the following:

- Energy, mass, and salt balances for each flashing stage.
- Heat transfer equations for condenser/preheater tubes in heat recovery and heat rejection section.
- Ejector equations.

It should be noted that all the model equations and correlations are given in the previous section for the brine circulation system. Also, the model equations for the ejector are given in the single and multiple effect systems.

6.6.3 System Performance

Performance of the MSF-TVC modes (a) and (b) are shown in Figs. 39-42, which includes the temperature profiles, the performance ratio, the specific heat transfer area, the specific flow rate of cooling water, and the piping size.

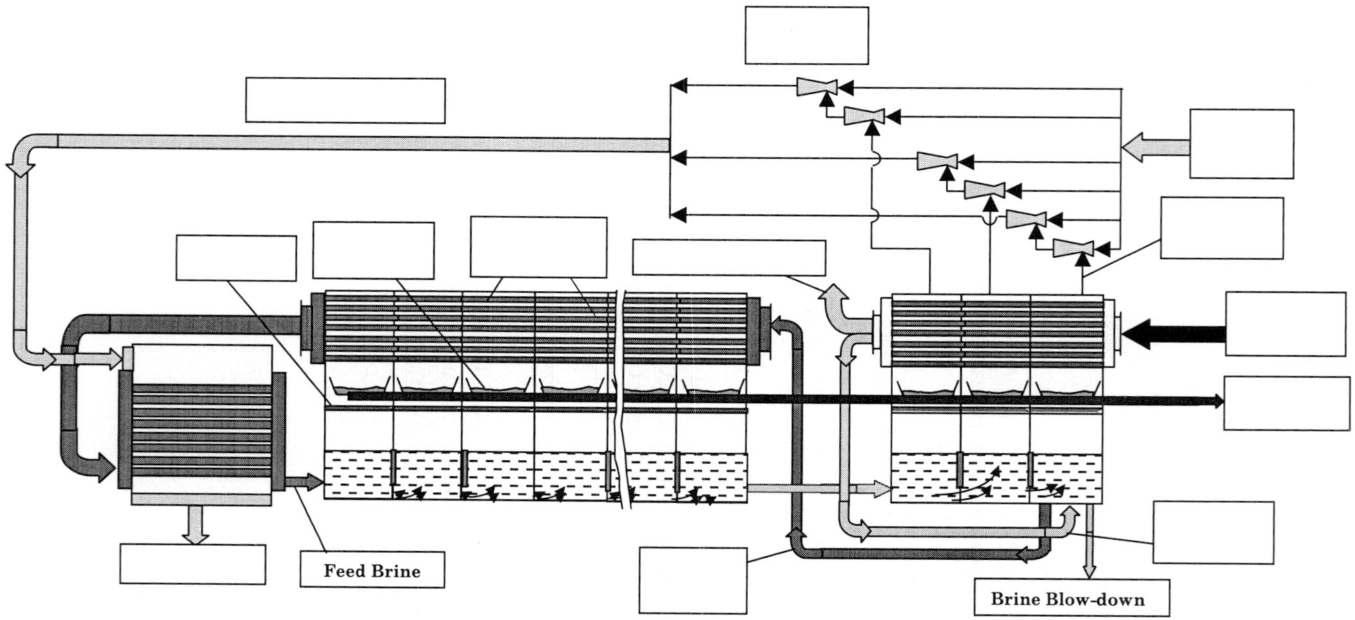

Fig. 37. MSF with thermal vapor compression. Mode (a) Vapor entrainment from the heat rejection stages.

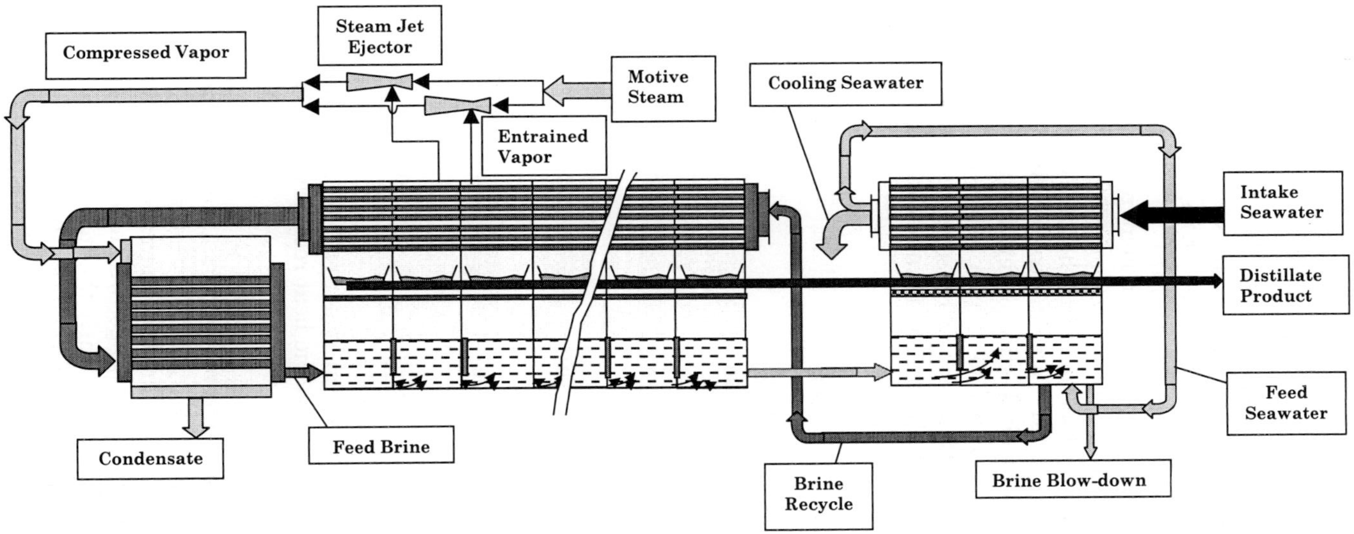

Fig. 38. MSF with thermal vapor compression. Mode (b) Vapor entrainment from the heat recovery stages.

The temperature profiles shown in Fig. 39 are obtained for the flashing brine, the brine recycle, and the feed seawater. In both configurations the temperature profiles for the flashing brine and flashed off vapor are identical to that of conventional MSF. This is because the flashing range and the temperature drop per stage are kept constant for all systems. Similarly, the temperature profiles of the feed seawater and brine recycle stream in configuration (a) is the same as conventional MSF. This is because the temperature differences across the heat recovery or heat rejection sections are the same as those for the MSF system. Therefore, vapor entrainment in the heat rejection section would only result in reduction of the thermal load, which implies reduction of the flow rate of cooling water. In configuration (b) the temperature profile of brine recycle in the heat recovery section differs from those for the MSF system. The difference is found in stages 1-6, where vapor is entrained. As a result, the recycle brine temperature does not increase to the same value as it leaves the heat recovery section as in configuration (a) and conventional MSF.

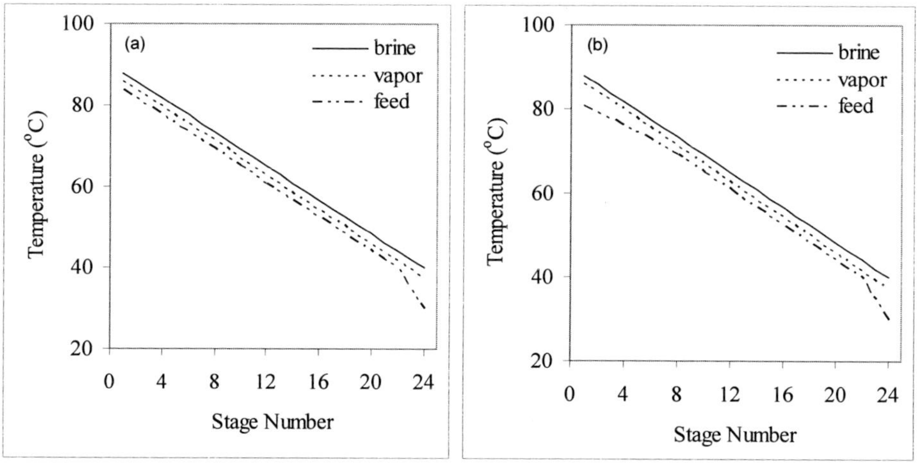

Fig. 39. Temperature profiles in MSF-TVC. (a) Vapor entrainment from the heat rejection section. (b) Vapor entrainment from the heat recovery section.

Fig. 40 shows variations in the thermal performance ratio for modes (a) and (b) as well as the conventional MSF system. For the three systems, the thermal performance ratio increases upon the increase in the top brine temperature. This is because of the flashing range, which results in the increase of the temperature drop per stage and consequently the increase in the amount of flashed off vapor per stage. The highest performance ratio is obtained for the vapor compression mode (b) with values varying between 9 and 10.5 as the top brine temperature increases from 90 to 110 °C. Lower performance ratio is obtained for the vapor compression mode (a) and the lowest values are obtained for the conventional MSF system. The higher performance ratio for mode (b) is caused by reduction in the motive steam requirements for vapor compression at higher temperatures. As a result, the average percentage increase in the thermal performance ratio for mode (a) over mode (b) and the conventional MSF is 6 and 14%, respectively.

Results for the specific heat transfer area for configuration (a), (b) and the MSF system are shown in Fig. 41. The specific heat transfer area for the three configuration decreases upon the increase in the top brine temperature. This is because of the increase in the driving force for heat transfer. It should be noted that the heat transfer area for mode (a) and the conventional MSF system are similar. This is because of the same temperature profiles for both system in the heat recovery and rejection sections. Therefore, the temperature drop per stage in each system and consequently, the driving force for heat transfer are identical. As for mode (b) it has a smaller specific heat transfer area than the other two systems because of the increase in the temperature driving force between the flashing vapor and the brine recycle stream, Fig. 40. The average percentage reduction in the specific heat transfer area for mode (B) is 8% lower than mode (a) and conventional MSF.

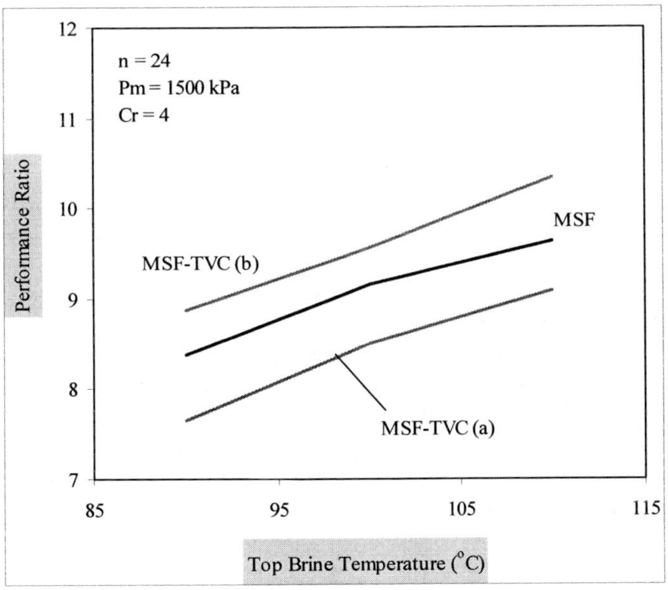

Fig. 40. Variation in the performance ratio as a function of the compression mode and the top brine temperature

Variations in the specific flow rate of cooling water are shown in Fig. 42 for the three configurations. As is shown, the specific flow rate of cooling water decreases with the increase in the top brine temperature. This is caused by reduction in the amount of heating or motive steam per unit mass of product water, or the increase in the system thermal performance ratio. Therefore, the thermal load of the system decreases at higher top brine temperatures and consequently the specific flow rate of the cooling water decreases. The lowest specific flow rate for the cooling water is obtained for mode (a), followed by mode (b), and then the MSF system. The average percentage reduction in the specific flow rate of cooling water in mode (a) over mode (b) and the conventional MSF are 44 and 56%, respectively. The large reduction in the specific flow rate of cooling water for mode is caused by vapor entrainment from the heat rejection section and simultaneous increase in the system thermal performance ratio. Both factor reduces the thermal load of the heat rejection section and consequently the specific flow rate of cooling water.

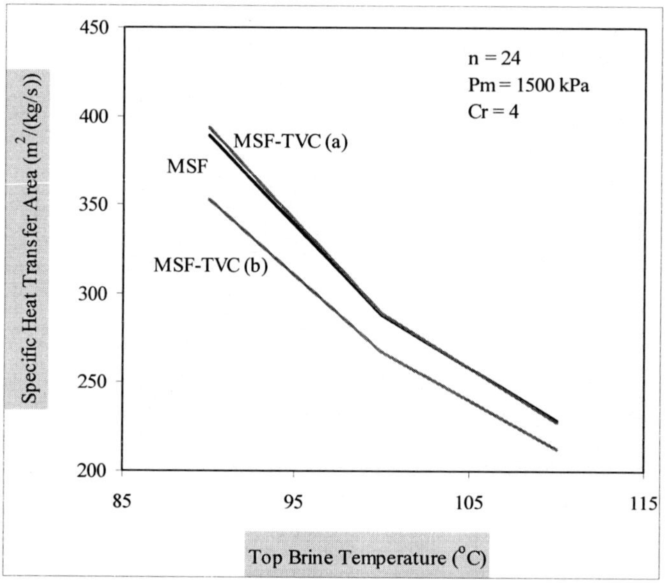

Fig. 41. Variation in the specific heat transfer area as a function of the compression mode and the top brine temperature

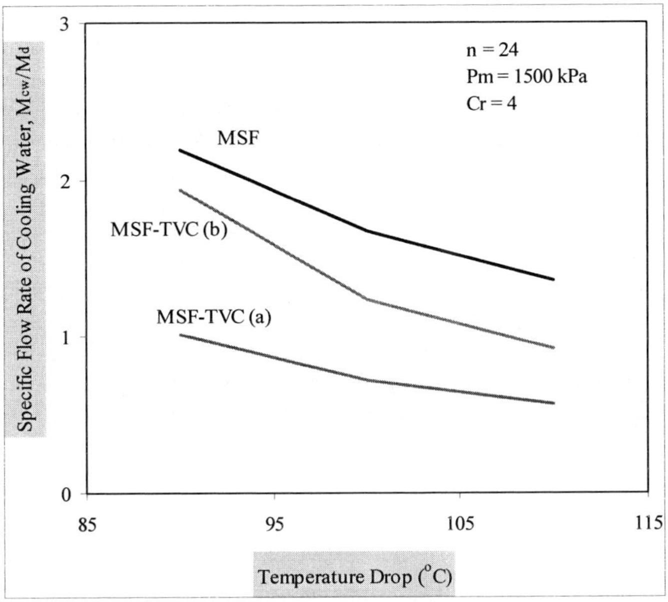

Fig. 42. Variation in the specific flow rate of cooling water as a function of the compression mode and the top brine temperature

6.6.4 Summary

A novel MSF configuration, which is based on thermal vapor compression, is proposed and evaluated as a function of the vapor compression mode and the operating conditions. Analysis of the results shows the following:

- Thermal vapor compression enhances the performance of the MSF system as a result of increase in the performance ratio and reduction in the specific flow rate of cooling water and the specific heat transfer area.
- Vapor compression from stages operating at higher temperatures in the heat recovery section give higher performance ratios than for vapor compression from the heat rejection section.
- The specific heat transfer area for the vapor compression mode from the heat recovery section gives lower specific heat transfer area from the heat rejection section.
- The pipe diameter for the motive steam in MSF-TVC is lower than the diameter for the heating steam by an average of 60%.

References

Darwish, M.A., Abdel-Jawad, M., Aly, G.S., Technical and economical comparison between large capacity multi stage flash and reverse osmosis desalting plants, Desalination, **72**(1989)367-379.

De Gunzbourg, J., and Larger, D., Cogeneration applied to very high efficiency thermal seawater desalination plants, A concept, Int. Desalination & Water Reuse Quart., **7**(1998)38-41

El-Dessouky, H.T., and Ettouney, H.M., Simulation of combined multiple effect evaporation - vapor compression desalination processes, 1st IDA Int. Desalination Conference in Egypt, Cairo, Egypt (Sept. 1997).

El-Dessouky, H.T., Alatiqi, I., Bingulac, S., and Ettouney, H.M., Steady-state analysis of the multiple effect evaporation desalination process, Chem. Eng. Tech., **21**(1998)15-29.

El-Dessouky, H.T., Ettouney, H.M., and Al-Juwayhel, F., Multiple effect evaporation/Vapor compression desalination process, Trans. I. Chem. E., 78(2000)662-676.

Genthner, K., and El-Allawy, M.M., Solutions for coupling a mechanical vapour compression distiller with a multi-stage-flash evaporator, Desalination, **45**(1983)143-152.

Glueckstern, P., Potential uses of solar energy for seawater desalination, Desalination, **101**(1995)11-20.

Power, B.R., Steam Jet Ejectors for Process Industries, McGraw-Hill, New York, 1994.

Temstet, C., and Laborie, J., Dual purpose desalination plant-high efficiency multiple-effect evaporator operating with a turbine for power production, Proc. IDA World Congress on Desalination and Water Science, Abu Dhabi, vol. 3, pp. 297-308, 1995.

6.7. MSF with Brine Mixing

This section presents a novel MSF process with design and operation advantages over the brine circulation MSF or the once through process MSF-OT. The proposed system is based on the essentials, features, and fundamentals of the MSF process. Therefore, implementation of the proposed system requires a limited number of modifications in MSF plants already in operation.

6.7.1 Process Description

The layout for the brine mixing process is shown in Fig. 43. Also, the temperature profiles of the unevaporated and flashing brine for the MSF, MSF-M, and MSF-OT are shown in Fig. 44. As illustrated, the configuration is similar, in some extent, to the heat recovery section in conventional MSF as well as the once through MSF. The system contains three main sections; these are the brine heater (or the heat input section), the flashing stages (the heat recovery section), and the brine mixing tank. As is shown the unevaporated brine recycle stream flows in a counter current direction to the flashing brine and distillate product in the flashing stages. The unevaporated brine flows from the cold side of the plant to the hot side, while, the distillate product and flashing brine flow in the opposite direction. In the brine heater, saturated steam at a flow rate of, M_s, is used to increase the temperature of the unevaporated brine to the top brine temperature, T_o. The temperature of the heating steam, T_s, is higher than the top brine temperature by a specified few degrees. The heated brine enters the first flashing chamber, where a small portion of distillate is formed. The flashing process converts the excess sensible heat of the unevaporated brine into latent heat required for vapor formation. The flashing process continues throughout the stages, where small amounts of distillate product, D_i, are formed in each stage. As a result of water evaporation, the brine salinity increases across the stages. The maximum permissible value of salt concentration is limited to 70,000 ppm in order to prevent formation of calcium sulfate scaling.

The formed vapor passes through wire mesh mist eliminator, known as the demister. The demister removes brine droplets entrained by the flashing vapor. Brine droplets are formed because of foaming, bursting of vapor bubbles at the brine surface, and brine splashing at the stage orifice. The separation of entrained brine droplets is necessary to prevent contamination of the distillate product and reduction of its quality; moreover, the separation prevents scale formation on the outside surface of the condenser tubes. The vapor condenses on the outside surface of the preheater/condenser tubes, where the unevaporated brine flows. The vapor releases its latent heat to the brine stream and as a result the temperature of the brine stream increases across the stages. The distillate product is collected in the distillate trays and accumulates across the stages.

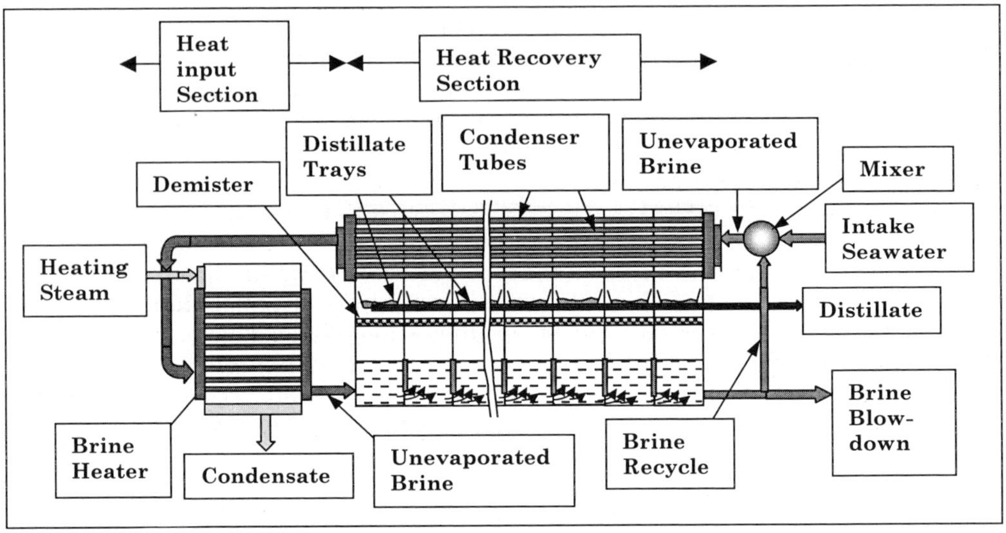

Fig. 43. Schematic of MSF with brine mixing

In each stage, the temperature of the formed vapor, T_{v_i}, is lower than the flashing brine temperature, T_i, by the boiling point elevation. The value of the boiling point elevation is affected by the salinity of the flashing brine and the boiling temperature. Further reduction in the vapor saturation temperature is caused by pressure drop in the demister and during condensation.

The brine leaving the last stage is divided into two parts; the first is rejected back to the sea and the second is recycled. The ratio between these parts can be controlled and depends mainly on the intake seawater temperature. The aim of brine rejection to the sea is to control salt concentration inside the plant. On the other hand, the purpose of brine recirculation is to decrease the flow rate of the feed seawater and recover a part of the heat added to the system in the brine heater. This lowers the chemical additive consumption rate and the size of the pretreatment facilities for the feed stream. Also, since the recycled brine contains higher energy than the feed seawater, the process thermal efficiency will improve. The portion of the brine stream leaving the last stage, M_r-M_f is mixed with the intake seawater stream, M_f. The resulting mixture, M_r, has a higher salinity and temperature than the intake seawater. The mixing process is expected to cause a thermal shock because of differences in the temperatures of the intake seawater and the brine recycle. This would result in dissociation of the bicarbonate compounds and formation of carbon dioxide gas. Therefore, the mixing unit must be properly vented to avoid accumulation of non-condensable gases in the brine recycle stream.

The flashing process and vapor formation is limited by increase in the specific vapor volume at lower temperatures and difficulties encountered for operation at low pressures. Common practice limits the temperature of the last stage to values of 32 and 40 °C, for winter and summer operation, respectively. Further reduction in these temperatures results in drastic increase of the stage volume and its dimensions. In addition, since most of the stages operate at temperatures below 100 °C, the pressure within the stages is at vacuum conditions. This may result in leak of the outside air. At such conditions, air is non-condensable and its presence in the system may result in severe reduction in the heat transfer rates within the chamber, increase of the chamber pressure, and reduction of the flashing rates. Accumulation of non-condensable gases is also generated from trace amounts of dissolved gases, which are not removed in the dearation pretreatment. This condition necessitates proper venting of the flashing stages to enhance the flashing process and to improve the system efficiency.

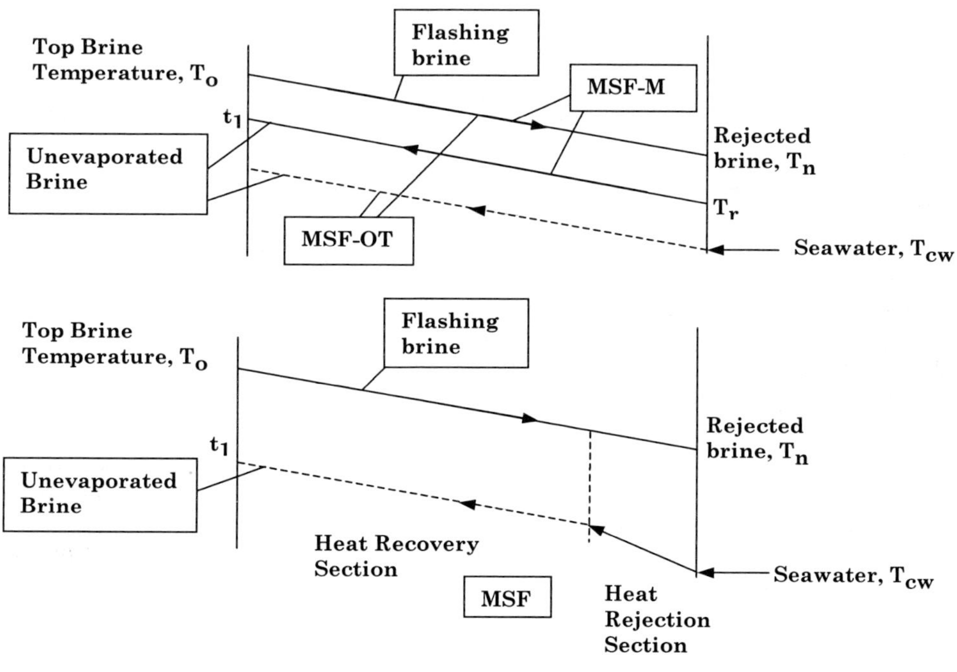

Fig. 44. Temperature profiles for the MSF-M, the MSF-OT, and MSF.

In summary, the main differences between the proposed system and conventional MSF with brine circulation are:

- Removal of the heat rejection section.
- Absence of the cooling water loop used in conventional MSF to control the flashing temperature in the last stage and to remove excess energy added to the system in the brine heater.
- Elimination of the cooling water recycle loop, which is used to adjust the flashing temperature of the seawater of the last flashing stage in the heat rejection section, when the seawater temperature becomes very at winter operation conditions.
- The mixing of the brine recycle and the feed seawater take place in an external mixing tank rather than inside the flashing stages.
- The salinity of the rejected brine can be less than the limiting value of 70,000; this depends on the temperature of feed seawater.
- The flow rate of feed seawater is not constant and is regulated subject to the temperature and salinity of the seawater.

Differences for the proposed system and the once through configuration are:

- Part of the brine flowing in the last stage is recirculated to the system.
- Brine recirculation reduces the flow rate of the feed seawater; consequently lower amounts of chemical additives are used and smaller size pretreatment plant is required, which include screening, filtration, and dearation.
- The use of the mixing tank for the feed stream and the brine recycle stream gives a better control on the temperature of the brine feed to the condenser tubes of the last flashing stage.
- The salinity of the recycle brine is higher than the feed seawater.
- Dearation of the feed seawater takes place outside the stage; this reduces the corrosion rate inside the stages.
- The system is less sensitive to variations in feed seawater temperature; because it can be controlled by the brine circulation rate.

6.7.2 Mathematical Model

The MSF-M mathematical model constitutes the same balance equations and correlations used in the previous MSF systems. The model includes stage energy and material balances as well as the heat transfer equations for condenser/preheater. The model also includes the material and energy equations for the brine mixer, Fig. 45. The salt balance is given by

$$X_f M_f + (M_r - M_f) X_n = M_r X_r$$

This simplifies to

$$X_r = X_n - M_f/M_r\,(X_n - X_f) \tag{99}$$

The energy balance is given by

$$(M_r - M_f)\,C_p\,(T_n - T_f) = M_r\,C_p\,(T_r - T_f)$$

This relation is rearranged into

$$T_r = T_n - M_f/M_r\,(T_n - T_f) \tag{100}$$

Solution procedure and system parameters used in performance evaluation are identical to those of the previous MSF systems. The solution procedure is based on Newton's method and it evaluates through iterative sequence the flow rates, temperature, and salinity of various streams.

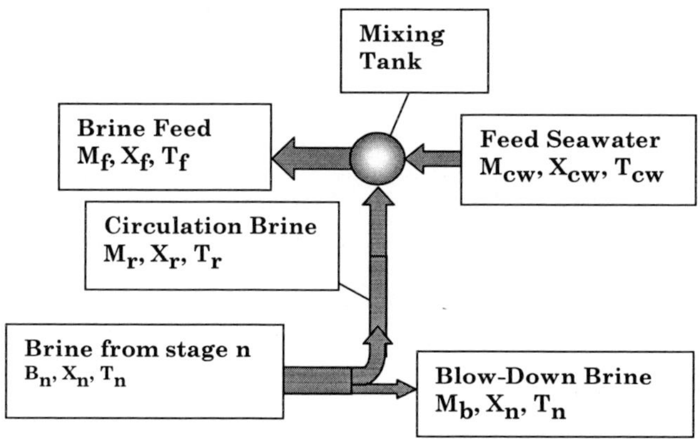

Fig. 45. Temperature profiles for the MSF-M, the MSF-OT, and MSF

6.7.3 System Performance

The performance of the proposed MSF-M system is analyzed as a function of the top brine temperature, the temperature of brine recycle, and the number of stages. Results shown in Figs. 46 and 47 include variations in the thermal performance ratio and the total specific heat transfer area as a function of the recycle temperature and the number of stages, respectively. The results shown in Fig. 46 are made for 24 flashing stages and seawater temperature of 32 °C. As is shown in Fig. 46 the thermal performance ratio increases at higher top brine temperatures and lower recycle temperatures. At these conditions, the flashing range increases resulting in higher temperature drop per stage. This increases the amount of flashing vapor per stage and consequently the total amount of product distillate increases. At the same conditions, the specific heat transfer decreases at higher top brine temperature and lower recycle temperatures. As discussed before, this decrease is caused by the increase in the amount of distillate product. Also, increase of the top brine temperature enhances the heat transfer rate across the condenser tubes. This is caused by the viscosity decrease and thermal conductivity increase for the seawater water and the distillate streams at higher temperatures. Effect of the number of stages on the system performance is shown in Fig. 47, at a recycle temperature of 37 °C and seawater temperature of 32 °C. Calculations are made for systems with 20, 24, 28, and 32 stages. As is shown the performance ratio and the specific heat transfer area increase at larger number of stages. Increase of the number of stages reduces the stage temperature drop. Therefore, the driving force for heat transfer is reduced. This increases the heat transfer area of the condensers in each stage and consequently the amount of distillate condensate per stage.

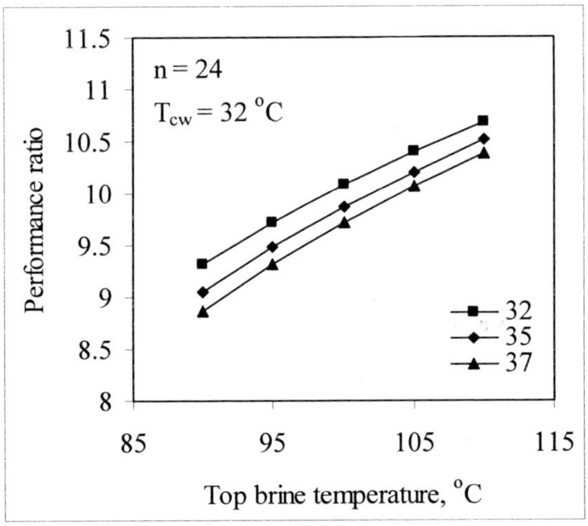

Fig. 46. Effect of the top brine temperature and brine recycle temperature on the performance ratio

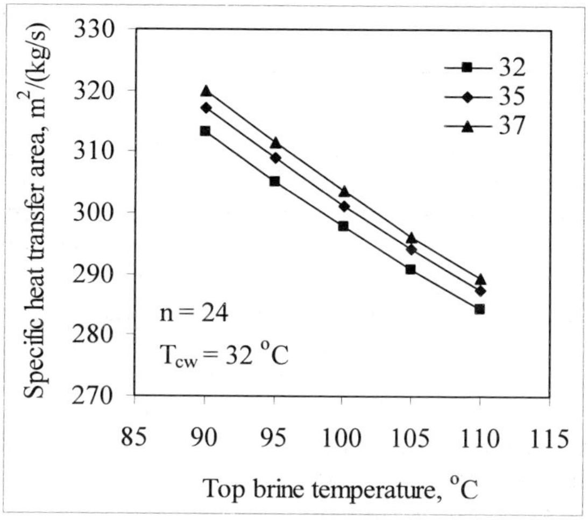

Fig. 47. Effect of the top brine temperature and brine recycle temperature on the specific heat transfer area.

6.7.4 Modification of Existing MSF Plants

Conversion of the MSF with brine recirculation to the MSF-M configuration is simple and primarily involves elimination of the brine circulation and the cooling seawater streams. The conversion includes the following:

- Removal of the cooling seawater loop as well as the temperature control loop on the feed seawater temperature.
- Modify the brine circulation loop to recycle the brine to the storage tank instead of the last stage.
- Addition of the accumulation tank for the recycled brine stream.
- Connect the preheater tubes of the first stage in the heat rejection section and the last stage in heat recovery section.
- Replace the intake seawater pump with a smaller capacity pump; this is necessary, since in the cooling seawater stream in the MSF-M system is eliminated.
- It should be stressed that conversion of the MSF system into MSF-M is feasible since the temperature of the brine blow down and the recycle flow rates, during the whole year, are similar in both systems. Therefore, the two ratios $M_r/\Delta P$ and M_d/ρ_v will have similar value and the design features of the MSF, which includes stage dimensions, brine flow area across the stage, and the non-condensable gases vacuum system, will be suitable for the MSF-M system.

6.7.5 Summary

A novel system is presented for the multi-stage flash desalination. The system is based on the conventional MSF configuration, where it includes two of its basic elements; the brine heater and the heat recovery section. The cooling water stream, typical of the MSF system, is also eliminated. The new system adopts a direct brine recycle stream from the brine stream leaving the last flashing stage. This stream is mixed with the intake seawater stream in an insulated and vented tank. Accordingly, the temperature of the feed stream is adjusted to meet summer and winter operating conditions. In the light of above, the following conclusions are made:

- Control of the intake seawater temperature is an essential feature in all thermal desalination process. This is found in the mixing tank of the MSF-M system and in the heat rejection section of the conventional MSF.
- This temperature control feature is essential in winter operation. Lack of this control, as in the MSF-OT system, reduces the brine temperature in the last stage and results in large increase the pressure drop across the stages.
- Reduction of the brine temperature in the last stage during winter operation for the MSF-OT system, necessitates increase of the stage dimensions to accommodate the large increase in the vapor specific volume. Also, larger flow area for the brine stream is necessary to meet the high pressure drop across the stages. Additionally, increase in the capacity of the non-condensable gases vacuum system, wall thickness, and leakage rate of the outside air are dictated by lack of control of the intake seawater temperature.
- Operation of the MSF-M system with no brine recycle reduces the system to the MSF-OT configuration. This condition can be adopted during the summer period, where the seawater temperature is high enough to ensure high brine temperatures in the last stage.
- In actual operation, brine circulation in the MSF-M system would be favored, even during summer operation. This condition reduces the amount of intake seawater and the amount of chemicals used to control scaling, corrosion, and foaming.
- The MSF-M has a similar consumption rate of the antiscalent to the MSF system. Since MSF-OT system has no brine recycle stream, its antiscalent consumption rate is 3-4 times higher than the other two systems.

Problems

Problem 1

An MSF-M system has the following design data:

Plant capacity:	Unknown
Seawater temperature:	32 °C
Brine recycle temperature:	36 °C
Seawater salinity:	49400 ppm
Top brine temperature:	100 °C
Performance ratio:	8
Number of flashing stages	22
Maximum brine concentration:	68600 ppm
Ratio of demister cross Sectional area to the chamber area:	0.75
Maximum vapor velocity in demister:	4 m/s
Maximum allowable flow rate of brine per chamber width:	600×10^3 kg/m hr
Overall heat transfer coefficient in brine heater:	2 kW/m^2 °C
Overall heat transfer coefficient in flashing stages:	2.4 kW/m^2 °C

Calculate the following:

- The specific heat transfer area
- The thermal performance ratio
- The brine recycle flow rate

Problem 2

An MSF-M plant has the following design data:

Plant capacity:	Unknown
Top brine temperature:	Unknown
Brine flow rate per chamber width:	Unknown
Number of stages:	20
Brine recycle temperature:	36 °C
Boiling temperature in last stage:	40 °C
Heat transfer area in the brine heater:	1000 m^2
Overall heat transfer coefficient	2.527 kW/m^2 °C
Mass flow rate of heating steam:	16.782 kg/s

Heating steam temperature: 120 °C
Number of tubes in the brine heater: 1000 tube
Specific flow rate of brine circulation: 8.422
Diameter of tubes used in brine heater: 31.8 mm

Calculate the following:

- The plant performance ratio
- The specific heat transfer area

Problem 3

An MSF-M plant has the following design data:

Seawater temperature: 34 °C
Brine recycle temperature: 36 °C
Seawater salinity: 42000 ppm
Top brine temperature: 100 °C
Temperature in the last stage: 40 °C
Temperature of heating steam: 110 °C
Specific flow rate of brine circulation: 8.478
Heat transfer area in the brine heater: 80 m^2
Overall heat transfer coefficient
in brine heater: 1.5 kW/m^2 °C

Calculate the following

- The plant performance ratio
- The terminal temperature difference of the first stage.
- The specific heat transfer area.

Problem 4

An MSF-M plant has the following design data:

Distillate flow rate: 5000 m^3/d
Brine recycle temperature: 34 °C
Seawater temperature: 30 °C
Seawater salinity: 44000 ppm
Top brine temperature: 112 °C
Temperature in the last stage: 40 °C
Temperature of heating steam: 116.7 °C
Terminal temperature difference: 3 °C
Number of flashing stages: 18

The overall heat transfer coefficient in the brine heater or the flashing stages is given by the relation

$$U = 6.5 - 0.03(115\text{-}T)$$

With U in $kW/m^2\,^oC$ and T in ^oC.

Calculate the following

- Thermodynamic losses in the first stage
- Plant performance ratio
- Brine heater surface area
- Flow rate of make up water.

<u>Problem 5</u>

An MSF-M plant has the following design data:

Distillate flow rate:	22750 m^3/d
Brine recycle temperature:	32 ^oC
Seawater temperature:	28 ^oC
Seawater salinity:	45000 ppm
Top brine temperature:	90 ^oC
Temperature in the last stage:	40 ^oC
Temperature of heating steam:	100 ^oC
Terminal temperature difference:	3 ^oC
Number of stages:	40
Width of stage 10:	16 m
Length of stage 10:	3.5 m

Calculate the following

- The temperature profile of flashing brine
- The temperature profile of seawater flowing in the preheaters
- The flow rates of heating steam, makeup water, brine circulation, cooling water, and brine blowdown.
- The pressure in stages 9, 10, and 11.
- The distillate product in stage 10.
- The vapor velocity in stage 10.

Chapter 7

Reverse Osmosis

Objectives

The objective of this chapter is to present elements of membrane separation processes and to explain principles of membrane separation. The material also defines ranges for application of different membrane processes and summarizes performance parameters of membrane separation processes. In addition, an outline is made for different construction materials and module configurations. This also includes components of the RO desalination process.

7.1 Historical Background

Membranes are an intimate part of being alive. Several examples are simple to cite:

- The skin in all mammals is a very efficient and highly selective type of membrane controlling release of sweat to cool off the bodies through evaporation of tinny water droplets during hot weather. Skin selectivity as apparent, when its cut the fine blood cells and vessels that runs underneath the skin are broken and releases its blood content. A healthy and intact skin does not release blood.
- The lungs are also a good example of effective membranes, where fine cells within the lungs allow passage of oxygen from the inhaled air and release carbon dioxide into the same stream. The lungs as a membrane prevents permeation of the nitrogen in the inhaled air, irrespective of its high content.
- The kidney membranes regulate the water, salt ions, proteins, and other nutrient within the body. The kidneys are extremely efficient that a healthy body can survive with a quarter of both kidneys.
- On a much smaller scale, membrane walls in single cells within mammals, bacteria, and other microorganisms maintains the cell contents intact and regulate the input/output rates of nutrients or products.

Since the early days of civilization mankind have adopted simple forms of membranes. In early agriculture communities, household sieves were invented and developed to separate fin grain ground from coarse grain particles and shells. Similarly, cheesecloth was made from cotton fibers and used to manufacture cheese. Both forms of separation are based on differences in particle size. However, developments in membrane technology have focused on adoption of other separation mechanisms, such differences in solution and diffusion rates of various species across the membrane material.

Other than the sieve type membrane use of artificial membranes is rather new. Major landmarks in use of artificial membranes are summarized in the following points:

- In 1823, Dutrochet gave correct explanation of osmosis (passage of solvent across a membrane from low to high concentration) and dialysis (passage of solute across a membranes from high to low concentration).
- In 1867, Traube and Pfeffer performed one of the first quantitative studies on performance of artificial membranes.
- Moritz Taube, 1867, prepared the first synthetic membrane.
- In the late 1800's Graham discovered that arranging a membrane between a reservoir of pressurized air and another reservoir of unpressurized air could produce oxygen-enriched air.
- Early use of membranes was applied to recovery of NaOH by dialysis from wastewater containing hemicellulose from the viscose-rayon industry.
- Also, uranium isotopes (235 and 238) are separated in the vapor phase through porous membranes.

As for the RO membranes, their history started with the following two studies

- Reid and Breton, 1959, at the University of Florida developed cellulose acetate RO membranes.
- Loeb and Sourirajan, 1963, from the University of California, Los Angeles developed the first asymmetric cellulose acetate membrane, with higher salt rejection and water flux.

Subsequently, a huge amount of research studies were conducted with focus on development of new membrane materials and performance evaluation of these membranes. As for commercialization of the RO membranes it is summarized in the following points:

- In the late 1960s, the Gulf General Atomics and Aerojet General used Loeb-Sourirajan cellulose acetate membranes for constructing spiral wound modules.
- In 1971, Dupont introduced the Permasep B-9 permeator for brackish water desalination. The permeator contains millions of asymmetric aromatic polyamide (aramid) hollow fine fibers.
- In late 1973, Dupont introduced the Permasep B-10 permeator, also using asymmetric aramid fibers, capable of producing potable water from seawater in a single pass.
- In the mid-1970s, cellulose triacetate hollow fiber permeators were introduced by Dow Chemical Company, followed by Toyobo of Japan
- During the same period, Fluid Systems and FilmTec introduced the spiral wound polyamide thin film composite membranes.
- Throughout the 1980s, improvements were made to these membranes to increase water flux and salt rejection with both brackish water and seawater.
- Today the predominate membrane materials are still aramids, polyamides, and cellulose acetate and triacetate in spiral wound and hollow fiber configurations.

- Applications of the RO membranes include potable water production, waste recovery, food applications, kidney dialysis, high-purity water for boiler feed, and ultrapure water electronics applications.
- In 2000, the RO technology was used to treat more than two billion gallons of water per day, and this market is expected to continue growing during the first half of the 21st century.

7.2 Elements of Membrane Separation

A number of membrane-based desalination processes are used on industrial scale. As is shown in Fig. 1, the membrane-based processes include reverse osmosis, nanofiltration, ultrafiltration, and microfiltration. Differences among these processes is shown in Fig. 1, where

- Microfiltration operates on a particle size range of 0.15 μm to 0.15 μm.
- Ultrafiltration operates on a particle size range of 0.15 μm to $5x10^{-2}$ μm
- Nanofiltration operates on a particle size range of $5x10^{-2}$ μm to $5x10^{-3}$ μm
- Reverse osmosis operates on a particle size range of $5x10^{-3}$ μm to 10^{-4} μm.

There is an inherent difference in the separation mechanism in all filtration processes and the reverse osmosis process. In filtration, separation is made by a sieving mechanism, where the membrane passes smaller particles and retains larger ones. In osmosis or reverse osmosis processes the membrane permeates only the solvent and retains the solute. Further distinction of the four membrane processes is shown in Fig. 1. As is shown, the microfiltration, ultrafiltration, and nanofiltration processes are used to separate the suspended material. On the other hand, the reverse osmosis process is used to separate dissolved solids. Nanofiltration is used for partial softening of brackish water.

A schematic for the osmosis and reverse osmosis phenomenon are shown in Fig. 2. In this configuration, the direction of solvent flow is determined by its chemical potential, which is a function of pressure, temperature, and concentration of dissolved solids. Pure water in contact with both sides of an ideal semi-permeable membrane at equal pressure and temperature has no net flow across the membrane because the chemical potential is equal on both sides. If a soluble salt is added on one side, the chemical potential of this salt solution is reduced. Osmotic flow from the pure water side across the membrane to the salt solution side will occur until the equilibrium of chemical potential is restored. Equilibrium occurs when the hydrostatic pressure differential resulting from the volume changes on both sides is equal to the osmotic pressure. This is a solution property independent of the membrane. Application of an external pressure to the salt solution side equal to the osmotic pressure will also cause equilibrium. Additional pressure will raise the chemical potential of the water in the salt solution and cause a solvent flow to the pure water side, because it now has a lower chemical potential.

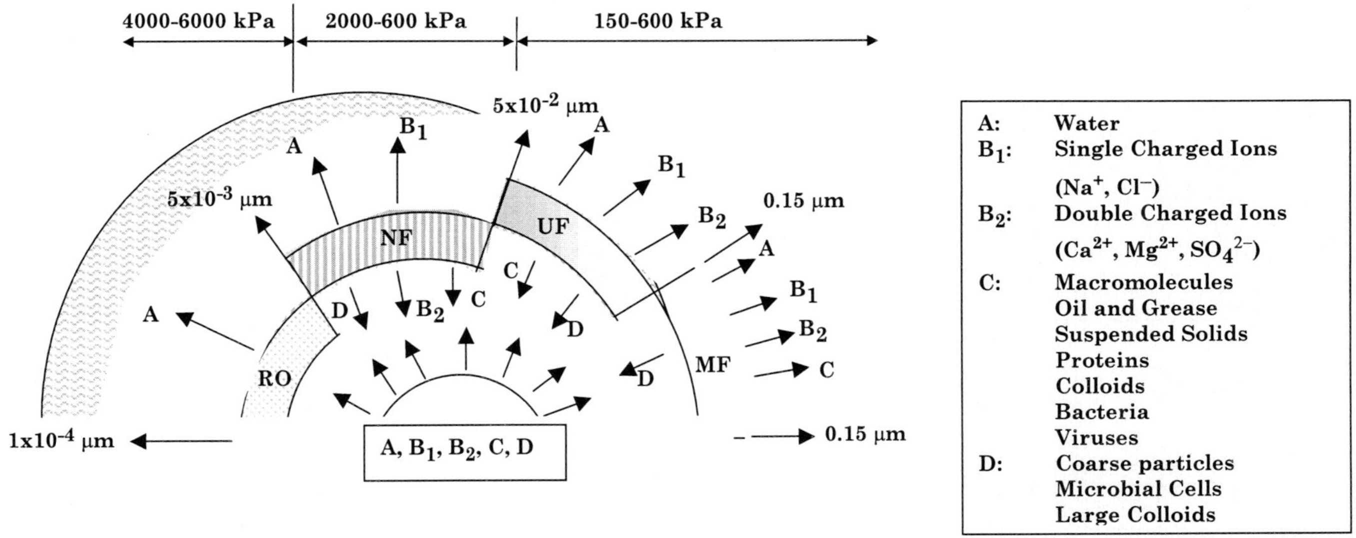

Fig. 1. Membrane separation processes and corresponding particle sizes

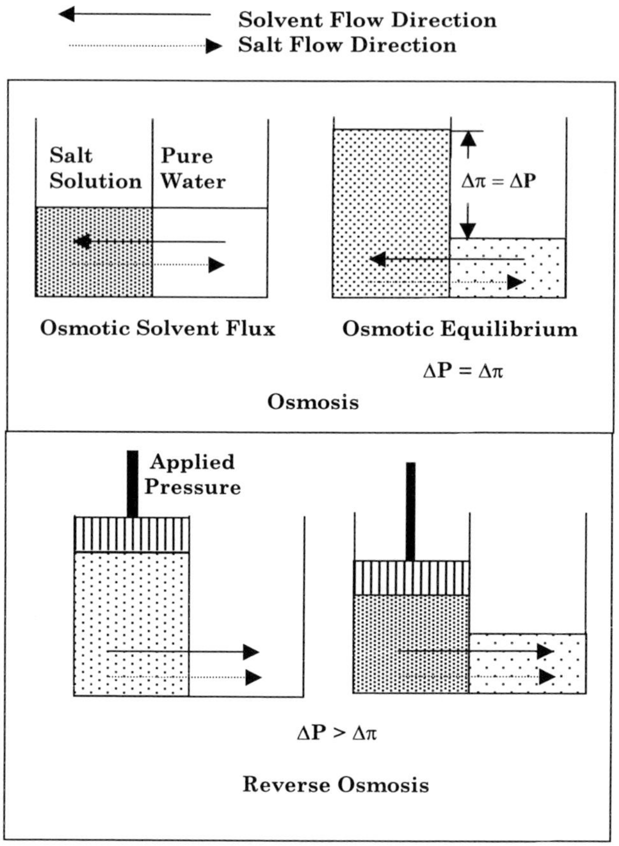

Fig. 2. Osmosis and reverse osmosis processes.

7.3 Performance Parameters

The RO process is defined in terms of a number of variables, which includes:

- Osmotic and operating pressure
- Salt rejection
- Permeate recovery

Membrane manufacturing companies define system specifications in terms of the feed quality, which includes salinity and temperature.

7.3.1 Osmotic and Operating Pressure

The osmotic pressure, π, of a solution can be determined experimentally by measuring the concentration of dissolved salts in the solution. The osmotic pressure is obtained from the following equation

$$\pi = R\,T\,\sum X_i \tag{1}$$

where

π is the osmotic pressure (kPa).
T is the temperature (K).
R is the universal gas constant, 8.314 kPa m^3/kgmol K
$\sum X_i$ is the concentration of all constituents in a solution (kgmol/m^3).

An approximation for π may be made by assuming that 1000 ppm of Total Dissolved Solids (TDS) equals to 75.84 kPa of osmotic pressure.

Operating pressure is adjusted to overcome the adverse effects of the following:

- Osmotic pressure
- Friction losses
- Membrane resistance
- Permeate pressure

If the operating pressure is set equal to the sum of the above resistances the net permeate flow rate across the membrane would be minimal or equal to zero; therefore, the operating pressure is set at higher value in order to maintain economical permeate flow rate.

7.3.2 Salt Rejection

Salt rejection is defined by

$$SR = 100\%\,(1 - (X_p/X_f)) \tag{2}$$

where SR is the salt rejection. For example, a feed seawater with 42,000 ppm and a permeate with a salinity of 150 ppm gives a percentage salt passage of 99.64%. Similarly, for a brackish water feed with salinity of 5000 ppm and a permeate salinity of 150 ppm gives a percentage salt passage of 97%. The two cases indicate the dramatic difference between the seawater and brackish water desalination membranes. Current membrane technology provides salt rejection values above 99% for both seawater and brackish water membranes.

7.3.3 Permeate Recovery

Permeate recovery is another important parameter in the design and operation of RO systems. Recovery or conversion rate of feed water to product (permeate) is defined by

$$R = 100\% \, (M_p/M_f) \tag{3}$$

where R is recovery rate (in %), M_p is the permeate water flow rate, and M_f is the feed water flow rate. The recovery rate affects salt passage and product flow. As the recovery rate increases, the salt concentration on the feed-brine side of the membrane increases, which causes an increase in salt flow rate across the membrane. Also, a higher salt concentration in the feed-brine solution increases the osmotic pressure, reducing the ($\Delta P - \Delta\pi$) and consequently reducing the product water flow rate.

Membrane recovery for RO systems have increased over the years from lower values of 10-20% to current higher values up to 50%. This is achieved in part by proper system design and use of multiple modules of spiral wound membranes within the same pressure vessel. As for the hollow fiber membranes it common to use a single module within the same pressure vessel.

7.4 RO Membranes

Features of the RO membranes include the following:

- The membranes are formed of thin film of polymeric material several thousand Angstroms thick cast on polymeric porous material.
- Commercial membranes have high water permeability and a high degree of semi-permeability; that is, the rate of water transport must be much higher than the rate of transport of dissolved ions.
- The membrane must be stable over a wide range of pH and temperature, and have good mechanical integrity.
- The life of commercial membranes varies between 3-5 years. On average annual membrane replacement rates stand at 5-15%; this depends on the feed water quality, pretreatment conditions, and stability of operation.
- Major types of commercial reverse osmosis membranes include cellulose acetate (CA) and polyamide (PA).
- It should be noted that membrane choice is often governed by compatibility considerations rather than separation performance and flux related characteristics.

7.4.1 Cellulose Acetate Membranes

The original CA membrane, developed in the late 1950's by Loeb and Sourirajan, was made from cellulose diacetate polymer. Current CA membrane is usually made from a blend of cellulose diacetate and triacetate. The membrane preparation process includes thin film casting, cod bath leaching, and high temperature annealing. The casting process is associated with partial removal of the solvent material by evaporation. The cold bath process removes the remaining solvent and other leachable compounds. The annealing process is made in a hot water bath at a temperature of 60-90°C. The annealing step improves the semipermeability of the membrane with a decrease of water transport and a significant decrease of salt passage.

The CA membranes have an asymmetric structure with a dense surface layer of about 1000-2000 A (0.1-0.2 micron) which is responsible for the salt rejection property. The rest of the membrane film is spongy and porous and has high water permeability. Salt rejection and water flux of a cellulose acetate membrane can be controlled by variations in temperature and duration of the annealing step.

7.4.2 Composite Polyamide Membranes

The composite polyamide membranes are formed of two layers, the first is a porous polysulfone support and the second is a semi-permeable layer of amine and carboxylic acid chloride functional groups. This manufacturing procedure enables independent optimization of the distinct properties of the membrane support and salt rejecting skin. The resulting composite membrane is characterized by higher specific water flux and lower salt passage than cellulose acetate membranes.

Polyamide composite membranes are stable over a wider pH range than CA membranes. However, polyamide membranes are susceptible to oxidative degradation by free chlorine, while cellulose acetate membranes can tolerate limited levels of exposure to free chlorine. Compared to a polyamide membrane, the surface of cellulose acetate membrane is smooth and has little surface charge. Because of the neutral surface and tolerance to free chlorine, cellulose acetate membranes will usually have a more stable performance than polyamide membranes in applications where the feed water has a high fouling potential, such as with municipal effluent and surface water supplies.

7.5 Membrane Modules

The two major membrane module configurations used for reverse osmosis applications are hollow fiber and spiral wound. Other configurations, which include tubular and plate and frame, are used in the food and dairy industry.

7.5.1 Hollow Fine Fiber

This configuration uses membrane in the form of hollow fibers, which have been extruded from cellulosic or non-cellulosic material. The fiber is asymmetric in structure and is as fine as a human hair, about 42 μm ID and 85 μm OD, Fig. 3. Millions of these fibers are formed into a bundle and folded in half to a length of approximately 120 cm. A perforated plastic tube, serving as a feed water distributor is inserted in the center and extends the full length of the bundle. The bundle is wrapped and both ends are epoxy sealed to form a sheet-like permeate tube end and a terminal end which prevents the feed stream from bypassing to the brine outlet.

The hollow fiber membrane bundle, 10 cm to 20 cm in diameter, is contained in a cylindrical housing or shell approximately 137 cm long and 15-30 cm in diameter. The assembly has the highest specific surface area, defined as the total area per unit volume, among all module configurations. The pressurized feed water enters the permeator feed end through the center distributor tube, passes through the tube wall, and flows radially around the fiber bundle toward the outer permeator pressure shell. Water permeates through the outside wall of the fibers into the hollow core or fiber bore, through the bore to the tube sheet or product end of the fiber bundle, and exits through the product connection on the feed end of the permeator.

In a hollow fiber module, the permeate water flow per unit area of membrane is low, and therefore, the concentration polarization is not high at the membrane surface. The net result is that hollow fiber units operate in a non-turbulent or laminar flow regime. The hollow fine fiber membrane must operate above a minimum reject flow to minimize concentration polarization and maintain even flow distribution through the fiber bundle. Typically, a single hollow fiber permeator can be operated at up to 50-percent recovery and meet the minimum reject flow required. The hollow fiber unit allows a large membrane area per unit volume of permeator that results in compact systems. Hollow fiber perimeters are available for brackish and seawater applications.

Membrane materials are cellulose acetate blends and polyamide type material. Because of very close packed fibers and tortuous feed flow inside the module, hollow fiber modules require feed water of better quality (lower concentration of suspended solids) than the spiral wound module configuration.

7.5.2 Spiral Wound

In a spiral wound configuration two flat sheets of membrane are separated with a permeate collector channel material to form a leaf. This assembly is sealed on three sides with the fourth side left open for permeate to exit. A feed/brine spacer material sheet is added to the leaf assembly. A number of these assemblies or leaves are wound around a central plastic permeate tube. This tube is perforated to collect the permeate from the multiple leaf assemblies. The typical industrial spiral wound membrane element is approximately 100 or 150 cm long and 10 or 20 cm in diameter, Fig. 4.

The feed/brine flow through the element is a straight axial path from the feed end to the opposite brine end, running parallel to the membrane surface. The feed channel spacer induces turbulence and reduces concentration polarization. Manufacturers specify brine flow requirements to control concentration polarization by limiting recovery (or conversion) per element to 10 – 20%.

Therefore, recovery (or conversion) is a function of the feed-brine path length. In order to operate at acceptable recoveries, spiral systems are usually staged with three to six membrane elements connected in series in a pressure tube. The brine stream from the first element becomes the feed to the following element, and so on for each element within the pressure tube.

The brine stream from the last element exits the pressure tube to waste. The permeate from each element enters the permeate collector tube and exits the vessel as a common permeate stream. A single pressure vessel with four to six membrane elements connected in series can be operated at up to 50-percent recovery under normal design conditions. The brine seal on the element feed end seal carrier prevents the feed/brine stream from bypassing the following element.

Spiral wound elements are most commonly manufactured with flat sheet membrane of either a cellulose diacetate and triacetate (CA) blend or a thin film composite. A thin film composite membrane consists of a thin active layer of one polymer cast on a thicker supporting layer of a different polymer. The composite membranes usually exhibit higher rejection at lower operating pressures than the cellulose acetate blends. The composite membrane materials may be polyamide, polysulfone, polyurea, or other polymers.

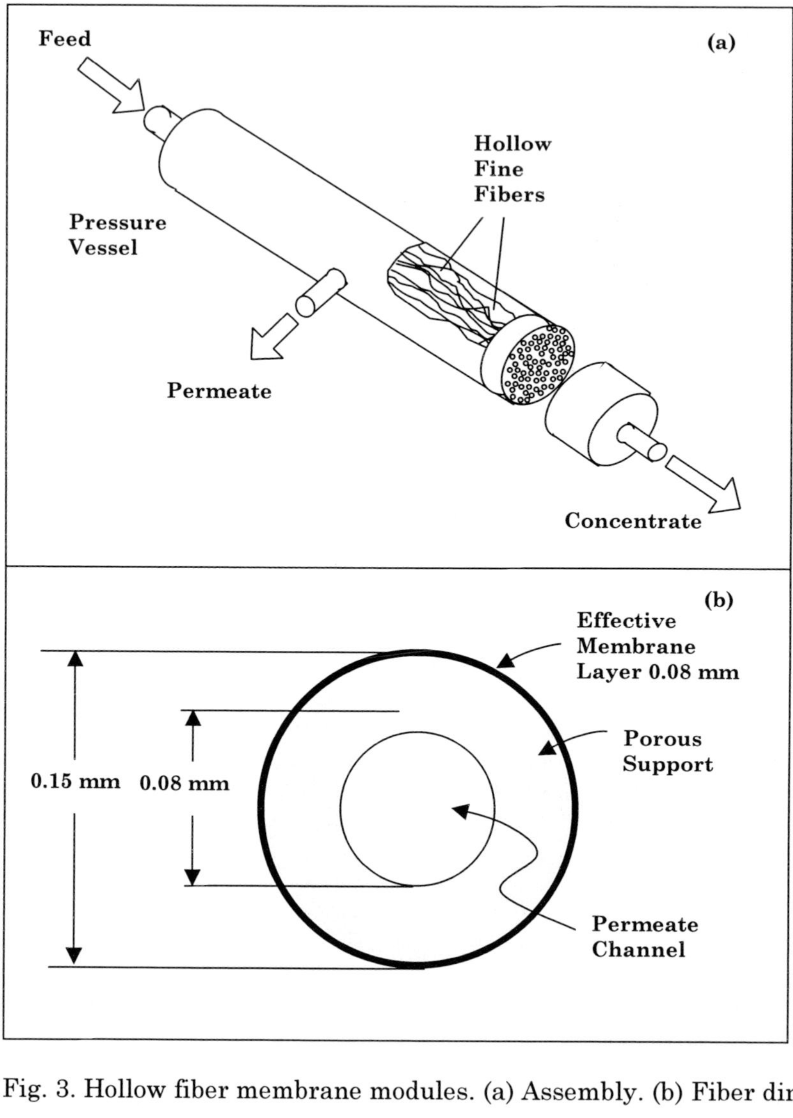

Fig. 3. Hollow fiber membrane modules. (a) Assembly. (b) Fiber dimensions

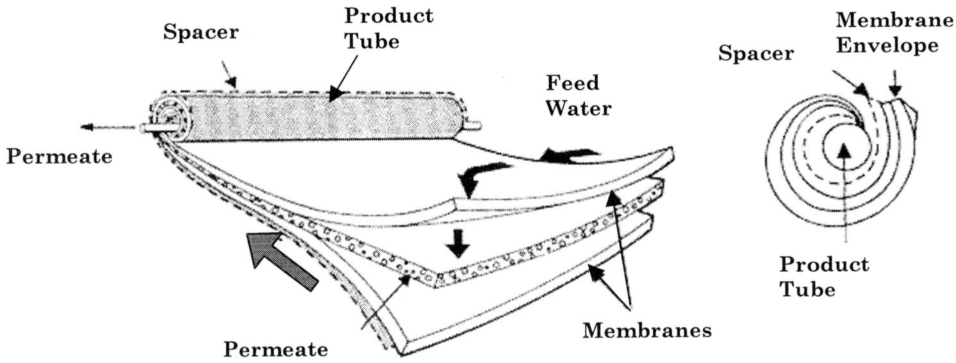

Fig. 4. Spiral wound membrane modules

7.6 RO Systems

The RO systems may consist of the following basic components:

- Feed water supply unit
- Pretreatment system
- High pressure pumping unit
- Membrane element assembly unit
- Instrumentation and control system
- Permeate treatment and storage unit
- Cleaning unit

Figure 5 shows a typical process diagram for the RO process. It should be noted that the system in Fig. 5 is only an example, where the level of feed pretreatment depends strongly on the quality of the feed water. Features of the RO process includes the following:

- Large particles are removed from the feed water using mesh strainers or traveling screens. Mesh strainers are used in well water supply systems to stop and remove sand particles, which may be pumped from the well. Traveling screens are used mainly for surface water sources, which typically have large concentrations of biological debris.
- It is common practice to disinfect surface feed water in order to control biological activity. Biological activity in well water is usually very low, and in majority of cases, well water does not require chlorination. In some cases, chlorination is used to oxidize iron and manganese in the well water before filtration. Well water containing hydrogen sulfide should not be chlorinated or

exposed to air. In presence of an oxidant, the sulfide ion can oxidize to elemental sulfur, which eventually may plug membrane elements.

- Settling of surface water in a detention tank results in some reduction of suspended particles. Addition of flocculants, such as iron or aluminum salts, results in formation of corresponding hydroxides; these hydroxides neutralize surface charges of colloidal particles, aggregate, and adsorb to floating particles before settling at the lower part of the clarifier. To increase the size and strength of the flock, a long chain organic polymer can be added to the water to bind flock particles together. Use of lime results in increase of pH, formation of calcium carbonate and magnesium hydroxide particles. Lime clarification results in reduction of hardness and alkalinity, and the clarification of treated water.
- Well water usually contains low concentrations of suspended particles, due to the filtration effect of the aquifer. The pretreatment of well water is usually limited to screening of sand, addition of scale inhibitor to the feed water, and cartridge filtration.
- Surface water may contain various concentrations of suspended particles, which are either of inorganic or biological origin. Surface water usually requires disinfection to control biological activity and removal of suspended particles by media filtration. The efficiency of filtration process can be increased by adding filtration aids, such as flocculants and organic polymers. Some surface water may contain high concentrations of dissolved organics. Those can be removed by passing feed water through an activated carbon filter. Depending on composition of the water, acidification and addition scale inhibitor may be required. The flow diagram of pretreatment system for surface water is shown below.
- Cartridge filters, almost universally used in all RO systems prior to the high pressure pump, serve as the final barrier to water born particles. The nominal rating commonly used in RO applications is in the range of 5 - 15 microns. Some systems use cartridges with micron ratings as low as 1 micron. There seems to be little benefit from lower micron rated filters as such filters require a high replacement rate with relatively small improvement in the final feed water quality.
- Recently, new pretreatment equipment has been introduced to the RO market. It consists of backwashable capillary microfiltration and ultrafiltration membrane modules. This new equipment can operate reliably at a very high recovery rates and low feed pressure. The new capillary systems can provide better feed water quality than a number of conventional filtration steps operating in series. The cost of this new equipment is still very high compared to the cost of an RO unit.

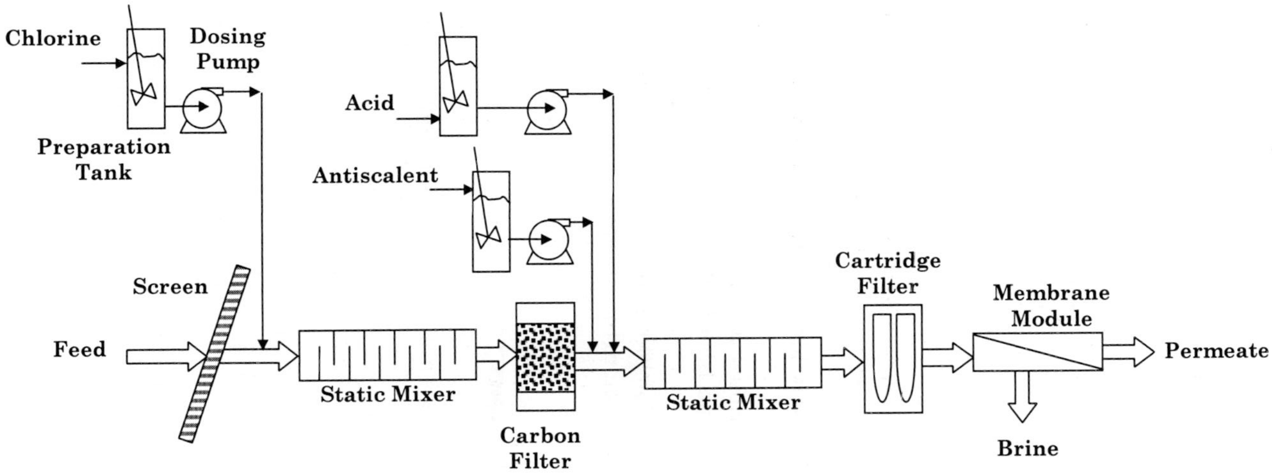

Fig. 5. Typical RO process with screening, chlorination, filtration, acidification and scale inhibition

7.7 RO Model and System Variables

The RO process is defined in terms of a number of variables, which includes:

- Osmotic pressure
- Water transport
- Salt transport
- Salt passage
- Salt rejection
- Permeate recovery
- Concentration polarization

The following sections include the equations and terms forming the RO simple model. The model is based on the following assumptions:

- Steady state and isothermal operation.
- Permeability coefficients of various salt ions or water are independent of temperature and concentration.
- Similar permeability coefficient for various salt ions.
- The salt flow rate across the membrane is negligible in comparison with the water permeate flow rate.
- Complete mixing within the permeate compartment.
- Salt concentration within the feed compartment varies linearly along the membrane area.

7.7.1 Permeator Mass and Salt Balances

The permeator mass and salt balances are given by the following relations:

$$M_f = M_p + M_b \quad (4)$$

$$X_f M_f = X_p M_p + X_b M_b \quad (5)$$

where

M_f is the feed flow rate, kg/s.

M_p is the permeate flow rate, kg/s

M_b is the brine flow rate, kg/s

X_f is the feed salinity, kg/m^3

X_p is the permeate salinity, kg/m^3.

X_b is the brine salinity, kg/m^3.

7.7.2 *Water Transport*

The following relation defines the rate of water passage through a semipermeable membrane

$$M_p = (\Delta P - \Delta\pi)\, K_w\, A \tag{6}$$

where

M_p is the rate of water flow through the membrane, m^3/s.
$\Delta\pi$ is the osmotic pressure differential across the membrane, kPa.
K_w is the water permeability coefficient, m^3/m^2 s kPa
A is the membrane area, m^2.

In Eq. (6) the terms ΔP and $\Delta\pi$ represent the net hydraulic and osmotic pressure differential across the membrane, respectively, or

$$\Delta P = \overline{P} - Pp \tag{7}$$

$$\Delta\pi = \overline{\pi} - \pi p \tag{8}$$

where

Pp and πp are the permeate hydraulic and osmotic pressure, respectively.
$\overline{P}$ and $\overline{\pi}$ are the average hydraulic and osmotic pressures on the feed side and are given by:

$$\overline{P} = 0.5\,(P_f + P_b) \tag{9}$$

$$\overline{\pi} = 0.5\,(\pi_f + \pi_b) \tag{10}$$

where P_f and π_f are the hydraulic and osmotic pressures of the feed stream, respectively. While P_b and π_b are the hydraulic and osmotic pressures of the reject stream, respectively.

7.7.3 *Salt Transport*

The rate of salt flow through the membrane is defined by

$$M_s = (\overline{X} - Xp)\, K_s\, A \tag{11}$$

where

M_s is the flow rate of salt through the membrane, kg/s.

K_s is the membrane permeability coefficient for salt, m^3/m^2 s.

Xp is the permeate total dissolved solids concentration, kg/m^3.

A is the membrane area, m^2.

In Eq. (11) the term $\overline{X}$ is defined by

$$\overline{X} = (M_f X_f + M_b X_b)/(M_f + M_b) \tag{12}$$

where X_f and X_b are the feed and reject salt concentrations, respectively.

Equations 6 and 11 show that for a given membrane:

- Rate of water flow through a membrane is proportional to net driving pressure differential $(\Delta P - \Delta\pi)$ across the membrane.
- Rate of salt flow is proportional to the concentration differential across the membrane $(\overline{X} - Xp)$ and is independent of applied pressure.

Salinity of the permeate, Xp, depends on the relative rates of water and salt transport through reverse osmosis membrane:

$$Xp = M_s/M_d \tag{13}$$

The fact that water and salt have different mass transfer rates through a given membrane creates the phenomena of salt rejection. No membrane is ideal in the sense that it absolutely rejects salts; rather the different transport rates create an apparent rejection. Equations 6 and 11 show that increasing the operating pressure will increase water flow without changing salt flow, thus resulting in lower permeate salinity.

7.7.4 Semi-Empirical Models

Several approaches are adopted in evaluation of experimental models, which includes the following:

- Irreversible thermodynamic model.
- Frictional model.
- Solution-diffusion model.
- Solution-diffusion-imperfection model.
- Preferential adsorption capillary flow model.
- Diffusion viscous flow model
- Finely porous model.

All these models are special cases of the statistical-mechanical model, Mason and Lonsdale (1990).

The statistical-mechanical model includes the following equations:
Salt rejection

$$SR = (C_1 \frac{1}{(M_p / A)} + C_2)^{-1} \tag{14}$$

Permeate Flux

$$M_p / A = (D_1 c_w + D_2)(\Delta p - \sigma\Delta\pi) \tag{15}$$

Membrane wall concentration

$$c_w = c_b + (c_b - c_d)(e^{M_p / Ak} - 1) \tag{16}$$

Variables in the above model are defined as follows:
- C_1, C_2, D_1, and D_2 are the fitting constants of Eqs. 14 and 15 with experimental data.
- SR is the salt reject.
- Mp is the permeate flow rate, m^3/s.
- A is the membrane area, m^2.
- ΔP is the pressure difference across the membrane, kPa.
- $\Delta\pi$ is the osmotic pressure difference across the membrane, kPa.
- cw is the salt concentration in the membrane wall, kg/m^3.
- c_b is the salt concentration in the bulk of the feed compartment, kg/m^3.
- cd is the salt concentration in the permeate stream, kg/m^3.
- k is the mass transfer coefficient, m/s.
- σ is the reflection coefficient.

The finely-porous model includes the following equations:
- Salt rejection

$$SR = 1 - (A_1 - (1 - A_1)\, e^{(-A_2 M_p /(AD_{sw})})^{-1} \tag{17}$$

- Permeate Flux

$$M_p / A = \frac{\Delta p - \sigma\Delta\pi}{B_1 c_d + B_2\, \mu} \tag{18}$$

- Membrane wall concentration

$$c_w = c_b + (c_b - c_d)(e^{M_p / Ak} - 1) \tag{19}$$

The variables in the above model are similar to those of the statistical mechanical model, except for the following:

- A_1, A_2, B_1, and B2 are the fitting constants of Eqs. 17 and 18 with experimental data.
- μ is the water dynamic viscosity, kg/m s.
- D_{sw} is the solute diffusion coefficient in solution, m^2/s.

Both models are nonlinear and it requires iterative solution to determine the permeate flow rate, salt rejection, and membrane wall concentration.

7.7.5 Concentration Polarization

As water flows through the membrane and the membrane rejects salts, a boundary layer is formed near the membrane surface in which the salt concentration exceeds the salt concentration in the bulk solution. This increase of salt concentration is called concentration polarization. The effect of concentration polarization is to reduce actual product water flow rate and salt rejection versus theoretical estimates. The effects of concentration polarization are as follows:

- Greater osmotic pressure at the membrane surface than in the bulk feed solution, $\Delta\pi$, and reduced Net driving pressure differential across the membrane $(\Delta P - \Delta\pi)$.
- Reduced water flow across membrane (M_d).
- Increased salt flow across membrane (M_s).
- Increased probability of exceeding solubility of sparingly soluble salts at the membrane surface, and the distinct possibility of precipitation causing membrane scaling.

The Concentration Polarization Factor (CPF) can be defined as a ratio of salt concentration at the membrane surface (C_s) to bulk concentration (C_b), where

$$CPF = C_s/C_b \tag{20}$$

An increase in permeate flux will increase the delivery rate of ions to the membrane surface and increase C_s. An increase of feed flow increases turbulence and reduces the thickness of the high concentration layer near the membrane surface. Therefore, the CPF is directly proportional to permeate flow (M_d), and inversely proportional to average feed flow (M_f), where

$$CPF = K_3 \exp(M_d/M_f) \qquad (21)$$

where K_3 is a proportionality constant depending on system geometry.

Using the arithmetic average of feed and concentrate flow as average feed flow, the CPF can be expressed as a function of the permeate recovery rate a of membrane element (R):

$$CPF = K_3 \exp(2R/(2-R)) \qquad (22)$$

The value of the Concentration Polarization Factor of 1.2, corresponds to 18% permeate recovery.

7.8 *Case Studies*

Example on Mathematical Model

Design a single stage RO desalination system by calculating the permeate salinity, the brine salinity, the brine flow rate, and the membrane area.

Data

Water permeability is 2.05×10^{-6} m^3/m^2 s kPa
Salt permeability is 2.03×10^{-5} m^3/m^2 s
Feed salinity is 42,000 ppm
Feed flow rate: 2.5 kg/s
Permeate flow rate: 1 kg/s
Feed pressure: 8000 kPa
Reject pressure: 7800 kPa
Permeate pressure: 101 kPa

Solution

The material balance on the permeator is given by

$$M_b = M_f - M_d$$
$$= 2.5 - 1$$
$$= 1.5 \text{ kg/s}$$

The remaining model equations include the salt balance and the water and salt permeation. Solution of these equations is iterative because of the dependence of the osmotic pressure on the salinity of the permeate and brine streams. The iteration sequence is as follows:

- Assume salinity of the permeate stream is equal to 145 ppm or 0.145 kg/m^3.
- This value is used to calculate the brine salinity by solving the salt balance equation, which is given by

$$M_f X_f = M_b X_b + M_d X_d$$

which gives

$$(2.5)(42) = (1.5)(X_b) + (1)(0.145)$$

or $X_b = 69.903$ kg/m^3

- The average salinity on the feed side is calculated

$$\begin{aligned}\overline{X} &= (M_f X_f + M_b X_b)/(M_f + M_b) \\ &= ((2.5)(42) + (1.5)(69.903))/(2.5+1.5) \\ &= 52.46 \text{ kg/m}^3\end{aligned}$$

- The salinity of various streams is used to calculate the osmotic pressure of each stream

$$\pi_f = 75.84\ X_f = (75.84)(42) = 3185.28$$

$$\pi_b = 75.84\ X_b = (75.84)(69.903) = 5301.452 \text{ kPa}$$

$$\pi_p = 75.84\ X_p = (75.84)(0.145) = 10.9968 \text{ kPa}$$

- The average osmotic pressure on the feed side is then calculated

$$\overline{\pi} = 0.5\ (\pi_f + \pi_b) = 0.5\ (3185.28 + 5301.452) = 4243.366 \text{ kPa}$$

- Therefore, the net osmotic pressure across the membrane is given by

$$\Delta\pi = \overline{\pi} - \pi_p = 4243.366 - 10.9968 = 4232.3 \text{ kPa}$$

- The net pressure difference across the membrane is given by

$$\Delta p = 0.5(P_f + P_b) - P_p = (0.5)(8000+7800) - 101 = 7799 \text{ kPa}$$

- The permeate flux equation is used to calculate the membrane area, or,

$M_d = A\ k_w\ (\Delta P - \Delta\pi)$
$(1) = A\ (2.05x10^{-6})\ (7799 - 4232.3)$

which gives $A = 136.76\ m^2$

– The average salinity on the feed side is calculated from

$\overline{X} = (M_f\ X_f + M_b\ X_b)/(M_f + M_b)$
$= ((2.5)(42) + (1.5)(69.903))/(2.5+1.5)$
$= 52.46\ kg/m^3$

– The salt transport equation is used to calculate the permeator area

$M_s = A\ k_s\ (\Delta X - X_p)$
$(1)\ (0.145) = A\ (2.03x10^{-5})\ (52.46 - 0.145)$

which gives $A = 136.5\ m^2$

The difference in the areas obtained by the two flux equations is small, which implies correctness of the initial assumption.

Example on Manufacturer Specifications

Design a single stage RO desalination system by calculating the permeate salinity, the brine salinity, the brine flow rate, and the membrane area.

Module Data

Membrane recovery ratio:	10%
Membrane module area:	2.6 m^2
Salt rejection:	99.5%
Module pressure drop:	69 kPa
Maximum applied pressure:	5500 kPa
Maximum feed flow rate:	32.7 m^3/d
Permeate flow rate:	1.5 m^3/d
Minimum brine flow rate:	7.5 m^3/d

Design Data

Feed salinity is 42,000 ppm
Permeate flow rate: 5000 m^3/d
Permeate pressure: 101 kPa

Solution

The solution includes the following steps:

- The salt rejection definition is used to calculate the product salinity, where

$SR = 1 - X_p/X_f$
$0.995 = 1 - X_p/42000$

which gives $X_p = 210$ ppm

- The permeate recovery equation is used to calculate the feed flow rate for each module, or

$R = Q_d/Q_f$

Where Q_d and Q_f are the module flow rates for permeate and feed streams. This gives

$0.1 = 1.5/Q_f$
or $Q_f = 15 \text{ m}^3/\text{d}$

This value is lower than the maximum limit per module, which is equal to 32.7 m^3/d

- Therefore the total amount of feed seawater is obtained by performing a simple balance on the total production capacity and the module production rate, where

$M_f = (M_d/Q_d)(Q_f)$
$= (5000/1.5)(15)$
$= 50000 \text{ m}^3/\text{d}$

Also this gives the total number of modules, which is given by

$N = 5000/1.5$
$= 3333$ modules

- The feed and permeate flow rates per module are used to calculate the brine flow rate, where

$Q_f = Q_d + Q_b$

$15 = 1.5 + Q_b$

which gives Q_b = 13.5 m^3/d. This value is higher than minimum brine flow rate per module, or 7.5 m^3/d.

- The salt balance for each module gives the salinity in the brine stream, where

$Q_f X_f = Q_d X_d + Q_b X_b$
$(15)(42000) = (1.5)(210) + (13.5)(X_b)$

which gives X_b = 46643.3 ppm

- The salinity of various streams is used to calculate the osmotic pressure of each stream

$\pi_f = 75.84 X_f = (75.84)(42) = 3185.28$ kPa

$\pi_b = 75.84 X_b = (75.84)(46.6433) = 3537.4$ kPa

$\pi_p = 75.84 X_p = (75.84)(0.21) = 15.9264$ kPa

- The resulting average osmotic pressure on the feed side is then calculated

$\overline{\pi} = 0.5 (\pi_f + \pi_b) = 0.5 (3185.28 + 3537.4) = 3361.3$ kPa

- Therefore, the net osmotic pressure across the membrane is given by

$\Delta\pi = \overline{\pi} - \pi_p = 3361.3 - 15.9264 = 3345.4$ kPa

- Assuming that the feed pressure is equal to the maximum module pressure, or 5500 kPa. Also, considering the module pressure drop gives a pressure of 5431 kPa in the brine stream.
- The net pressure difference across the membrane is given by

$\Delta p = 0.5(P_f+P_b) - P_p = 0.5(5500+5431) - 101 = 5364.5$ kPa

which is higher than the exerted osmotic pressure

Example on Semi-Empirical Model

Use the statistical mechanical model to obtain the permeate flow rate and membrane area at the following conditions:

- Membrane constants C_1, C_2, D_1, and D_2 with

$C_1 = -3.01\text{x}10^{-1}$

$C_2 = 1.195164$

$D_1 = -1.56\text{x}10^{-4}$

$D_2 = 1\text{x}10^{-2}$

- Other data includes:
  Membrane salt rejection = 99%
  Feed salinity = 34,000 ppm
  Feed flow rate = 1000 m^3/d
  Permeate flow rate = 325 m^3/d
  Feed pressure = 6000 kPa
  Pressure of brine reject = 5900 kPa
  Permeate pressure = 101 kPa
  Salt concentration in membrane = 1.76 C_f

- The salt rejection definition is used to calculate the product salinity, where

$SR = 1 - X_p/X_f$
$0.99 = 1 - X_p/34000$

which gives $X_p = 340$ ppm

- From the above the pressure drop across the membrane is given by

$$\Delta P = 0.5\,(Pf + Pb) - Pp$$
$$= 0.5\,(6000+5900) - 101$$
$$= 5849 \text{ kPa}$$

- The osmotic pressure is calculated for the feed, brine, and permeate

$\pi_f = 75.84\ X_f = (75.84)(34) = 2578.6$ kPa

$\pi_b = 75.84\ X_b = (75.84)(50.2) = 3806.3$ kPa

$\pi_p = 75.84\ X_p = (75.84)(0.34) = 25.8$ kPa

– The resulting average osmotic pressure on the feed side is then calculated

$\bar{\pi} = 0.5\ (\pi_f + \pi_b) = 0.5\ (2578.6 + 3806.3) = 3192.4$ kPa

– Therefore, the net osmotic pressure across the membrane is given by

$\Delta\pi = \bar{\pi} - \pi_p = 3192.4 - 25.8 = 3166.7$ kPa

– The permeator area is calculated from the following equation

$$M_p / A = (D_1\ c_w + D_2)(\Delta p - \sigma\Delta\pi)$$

$$325 / A = (-0.000156\ (60.2) + 0.01)(5849 - 3166.6)$$

$$A = 199\ m^2$$

– The salt rejection is then calculated from the following relation

$$SR = (C_1 \frac{1}{(M_p / A)} + C_2)^{-1}$$

$$= ((-0.301)\frac{1}{(325/199)} + 1.19514)^{-1}$$

$$= 0.99$$

Problems

Problem 1

Design a single stage RO desalination system by calculating the permeate salinity, the brine salinity, the brine flow rate, and the membrane area. Analyze the following cases:

Data

Water permeability is $2.05\text{x}10^{-6}\ m^3/m^2$ s kPa
Salt permeability is $2.03\text{x}10^{-5}\ m^3/m^2$ s
Feed salinity is 34,000 ppm

Feed flow rate: 2.5 kg/s
Permeate flow rate: 1 kg/s
Feed pressure: 6000 kPa
Reject pressure: 5800 kPa
Permeate pressure: 101 kPa

Problem 2

The following module data are those for the DuPont B-10 permeator

Membrane recovery ratio:	35%
Salt rejection:	99.2%
Maximum applied pressure:	6895 kPa
Minimum brine flow rate:	1.584 m^3/d
Maximum brine flow rate:	14.256 m^3/d
Permeate flow rate:	2.46 m^3/d

Design Data

Feed salinity is 42,000 ppm
Permeate flow rate: 5000 m^3/d
Permeate pressure: 101 kPa

Design a single stage system by calculating the number of modules, osmotic pressure, permeation pressure difference, flow rates, and stream salinity.

Problem 3

A three stage RO system is designed using the following module data of the Osmonics permeator

Module Data

Membrane recovery ratio:	7%
Salt rejection:	99.5%
Maximum applied pressure:	5500 kPa
Minimum brine flow rate:	1.584 m^3/d
Maximum brine flow rate:	14.256 m^3/d
Permeate flow rate:	1.43 m^3/d

Design Data

Feed salinity is 34,000 ppm
Plant capacity: 24000 m^3/d

Permeate pressure: 101 kPa

Calculate the number of modules in each stage for the following configuration, flow rates and salinity of all stream, and the specific power consumption.

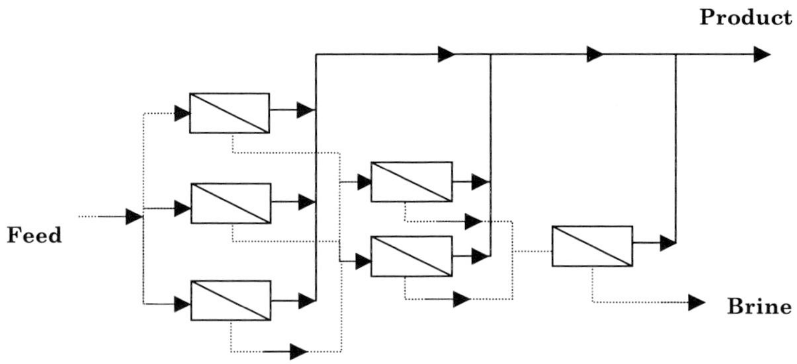

Chapter 8

Reverse Osmosis Feed Treatment, Biofouling, and Membrane Cleaning

Objectives

The objective of this chapter is to outline characteristics of major features in RO operation, which includes feed pretreatment, biofouling, and membrane cleaning. The analysis include description of processes, procedures, troubleshooting, selection criteria

8.1 Need for Pretreatment Processes in RO

The RO feed water may contain various concentrations of suspended solids and dissolved matter. Suspended solids may consist of the following:
- Inorganic particles.
- Colloids.
- Biological matter, which includes microorganisms and algae.

On the other hand, Dissolved matter may consist of highly soluble salts, such as chlorides, and sparingly soluble salts, such as carbonates, sulfates, and silica.

Reduction in the feed water volume during the RO process results in increase of the concentration of suspended particles and dissolved ions. Settling of the suspended particles or scale formation from the sparingly soluble salts would result in the following:
- Blocking of the flow channels that would increase the pressure drop in the feed channels in the membrane module. This would require increasing the pumping power, which is limited by the original design value that takes into account clean operation in addition to some finite level of blockage.
- Reducing the permeation rate across the membrane and increase in the amount of salt passage through the membrane. This is caused by scale formation on the membrane surface.

In addition to physical masking of the membrane surface area and blockage of the membrane module, membrane damage can be caused by system operation at excessively low pH values, high chlorine concentration, or presence of other aggressive chemical compounds that would react and destruct the membrane material.

Depending on the raw water quality, the pretreatment process may consists of all or some of the following treatment steps:

- Removal of large particles using a coarse strainer.
- Water disinfection with chlorine or other biocides.
- Media filtration.
- Reduction of alkalinity by pH adjustment.
- Addition of scale inhibitor.

- Reduction of free chlorine using sodium bisulfite or activated carbon filters.
- Final removal of suspended particles using cartridge filters.

This chapter provides a summary for a large number of studies found in literature that covers the period from 1980 to 2001, see references on page 452.

8.2 Testing Methods

Testing methods of feed water include the following:

- Ionic constituents.
- Turbidity.
- Suspended solids.
- Total organic carbon.
- Silt Density Index (SDI).
- Dissolved gases such as CO_2, O_2, and Cl_2.
- PH.
- Temperature.

The RO operators use the SDI test as measure of the feed quality and operation ease. The guidelines for RO operation using the SDI test is as follows:

- SDI < 1, implies high quality feed water that would provide trouble free operation for years.
- 1 < SDI < 3, implies moderate to low quality feed water that would allow for few months of operation before need for membrane cleaning.
- 3 < SDI < 5 implies low quality feed water, which would require frequent cleaning.
- SDI > 5 implies very poor water quality and operation at these conditions is not acceptable.

The SDI test involves filtration of the water sample. A schematic of the SDI test unit is shown in Fig. 1. As is shown, the system includes the following elements:

- Feed pressure line.
- Pressure regulator.
- Filter paper assembly.
- Sample collection.
- Pressure gauge.
- Stop watch.

The SDI test procedure includes the following:

- A 500 ml (or 100 mℓ) volume feed water is forced through a standard filter under specified test conditions.

- After 15 min (or 5 min for the 100 mℓ sample) the test is repeated using the same filter.
- The filtration times t_1 and t_2 are used to determine the SDI, where

$$SDI = \frac{(1 - \frac{t_1}{t_2})(100)}{T} \quad (1)$$

where, T is 5 (or 15 minutes for 100 mℓ sample).

Other test conditions include the following:
- Applied pressure difference across the membrane is 2 bar.
- Filter pore diameter is 45 μm.
- Filter area is 1350 mm^2.

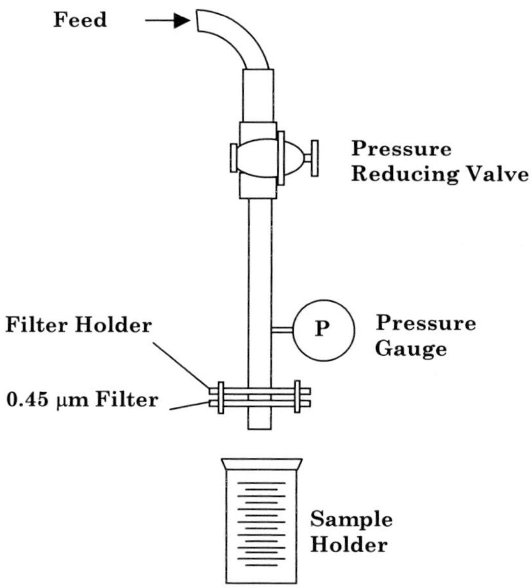

Fig. 1. SDI System.

8.3 Suspended Solids and Silt Reduction

Typical examples of suspended solids include the following:
- Mud and silt
- Organic colloids
- Iron corrosion products
- Precipitated iron

- Algae
- Bacteria
- Rocks
- Silica/Sand
- Precipitated Manganese
- Precipitated Hardness
- Aluminum hydroxide flock

8.3.1 Media Filters

Features of media filters include the following:

- Formed of single- or multi-layered particles.
- The layers contain gravel, activated carbon or anthracite.
- The top layer of the bed consists of coarsely graded material, whereas the finely graded material is layered on the bottom
- The thickness of the filter layer is about 1-3 m.
- Compressed air or water is used for back-flushing.
- The particle size of the filter material is 0.5 – 3 mm.
- The filtration rate is 10-20 m h^{-1}.
- Most of the filtration process takes place in the top layer of the bed.

8.3.2 Cartridge Filters

Features of cartridge filters include the following:

- Cartridge filters are used in various RO configurations.
- The last pretreatment step prior to the RO process.
- The most common size is the 5 μm filter.
- The separation media can be made of polymer materials, such as polypropylene. Use of fiber shedding compounds, such as cotton and hemp, is not recommended since it may
- Combination of media and cartridge filters may be the most optimum and economic solution for particle removal.

8.4 Fouling and Scale Control

Scale forming compounds, in order of occurrence, include the following:

- Calcium carbonate
- Calcium sulfate
- Silica complexes
- Barium sulfate
- Strontium sulfate

- Calcium fluoride

It should be stressed that the formed scale is made of several compounds. This is because that the first compound being precipitated would provide nucleation sites for other compounds.

8.4.1 Calcium Carbonate

The calcium carbonate equilibrium reaction is given by:

$Ca^{+2} + 2\ HCO^{-3} \rightarrow CaCO_3 + CO_2 + H_2O$

Precipitation of calcium carbonate is favored by:
- Increasing calcium or bicarbonate concentration.
- Decreasing the carbon dioxide concentration.
- Increasing the temperature.
- Increasing the pH (more alkaline solution).

Common methods to prevent calcium carbonate scaling include the following:
- Removal of all or some of the bicarbonate alkalinity by feeding acid.
- Use of scale-control agents

The choice between acid dosing and anti-scale control depends on:
- Type of membrane, especially compatibility for long-term operation at low pH.
- Process economics, which would optimize the purchasing cost of antiscalent and acid as well as the capital of handling and dosing equipment for the acid and the antiscalent. In this regard, sulfuric acid is very inexpensive when compared to other acids or antiscalent compounds. However, its use should be handled properly. Otherwise, the presence of the sulfate group in the acid may enhance formation of the calcium, barium, or strontium sulfate scale.
- Final selection would depend on the feed salinity and product quality, which may call for use of each method separately or the combined use of both techniques. For example, seawater desalination would call for simultaneous use of acid and antiscalent dosing.
- Another problem related to acid dosing is the generation of large amounts of CO_2. As a result, the CO2 concentration increases in the permeate water because of its high permeability of CO_2 across various types of RO membranes. This reduces the permeate pH and imposes an overload on ion exchange units used for polishing of the boiler makeup water.

8.4.2 Calcium Sulfate

Various forms $CaSO_4$ crystals include the following:

- Dehydrate calcium sulfate, $CaSO_4.2H_2O$.
- Anhydrous calcium sulfate, $CaSO_4$.
- Hemihydrate calcium sulfate, $CaSO_4.1/2H_2O$.

All of these compounds have reverse solubility, where it precipitates at high temperature. However, increase in the ionic concentration of Ca^{2+} and SO_4^{2-} beyond the solubility limit ions in the brine stream may result in severe scaling.

Prevention of $CaSO_4$ scaling in RO includes the following:

- Addition of antiscaling agents, which includes polyphosphates, polycarboxylates or sodium hexametaphosphate to prevent precipitation of $CaSO_4$. Proper dosing would allow for safe operation even if the brine stream has higher concentration than the saturation limit.
- Lowering the permeate recovery rate to prevent increase in the concentration beyond the saturation limit.

8.4.3 Silica

Silica scaling has the following features:

- The solubility of silica depends on the system pH and temperature.
- Low temperature operation, i.e., below 10 °C, and silica saturation below 120 ppm allows for operation with brine solutions supersaturated in silica with little or no silica scaling.
- Operation at higher concentrations or temperature enhances the silica scaling process.
- Silica forms complex precipitates with iron, aluminum, and magnesium hydroxides.
- The most effective method to prevent silica scaling is to maintain its concentration below the saturation limit, which is strongly affected by the system temperature.

8.4.4 Organics

A wide range of organic compounds can be found in the RO feed water. This depends on the water source and if it has been contaminated with industrial, municipal, or agricultural effluents. Variations in the organics physical and chemical properties give different effects on the RO process, which includes the following:

- Organic compounds reacts differently with the membrane separation process, where the membrane may reject or permeate the compound. Also, rejected or permeating organics may foul the membrane and reduce its performance.
- Organic compounds can be removed from the feed water through degasification of low boiling compounds, adsorption on activated carbon, coagulation and filtration, or removal by ion exchange resins.
- Final selection of the pretreatment process requires measuring the TOC of the feed water and chemical analysis to identify major compounds.

8.5 Biofouling in RO

Biofouling in RO is combined result of the following factors:

- Presence of microorganisms in the feed water.
- Availability of the membrane surface area, which can be colonized by the microorganisms.
- The RO membrane rejects all microorganisms found in the feed water. As a result, part of the rejected cells remains adhered to the membrane and initiates the process of biofilm formation.
- All pretreatment operations prior to the RO module may provide sufficient surface area for microorganisms and bacterial growth. For example, the large surface areas found in media filters, activated carbon beds, or even pipelines connecting various units.
- The biofilm growth mode, on the membrane surface or any other surface, requires the minimum possible amount of nutrients in the feed water. In addition, diffusion and penetration of biocides from the feed water bulk into the bacterial film is less effective than attacking single cells suspended in the feed water.

8.5.1 Effects of Biofouling on RO Performance

Biofouling effects on RO performance are characterized by gradual deterioration in the system performance. This includes a period of rapid decline followed by an asymptotic limit. Performance deterioration includes the following:

- Decrease in the permeate flux.
- Increase in pressure drop.
- Decrease in salt rejection.

The results of the above effects on the overall performance include the following:

- Increased cleaning and maintenance costs. This is necessary to maintain economic operation that calls for a specific range for production rate and product quality. Membrane cleaning restores the system production rate within the specified limits as well as the product quality

- Reduction in membrane lifetime. This is a complex phenomenon, since exposure of the membrane to biofilm formation would result in several side effects, which includes increase in the pressure drop, increase in the flow velocity, exposure of the membrane to extracellular compounds, entrapment of settled solids in the biofilm, precipitation of scaling compounds on the membrane and within the biofilm.
- Other than reducing the flow area, the biofilm may increase friction and drag. Both factors increase the pressure drop within the system, which in turn increase the demand load on the high pressure feed pumps. This is necessary to maintain sufficient feed pressure and permeation rate.
- Uneven growth of the biofilm is also quiet common and that results in uneven flow within the system. As results, sections with lower flow rate may result in enhancement of scaling and settling of solids.

8.5.2 Biofouling Mechanism

The biofouling mechanism is formed of the following elements:

- Bifouling potential depends on feed water conditions, system design, and operating conditions. Feed water temperature is a very important parameter; where in the Gulf and equatorial regions the seawater temperature during the long summer period remains close to 30 °C. Such condition would highly favor bacterial growth and biofilm formation. In addition, presence of dead ends, rough piping, or non-disinfected water tanks in the system would enhance the biofilm formation process. Poor housekeeping may also contribute to microbial contamination and formation of the biofilm.
- Formation of the conditioning film, which involves adsorption of macromolecules on the membrane surface. This step may last for few seconds to minutes after the membrane is exposed to the feed water. The conditioning film has different surface properties than the original membrane surface. Therefore, the electrostatic charge of the conditioning film may considerably affect microbial adhesion.
- Microbial adhesion is achieved by hydrodynamic forces, motility, and diffusion. The microbial layer is protected to a large extent from the shearing forces of the bulk flow because of the small thickness of the biofilm in comparison with the boundary layer thickness. In addition, the RO membranes reject all the bacterial cells and microorganisms arriving at membrane surface. The adhesion process is also affected by the number of cells in the bulk, nutrient concentration, temperature, pH, flow hydrodynamics, and surface charge. Recent membrane development focus on construction of membranes with surface charge that inhibit microbial adhesion and formation of the conditioning film.

- Biofil 8.2.4 Assessment of Biofouling Potential ; are irreversibly attached to the membrane surface. As mentioned before, the biofilm formation is favored since it is protected against biocide effects (because of diffusion limitations) and it consumes much smaller amounts of nutrients than cells in suspension. Therefore, the biofilm grows rapidly until it reaches an asymptotic limit, where the growth rate (controlled by nutrient concentrations and temperature) reaches equilibrium with the removal rate by shear forces.

8.5.3 Biofouling Assessment

Biofouling is recognized by indirect effects on the system performance: permeate decline, decrease of salt rejection, or increase of the pressure drop on the feed-side. Further assessment of membrane biofouling is achieved by any of the following techniques:

- Analysis of the pretreatment filter media for biofouling. Presence of biofouling in these systems may give strong indication for membrane biofouling.
- Use of on-line membrane testing elements, which have smaller size than the actual membrane module. These elements can be removed and replaced on frequent basis. The removed elements are then dismantled and properly tested for biofouling and other forms scaling.

8.5.4 Biofouling Treatment

Addition of biocides to the feed water at an appropriate dosing rate would kill the microorganisms and bacterial cells suspended in the water and forming the biofilm on various parts of the system. Biocide treatment would prevent further growth of existing biofilms or formation of new biofilms. However, existing biofilms, formed of dead bacterial cells, would remain to affect the system performance in various aspects, i.e., higher pressure drop, lower recovery, lower salt rejection, etc. In other words, biocide treatment kills the microorganisms and bacteria forming the biofilm, but has no effect on removal of the biofilm. Moreover, a dead biofilm might become a substrate for new bacterial cells, which may get attached to the dead biofilm and consume its nutrient content. The same concept applies for the dead and suspended cells in the feed stream, which may ends up being attached to the membrane surface and causes reduction in recovery rate and salt rejection. Features of the biofouling treatment processes are summarized below:

- Effectiveness of the biocide depends on several operating and design factors, which includes; type of biocide, concentration of biocide, side-reactions of biocide and compounds other than bacteria or microorganisms, pH, temperature, residence time, type of microorganism, growth state, physical nature of the biofilm

- Increase in the biocide concentration, temperature, and residence time would increase the percent kill of the microorganisms.
- Chlorine is one of the most effective and common biocide agents. However, it requires a de-chlorination unit placed ahead of the RO to protect membranes sensitive to chlorine attack. Another disadvantage of chlorine use is the safety precautions required for storage and handling.
- Other biocides include formaldehyde, ozone, peracetic acid, hydrogen peroxide, etc. Use of these is found much smaller scale than chlorine. In addition, some these have lower efficiency than chlorine and have harmful effects on the environment.

8.6 Membrane Cleaning

Membrane cleaning is dictated by increase in pressure drop, decrease in permeate recovery, and decrease in salt rejection. It should be stressed that normalized data should be used to correct for temperature effects on system performance. In addition, identification of foulants as well as analysis of feed and outlet water are important factors in determining the proper cleaning solution. The following is a summary of cleaning methods of various fouling and scaling compounds.

Calcium Carbonate and Metal Oxides Scale: Clean with low pH water. The water pH is adjusted to 3-4 and sulfuric, hydrochloric, or citric acids are used.

Calcium Sulfate Scale: Clean with a solution that include sodium tripoly phosphate or sodium salt of ethylene diamine tetra acetic acid.

Silica: Detergents and hydraulic cleaning.

Organics and biofouling: similar solution to calcium sulfate cleaning. In addition, use of detergents and biocides are recommended.

8.6.1 Membrane Cleaning Procedure

Generally, low pH solutions are used to clean metallic scales while alkaline solutions are used to clean biological and organic fouling. Relatively high flow with low pressure is recommended. System cleaning follows the following basic steps:

- Preparation of the cleaning solution and adjustment of temperature and pH.
- Displacement of the solution in RO modules by pumping the cleaning solution.
- Recycling and soaking of the element. Soaking time may vary from few hours to overnight depending on the fouling level.

- Flushing the unit with RO permeate water. The flushing procedure continues until foaming disappears and the pH and conductivity of the effluent solution approach those of the feed water.

Other features of the cleaning process include the following:

- Use of chlorine or other strong oxidants on polyamide membranes can cause irreversible damage to the membrane.
- Warm water at a temperature of 32 °C to 37 °C gives better cleaning results than lower temperature solutions.
- If the pH of an acid solution increases during recirculation, add more acid to return the pH back to the target value. This is because of acid consumption in dissolving inorganic scale.
- Use of sulfuric acid in low pH solutions may result in forming sulfate scale.
- Permeate water is preferred for mixing cleaning solutions.
- Use of filtered tap water for high pH solutions can result in carbonate fouling if the water is hard.

8.6.2 Membrane Sterilization

Membrane sterilization is necessary if the system is shut down for period of more than 2 days. Sterilization compounds include the following:

- Hydrogen peroxide (0.25 wt %).
- Sodium bisulfite/glycerin (0.2 to 1 wt %/16 to 20 wt %).
- Sodium bisulfite (0.5 to 1 wt %).
- Formaldehyde (0.25 to 1 wt%).
- Copper sulfate (0.1 to 0.5 ppm)

Procedure for membrane sterilization includes: preparation of the sterilizing solution, initial flushing with RO permeate water, circulation of the sterilizing solution, drainage, and tight closure during the storage period. Placing a sterilized unit back into operation requires flushing permeate water.

8.6.3 Cleaning System Specifications

The RO cleaning system is formed of a tank, 5 μm cartridge filter, pumping unit, and instrumentation, Fig. 2. Cleaning solution is pumped from the tank through the cartridge filter to the RO array. Solution is then recycled back to the tank. The volume of solution should be sufficient to fill the volume of the vessels, filters and piping. Instrumentation includes pH, temperature, flow rate, pressure, and level controllers.

Choice of the construction material for the cleaning system includes the following:

- Tank: Fiberglass reinforced plastic or polypropylene. Tank should have a removable cover.
- Piping: PVC schedule 80 or Nylon reinforced flex hose.
- Valves: Stainless Steel.
- Pump: Stainless Steel or Non-metallic composite polyesters. Pump should be centrifugal type.

Material selection should be made to withstand extremes in pH, temperatures up to (45 °C), and electrical sources/switches should be protected and well grounded.

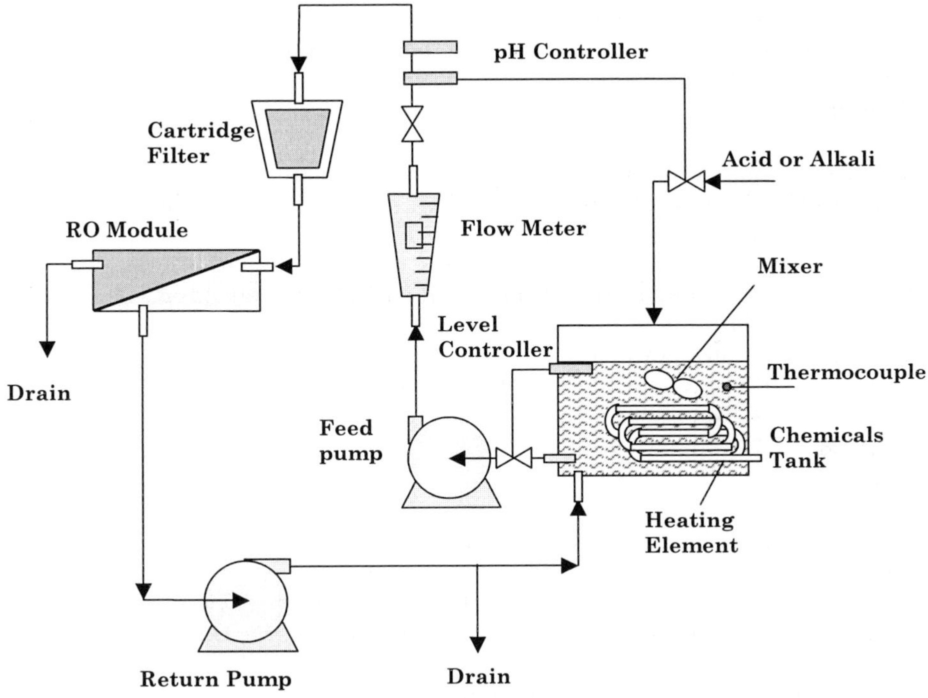

Fig. 2. RO cleaning system.

8.7 Membrane Storage

Various forms of membrane storage include the following:

- Short-term storage of membrane elements in place inside the pressure tubes.

- Long-term storage of membrane elements in place inside the pressure tubes.
- Dry storage of membrane elements as spares or before start-up of the plant.

Short-term storage ranges between 5 to 30 days. On the other hand the long term storage remains for more than 30 days. For both modes of storage, the membrane elements remain in place inside the pressure tubes. Both storage modes have similar sequence and are summarized below:

- Cleaning of the membrane elements.
- Flushing the membrane element with acidified water to prevent precipitation of scaling material.
- A suitable biocide is added to the flushing water to prevent microorganism growth.
- The pressure tubes are filled with the above solution.
- The cleaning, flushing, and filling sequence is repeated on frequent basis subject to the effectiveness of the biocide.
- Before, returning the system to service, the membrane elements are flushed with feed water. The flushing process starts at low pressure and then proceeds to high pressure operation. During the flushing process, the brine and product streams are rejected. Also, water samples are collected from both streams to check for presence of biocides.

Storage of membrane elements during transportation or prior to installation should provide protection from direct sunlight. The elements should be stored in a cool, dry place, and at a temperature range of 20°C to 35°C.

References

American Water Works Association, Reverse Osmosis and Nanofiltration, American Water Works Association, 2001.

Amjad, Z., Reverse Osmosis: Membrane Technology, Water Chemistry, and Industrial Applications, Ed., Chapman & Hall, New York, 1992.

Fluid Systems Corporation, RO PRO, version 6.1, 1997.

Hydranautics, Hydranautics RO system design software, version 6.4, 1998.

Parekh, B.S., Reverse Osmosis Technology: Applications for High-Purity-Water Production, Edt., Marcel Dekker, New York, 1988.

Rautenbach, R., Membrane Processes, Wiley, New York, 1989.

Chapter 9

Associated Processes

Objectives

The objective of this chapter is to analyze and evaluate the characteristics of major associated processes in thermal desalination, which includes.

- Venting
- Steam Jet Ejectors
- Demisters
- MSF Weirs

Pumps and valves play a very important role in desalination plant; however, analysis of these unit processes is not covered here since several textbooks are available on their design details.

9.1 Venting of Non-Condensable Gases

Venting in the thermal desalination is driven by steam ejectors for removal of the non-condensable gases during startup and operation. During maintenance and other shut down procedures, the flashing chambers in MSF or the evaporation effects in MEE are opened to ambient air. Therefore, air removal is one of the main activities in the startup procedure. The startup ejectors have higher capacity and are capable of processing large air volumes over a short period of time.

During steady state operation, the ejector for removal of the non-condensable gases has smaller capacity and their main function is to reduce the concentration of these gases, which would accumulate around the condenser tubes. Effects of the non-condensable gases include the following:

- Reduce the heat transfer coefficient because of their low thermal conductivity. Designers view this reduction as an additional thermal resistance to heat transfer. Therefore, additional heat transfer area is incorporated to handle the presence of non-condensable gases.
- In the MSF process, gas accumulation beyond the design specifications would reduce the brine recycle temperature entering the brine heater. This will require use of a larger amount of the heating steam, which will reduce the process thermal performance ratio.
- In the MVC process presence of non-condensable would affect the performance of the compressor. Their presence would reduce the amount of compressed vapor and consequently the thermal load would decrease as well as the amount of product water.
- Reduce the partial pressure of the condensing vapor and hence the condensation temperature. This also reduces the brine recycle temperature entering the brine and would require use of larger amounts of heating steam.
- Enhance corrosion reactions due to presence of O_2 and CO_2 gases.

9.1.1 Venting System in MSF

The venting system in the flashing stage is attached between the partition wall and the confining plate of the distillate tray. Schematic for venting line between two stages is shown in Fig. 1. As is shown, the inlet of the venting line is protected by an air baffle, which routs the air and other non-condensable gases into the venting line and minimizes entrainment of condensate droplets or escape of vapor in the venting line.

The venting lines cascades through the stages. Connection to the ejector unit depends on the system designer. In a conventional 24 stage MSF system, venting occurs at three points within the system. In this regard, vent lines are located in stages 7, 12, and 24. Therefore, venting of the non-condensable gases is cascaded through the seven stages, then through stage 8-12, and from stage 13 into stage 24.

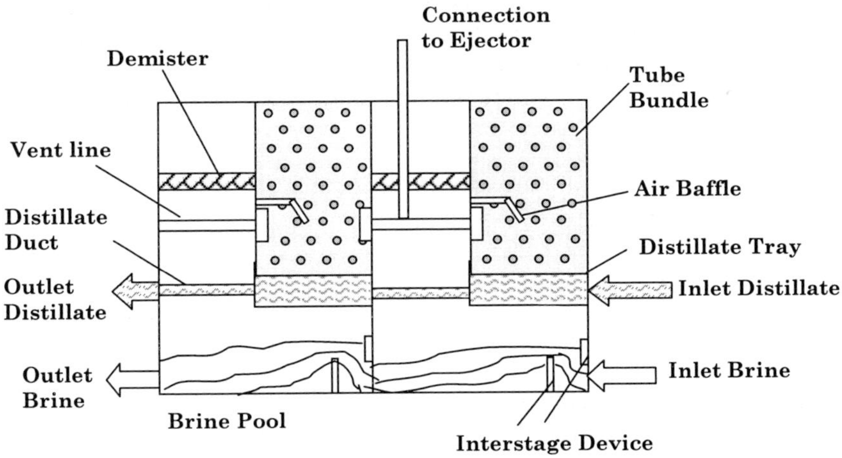

Fig. 1. Venting lines and other parts of the flashing stage.

Design considerations for the venting include the following:

- The venting point is located at the coldest location inside the flashing stage. This is to allow for condensation of the water vapor and to reduce the load on the venting system.
- The temperature of the gas emerging from the condenser vent must be lower than the dew point of the gas entering the condenser.
- An air baffle is placed near the vent entrance. This is to guide the air flow and to adjust the gas velocity to achieve the most effective cooling. The baffle

geometry is designed to direct the air flow through the tube bundle and towards the vent opening. This is important to prevent gas bypassing.

- Near the vent entrance, the tube spacing is reduced as well as the superficial flow area. This is to maintain high gas velocity and to increase the heat transfer rates.
- The vent entrance zone should avoid creation of dead zones or stagnation point, which may cause accumulation of the non-condensable gases.
- The vent opening should be located above the water level in the condensate tray. This is to prevent condensate flow into the vent line or droplet entrainment into the gas stream.
- On average five vent lines are connected between the flashing chambers. This is necessary to obtain the desired flow rate, velocity, and pressure drop in the vent line.

9.1.2 Design of Vent Line Orifice

The main function of the vent line orifice is to regulate the vapor flow rate between the stages. The vapor flow through the vent line is limited to a maximum of 2% of the total flow rate of the flashed off vapor. In presence of non-condensable gases, the amount of the vapor flowing through the vent line is reduced. The vent line orifice is a sharp-edged hole with straight walls perpendicular to the flow direction, Fig. 2. The pressure drop in the orifice is given by

$$\Delta P = (Kc + Ke + Ko)\, \rho\, v^2 \qquad (1)$$

where

v is the vapor velocity in the vent line
Kc is the coefficient for sudden contraction losses, Kc = 0.5
Ke is the coefficient for sudden expansion losses, Ke = 1
Ko is the orifice coefficient, $Ko = (\frac{1}{(d_2/d_1)} - 1)(\frac{2.75}{(d_2/d_1)} - 1.56)$
d_1 is the diameter of the vent line
d_2 is the opening diameter of the orifice

The total mass flow rate of the vapor in the vent lines is given by

$$m = n\, \rho\, v\, A_2 \qquad (2)$$

where

n is the number of vent lines
A_2 is the cross section area of each vent line

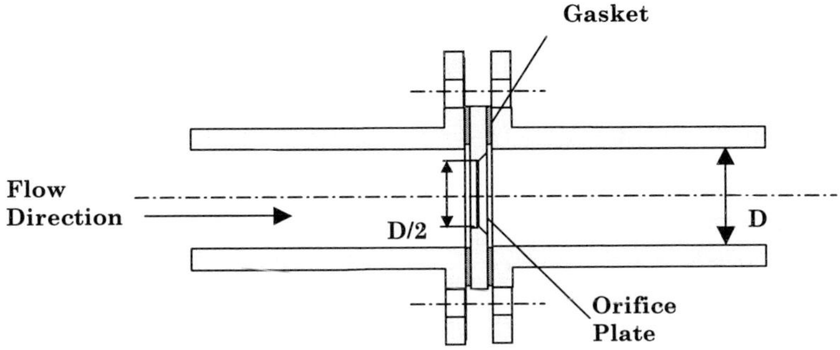

Fig. 2. Square orifice in venting lines

Example

The following data is typical for a flashing stage in the MSF system and is used to illustrate iterative calculations to determine the cross section area (A_2) of the vent line

Data:

- Mass flow rate of flashing vapor in the stage = 13.83 kg/s
- Number of vent lines = 5
- Diameter of the vent line = 100 mm
- Vapor temperature inside the stage= 69.38 °C
- Vapor temperature in the next stage = 67.18 °C
- Vapor density = 0.1932 kg/m3
- Vapor pressure inside the stage = 30.34 kPa
- Vapor pressure in the next stage = 27.58 kPa

Solution:

- Assume that the orifice diameter (d_2) is equal to 54 mm, which gives a cross section area (A_2) of 0.0079 m^2
- Orifice coefficient

$$Ko = (\frac{1}{(54/100)^2} - 1)(\frac{2.75}{(54/100)^2} - 1.56) = 19.1$$

- Vapor velocity in the vent line

$v = (2)(105)(27.6)/((0.5+1+19.1)(0.1932))^{0.5} = 36.8$ m/s

- Vapor mass flow rate in the vent lines

$m = (5)(0.1932)(36.8)(0.0079) = 0.28$ kg/s

- Calculate the ratio of mass loss in the vent line, or (0.28)/(13.83) = 2.03%. This value is within the design limits. Therefore, it is not necessary to attempt a new value for the orifice diameter.

9.2 Steam Jet Ejectors

Conventional jet ejector has three main parts: 1) the nozzle, 2) the suction or mixing chamber, and 3) the diffuser, Fig. 3. The nozzle and the diffuser have the geometry of converging/diverging venturi. The diameters and lengths of various parts forming the nozzle, the diffuser, and the suction chamber together with the stream flow rate and properties define the ejector capacity and performance. The ejector capacity is defined in terms of the flow rates of the motive fluid and the entrained stream. Their sum gives the mass flow rate of the compressed gases. As for the ejector performance it is defined in terms of entrainment, expansion, and compression ratios. The entrainment ratio (w) is the flow rate of the entrained stream divided by the flow rate of the motive fluid. As for the expansion ratio (Er) it is defined as the ratio of the motive steam pressure to the entrained gas pressure. The compression ratio (Cr) gives the pressure ratio of the compressed gas to the entrained gas.

The velocity and pressure profiles of various stream inside the ejector and as a function of location, Fig. 3, are summarized below:

- The motive fluid enters the ejector at point (p) with a subsonic velocity.
- As the stream flows in the converging part of the ejector its pressure is reduced and its velocity increases. The stream reaches sonic velocity at the nozzle throat, where its Mach number is equal to one.
- In the diverging part of the nozzle, the increase in the cross section area results in decrease of the shock wave pressure and increase in its velocity to supersonic conditions.
- At the nozzle outlet plane, point (2), the motive steam pressure becomes lower than the entrained gas pressure and its velocity ranges between 900 and 1200 m/s.
- The entrained gas at point (e) enters the ejector, where its velocity increases and its pressure decreases to that of point (3).
- The motive fluid and entrained gas streams may mix within the suction chamber and the converging section of the diffuser or it may flow as two separate streams as it enters the constant cross section area of the diffuser, where mixing occurs.

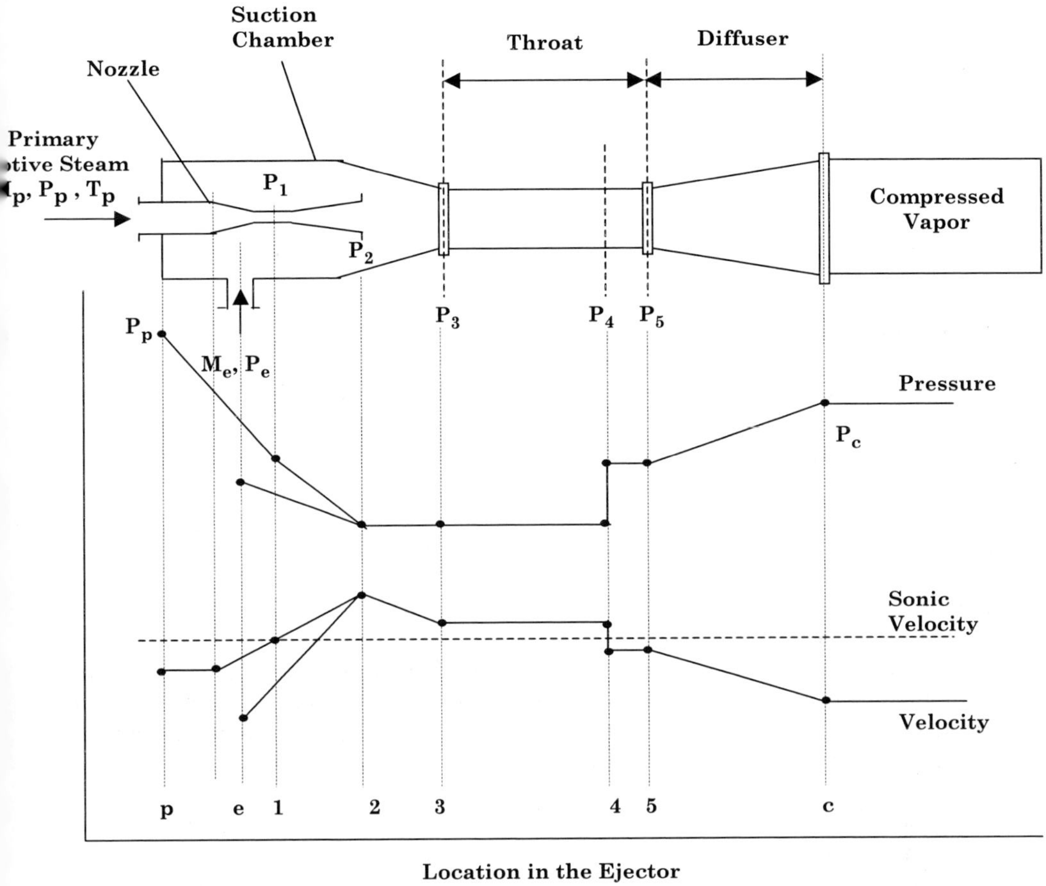

Fig. 3. Ejector parts and profiles of velocity and pressure inside the ejector

- In either case, the mixture goes through a shock inside the constant cross section area of the diffuser. The shock is associated with increase in the mixture pressure and reduction of the mixture velocity to subsonic conditions, point (6). The shock occurs because of the back pressure resistance of the condenser.
- As the subsonic mixture emerges from the constant cross section area of the diffuser, further pressure increase occurs in the diverging section of the diffuser, where part of the kinetic energy of the mixture is converted into pressure. The pressure of the emerging fluid is slightly higher than the condenser pressure, point (c).

9.2.1 Ejector Analysis

Summary for a number of literature studies on ejector design and performance evaluation is shown in Table 1. The following outlines the main findings of these studies:

- Optimum ejector operation occurs at the critical condition. The condenser pressure controls the location of the shock wave, where increase in the condenser pressure above the critical point results in rapid decline of the ejector entrainment ratio since the shock wave moves towards the nozzle exit. Operating at pressures below the critical points has negligible effect on the ejector entrainment ratio.
- At the critical condition, the ejector entrainment ratio increases at lower pressure for the boiler and condenser. Also, higher temperature for the evaporator increases the entrainment ratio.
- Use of a variable position nozzle can maintain the optimum conditions for ejector operation. As a result, the ejector can be maintained at critical conditions even if the operating conditions are varied.
- Multi-ejector system increases the operating range and improves the overall system efficiency.
- Ejector modeling is essential for better understanding of the compression process, system design, and performance evaluation. Models include empirical correlations, such as those by Ludwig (1977), Power (1964), El-Dessouky and Ettouney (1999), and El-Dessouky et al. (2001). Such models are limited to the range over which it was developed which limits their use in investigating the performance of new ejector fluids, designs, or operating conditions. Semi-empirical models give more flexibility in ejector design and performance evaluation, Munday and Bagster (1977) and Henzler (1983). Other ejector models are based on fundamental balance equations, Keenan and Neuman (1942).

Table 1
Summary of literature studies on ejector design and performance.

Reference	Fluid	Boiler Evaporator Condenser Temperature	Features
Al-Khalidy (1997)	R-113	60–100 °C 5–18 °C 40–50 °C	Basis for refrigerant selection for solar system, system performance increased with increasing boiler and evaporator temperatures and decreasing condenser temperature.
Chen et al. (1998)	R-113 R-114 R-142b R-718	80–95 °C 5–13 °C 25–45 °C	Comparison of ejector and refrigerant performance. Dry, wet, and isentropic fluids. Wet fluid damage ejectors due phase change during isentropic expansion. R-113 (dry) has the best performance and R142b (wet) has the poorest performance.
Sokolov and Hershgal (1990a and 1990b)	R-114	86 °C –8 °C 30 °C	Increase in ejector performance using mechanical compression booster.
Eames et al. (1995)	Water	120–140 °C 5–10 °C 30–65 °C	Choking of the entrained fluid in the mixing chamber affects system performance. Maximum COP is obtained at the critical flow condition.
Aphornratana and Eames (1997)	Water	120–140 °C 5–10 °C 30–60 °C	Effect of varying the nozzle position to meet operating condition. Increase in COP and cooling capacity by 100%
Al-Khalidy and Zayonia (1995)	R-113	70–100 °C 6-25 °C 42-50 °C	Entrainment ratio is highly affected by the condenser temperature especially at low evaporator temperature.
Sokolov and Hershgal (1993a and 1993b)	R-114	90 °C 4 °C 30 °C	Combined solar generator and ejector air conditioner. More efficient system requires multi-ejector and cold energy storage (cold storage in either phase changing materials, cold water or ice).
Tomasek and Radermacher (1995)	R134a	5 °C –12, –18 °C 40 °C	Combined ejector and mechanical compressor for operation of domestic refrigerator-freezer increases entrainment ratio from7% to12.4%. The optimum throat diameter depends on the freezer temperature

Table 1 (continued)
Summary of studies on ejector cooling/heating systems.

Reference	Fluid	Boiler Evaporator Condenser Temperature	Features
Menegay and Kornhauser (1994)	R-134A	 −15 °C 30 °C	Modeling the effect of motive nozzle on system performance, in which ejector is used to recover part of the work that would be lost in the expansion valve using high-pressure motive liquid.
Abdel-Aal et al. (1990)	Water	100–165 °C 10 °C 30–45 °C	Combined solar collector, refrigeration, and seawater desalination system. Performance depends on steam pressure, cooling water temperature, and suction pressure.
Munday and Bagster (1977)	Water		Developed a new ejector theory in which the entrained fluid is choked, the plant scale results agree with this theory. Steam jet refrigeration should be designed for the most often prevailing conditions rather than the most severe to achieve greater overall efficiency.
Grazzini and Mariani (1998)	Water	–	Model of multistage steam ejector refrigeration system using annular ejector in which the primary fluid enters the second stage at annular nozzle on the sidewall. This will increase static pressure for low-pressure stream and the mixture and reduce the velocity of motive stream and reduce jet mixing losses shock wave formation losses.
Chen and Hsu (1987)	R11 R113 R114	93.3 °C 10 °C 43.3 °C	Measure and calculate ejector entrainment ratio as a function of boiler, condenser, and evaporator temperatures. Entrainment ratio decreases for off design operation and increases for the two stage ejectors.
Huang and Chen (1996)	R113 R114 R142b	120–140 °C 65–80 °C	Effect of throat area, location of main nozzle, and length of the constant area section on backpressure, entrainment ratio, and compression ratio.
Henzler (1983)			Mathematical model use empirical parameters that depend solely on geometry. The parameters are obtained experimentally for various types of ejectors.
Sun and Eames (1996)	R11 HR-123	80 °C 5 °C 30 °C	Performance of HR-123 is similar to R-11 in ejector refrigeration. Optimum performance is achieved by the use of variable geometry ejector when operation conditions change.

9.2.2 Ejector Models

The review by Sun and Eames (1995) outlined developments in mathematical modeling and design of jet ejectors. The review shows that there are two basic approaches for ejector analysis. These include mixing of the motive steam and entrained vapor either at constant pressure or at constant area. Design models of stream mixing at constant pressure are more common in literature because the performance of the ejectors designed by this method is more superior to the constant area method and it compares favorably against experimental data. The basis for modeling the constant pressure design procedure was developed first by Keenan and Neuman (1942). Subsequently, several investigators have used the model for design and performance evaluation of various types of jet ejectors. This involved a number of modifications in the model, especially, losses within the ejector and mixing of the primary and secondary streams.

9.2.3 Constant Pressure Model

In this section, the constant pressure ejector model is developed. The developed model is based on a number of literature studies that include Eames et al. (1995), Sun and Eames (1996), Aly et al. (1999), Huang et al. (1999), and El-Dessouky et al. (2001). Assumptions invoked in this model include:

- The motive steam expands isentropically in the nozzle. Also, the mixture of the motive steam and the entrained vapor compresses isentropically in the diffuser.
- The motive steam and the entrained vapor are saturated and their velocities are negligible.
- Velocity of the compressed mixture leaving the ejector is insignificant.
- Constant isentropic expansion exponent and the ideal gas behavior.
- The mixing of motive steam and the entrained vapor takes place in the suction chamber.
- The flow is adiabatic.
- Friction losses are defined in terms of the isentropic efficiencies in the nozzle, diffuser, and mixing chamber.
- The motive steam and the entrained vapor have the same molecular weight and specific heat ratio.
- The ejector flow is one-dimensional and at steady state conditions.

The model equations include the following:

- Overall material balance

$$m_p + m_e = m_c \tag{3}$$

where m is the mass flow rate and the subscripts c, e, and p define the compressed vapor mixture, the entrained vapor, and the primary stream.

– Entrainment ratio

$$w = m_e/m_p \tag{4}$$

– Compression ratio

$$Cr = P_c/P_e \tag{5}$$

– Expansion ratio

$$Er = P_p/P_e \tag{6}$$

– Isentropic expansion of the primary fluid in the nozzle is expressed in terms of the Mach number of the primary fluid at the nozzle outlet plane

$$M_{p_2} = \sqrt{\frac{2\,\eta_n}{\gamma-1}\left[\left(\frac{P_p}{P_2}\right)^{\frac{\gamma-1}{\gamma}}-1\right]} \tag{7}$$

where M is the Mach number, P is the pressure, and γ is the isentropic expansion coefficient. In the above equation η_n is the nozzle efficiency and is defined as the ratio between the actual enthalpy change and the enthalpy change undergone during an isentropic process.

– Isentropic expansion of the entrained fluid in the suction chamber is expressed in terms of the Mach number of the entrained fluid at the nozzle exit plane

$$M_{e_2} = \sqrt{\frac{2}{\gamma-1}\left[\left(\frac{P_e}{P_2}\right)^{\frac{\gamma-1}{\gamma}}-1\right]} \tag{8}$$

– The mixing process is modeled by one-dimensional continuity, momentum and energy equations. These equations are combined to define the critical Mach number of the mixture at point 5 in terms of the critical Mach number for the primary and entrained fluids at point 2

$$M_4^* = \frac{M_{p_2}^* + w\,M_{e_2}^*\sqrt{T_e/T_p}}{\sqrt{(1+w)(1+w\,T_e/T_p)}} \tag{9}$$

where w is the entrainment ratio, and M^* is the ratio between the local fluid velocity to the velocity of sound at critical conditions.

- The relationship between M and M^* at any point in the ejector is given by this equation

$$M^* = \sqrt{\frac{M^2(\gamma+1)}{M^2(\gamma-1)+2}} \tag{10}$$

Equation (10) is used to calculate $M^*_{e_2}$, $M^*_{p_2}$, M_4.

- Mach number of the mixed flow after the shock wave

$$M_5 = \frac{M_4^2 + \frac{2}{(\gamma-1)}}{\frac{2\gamma}{(\gamma-1)} M_4^2 - 1} \tag{11}$$

- Pressure increase across the shock wave at point 4

$$\frac{P_5}{P_4} = \frac{1+\gamma M_4^2}{1+\gamma M_5^2} \tag{12}$$

In Eq. (12) the constant pressure assumption implies that the pressure between points 2 and 4 remains constant. Therefore, the following equality constraint applies $P_2 = P_3 = P_4$.

- Pressure lift in the diffuser

$$\frac{P_c}{P_5} = \left[\frac{\eta_d(\gamma-1)}{2} M_5^2 + 1\right]^{\frac{\gamma}{\gamma-1}} \tag{13}$$

where η_d is the diffuser efficiency.

- The area of the nozzle throat

$$A_1 = \frac{m_p}{P_p}\sqrt{\frac{RT_p}{\gamma\eta_n}\left(\frac{\gamma+1}{2}\right)^{(\gamma+1)/(\gamma-1)}} \tag{14}$$

- The area ratio of the nozzle throat and diffuser constant area

$$\frac{A_1}{A_3}=\frac{P_c}{P_p}\left(\frac{1}{(1+w)(1+w(T_e/T_p))}\right)^{1/2}\frac{\left(\frac{P_2}{P_c}\right)^{1/\gamma}\left(1-\left(\frac{P_2}{P_c}\right)^{(\gamma-1)/\gamma}\right)^{1/2}}{\left(\frac{2}{\gamma+1}\right)^{1/(\gamma-1)}\left(1-\frac{2}{\gamma+1}\right)^{1/2}} \tag{15}$$

– The area ratio of the nozzle throat and the nozzle outlet

$$\frac{A_2}{A_1}=\frac{1}{M_{p_2}^2}\left(\frac{2}{(\gamma+1}\left(1+\frac{(\gamma-1)}{2}M_{p_2}^2\right)\right)^{(\gamma+1)/(\gamma-1)} \tag{16}$$

9.2.4 Solution Procedure of the Constant Pressure Model

Solution of the constant pressure model requires an iterative procedure. Also, it is necessary to define values of η_n and η_d. The values of these efficiencies widely differ from one study to the other as shown in Table 2. Two solution procedures are shown in Fig. 4. Either procedure requires iterative calculations. The first procedure is used for system design, where the system pressures and the entrainment ratio are defined. Iterations are made to determine the pressure of the motive steam at the nozzle outlet (P_2) that gives a specified value for the back pressure (P_c). The iteration sequence for this procedure is shown in Fig. 4a. The second solution procedure is used for performance evaluation or ejector rating, where the cross section areas and the pressures of the entrained vapor and motive fluid are specified. Iterations are made to calculate the entrainment ratio and the ejector capacity. The iteration sequence for this procedure is shown in Fig. 4b.

Table 2
Examples of ejector efficiencies used in literature studies

Reference	η_n	η_d	η_m
Menegay and Kornhauser (1994)	0.9	0.75	
Rogdakis and Alexis (2000)		0.8	0.8
Sun (1996)	0.85	0.85	
Tomasek and Radermacher (1995)	0.7-1	0.7-1	
Aly et al. (1999)	0.8-1	0.8-1	
Chen and Hsu (1987)	0.85-0.98	0.65-0.85	
Eames et al. (1995)	0.85	0.85	0.95
Gupta et al. (1979)	0.75	0.9	

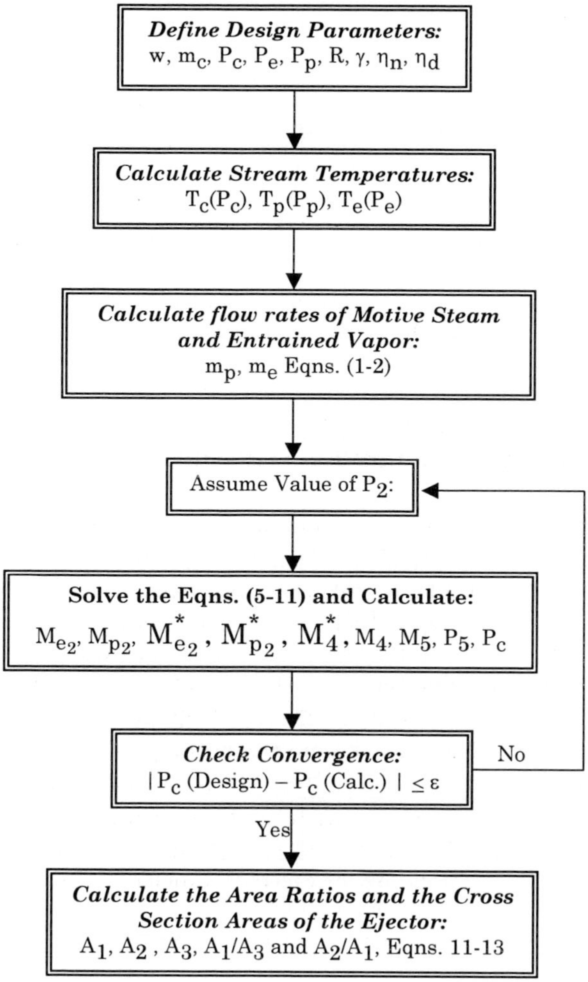

Fig. 4a. Solution algorithm of the ejector mathematical model. Design procedure to determine the area ratios.

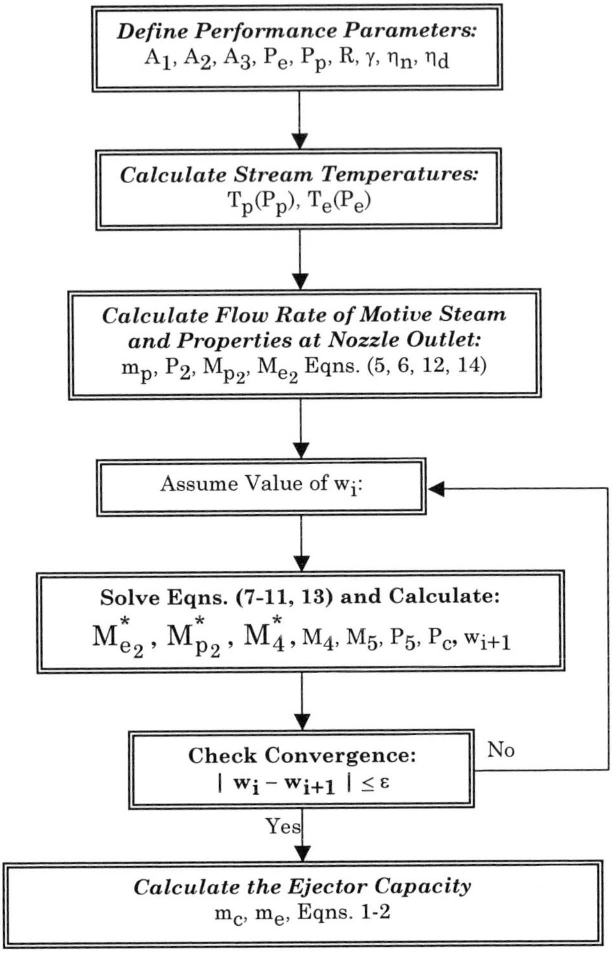

Fig. 4b. Solution algorithm of the ejector mathematical model. Performance evaluation to calculate the entrainment ratio.

9.2.5 Semi-Empirical Model

Development of the semi-empirical model is thought by a number of investigators that include Ludwig (1977), Power (1964), El-Dessouky and Ettouney (1999), and El-Dessouky et al. (2001). This is to provide a simple method for designing or rating of steam jet ejectors. The semi-empirical model presented here is based on the development by El-Dessouky et al. (2001). The model is based on three sets of design data acquired from major ejector manufacturers, which includes Croll Reynolds, Graham, and Schutte-Koerting. Also, several sets of experimental data are extracted from the literature and are used in the development of the empirical model. The semi-empirical model includes a number of correlations to calculate the entrainment ratio (w), the pressure at the nozzle outlet (P_2), and the area ratios in the ejector (A_2/A_1) and (A_1/A_3). The correlation for the entrainment ratio is developed as a function of the expansion ratio and the pressures of the motive steam, the entrained vapor, and the compressed vapor. The correlation for the pressure at the nozzle outlet is developed as a function of the evaporator and condenser pressures. The correlations for the ejector area ratios are defined in terms of the system pressures and the entrainment ratio. Table 3 shows summary for the ranges of the experimental and the design data. The table also includes the ranges for the data reported by Power (1994). Experimental data used in developing the semi-empirical model are summarized in Table 4.

The correlation for the entrainment ratio of choked flow or compression ratios above 1.8 is given by

$$w = a\ Er^b\ P_e^c\ P_c^d\ \frac{(e + fP_p^g)}{(h + iP_c^j)} \tag{17}$$

Similarly, the correlation for the entrainment ratio of un-choked flow with compression ratios below 1.8 is given by

$$w = a\ Er^b\ P_e^c\ P_c^d\ \frac{(e + f\ \ln(P_p))}{(g + h\ \ln(P_c))} \tag{18}$$

470

Table 3
Range of design and experimental data used in model development.

Source	Cr	Er	P_e(kPa)	P_c(kPa)	P_p(kPa)	w
Experimental	1.4-6.19	1.6-526.1	0.872-121.3	2.3-224.1	38.6-1720	0.11-1.132
Schutte-Koerting	1.008-3.73	1.36-32.45	66.85-2100.8	790.8-2859.22	84.09-2132.27	0.1-4
Croll-Rynolds	1.25-4.24	4.3-429.4	3.447-124.1	446.06-1480.27	6.2-248.2	0.1818-2.5
Graham	1.174-4.04	4.644-53.7	27.58-170.27	790.8-1480.27	34.47-301.27	0.18-3.23
Power	1.047-5.018	2-1000	2.76-172.37	3.72-510.2	344.74-2757.9	0.2-4

Table 4
Experimental data used in development of empirical model

Reference	Compression ratio	Expansion ratio	Entrainment ratio	Area ratio
Eames et al. (1995)	3-6	160-415	0.17-0.58	90
Munday and Bagster (1977)	1.8-2	356-522	0.57-0.905	200
Aphornratana and Eames (1997)	4.6-5.3	309.4	0.11-0.22	81
Bagster and Bresnahan (1983)	2.4-3.4	165-426	0.268-0.42	145
Sun (1996)	2.06-3.86	116-220	0.28-0.59	81
Chen and Sun (1997)	1.77-2.76,	1.7-2.9	0.37-0.62	79.21
Arnold et al. (1982)	2.47-3.86	29.7-46.5	0.27-0.5	
Everitt and Riffat (1999)	1.37-2.3	22.6-56.9	0.57	

Fitting results against the design and experimental data are shown in Figs. 5 and 6, respectively. The results shown in Fig. 5 cover the most commonly used range for steam jet ejectors, especially in vacuum and vapor compression applications. As is shown in Fig. 5 the fitting result are very satisfactory for entrainment ratios between 0.2 and 1. This is because the major part of the data is found between entrainment ratios are clustered over a range of 0.2-0.8. Examining the experimental data fit shows that the major part of the data fit is well with the correlation predictions, except for a small number of points where the predictions have large deviations.

The correlations for the motive steam pressure at the nozzle outlet and the area ratios are obtained semi-empirically. In this regard, the design and experimental data for the entrainment ratio and system pressures are used to solve the mathematical model and to calculate the area ratios and motive steam pressure at the nozzle outlet. The results are obtained for efficiencies of 100% for the diffuser, nozzle, and mixing and a value of 1.3 for γ. The results are then correlated as a function of the system variables. The following relations give the correlations for the choked flow:

$$P_2 = 0.13 \; P_e^{0.33} \; P_c^{0.73} \tag{19}$$

$$A_1/A_3 = 0.34 \; P_c^{1.09} \; P_p^{-1.12} \; w^{-0.16} \tag{20}$$

$$A_2/A_1 = 1.04 \; P_c^{-0.83} \; P_p^{0.86} \; w^{-0.12} \tag{21}$$

The R^2 for each of the above correlations is above 0.99. Similarly, the following relations give the correlations for the un-choked flow:

$$P_2 = 1.02 \; P_e^{-0.000762} \; P_c^{0.99} \tag{22}$$

$$A_1/A_3 = 0.32 \; P_c^{1.11} \; P_p^{-1.13} \; w^{-0.36} \tag{23}$$

$$A_2/A_1 = 1.22 \; P_c^{-0.81} \; P_p^{0.81} \; w^{-0.0739} \tag{24}$$

The R^2 values for the above three correlations are above 0.99.

The constants in Eqs. 17 and 18 are given in the next table

Constant	Entrainment Ratio Correlation Choked flow (Eq. 17 - Fig. 5)	Entrainment Ratio Correlation Non-Choked Flow (Eq. 18 -Fig. 6)
a	0.65	$-1.89\text{x}10^{-5}$
b	-1.54	-5.32
c	1.72	5.04
d	$6.79\text{x}10^{-2}$	$9.05\text{x}10^{-2}$
e	22.82	22.09
f	$4.21\text{x}10^{-4}$	-6.13
g	1.34	0.82
h	9.32	$-3.37\text{x}10^{-5}$
j	$1.28\text{x}10^{-1}$	–
j	1.14	–
R^2	0.85	0.79

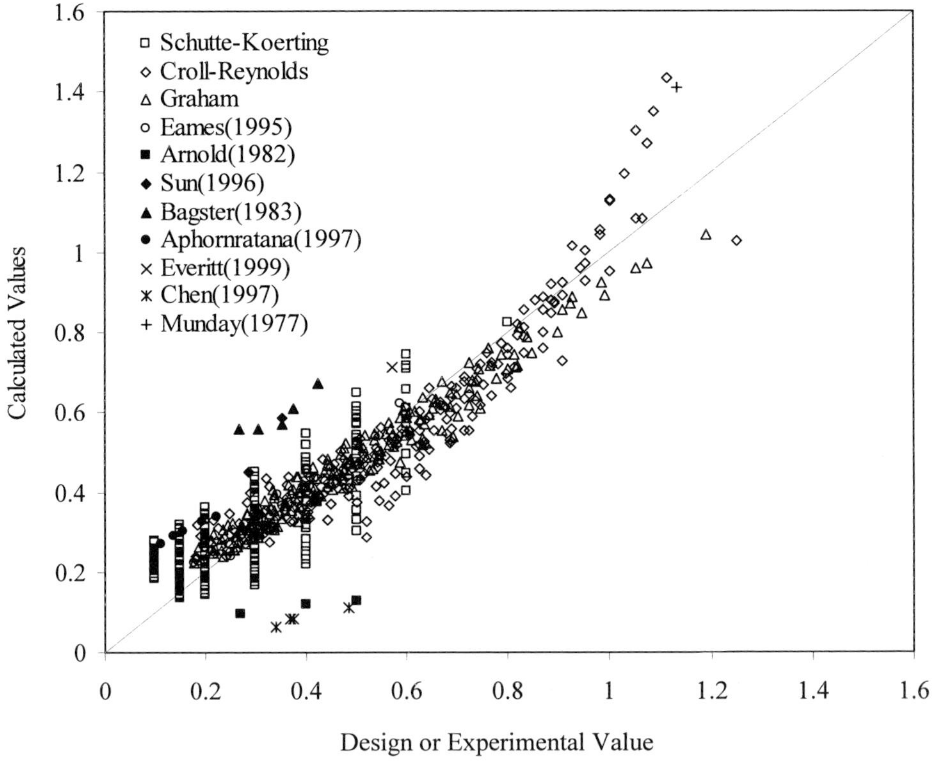

Fig. 5. Fitting of the entrainment ratio for compression ratios higher than 1.8.

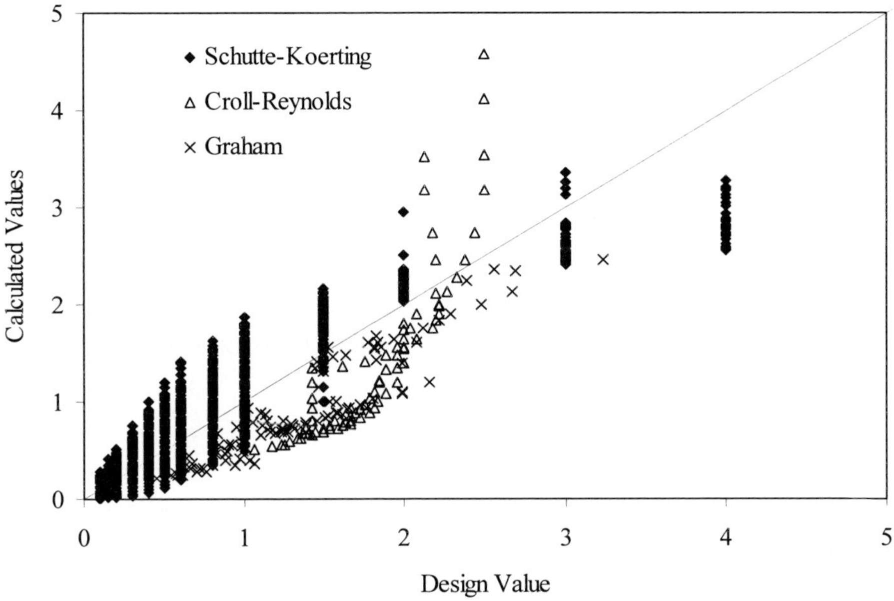

Fig. 6. Fitting of the entrainment ratio for compression ratios lower than 1.8.

Example

Calculate the entrainment ratio, the area ratios, areas, and flow rates of the following operating conditions:

- Pressure of motive steam = 800 kPa
- Pressure of the entrained vapor = 36.63 kPa
- Pressure of the compressed vapor = 101.325 kPa
- Flow rate of the entrained vapor = 1 kg/s
- Nozzle diameter = 0.02 m

Solution:

- Calculate the compression ratio

Cr = Pc/Pe = 101.325/36.63 = 2.77

- Calculate the expansion ratio

Er = Pp/Pe = 800/36.63 = 21.84

- Calculate the nozzle cross section area

$A_1 = \pi\,(d_1)^2/4 = (3.14)(0.02)^2/4 = 3.14\text{x}10^{-4}\ m^2$

– Calculate the entrainment ratio

$$w = a\ Er^b\ P_e^c\ P_c^d\ \frac{(e + fP_p^g)}{(h + iP_c^j)}$$

$$= 0.65\ (21.84)^{-1.54}\ (36.63)^{1.72}\ (101.325)^{0.0679}\ \frac{(22.82 + (4.21\text{x}10^{-4})(800)^{1.34})}{(9.32 + (0.128)(101.325)^{1.14})}$$

$$= 2.89$$

– Calculate the area ratios and the areas of the nozzle outlet and the diffuser

$A_1/A_3 = 0.34\ P_c^{1.09}\ P_p^{-1.12}\ w^{-0.16}$
$= 0.34\ (101.325)^{1.09}\ (800)^{-1.12}\ (2.89)^{-0.16}$
$= 0.0247$

$A_2/A_1 = 1.04\ P_c^{-0.83}\ P_p^{0.86}\ w^{-0.12}$
$= 1.04\ (101.325)^{-0.83}\ (800)^{0.86}\ (2.89)^{-0.12}$
$= 6.22$

$A_3 = A_1/(A_1/A_3)$
$= 3.14\text{x}10^{-4}/0.0247$
$= 1.27\text{x}10^{-2}\ m^2$

$A_2 = A_1\ (A_2/A_1)$
$= 3.14\text{x}10^{-4}\ (6.22)$
$= 1.95\text{x}10^{-3}$

– Combine Eqs. (17) and (18) to calculate the flow rates of the motive steam and the entrained vapor

$w = m_e/(m_c - m_e)$
$2.89 = 1/(m_c - 1)$
$m_c = 1.35$ kg/s

– Calculate the flow rate of the motive steam

$$m_p = m_c - m_e$$
$$= 1.35 - 1 = 0.35 \text{ kg/s}$$

9.3 Wire Mesh Demisters

Wire mesh mist eliminator or demisters is a simple porous blanket of metal or plastic wire that retains liquid droplets entrained by the gas phase. Demisters are used in distillation or fractionation, gas scrubbing, evaporative cooling, evaporation and boiling, and trickle filters. Removal of mist droplets is desirable or even mandatory for different reasons, which includes:

- Recovery of valuable products.
- Improving emission control.
- Protection of downstream equipment.
- Improving product purity

In thermal desalination processes the flashed or boiled off fresh water vapor entrains brine in the form of fine mist droplets. The brine droplets must be removed before vapor condensation over the condenser tubes. If the demister does not operate efficiently, the entrained brine droplets will reduce the quality of the distilled water and will form salt scale on the outer surface of the condenser tubes. The first effect results in disposal of the distillate product because of limits imposed by the end user, especially if the product water is used as a makeup for boilers. The last effect is very harmful because it reduces the heat transfer coefficient and enhances the corrosion of the tube material.

Vane type separators were common in earlier designs of evaporators; however, the system suffers from the following drawbacks: 1) high pressure drop (which could result in the total loss of temperature driving force between stages) and 2) excessive brine carry over.

Today, the wire mesh mist eliminator is widely used in thermal desalination plants. The main features of wire mesh mist eliminators are low pressure drop, high separation efficiency, reasonable capital cost, minimum tendency for flooding, high capacity, and small size. The performance of wire mesh eliminators depends on many design variable such as supporting grids, vapor velocity, wire diameter, packing density, pad thickness, and material of construction. Because the wire mesh is not rigid, it must be supported on suitable grids. To obtain minimum pressure drop, maximum throughput, and maximum efficiency, the support grids must have a high percentage of free passage. To take full advantage of the 98% or so free volume in the wire-mesh, the free passage through the support grids should be greater than 90%. If the free passage through the support grids is much lower than 90%, the accumulated liquid is prevented from draining back through the support grids, causing premature flooding, Fabian et al. (1993).

Typically, maximum allowable velocity in demisters is limited by the ability of the collected liquid to drain from the unit. In vertical flow, when the gas velocity increases past design levels, liquid begins to accumulate in the bottom of the unit. The liquid buildup results in re-entrainment of the downstream. This is because the inertia of the incoming gas prevents the liquid from draining out of the unit. In horizontal flow, the gas inertia pushes the captured liquid toward the downstream face, Holmes and Chen (1984). As a rule, smaller diameter wire targets collect smaller liquid droplets more efficiently. For example, a 10 μm wire removes smaller droplets than a 200 μm wire. However, a bed of 10 μm wires normally has the tendency to flood and re-entrain at much lower gas velocities than a bed of 200 μm wire. This is because the thinner wires provide dense packing that can trap the liquid by capillary action between the wires, Fabian et al. (1993). Interweaving of small diameter wires with larger diameter wire has been used often to tackle some of the most difficult mist removal problems. This design uses metallic or plastic wires as a support structure to hold the fine wires apart. Even with this approach, the throughput capacity of the unit is limited, compared to that possible with conventional mesh. Special internal mesh geometry modifications are now available that allow these bi-component (that is, small-fiber and large-diameter wire mesh) configurations to operate at velocities essentially the same as conventional mesh designs. These ultra-high-efficiency designs can replace conventional wire mesh. For example, in the dehydration towers of natural gas production plants, where even small losses of absorption chemicals, such as ethylene glycol, can be a significant operating expense, Lerner and Woinsky (1986).

Another important design parameter for proper operation of wire mesh mist eliminators is the distance between the mist eliminator and the surface of the flashing liquid. This distance must be optimized to minimize the volume of the flashing chamber and to provide sufficient distance for settling of the liquid droplets falling from the mist eliminator into the flashing liquid. A short distance will not allow for settling of these droplets and a long distance would come at the expense of increase in the total volume of the flash chamber.

9.3.1 Separation Mechanism

The separation processes in the demisters consist of three successive steps, which are:

- **Accumulation:** Droplet accumulation occurs by three main mechanisms, impaction, diffusion, and interception.

  - **Inertia impaction:** As the gas phase flows past the surface or around wires in the mesh pad the streamlines are deflected, but the kinetic energy of the liquid droplets associated with the gas stream may be too high to follow the streamline of the gas. As the droplets impact against the wires their momentum drops to zero.
  - **Diffusion:** The diffusion mechanism, sometimes called Brownian motions, is significant only for the capture of sub-micron droplets at a very low gas velocity.
  - **Interception:** This mechanism occurs for droplets with dimensions similar to or higher than the wire diameter.
- **Coalescence:** The droplets impinging on the surface of the wires coalesce to form larger size drops.
- **Detachment:** In the vertical-flow installations, detached liquid drops drain back from the upstream face of the wire mesh pad. In the horizontal flow systems, collected liquid droplets drain down through the vertical axis of the mesh pad in a cross flow fashion.

9.3.2 Materials of Wire Mesh Mist Eliminator

Construction materials for the wires include metal, fiberglass, plastics or polymers such as polypropylene or Teflon. Commercial alloys have been developed specifically for construction of wire mesh demisters, which provide three to five times the service lives of the traditional materials. Also, alloy wires offer improved service depending on the temperature and acid concentration of the gas stream, Fabian et al. (1993).

9.3.3 Demister Developments

Literature studies of wire mesh mist eliminator are rather limited. Table 5 shows a brief summary for some of the main studies, which includes the following:

- Buerkholz (1989) presented a complete review of studies on demisters. He reports that preventing droplets re-entrainment captured in the demister, should limit the gas phase velocity to 4- 5 m/s.
- Buerkholz (1986) presented a simple approximation formula for the fractional degree of precipitation and the limiting droplet size for all types of separators.
- Bradie and Dickson (1969) discussed the factors governing the wire mesh demister, and in particular with their application to entrainment removal in pool boiling systems.
- Feord et al. (1993) proposed a mathematical model to specify the outlet concentration and droplet size distribution for wire mesh demisters. The model offers the prospect of optimizing the pad construction to maximize the

separation efficiency at a target pressure drop or designing to a maximum pressure drop.

- The main drawback of the models developed by Feord et al. (1993) and Bradie and Dickson (1969) is the need for complete information on the entrainment level and the droplet size spectra. This type of data is not always available in practical units. Measuring droplet size distribution, especially in the small size range, is quite difficult, costly, and prone to inaccuracies.
- Robinson and Homblin (1987) presented detailed experimental work undertaken to demonstrate the collection of liquid aerosols in a helical coil as a function of aerosol size, gas velocity, tube diameter and number of coils. They compared the performance of this helical coils with that of a knitted mesh, a cyclone and a packed bed. They found that the performance of the helical coil demister is superior to the other systems. Nevertheless, the pressure drop for the knitted mesh system was to some extent lower than that of any one of the considered systems.
- Capps (1994) presented a guide for selecting mist-eliminating devices that include knockout drums, wire mesh pads, impingement separators, cyclones, filters, wet scrubbers, and electrostatic precipitators.
- Bayley and Davies (1973) developed design nomogram, which allows the process design engineers to quickly and accurately assess size and capacity of the required demister.
- Lerner (1986) stated that the width of the flashing chambers in the multistage flash desalination process (MSF) process is usually dictated by mist eliminator area requirements. They presented performance evaluation of two new types of commercial mist eliminators with capacities two to three times those of conventional mesh pads.
- El-Dessouky et al. (2000) performed a fundamental experimental investigation on the performance of wire mesh mist eliminator for removal of entrained water droplets from a water vapor stream. They used wire mesh mist eliminator made of Stainless steel 316L wires with 0.2 to 0.32 mm in diameter, which has the same specifications found in the MSF process. The demister was formed of several layers of knitted wires. They measured the separation efficiency, pressure drop, loading, and flooding points. They developed correlations for the above design and operating parameters as a function of the wire diameter, pad thickness, vapor velocity, and droplet size.

In summary, the following is concluded:

- Mathematical models of demisters are difficult to apply because it is necessary to have details of the droplet size distribution, which depend on the system temperature.
- Empirical correlations are simple to use and useful for design and performance evaluation within the range of experimental parameters.

Table 5
Summary of literature studies on wire mesh mist eliminator

Reference	Study	Remarks
Lerner (1986)	Develop two new types of commercial mist eliminators	Demister capacity is two to three times higher than conventional demisters
Buerkholz (1989)	Review studies on mist eliminator	Vapor velocity should be limited to 4-5 m/s to prevent droplets re-entrainment
Buerkholz (1986)	Develop mathematical models	Models are simple to use and apply to several types of mist elimination.
Bradie and Dickson (1969)	Discuss performance parameters of wire mesh mist eliminators	Focus is made on entrainment removal in pool boiling
Feord et al. (1993)	Develop mathematical model	Optimize pad dimensions and maximize separation efficiency as a function of pressure drop
Robinson and Homblin (1987)	Measure removal of liquid aerosols in helical coil demisters	Performance of helical coil demisters is superior to the other systems.
Capps (1994)	Develop guides for selection of demisters	Several types of demisters is included in the study
Bayley and Davies (1973)	Develop design nomograms	Simple to use for design and performance evaluation
El-Dessouky et al. (2000)	Measure removal rates in industrial type wire mesh mist eliminator.	Correlations cover the practical range for design and operation of wire mesh mist eliminators in desalination processes
Carpenter and Othmer (1955)	Develop separation efficiency model	
Brunazzi and Paglianti (1998)	Measure and develop and new separation efficiency model	

9.3.4 Design Parameters

Wire mesh demisters are usually specified in terms of the specific area A_s, packing density, ρ_p, and void fraction ε. These parameters are defined as:

$$A_s = \frac{\text{Surface area of wires}}{\text{Volume of demisters}} \tag{25}$$

$$\rho_p = \frac{\text{Mass of wire}}{\text{Volume of demister}} \tag{26}$$

$$\varepsilon = 1 - \frac{\text{Volume occupied by wires}}{\text{Volume of demister}} \tag{27}$$

For industrial demisters A_s ranges from 140 to 300 m^2/m^3, ρ_p from 80 to 268 kg/m^3, and ε from 0.979 to 0.99 (2000).

Separation efficiency is a measure to the fraction of droplets in the vapor swept out by the wire mesh mist eliminator and is given by

$$\eta = \frac{M_{in} - M_{out}}{M_{in}} \times 100 \tag{28}$$

where M_{in} and M_{out} are the mass of entrained water droplet by the vapor up and down stream the mist eliminator, respectively. The capacity of a wire mesh mist eliminator is determined by the conditions of loading and flooding. Beyond the loading point, the liquid holdup is high enough to improve the separation efficiency. The demister should operate at a velocity higher than the loading velocity. Flooding occurs when the vapor velocity exceeds a critical value. To prevent flooding, the mist eliminator must be designed and sized so that the design velocity is below the critical flooding velocity.

9.3.5 Demister Performance

Demister performance is illustrated in Figs. 7-13. Performance features are summarized below:

- Figure 7 shows the measured separation efficiency as a function of the maximum diameter of captured droplets. As is shown, the removal efficiency increases with the increase of maximum diameter of collected droplets and the vapor velocity. As the vapor and entrained liquid droplets pass through the mist eliminator, the vapor phase moves freely, but the liquid, due it's greater inertia, is unable to make the required sharp turns. Therefore, the droplets are impacted and collected on the surface of the mesh wires. The droplet momentum or inertia is proportional to their velocity, mass, and diameter.

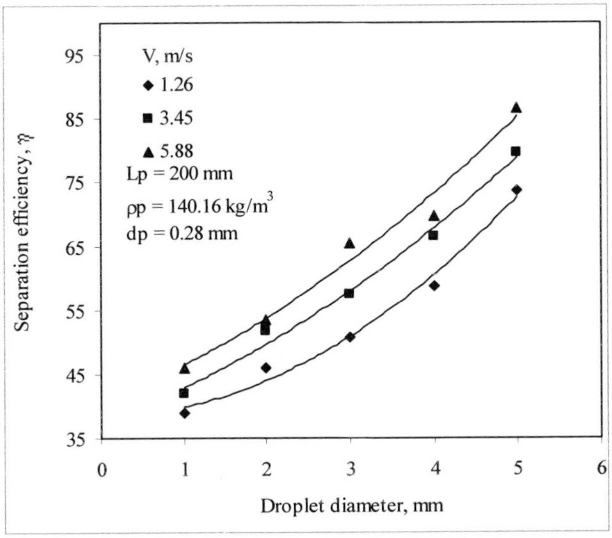

Fig. 7. Effect of droplet size on the separation efficiency at different values of vapor velocity

– Figure 8 displays the variation of the separation efficiency with the vapor velocity at different values of maximum size of collected droplets. The separation efficiency increases steadily with the vapor velocity until it reaches a peak value and thereafter diminishes with further increase in vapor velocity. This observation can be attributed to re-entrainment of the water droplet with the vapor flowing inside the pad. There are three different forces controlling the movement of the water droplets accumulated within the demister pad. These are the drag, the gravity, and the surface tension forces. When the gravity force is dominant, the droplets are detached from the wire and drained by gravity. Re-entrainment of the collected water droplets with the vapor flowing in the demister occurs because the shearing energy per unit volume of liquid droplet is sufficient to create liquid droplets having a large ratio of surface to volume. This will increase the drag force exerted by the vapor on the droplets external surface. The drag force is proportional to the vapor interstitial velocity and droplets surface area and density. The vapor velocity in the pad increases at larger liquid hold up. This is because the liquid occupies part of the void fraction, thus, decreasing the available area for vapor flow. The liquid hold up in the pad is related to the rates of draining, re-entrainment, and collection. The increase in vapor velocity reduces the rate draining and increases the rates of collection and re-entrainment. Moreover, at high vapor velocities, the re-entrained liquid droplets can re-impact on the surfaces of subsequent wires. This may lead to the atomization of the water droplet by the force of the impact. Consequently, a very fine spray of droplets is generated, which is difficult to recapture by subsequent separators. This may explain the decrease in the separation efficiency at large vapor velocities.

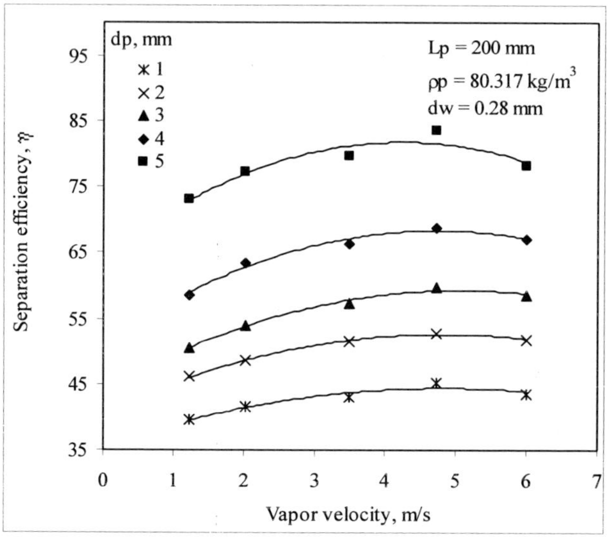

Fig. 8. Effect of vapor velocity on the separation efficiency at different values of droplet size

– The effect of the wire mesh diameter on the separation efficiency at different values of vapor velocity is displayed in Fig. 9. As displayed, the droplet separation efficiency improves with the decrease of the wire diameter. The influence of the wire diameter on the separation efficiency is more pronounced at lower vapor velocities. This is caused by the fact that the surface area of the wires at constant packing density and depth is directly related to the wire diameter. As a result, more droplets with smaller size can be trapped for mesh wires with smaller diameters, where the number of liquid droplets touching the wire is primarily determined by the ratio between the wire diameter and the droplet size. Moreover, as the wire diameter is reduced, the surface area is increased. Therefore, a larger amount of entrained droplets are caught within the demister pad due to capillary action. At high vapor velocity, the shearing energy per unit volume of the liquid droplets captured in the pad progressively increases as the wire diameter decreases. As a result, the liquid re-entrainment becomes excessive. Although, the results show the superior performance of demisters with smaller diameter wires; however, use of larger diameter wires is necessary to facilitate demister washing and cleaning. Also, use of larger diameter wires gives the demister adequate mechanical strength and operational stability.

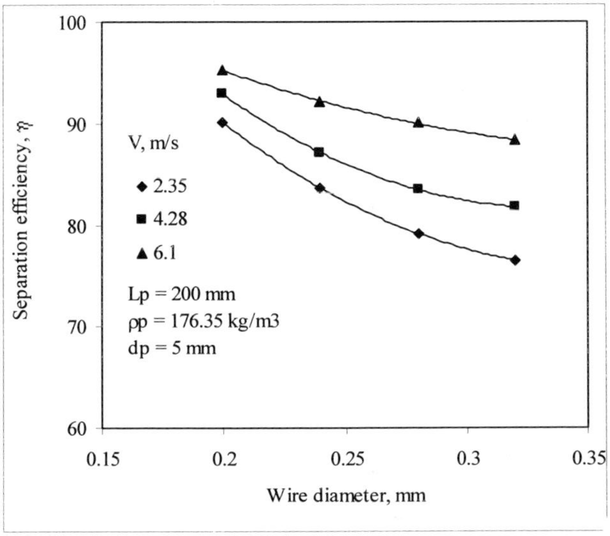

Fig. 9. Effect of wire diameter on the separation [illegible] vapor velocity

- The wire surface area for the wire mesh mist eliminator is directly related to the demister packing density. The increase of the pad packing density is associated with the reduction of the porosity of the pad. As a result, the number of entrained droplets which approaching the wires and the amount of captured droplets increases at larger surface area for the wires. This fact can be employed to explain the steady augmentation of the separation efficiency with the increase of the demister packing density as illustrated in Fig. 10. It is interesting to note that the effect of the pad density on the separation efficiency is more pronounced at low vapor velocities. This is mainly due to the increase in droplet re-entrainment and liquid hold up at higher vapor velocities.
- Figure 11 elucidates the effect of the pad thickness on the separation efficiency. As it can be seen, the separation efficiency is insensitive to the increase in the pad thickness. The pad thickness is a measure of how many times the vapor will consecutively impinge on a wire during its passage through the mesh pad. It was observed visually that a thin layer of water was

accumulated on the bottom surface of the wires facing the vapor. This thin film acts as an extra droplet collection media. However, this layer does not penetrate through the wire mesh pad and is not affected by the pad thickness. This explains the insensitivity of the separation efficiency with the increase in the pad thickness.

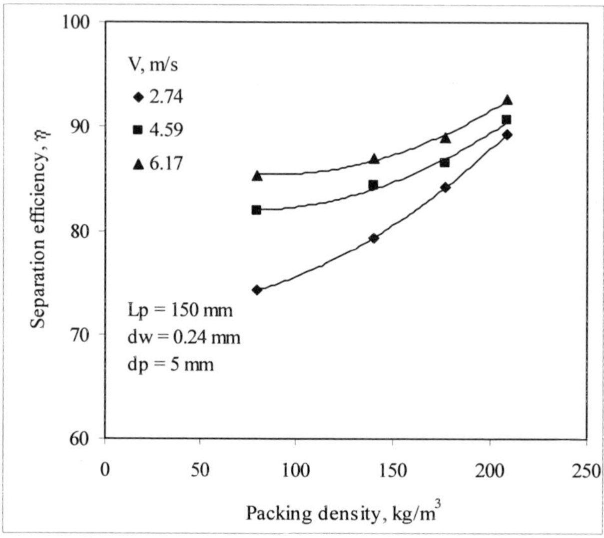

Fig. 10. Effect of packing density on the separation efficiency at different values of vapor velocity

- The variation in the specific pressure drop for dry and wet demisters at different packing densities is shown in Fig. 12. The dry demister refers to the condition at which the pad is free of water droplets. On the other hand, the wet demister assigns the state at which the water droplets are retained inside the pad. As it can be seen, the pressure drop for the dry demister, in general, is relatively low and increases linearly with the vapor velocity. The pressure drop for the dry demister is a measure of the relative resistance of the fluid flow through the pad. This arises from the viscous drag between the vapor and the wires forming the demister pad and also because of kinetic energy loss due to the changes of the flow direction. The viscous drag force and the kinetic energy loss are directly related to the vapor velocity and the total surface area of the wires (wetted area). The flow resistance is comparatively low because of the high voidage of the demister. Measurements show that the specific pressure drop for the dry demister is nearly independent of the demister thickness, density, and the wire diameter. The profiles for the specific pressure drop of the wet demister can be divided into three regions.

The first occurs at low vapor velocities, where the specific pressure drop varies linearly with the vapor velocity and is parallel to the specific pressure drop of the dry demister. In the second region, the specific pressure drop increases more rapidly with the increase in the vapor velocity. In the third region, the pressure drop rises very steeply even with the slight increase of vapor velocity. The specific pressure drop in wet demisters is caused by the dry pad and due to the presence of water droplets. The pressure drop of the wet demister is complicated and there are three contributions to this pressure drop. The first term represents the frictional pressure drop because of the slip between the vapor phase and the demister pad. The second term represents the pressure drop due to the vapor phase acceleration. The last term accounts for the gravitational effects, which is smaller than the other two terms and can be safely neglected. The frictional and acceleration pressure drops are strongly dependent on the vapor velocity. The vapor velocity inside the demister is changed as a result of variations in the system operating parameters or due to the holdup of the liquid phase. As the liquid holdup progressively increases, the free space area available for the vapor flow decreases and results in rapid increase in the flow resistance. The liquid holdup may be either static or dynamic. Capillary action causes the static holdup and occurs at high retention of the liquid within the demister pad. The dynamic holdup takes place, as the settling velocity of the falling droplets becomes lower than the upward vapor velocity. The static holdup increases with the increase of the wire surface area, which is directly related to the packing density.

- Figure 13 depicts variations of the specific pressure drop with the vapor velocity at different values of wire diameter. The figure illustrates that the pressure drop is inversely related to the wire size, which is caused by the increase of the wires surface area at smaller wire diameters. The static liquid holdup augments with the increase in the wires surface area. It is interesting to note that, the effect of wire diameter on the pressure drop is more pronounced at higher vapor velocities.
- Variations in the specific pressure drop at loading and flooding conditions are shown in Figs. 12 and 13. Below the loading point the liquid holdup is relatively low and the vapor velocity does not significantly affect the pressure drop. Above the loading point, the liquid begins to accumulate or load the demister progressively causing the decrease of the free space for vapor flow. In this region, the pressure drop increases more rapidly with the increase in the vapor velocity. The flooding point represents the maximum loading the demister can accommodate. The flooding and loading points occur at lower vapor velocities upon the increase of the demister packing density and the decrease of the wire diameter. All these variables increase the surface area of wires forming the demister pad and consequently the extent of liquid holdup.
- It is important to emphasize that the design capacity of most mist eliminators is determined by the phenomenon of re-entrainment. However, in thermal desalination processes, especially MSF and MEE, the design capacity of the

demister is obtained as a function of the flooding points. The temperature difference available per stage in MSF plants is relatively small particularly when the number of stages is high. However, the temperature depression corresponding to the pressure drop in the demister can considerably influence the plant thermal performance, specific heat transfer surface area and the specific cooling water flow rate. All these variables control the cost of desalinated water. It is worth mentioning that the flooding velocity is relatively lower than the re-entrainment velocity. Thus limiting the capacity of wire mesh mist eliminator by the flooding velocities will protect the heat transfer surface areas downstream the demister, prevent the diminishing of the product water quality, and minimize the thermodynamic losses. However, the vapor velocity must be higher than the loading point to ensure reasonable liquid holdup, which improves the separation efficiency.

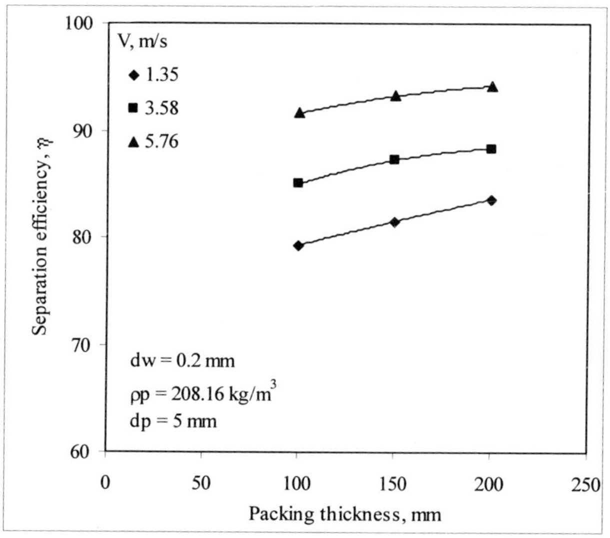

Fig. 11. Effect of packing thickness on the separation efficiency at different values of vapor velocity

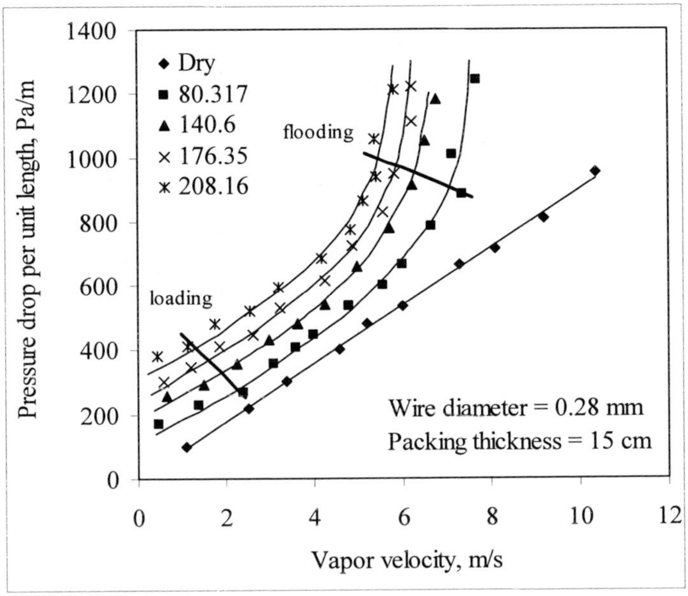

Fig 12. Variation in pressure drop as a function of vapor velocity for different packing density.

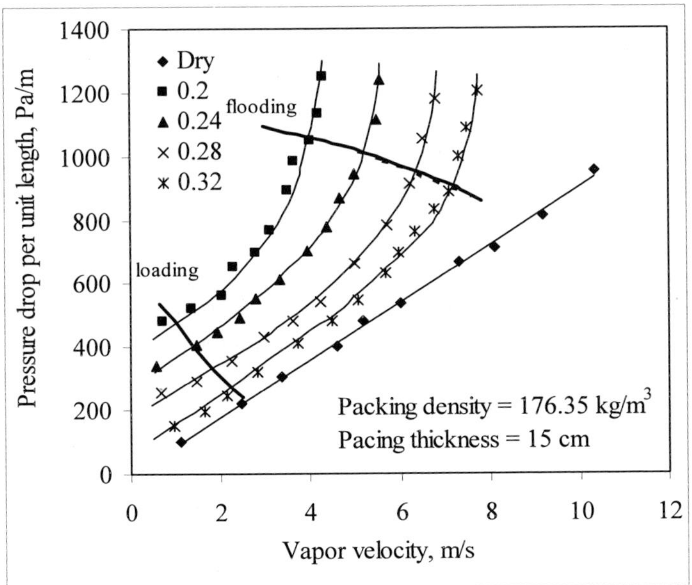

Fig 13. Variation in pressure drop as a function of vapor velocity for different wire diameter.

9.3.6 Empirical Correlations

In terms of the forgoing discussion, the separation efficiency of the wire mesh mist eliminator is affected by vapor velocity (V), wire diameter (d_w), droplet size (d_p), and packing density (ρ_p). On the other hand, the wet pressure drop is affected by the packing density (ρ_p), wire diameter (d_w), and vapor velocity (V). The flooding and loading velocities (V_ℓ and V_f) are dependent on the packing density (ρ_p) and wire diameter (d_w).

The least square fitting of the experimental data gives the following empirical correlations:

$$\eta = 17.5\,(d_w)^{-0.28}(\rho_p)^{0.09}(V)^{0.11}(d_d)^{0.38} \tag{29}$$

$$\Delta P = 3.9\,(\rho_p)^{0.38}(V)^{0.81}(d_w)^{-1.56} \tag{30}$$

$$V_\ell = 192.7\,(\rho_p)^{-0.47}(d_w)^{1.76} \tag{31}$$

$$V_f = 128.4\,(\rho_p)^{-0.29}(d_w)^{1.22} \tag{32}$$

The ranges of the experimental variables were V (0.98-7.5 m/s), ρ_p (80.32-208.16 kg/m^3), L (100-200 mm), d_w (0.2-.32 mm), and d_p (1-5 mm).

The pressure drop correlation is used to calculate the pressure drop and the associated temperature losses in the wire mesh demisters in the MSF process. Calculations are compared against the wire mesh design data of the MSF process. The parameters used in the calculations include a demister thickness of 150 mm, a packing density of 180.518 kg/m^3, and a wire diameter of 0.28 mm. Figure 14 shows variations in the design and predicted temperature loss in the demister as a function of the stage number and operating temperature. As is shown, good agreement is obtained between the design and the predictions of the empirical correlation.

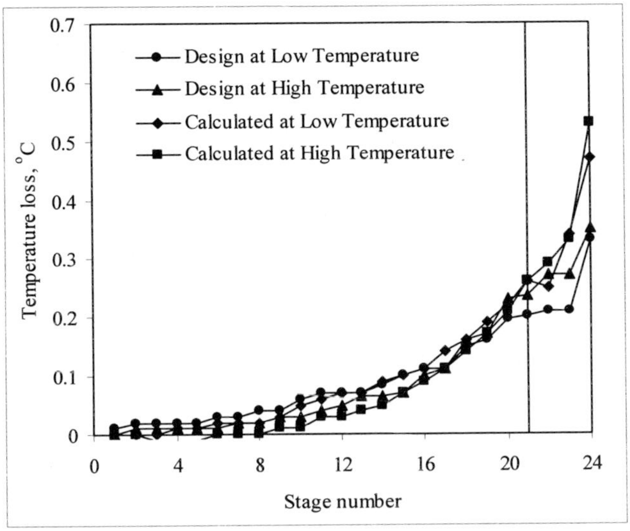

Fig. 14. Comparison of design and calculated temperature losses in wire mesh mist eliminator in the MSF process

9.3.7 Semi-Empirical Model

Two semi-empirical models are presented here. Carpenter and Othmer (1955) proposed the first model and the second model is developed by Brunazzi and Paglinati (1998). Common assumption for the two models include:

- No re-entrainment.
- No buildup of liquid

As for the model by Brunazzi and Paglinati (1998) they also assumed no mixing after passage through each layer of the demister. Carpenter and Othmer (1955) suggested the following semi-empirical equations

$$\eta_n = 1 - (1 - (2/3)\, A_e \eta_{ST}\, (z/\pi))^n \tag{33}$$

$$\eta_{ST} = St \text{ for } St \leq 1 \tag{34}$$

$$\eta_{ST} = 1 \text{ for } St \geq 1 \tag{35}$$

where

A_e is the specific area of the demister
η_n is the removal efficiency of the demister
η_{ST} is the removal efficiency of a single target
z is distance between two successive layers in the demister
n is the number of layers in the demister

St is the stokes number defined by $St = \dfrac{\rho_\ell u d_d^2}{18\,\mu_g d_w}$

d_w is the wire diameter
d_d is the droplet size
ρ_ℓ is the liquid density
u is the gas velocity
μ_g is the gas viscosity

The following equations is proposed by Brun

$$\eta_n = 1-(1-\eta_{ST})^M\left(\frac{\bar{n}-n'}{\bar{n}}+\frac{n'}{\bar{n}}(1-\eta_{ST})\right) \tag{36}$$

where

$\bar{n}$ is the number of layers in each cell, $\bar{n} = d_{eq}/d_w$
d_{eq} is the equivalent diameter, $d_{eq} = (4\,\pi\,\varepsilon\, d_w)/(A_a\, z)$
M is the number of cells in the pad, $M = \text{int}(n/\bar{n})$
n' is the number of layer not sufficient for form a complete cell, $n' = n - \bar{n}M$.
ε is the void space in the demister

Table 6 shows comparison of the separation efficiency as predicted by the above semi-empirical models.

Table 6
Comparison of separation efficiency for semi-empirical models*

Type of demister	A	B	C	D	E
Wire diameter (mm)	0.27	0.27	0.27	0.15	0.15
Packing density (kg/m^3)	145	190	200	128	190
Specific area (m^2/m^3)	267	360	363	459	643
Void fraction	0.98	0.975	0.974	0.984	0.975
Pad thickness (mm)	40	100	20	150	65
No. of layers	8	28	7	45	46
Equivalent mesh diameter (mm)	2.35	2.64	3.16	1.23	2.08
η_n (Carpenter and Othmer (1955))	0.89	0.99	0.77	1.	0.99
η_n (Brunazzi and Paglianti (1998))	0.75	0.97	0.51	1.	1.00

* Demister properties are obtained from Brunazzi and Paglianti (1998)

9.4 Interstage Brine Transfer Devices

Design and analysis of the interstage brine transfer devices is primarily a process of trail and error. Development of fully predictive model is rather a very difficult task. This is because of the complexity of flow, which is associated with formation, growth, and release of vapor bubbles as well as self entrainment of released vapor and non-condensables. Therefore, design of the interstage devices is based on models of the isothermal flow. Actual tuning of the system takes place in field equipment through a trial and error procedure, until the proper flow and flashing rates are achieved.

The interstage brine transfer devices play a major role in the MSF flashing stage. Their functions include the following:

- Provide stable operation at large load variation of 60-120% of the full load. The upper load is limited by the maximum pump capacity, maximum brine velocity in the tubes, erosion of mechanical parts, demister flooding. The minimum load is limited by minimum brine level that generate vapor lock between stages, minimum brine velocity inside the tubes,
- Maintain the minimum/optimum brine level throughout the stages. This is necessary to reduce the head loss caused by higher brine levels, which would reduce flashing rate, increase pressure drop, increase entrainment rate, and reduce the stage temperature.
- Provide vapor lock between stages to prevent vapor blow through across the stages. The liquid level is adjusted to 0.1-0.2 m higher than the gate height and the gate height varies between 0.2-0.3 m, however; adjustments might be necessary to prevent blow through and control of the liquid level within the stages.
- Provide sufficient turbulence that allow for liquid mixing and flashing from the brine pool.

- Controls the brine flow path and rate per unit width of the flashing chamber. This has a strong effect on the flashing rate of distillate product

9.4.1 Stage Orifice

Review of major MSF plant data show two main types of orifices, which includes:

- Low temperature weir/orifice configuration.
- High temperature box orifice.

Schematics of the two orifices are shown
configuration has simple geometry and it is formed of sluice made through the partition wall. A weir plate, known also as baffle, jump plate, or sill, is placed at distance upstream of the sluice gate. Properties of the weir orifice depends on the following parameters:

- The effective width and height of the gate.
- Distance between the gate and the weir.
- Height and width of the weir.

As is shown in Fig. 15a, the free-surface flow accelerates in the constriction posed by the gate between the two stages. Also, partial recirculation occurs due to stagnation in the upstream region of the gate. The stream leaving stage (i) and entering stage (i+1) expands and decelerates because of reduction in pressure. This expansion forms a hydraulic jump that prevent interstage blow-through of vapor. Formation and location of the hydraulic jump is affected by the flow velocity, brine levels, and pressure drop. It should be stressed that flow configuration is very complex because of stagnation, recirculation, rapid acceleration, strong turbulence, and presence of the hydraulic jump. This is further complicated by generation, growth, and translation of bubbles as well as self entrainment or aeration of vapor and non-codensable gases

As for the box orifice its flow properties depend on the dimensions of the box and it's opening. In this configuration bubble formation is observed to form inside the flashing box. Bubbles continue to grow inside the box. Also, a vapor film is formed at the undersurface of the box upper platen (1987).

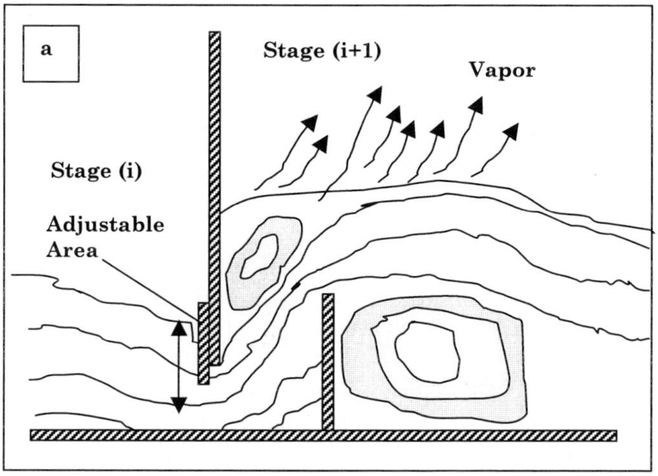

Fig. 15a. MSF weir orifice

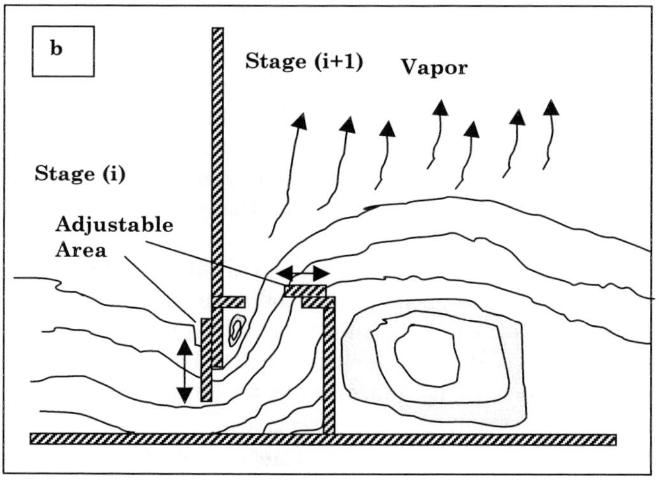

Fig. 15b. MSF box orifice

9.4.2 Orifice/Weir Analysis

Table 7 gives a summary for studies on the MSF orifices and weirs, which include the following:

- Lior (1999) reviewed the hydrodynamics of non-flashing flow in different regimes, the flashing stage flow, and the heat transfer models. The study includes quantitative explanation for the role of the hydraulic jump in the flash evaporation process. The review confirms the limited number of studies on orifice configuration in the MSF process. On the other hand, a larger number of correlations and design equations are found for the isothermal orifice flow in civil engineering applications.
- Bodendieck et al. (1997) presented a semi-empirical model for two types of MSF orifices, which include a single slot and the orifice/weir configuration. The model is integrated in an MSF model to determine the brine heights across the stages. Analysis is based on strategies of stable operation for various conditions of summer, winter, and partial load operation. The analysis shows suitable operation for the two orifices. However, a larger operating range is predicted for the orifice/weir configuration.
- Miyatake et al. (1992) presented a hydrodynamic model for the isothermal flow through the simple orifice and the orifice/weir configurations. The model results are validated against experimental measurements for loading range of 4.3x105-8.7x105 kg/(m hr) and liquid level of 0.4 m. The weir is found to promote the evaporation rates through propelling the entering liquid to the free surface and in generating low pressure regions near the top edge of the weir.
- Seul and Lee (1992) developed a two dimension, non-isothermal, two phase hydrodynamic model. An Eulerian approach for the continuous liquid phase and a Lagrangian approach for the dispersed vapor phase. In addition, the model takes into considerations bubble interaction. The model is applied to the simple orifice configuration. Results show that the evaporation rate increases at lower brine levels.
- The previous study was followed by a semi-empirical evaluation for the non-isothermal system, Miyatake et al. (1993). In their study, an empirical model is used for the temperature field and is combined with the two-dimensional hydrodynamic model developed by Miyatake et al. (1992). The results show good agreement with the experimental measurements as long as the flow is not disturbed by ebullition or separation.
- Reddy et al. (1995) proposed a model for simple orifice and orifice/weir configuration. The model is based on a set of empirical correlations previously developed in the ORNL reports and by Chow (1959), Subramanya (1986). A solution procedure for the proposed model is outlined for steady-state and transient conditions. However, no results are reported for the proposed model.
- Rautenbach and Schafer (1999) constructed a full scale experimental system to test various design configurations for the MSF orifices. Several configurations are tested, which include rectangular orifice, orifice/weir, siphon/sieve, and siphon with self-adjusting plate. The system allows for visual observations and measurements of pressure and temperature distribution Experimental measurements and resulting correlations show less than 10% deviations from the measured data for the pressure drop.

The most critical point in the above literature studies is the difficulty in full analytical or numerical solution for non-isothermal flow of orifice/weir configuration in the MSF system. As is shown, in the above citations the flow is highly turbulent and is associated with generation of low pressure regions and discontinuity in the flow regime near the orifice opening and the weir. As a result, problem solution is made through combination of empirical correlations and numerical solution for the isothermal flow regime.

Table 7
Summary of literature studies on MSF orifices

Reference	Remarks
Lior (1999)	Review of models for the isothermal models of simple orifice and orifice/weir configuration. Review shows a larger number of models for the isothermal applications and limited number of models for the MSF process.
Bodendieck et al. (1997)	Semi-empirical model for the simple slot and orifice/weir configuration. Model shows stable operation for both configuration with wider operating range for the orifice/weir configuration.
Miyatake et al. (1992)	Predictions of a two-dimensional isothermal analysis for simple orifice and orifice/weir configurations against experiment measurements. The results show better mixing and generation of low pressure regions for the orifice/weir system.
Seul and Lee (1990)	Based on the model developed by Seul and Lee (1990). Study effect of brine height. Results show higher evaporation rates at lower brine heights.
Miyatake et al. (1993)	A semi-empirical model that combines a correlation for the temperature field with the numerical model previously developed by Miyatake et al. (1992). Model predictions are limited to low loading rates with no separation or ebullition.
Reddy et al. (1995)	Developed iterative procedure for calculations of the brine levels in various flash stages at steady-state or transient conditions. No results are reported for the proposed model.
Rautenbach and Schafer (1999)	Designed a full scale test with a width of 0.1 m and weir loads up to $1.7x10^6$ kg/m hr. New weir design that allow for very low brine levels without vapor blow through is discussed.
Seul and Lee (1992)	Two phase model, which includes two phase flow and bubble interaction.

9.4.3 Submerged Orifice/Weir Design Formula

The following is a typical isothermal hydrodynamic orifice model. The model is based on mechanical energy balances as well as empirical correlations based on experimental measurements, Bodendieck et al. (1992). The model parameters are shown in Fig. 16.

$$u_1 = \frac{V}{BH_1} \tag{37}$$

$$u_2 = \frac{V}{B\,c\,H_o} \tag{38}$$

$$H_1 + \frac{P_1}{\rho g} + \frac{u_1^2}{2g} = H_2 + \frac{P_2}{\rho g} + \frac{u_2^2}{2g} \tag{39}$$

$$c = 0.6145 + 0.0592\,x - 0.5121x^2 + 2.5633\,x^3 - 4.1897\,x^4 + 2.4606\,x^5 \tag{40}$$

$$\frac{V}{B} = C\frac{2}{3}\mu\sqrt{2g}(H_3 - H_w)^{3/2} \tag{41}$$

$$x = \frac{H_o}{H_1 + h} \tag{42}$$

$$h = \frac{\Delta P}{9.81\,\rho} \tag{43}$$

$$\mu = 0.605 + \frac{1}{1000(H_3 - H_w)} + 0.08\frac{H_3 - H_w}{H_w} \tag{44}$$

where

$C = 1$ for crested weir, or $H_3 < H_w$

$$C = (1 - (\frac{H_3 - H_w}{H_w})^{1/2})^{0.385} \quad H_4 < H_o$$

Equations 37 to 44 can be solved iteratively to determine the brine level at various points up and down stream the orifice.

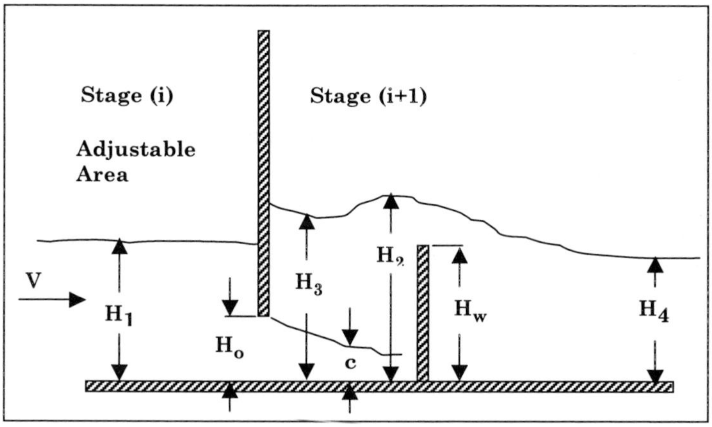

Fig. 16. Hydrodynamic parameter for flow through orifice/weir, Bodendieck et al. (1997)

References

Abdel-Aal, H.K., Al-Zakri, A.S., El-Sarha, M.E., El-Swify, M.E., Assassa, G.M., Other options of mass and energy input for steam jet refrigeration systems, Chem. Eng. J., 45(1990)99-110.

Aly, N.H., Karmeldin, A., and Shamloul, M.M., Modelling And Simulation Of Steam Jet Ejectors, Desalination, 123(1999)1-8.

Al-Khalidy, N., and Zayonia, A., Design and experimental investigation of an ejector in an air-conditioning and refrigeration system, ASHRAE Trans., 101(1995)383-391.

Al-Khalidy, N., Performance of solar refrigerant ejector refrigerating machine, ASHRAE Trans., 103(1997)56-64.

Aphornratana, S., and Eames, I.W., A Small Capacity Steam-Ejector Refrigerator: Experimental Investigation Of A System Using Ejector With Moveable Primary Nozzle, Int. J. Refrig., 20(1997)352-358.

Arnold, H.G., Huntley, W.R., Perez-Blanco, H., Steam ejector as an industrial heat pump, ASHRAE Trans., 88(1982)845-857.

Bagster, D.F., and Bresnahan, J.D., An examination of the two-stream theory of steam-jet ejectors, The 11th Australian Conference on Chemical Engineering, Brisbane, 4-7 Sept., 1983.

Bayley, R.D. and Davies, G.A. Process applications of knitted mesh mist eliminators, Chemical Process, May, 33-39, 1973.

Bodendieck, F., Genthener, K., Gregorzewski, A., The effect of brine orifice design on the range of operation and operation stability of MSF distillers, Proceeding of the IDA World Congress on Desalination and Water Sciences, Madrid, Spain, Vol. I, pp. 179-197, October, 1997.

Bradie, I.K. and Dickson, A.N. Removal of entrainment liquid droplets by wire mesh demisters, Proc. Inc. Mech. Engrs., 184(1969)195-203.

Brunazzi E. and Paglianti, A. Design of wire mesh mist eliminators, AIChE J., 44(1998)505-512.

Buerkholz, A. Die Beschreibung der partikelabscheidung durch tragheitskrafte mit hilfe einer dimensionsanalytisch abgeleiteten kennzahl, Chem.-Ing.-Tech. 58(1986)548-556.

Buerkholz, A. Droplet separation, VCH Publishers, New York, 1989.

Carpenter, C.L., and Othmer, D.F., Entrainment removal by a wire-mesh separator, AIChE J., 1(1955)549.

Capps, W., Properly specify wire-mesh mist eliminators, Chemical Engineering Progress, 90(1994)49-55.

Chen, F.C., and Hsu, C.T., Performance of ejector heat pumps, Energy Res., 11(1987)289-300.

Chen, Y.M., and Sun, C.Y., experimental study of the performance characteristics of a steam-ejector refrigeration system, Exper. Thermal and Fluid Sci., 15(1997)384-394.

Chen, S.L., Yen, J.Y., and Huang, M.C., An experimental investigation of ejector performance based upon different refrigerants, ASHRAE Trans., Vol. 104, No. 00, pp. 153-160, 1998.

Eames, I.W., Aphornaratana, S., and Haider, H., A Theoretical And Experimental Study Of A Small-Scale Steam Jet Refrigerator, Int. J. Refrig., Vol. 18, No. 00, pp. 378-385, 1995.

El-Dessouky, H.T., and Ettouney, H.M., Single Effect Thermal Vapor Compression Desalination Process: Thermal Analysis, Heat Transfer Eng., Vol. 20, No. 00, pp. 52-68, 1999.

El-Dessouky, H.T., Alatiqi, I.M., Ettouney, H.M., and Al-Deffeeri, N.S., Performance of wire mesh mist eliminator, Chem. Eng. Proc., Vol. 39, No. 2, pp. 129-139, 2000.

El-Dessouky, H.T., Ettouney, H.M., Alatiqi, I., and Al-Nuwaibit, G., Steam jet ejectors: Modeling and analysis, Chem. Eng. Proc., in print, 2001.

Everitt, P., and Riffat, S.B., Steam jet ejector system for vehicle air conditioning, Int. J. Ambient Energy, Vol. 20, No. 00, pp. 14-20, 1999.

Fabian, P., Cusack, R., Hennessey, P. and Neuman, M., Demystifying the selection of mist eliminators, Part I: The basics, Chemical Engineering, Vol. 100, No. 11, pp. 148-156, 1993.

Fabian, P., Van Dessel, P., Hennessey, P., and Neuman, M., Demystifying the selection of mist eliminators, Part II: The applications, Chemical Engineering, Vol. 100, No. 12, pp. 106-111, 1993.

Feord, D. Wilcock, E. and Davies, G.A. A stochastic model to describe the operation of knitted mesh mist eliminators, Computation of separation efficiency, Trans. IChemE, Research and Design, Vol. 71, No. A3, pp. 282-295, 1993.

Grazzini, G., and Mariani, A., A simple program to design a multi-stage jet-pump for refrigeration cycles, Energy Convers. Mgmt., Vol. 39, No. 00, pp. 1827-1834, 1998.

Gupta,S.K., Singh, R.P., and Dixit, R.S., A comparative parametric study of two theoretical models of a single-stage single-fluid, steam jet ejector, The Chemical Engineering J., Vol. 18, No. 00, pp. 81-85, 1979.

Henzler, H.J., Design Of Ejectors For Single-Phase Material Systems, Ger. Chem. Eng., Vol. 6, No. 5, pp. 292-300, 1983.

Holmes T.L., and Chen, G.K., Design and selection of spray/mist elimination equipment, Chemical Engineering, Vol. 91, No. 21, pp. 82-89, 1984.

Huang, M.C., and Chen, S.L., An experimental investigation of ejector performance characteristics in a jet refrigeration system, J. Chin. I. Ch. E., Vol. 27, No. 00, pp. 91-100, 1996.

Huang, B.J., Chang, J.M., Wang, C.P., Petrenko, V.A., A 1-D Analysis Of Ejector Performance, Int. J. Refrigeration, Vol. 22, No. 00, pp. 354-364, 1999.

Keenan, J.H., and Neumann, E.P., A Simple Air Ejector, J. Applied Mechanics, Vol. 64, No. 00, pp. 85-??, 1942.

Kishi, M., Mochizuki, Y., Matsubayashi, M., Hattori, K., Development of flashing flow model for box type orifice, Desalination, Vol. 65, No. 00, pp. 57-62, 1987.

Lerner, B.J., and Woinsky, S.G.
High-tech mist elimination in multi-stage evaporators, Plant Operations Progress, Vol. 5, No. 1, pp. 52-56, 1986.

Lior, N., Formulas for calculating the approach to equilibrium in open channel flash evaporators for saline water, Desalination, Vol. 60, No. 3, pp. 223-000, 1986.

Lior, N., The role of the hydraulic jump in MSF plant flow, Proceeding of the IDA World Congress on Desalination and Water Sciences, San Diego, USA, Vol. I, 155-166, August, 1999.

Ludwig, E.E., Applied Process Design for Chemical and Petrochemical Plants, Vol. 1, 2nd ed., Gulf, Houston, Texas, 1977.

Menegay, P., and Kornhauser, A.A., Ejector expansion refrigeration cycle with underexpanded motive nozzle, American Inst. Aeronautics and Astronautics, Vol. 2, No. 00, 915-920, 1994.

Miyatake, O., Hashimoto, T., and Lior, N., The liquid flow in multi-stage flash evaporators, Int. J. Heat Mass Transfer, Vol. 35, No. 12, pp. 3245-3257, 1992.

Miyatake, O., Hashimoto, T., and Lior, N., The relationship between flow pattern and thermal non-equilibrium in the multi-stage flash evaporation process, Desalination, Vol. 91, No. 1, pp. 51-64, 1993.

Munday, J.T., and Bagster, D.F, A New Ejector Theory To Steam Jet Refrigeration, I. & E.C., Vol. 16, No. 4, pp. 442-449, 1977.

Power, R.B., Hydrocarbon Processing, Vol. 43, No. 00, pp 138-000, 1964.

Power, R.B., Predicting Unstable-Mode Performance Of A Steam Jet Ejector, American society of mechanical engineers, FSPI, Vol. 1, No. 00, 11-15, 1994.

Rautenbach, R. and Schafer, S., Experimental and theoretical studies on interstage brine orifices for MSF desalination plants, Proceeding of the IDA

World Congress on Desalination and Water Sciences, San Diego, USA, Vol. I, pp. 47-63, August, 1999.

Reddy, K.V., Husain, A., Woldai, A., Nabi, S.M., Kurdali, A., Holdup and interstage orifice flow model for an MSF desalination plant, Proceeding of the IDA World Congress on Desalination and Water Sciences, Abu-Dhabi, UAE, Vol. IV, pp. 179-197, October, 1995.

Robinson, K.S. and Homblin, C. Investigating droplet collection in helices and a comparison with conventional demisters, Filtration & Separation, Vol. 24, No. 3, pp. 166-171, 1987.

Rogdakis, E.D., and Alexis, G.K., Design and parametric investigation of an ejector in an air-conditioning system, Applied Thermal Eng., Vol. 20, No. 00, pp. 213-226, 2000.

Seul, K.W., and Lee, S.Y., Numerical predictions of evaporative behaviors of horizontal stream inside a multi-stage-flash distillator, Desalination, Vol. 79, No. 00, pp. 13-35, 1990.

Seul, K.W., and Lee, S.Y., Effect of liquid level on flow behaviors inside a multi-stage flash evaporator – A numerical prediction, Desalination, Vol. 85, No. 2, pp. 161-177, 1992.

Sokolov, M., and Hershgal, D., Enhanced ejector refrigeration cycles powered by low grade heat. Part 1. Systems Characterization, Int. J. Refrig., Vol. 13, No. 00, pp. 351-356, 1990a.

Sokolov, M., Hershgal, D., Enhanced ejector refrigeration cycles powered by low grade heat. Part 2. Design procedures, Int. J. Refrig., Vol. 13, No. 00, pp. 357-363, 1990b.

Sokolov, M., and Hershgal, D., Optimal coupling and feasibility of solar powered year-round ejector air conditioner, Solar Energy, Vol. 50, No. 00, pp. 507-516, 1993a.

Sokolov, M., and Hershgal, D., Solar-powered compression-enhanced ejector air conditioner, Solar Energy, Vol. 51, No. 00, pp. 183-194, 1993b.

Sun, D.W., and Eames, I.W., Recent Developments In The Design Theories And Applications Of Ejectors – A Review, J. Inst. Energy, Vol. 68, No. 00, pp. 65-79, 1995.

Sun, D., Variable geometry ejectors and their applications in ejector refrigeration systems, Energy, Vol. 21, No. 00, pp. 919-929, 1996.

Sun, D.W., and Eames, I.W., Performance Characteristics Of HCFC-123 Ejector Refrigeration Cycle, Int. J. Energy Research, Vol. 20, No. 00, pp. 871-885, 1996.

Sun, D.W., Comparative study of the performance of an ejector refrigeration cycle operating with various refrigerants, Energy Conversion & Management, Vol. 40, No. 00, pp. 873-884, 1999.

Tomasek, M.L., and Radermacher, R., Analysis of a domestic refrigerator cycle with an ejector, ASHRAE Trans., Vol. 1, No. 00, pp. 1431-1438, 1995.

Chapter 10

Economic Analysis of Desalination Processes

Objectives

The objective of this chapter is to present elements desalination cost. The presentation outlines components of direct/indirect and operating cost. A number of case studies are presented for the calculation procedure. Also, a comprehensive review of literature cost data is presented.

10.1 Factors Affecting Product Cost

Unit product cost is affected by several design and operational variables, Ettouney et al. (2001), which includes the following:

- Salinity and quality of feed water: Lower feed salinity allows for higher conversion rates. As a result, the plant can operate with lower specific power consumption and dosing of antiscalent chemicals. Also, downtime related to chemical scaling is considerably reduced.
- Plant capacity: Larger plant capacity reduces the capital cost for unit product. Although, the increase in the plant capacity implies higher capital.
- Site conditions: Installation of new units as an addition to existing sites, would eliminate cost associated with facilities for feed water intake, brine disposal, and feed water pretreatment.
- Qualified manpower: Availability of qualified operators, engineers, and management would result in higher plant availability, production capacity, and lower down time caused by trips of devices.
- Energy cost: Availability of inexpensive sources for low cost electric power and heating steam have strong impact on the unit product cost.
- Plant life and amortization: Increase in plant life reduces the capital product cost.

10.2 Elements of Economic Calculations

Calculations of the unit product cost depend on the process capacity, site characteristics, and design features. System capacity specifies sizes for various process equipment, pumping units, and membrane area. Site characteristics have a strong effect on the type of pretreatment and postreatment equipment, and consumption rates of chemicals. In addition, design features of the process affect consumption of electric power, heating steam, and chemicals.

Figure 1 shows a summary for the economics of desalination processes. As is shown the production cost is divided into direct/indirect cost and annual operating cost. Elements forming both categories are explained in the following points:

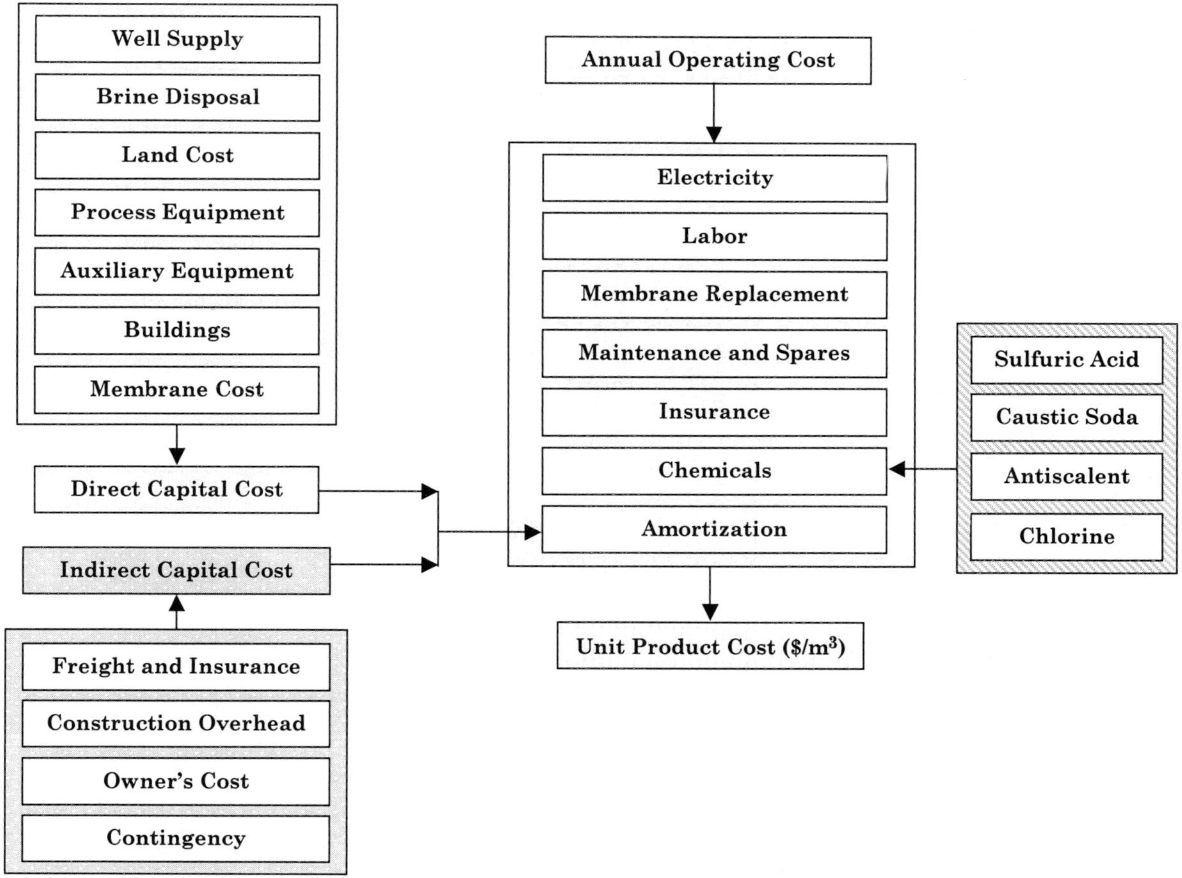

Fig. 1. Cost elements of desalination processes

10.2.1 Direct Capital Cost

The direct capital cost covers purchasing cost of various types of equipment, auxiliary equipment, land cost, construction, and buildings. The following gives brief description for various cost items with current cost estimates.

a. Land Cost

Land cost may vary considerably from zero charges to a total sum that depends on the site properties. Government owned plants normally have zero charges. Also, plants under BOOT contracts with governments or municipalities can zero or very highly reduced charges.

b. Well Supply

Recent estimates indicate \$650/m depth for construction. Average well capacity is estimated at 500 m^3/d.

c. Process Equipment

This is one of the most cost items and it depends on the process type and capacity. The cost of the process equipment may be less than \$1000; a good example is a laboratory scale RO unit operating on low salinity tap water. On the other hand, the equipment cost for 100,000 m^3/d RO system would be close to \$50x10^6. The equipment cost for the MSF and MEE is more expensive than the RO processes with current estimates of \$40x10^6 for a capacity of 27,000 m^3/d units. Item included under this category are listed below

- Process equipment
- Instrumentation and controls
- Pipelines and valves
- Electric wiring
- Pumps
- Process cleaning systems
- Pre and post-treatment equipment
- Seawater intake and brine discharge line
- Chlorination plant.

d. Auxiliary Equipment

The following auxiliary equipment is included:

- Open intakes or wells
- Transmission piping.
- Storage tanks
- Generators and transformers
- Pumps
- Pipelines and valves

e. Building Cost

Building cost varies over a wide range from \$100/m^2 to \$1000/m^2. This range is site specific and depends on the building type. Buildings include the following:

- Control room

- Laboratory
- Offices
- Workshop

f. Membrane Cost

Cost of membrane modules depends on the plant capacity and varies between \$500 to \$1000 per module, which have production rates of 50-100 m^3/d

10.2.2 Indirect Capital Cost

All cost items listed in this category are expressed as percentage of the total direct capital cost. Indirect capital costs include the following items:

a. Freight and Insurance

This cost is equal to 5% of total direct costs.

b. Construction Overhead

This cost is equal to 15% of direct material and labor cost and then adjusted for the size (total capital cost) of the plant. Construction overhead costs include the following:

- Fringe benefits
- Labor burden
- Field supervision
- Temporary facilities
- Construction equipment
- Small tools
- Miscellaneous
- Contractor's profit

c. Owner's Costs

Owner's costs are engineering and legal fees. This cost is equal to 10% of direct material and labor cost, and then adjusted for the size of the plant.

d. Contingency

Project contingency is taken at 10% of total direct costs.

10.2.3 Operating Cost

Operating cost covers all expenditure incurred after plant commissioning and during actual operation. These items include labor, energy, chemical, spare parts, and miscellaneous. The following gives brief description of each item and current cost estimates:

a. Electricity

This cost varies over a range of \$0.04-0.09/kWh. The upper limit is characteristic of European countries and the lower limit can be found in the Gulf States and the US.

b. Labor

This cost item is site specific and depends whether the plant is government or privately. In addition, recent trends in plant operation aims for contracting operation and maintenance duties. This reduces the full time manpower, which may include plant director and small team of experienced engineers and technicians.

c. Membrane Replacement

Replacement rate may vary between 5%-20% per year. The lower bound applies to low salinity brackish water supported by proper operation and pretreatment system and the upper would reflect high salinity seawater, similar to the Gulf area, in addition to relatively poor operation and inefficient pretreatment system.

d. Maintenance and Spares

This cost item can be assigned a value lower than 2% of the total capital cost was used as a yearly rate.

e. Insurance

Insurance is rated at 0.5% of the total capital cost.

f. Amortization or Fixed Charges

This item defines the annual payments that cover the total direct and indirect cost. This cost is obtained by multiplying the total direct and indirect cost by the amortization factor, which is defined by the following relation

$$a = \frac{i(1+i)^n}{(1+i)^n - 1}$$

where i is the annual interest rate and n is the plant life. Accumulated experience in the desalination industry indicates that an amortization life of 30 years is adequate. As for the interest rate, its average value is equal 5%, however, a range of 3-8% should be considered in economics analysis.

g. Chemicals

The chemicals used in feed treatment and cleaning include sulfuric acid, caustic soda, antiscalent, and chlorine. Cost of these items may be affected by availability of nearby manufacturing plants and prices in the global markets. Also, chemical treatment differs between thermal and membrane processes, where higher specific cost is obtained for the membrane processes. Also, treatment depends on the top brine temperature and feed salinity. Table 1 gives estimates for the unit cost of chemicals used in thermal and membrane desalination, dosing rates, and specific rates per unit volume of product water.

Table 1
Estimates of chemicals cost and dosing rates

Chemical	Unit Cost ($/kg of Chemical)	Dosing Rate (kg Chemical/kg water)	Specific Cost ($/m^3 water)
Sulfuric acid	0.504	2.42x10^{-5}	0.0122
Caustic soda	0.701	1.4x10^{-5}	0.0098
Antiscalent	1.9	4.99x10^{-6}	0.0095
Chlorine	0.482	4.00x10^{-6}	0.00193

10.3 Cost Evaluation

Table 2 shows a summary for unit product cost of the major desalination processes, which includes MSF, RO, MEE, and MVC. Table 2 also includes data for MEE-TVC and two novel schemes, which include MEE with absorption heat pump and vertical stack MEE. The data show the following trends:

- The unit product cost of RO process depends on the capacity. Recent field estimates give \$0.55/m^3 for the large RO project in Florida, USA, with a capacity of 113,652 m^3/d. Examining recent data smaller capacity units give \$0.83/m^3 and \$1.22/m^3 for capacities of 40,000 m^3/d and 20,000 m^3/d, respectively. Both price quotes are for existing RO units in Cyprus.
- The RO unit product cost has decreased over the years. This is evident by comparing the price quotes by Leitner (1998), Leitner (1992), and Wade (1993).
- The unit product cost for MSF, MEE, and MEE-TVC are higher than the RO process with values averaging \$1.5/m^3.
- The highest unit product cost is quoted for the MVC process. Veza (1995) quoted a field value of \$3.22/m^3, however, this high value attributed to thigh cost energy. This irrespective of the plant availability. Recent economic calculations by Morin (1999) remains to show similar and high unit product cost with an average of \$2.43/m^3.
- The lowest unit product quotes in Table 2 are those for novel configurations, which includes the MEE vertical stack and the MEE combined with absorption heat pump. The capacity of the vertical stack system is 340,965 m^3/d; the largest found in this table. The other configuration is a prototype unit, which has a performance ratio of 21, which is very high in comparison with the conventional performance ratio of range of 8-16 for the MEE system with/without thermal vapor compression.

The most critical parameters in cost evaluation are the fixed charges (amortization) and the energy cost. Other parameters that have lesser effect on the unit product cost include the cost of chemicals and labor. Tables 3, 4, and 5, and 6 include estimates for the cost of the process capital, energy, chemicals, and membrane replacement.

Table 2
Unit product cost for conventional and novel processes

Reference	$/m^3	Remarks
a) Novel Process		
Hammond et al. (1994)	0.48	MEE Vertical Stack, 340956 m^3/d, 30 effects, Aluminum alloy, Fluted tube
DeGunzbourg and Larger (1998)	0.35	MEE-ABS, 9600 m^3/d, Absorption Heat Pump and Gas Turbine
b) Mechanical Vapor Compression (MVC)		
Matz and Fisher (1981)	1.51	1000 m^3/d
Darwish and Al-Najim (1987)	0.89	750 m^3/d
Darwish et al. (1990)	5	100 m^3/d
Leitner (1992)	2.48	4000 m^3/d
Zimmerman (1994)	0.46	Operating cost, 20,000 m^3/d
Veza (1995)	3.22	500 m^3/d
Morin (1999)	2.43	4546 m^3/d
c) Reverse Osmosis		
Darwish and Al-Najim (1987)	0.64	Single Stage
Darwish and Al-Najim (1987)	0.76	Two Stage
Darwish et al. (1990)	2.37	100 m^3/d
Leitner (1992)	1.98	4000 m^3/d
Leitner (1992)	1.09	37,850 m^3/d
Wade (1993)	1.39	32000 m^3/d
Zimmerman (1994)	0.58	Operating cost, 20,000 m^3/d
Pappas (1997)	1.22	20,000 m^3/d
Leitner (1998)	0.83	40,000 m^3/d
Morin (1999)	1.29	4546 m^3/d
Morin (1999)	0.68	45460 m^3/d
Leitner (1999)	0.55	113,652 m^3/d
d) Multistage Flash Desalination (MSF)		
Darwish and Al-Najim (1987)	0.77	Dual purpose
Darwish and Al-Najim (1987)	1.84	Single Purpose
Leitner (1992)	1.25	37,850 m^3/d
Morin (1993)	1.61	45461 m^3/d
Wade (1993)	1.57	Steam Turbine
Wade (1993)	1.44	Gas Turbine, Waste Heat Boiler
Morin (1999)	1.498	45460 m^3/d

Table 2 (continued)
Unit product cost for conventional and novel processes

e) Multiple Effect Evaporation (MEE)		
Darwish and Al-Najim (1987)	0.87	Dual purpose
Darwish and Al-Najim (1987)	1.95	Single Purpose
Leitner (1992)	1.08	37,850 m^3/d
Wade (1993)	1.31	Gas Turbine, Waste Heat Boiler
Morin (1993)	1.24	22730 m^3/d
Morin (1999)	1.397	22730 m^3/d
f) Multiple Effect Evaporation – Thermal Vapor Compression		
Darwish and Al-Najim (1987)	2.34	Single Purpose
Darwish and Al-Najim (1987)	1.31	Dual purpose
Morin (1993)	1.55	22166 m^3/d

Table 3
Capital cost for various desalination processes

Reference	Process	$	$/(m^3/d)$	Capacity
Matz and Fisher (1981)	RO	$9.24x10^5$	924	1000 m^3/d
Leitner (1992)	RO	$49.7x10^6$	1313.1	37,850 m^3/d
Wade (1993)	RO	$53.3x10^6$	1665.6	32,000 m^3/d
Leitner (1999)	RO	$98x10^6$	1035	94,625 m^3/d low salinity feed (26,000 to 30,000 ppm)
Matz and Fisher (1981)	MVC	$8.94x10^5$	894	1000 m^3/d
Veza (1995)	MVC	$1.586x10^6$	1322	1200 m^3/d
Leitner (1992)	MSF	$60.5x10^6$	1598	37,850 m^3/d
Wade (1993)	MSF	$72.6x10^6$	2269	32,000 m^3/d
Morin (1993)	MSF	$76.817x10^6$	1690	45460 m^3/d
Leitner (1992)	MEE	$70.4x10^6$	1860	37,850 m^3/d
Wade (1993)	MEE	$67.2x10^6$	2100	32,000 m^3/d
Morin (1993)	MEE	$35.05x10^6$	1562	22730 m^3/d
Morin (1993)	MEE-TVC	$34.65x10^6$	1524	22730 m^3/d
Hammond et al. (1994)	MEE Vertical Stack	$187.1x10^6$	548	340956 m^3/d

Table 4
Energy cost for various desalination processes

Reference	Process	\$/yr	\$/m^3	Remarks
Matz and Fisher (1981)	RO	1.71x10^6	0.52	1000 m^3/d
Leitner (1992)	RO	4.3x10^6	0.35	37850 m^3/d
Wade (1993)	RO	6.261x10^6	063	32000 m^3/d
Zimmerman (1994)	RO	1.971x10^6	0.3	20000 m^3/d
Veza (1995)	MVC	1.68x10^5	1.057	500 m^3/d
Zimmerman (1994)	MVC	2.69x10^6	0.41	20000 m^3/d
Matz and Fisher (1981)	MVC	3.42x10^5	0.52	2000 m^3/d
DeGunzbourg and Larger (1998)	MEE-ABS	5.21x10^5	0.165	9600 m^3/d
Morin (1993)	MEE-TVC	5.658x10^6	0.758	22730 m^3/d
Hammond et al. (1994)	MEE Vertical Stack	13.65x10^6	0.129	340956 m^3/d
Leitner (1992)	MEE	1x10^6	0.08	37850 m^3/d
Morin (1993)	MEE	3.719x10^6	0.49	22730 m^3/d
Wade (1993)	MEE	1.2059x10^7	1.147	32000 m^3/d
Leitner (1992)	MSF	4.3x10^6	0.35	37850 m^3/d
Wade (1993)	MSF	1.1539x10^7	1.098	32000 m^3/d
Morin (1993)	MSF	1.2453x10^7	0.88	45461 m^3/d

Table 5
Chemicals cost for various desalination processes

Reference	Process	$/m^3$	Remarks
Matz and Fisher (1981)	RO	0.11	1000 m^3/d
Darwish et al. (1990)	RO	0.35	100 m^3/d Chloronation Acid Treatment Caustic soda Dechloronation
Leitner (1992)	RO	0.07	37850 m^3/d
Wade (1993)	RO	0.33	32000 m^3/d
Zimmerman (1994)	RO	0.07	20000 m^3/d
Zimmerman (1994)	MVC	0.05	20000 m^3/d
Darwish et al. (1990)	MVC	0.025	100 m^3/d Chlorination Antiscalent
Matz and Fisher (1981)	MVC	0.02	1000 m^3/d
DeGunzbourg and Larger (1998)	MEE-ABS	0.08	9600 m^3/d
Hammond et al. (1994)	MEE Vertical Stack	0.04	340956 m^3/d
Leitner (1992)	MEE	0.024	37850 m^3/d
Morin (1993)	MEE	0.0606	22730 m^3/d
Wade (1993)	MEE	0.207	32000 m^3/d
Leitner (1992)	MSF	0.024	37850 m^3/d
Wade (1993)	MSF	0.207	32000 m^3/d
Morin (1993)	MSF	0.058	45461 m^3/d

Table 6
Membrane replacement cost

Reference	$/yr	$/m^3$	Capacity
Matz and Fisher (1981)	2.673×10^5	0.81	1000 m^3/d
Darwish et al. (1990)	5600	0.17	100 m^3/d
Wade (1993)	2.722×10^6	0.27	32000 m^3/d
Leitner (1992)	25000	0.19	400 m^3/d
Leitner (1992)	1.9×10^6	0.153	37850 m^3/d

10.4 Case Studies

The following sections include four case studies for the major desalination processes. All calculations are based on recent economic data extracted from the field data and design studies in the literature. Common assumptions in these calculations include the following:

- Plant life (n) = 30 years
- Electric cost (c) = $0.05/m^3
- Average latent heat of heating steam (λ) = 2200 kJ/kg
- Steam heating cost (s) = $1.466/MkJ
- Performance ratio (PR) of MSF and MEE = 8 kg product/kg steam
- Specific cost of operating labor (ℓ) = $0.1/m^3
- Interest rate (i) = 5%
- Plant availability (f) = 0.9

10.4.1 MSF Product Cost

The MSF cost data includes the following:

- Direct capital cost (DC) = $64x10^6
- Plant capacity (m) = 32731 m^3/d
- Specific consumption of electric power (w) = 5 kWh/m^3
- Specific chemicals cost (k) = $0.025/m^3

The calculations proceed as follows:

- Calculate the amortization factor

$$a = \frac{i(1+i)^n}{(1+i)^n - 1} = \frac{0.05(1+0.05)^{30}}{(1+0.05)^{30} - 1} = 0.065051 \text{ yr}^{-1}$$

- Calculate the annual fixed charges

$$A_1 = (a)\,(DC) = (0.065051)(64\text{x}10^6) = \$4.163264\text{x}10^6/\text{yr}$$

- Calculate the annual heating steam cost

$$\begin{aligned} A_2 &= (s)\,(\lambda)\,(f)\,(m)\,(365)/((1000)(PR)) \\ &= (1.466)\,(2200)\,(0.9)\,(32731)\,(365)/((1000)(8)) \\ &= \$4.334723\text{x}10^6/\text{yr} \end{aligned}$$

– Calculate the annual electric power cost

$$A_3 = (c)\ (w)\ (f)\ (m)\ (365)$$
$$= (0.05)\ (5)\ (0.9)\ (32731)\ (365)$$
$$= \$2.688033\text{x}10^6/\text{yr}$$

– Calculate the annual chemicals cost

$$A_4 = (k)\ (f)\ (m)\ (365)$$
$$= (0.025)\ (0.9)\ (32731)\ (365)$$
$$= \$2.68803\text{x}10^5/\text{yr}$$

– Calculate the annual labor cost

$$A_5 = (\ell)\ (f)\ (m)\ (365)$$
$$= (0.1)\ (0.9)\ (32731)\ (365)$$
$$= \$1.075213\text{x}10^6/\text{yr}$$

– Calculate total annual cost

$$A_t = A_1 + A_2 + A_3 + A_4 + A_5$$
$$= 4.334723\text{x}10^6 + 4.334723\text{x}10^6 + 2.688033\text{x}10^6 + 2.68803\text{x}10^5 + 1.075213\text{x}10^6$$
$$= \$12.530065\text{x}10^6/\text{yr}$$

– Calculate unit product cost

$$A_s = A_t\ /((f)(m)(365))$$
$$= (12.530065\text{x}10^6)/((0.9)(365)(32731))$$
$$= \$1.165/\text{m}^3$$

– Calculate unit product cost

$$A_s = A_t\ /((f)(m)(365))$$
$$= (12.530065\text{x}10^6)/(32731)$$
$$= \$383/(\text{m}^3/\text{d})$$

Table 7
Summary of annual cost data of the MSF process

	\$/yr	\$/m^3	\$/(m^3/d)
Fixed Charges	4163291.845	0.387	127.2
Steam	4334722.621	0.4036	132.4
Electric Power	2688033.375	0.05	82.1
Chemicals	268803.3375	0.025	8.21
Operating Labor	1075213.35	0.1	32.85
Total	12530064.53	1.165	383

10.4.2 MEE Product Cost

The MEE cost data includes the following:

- Direct capital cost (DC) = $\$20x10^6$
- Plant capacity (m) = 12,000 m^3/d
- Specific consumption of electric power (w) = 3 kWh/m^3
- Specific chemicals cost (k) = \$0.025/m^3

The calculations proceed as follows:

- Calculate the amortization factor

$$a = \frac{i(1+i)^n}{(1+i)^n - 1} = \frac{0.05(1+0.05)^{30}}{(1+0.05)^{30} - 1} = 0.065051 \text{ yr}^{-1}$$

- Calculate the annual fixed charges

A_1 = (a) (DC) = (0.065051)(22x10^6) = \$1,301,029/yr

- Calculate the annual heating steam cost

A_2 = (s) (λ) (f) (m) (365)/((1000)(PR))
= (1.466) (2200) (0.9) (12,000) (365)/((1000)(8))
= \$1,589,217.3/yr

- Calculate the annual electric power cost

A_3 = (c) (w) (f) (m) (365)
= (0.05) (3) (0.9) (12,000) (365)

$= \$591{,}300/yr$

– Calculate the annual chemicals cost

$$A_4 = (k)\,(f)\,(m)\,(365)$$
$$= (0.025)\,(0.9)\,(12{,}000)\,(365)$$
$$= \$98{,}550/yr$$

– Calculate the annual labor cost

$$A_5 = (\ell)\,(f)\,(m)\,(365)$$
$$= (0.1)\,(0.9)\,(12{,}000)\,(365)$$
$$= \$394{,}200/yr$$

– Calculate total annual cost

$$A_t = A_1 + A_2 + A_3 + A_4 + A_5$$
$$= 1{,}301{,}029 + 1{,}589{,}217.3 + 985{,}500 + 98{,}550 + 394{,}200$$
$$= \$3{,}974{,}296/yr$$

– Calculate unit product cost

$$A_s = A_t\,/((f)(m)(365))$$
$$= (3{,}974{,}296)/((0.9)(365)(12{,}000))$$
$$= \$1.008/m^3$$

– Calculate unit product cost

$$A_s = A_t\,/((f)(m)(365))$$
$$= (3{,}974{,}296)/(12{,}000)$$
$$= \$331.2/(m^3/d)$$

Table 8
Summary of annual cost data of the MSF process

	\$/yr	\$/m^3	\$/(m^3/d)
Fixed Charges	1301029	0.330	108.4
Steam	1589217	0.403	132.4
Electric Power	985500	0.250	82.1
Chemicals	98550	0.025	8.2
Operating Labor	394200	0.100	32.9
Total	4368496	1.108	364.0

10.4.3 MVC Product Cost

The MVC cost data includes the following:

- Direct capital cost (DC) = \$400,000
- Plant capacity (m) = 3000 m^3/d
- Specific consumption of electric power (w) = 7 kWh/m^3
- Specific chemicals cost (k) = \$0.025/m^3

The calculations proceed as follows:

- Calculate the amortization factor

$$a = \frac{i(1+i)^n}{(1+i)^n - 1} = \frac{0.05(1+0.05)^{30}}{(1+0.05)^{30} - 1} = 0.065051 \text{ yr}^{-1}$$

- Calculate the annual fixed charges

A_1 = (a) (DC) = (0.065051)(400,000) = \$26,020/yr

- Calculate the annual electric power cost

A_3 = (c) (w) (f) (m) (365)
= (0.05) (7) (0.9) (3000) (365)
= \$344,925/yr

- Calculate the annual chemicals cost

A_4 = (k) (f) (m) (365)
= (0.025) (0.9) (3000) (365)
= \$24,638/yr

- Calculate the annual labor cost

A_5 = (ℓ) (f) (m) (365)
= (0.1) (0.9) (3000) (365)
= \$98,550/yr

- Calculate total annual cost

$A_t = A_1 + A_3 + A_4 + A_5$

$= 26{,}020 + 344{,}925 + 24{,}638 + 98{,}550$
$= \$494{,}133/yr$

- Calculate unit product cost

$A_s = A_t /((f)(m)(365))$
$= (494{,}133)/((0.9)(365)(3000))$
$= \$0.501/m^3$

- Calculate unit product cost

$A_s = A_t /((f)(m)(365))$
$= (494{,}133)/(3000)$
$= \$165/(m^3/d)$

Table 9
Summary of annual cost data of the MVC process

	\$	\$/m^3	\$/(m^3/d)
Fixed Charges	26,020	0.026	8.7
Electric Power	34,4925	0.35	115
Chemicals	24,638	0.025	8.2
Operating Labor	98,550	0.1	33
Total	494,133	0.501	165

10.4.4RO Product Cost

The RO cost data includes the following:

- Direct capital cost (DC) = $\$98x10^6$
- Membrane purchase cost (@60%DC) = $\$58.8x10^6$
- Membrane annual replacement cost (@10% of membrane purchase cost) = $\$5.88x10^5$
- Plant capacity (m) = 94,625 m^3/d
- Electric cost (c) = \$0.04/m^3
- Specific consumption of electric power (w) = 5 kWh/m^3
- Specific cost of operating labor (ℓ) = \$0.05/m^3
- Specific chemicals cost (k) = \$0.033/m^3

The calculations proceed as follows:

- Calculate the amortization factor

$$a = \frac{i(1+i)^n}{(1+i)^n - 1} = \frac{0.05(1+0.05)^{30}}{(1+0.05)^{30} - 1} = 0.065051 \text{ yr}^{-1}$$

– Calculate the annual fixed charges

$A_1 = (a)\ (DC) = (0.065051)(98x10^6) = \$\ 6,375,041/yr$

– Calculate the annual electric power cost

$A_2 = (c)\ (w)\ (f)\ (m)\ (365)$
$= (0.04)\ (5)\ (0.9)\ (94,625)\ (365)$
$= \$6216863/yr$

– Calculate the annual chemicals cost

$A_3 = (k)\ (f)\ (m)\ (365)$
$= (0.033)\ (0.9)\ (94,625)\ (365)$
$= \$1,039,129/yr$

– Calculate the annual membrane replacement cost

$A_4 = \$\ 5,880,000/yr$

– Calculate the annual labor cost

$A_5 = (\ell)\ (f)\ (m)\ (365)$
$= (0.05)\ (0.9)\ (94,625)\ (365)$
$= \$\ 1,554,216/yr$

– Calculate total annual cost

$A_t = A_1 + A_2 + A_3 + A_4 + A_5$
$= 6,375,041 + 6,216,863 + 1,039,129 + 5,880,000 + 1,554,216$
$= \$21,065,248/yr$

– Calculate unit product cost

$A_s = A_t /((f)(m)(365))$
$= (21,065,248)/((0.9)(365)(94,625))$
$= \$0.678/m^3$

– Calculate unit product cost

$$A_s = A_t /((f)(m)(365))$$
$$= (21{,}065{,}248)/(94{,}625)$$
$$= \$223.1/(m^3/d)$$

Table 10
Summary of annual cost data of the RO process

	$	$/m^3$	$/(m^3/d)$
Fixed Charges	6,375,041	0.205	67.4
Electric Power	6,216,863	0.200	65.7
Chemicals	1,039,129	0.033	11.0
Membrane Replacement	5,880,000	0.189	62.1
Operating Labor	1,554,216	0.050	16.4
Total	21,065,248	0.678	222.6

10.5 Summary

Economics of thermal and membrane desalination processes are evaluated in the light of field data and design studies. The analysis is presented for the major desalination processes, which includes MSF, MEE, MVC, and RO. In light of the above evaluation the following conclusions are made:

- Increase in plant capacity although increases the capital cost, but, it has a drastic effect on the product unit cost. This is illustrated in several examples for various desalination processes.
- A good estimate of the unit product cost of the RO process would have equal shares for the fixed charges, power, and membrane replacement. Inspection of data and results show drastic reduction in the chemicals and labor cost.
- Each of the fixed charges and the energy cost in the MSF and MEE processes represent 40-50% of the total unit product cost.
- Today, the RO process represents the optimum choice for desalination of low salinity water; this is irrespective of the plant capacity. The best example is the RO plant in Florida with a capacity of 94625 m^3/d and feed salinity of less than 30,000 ppm.
- The optimum choices for desalination of higher salinity depend on the required capacity, where the MSF process would be optimum for capacities higher than 25,000 m^3/d, the MEE process for capacities averaging 10,000 m^3/d, and the MVC process for capacities averaging 3000 m^3/d.

References

Darwish, M.A., and Al-Najem, N.M., Energy consumptions and costs of different desalting systems, Desalination, **64**(1987)83-96.

Darwish, M.A., Abdel Jawad, M., and Aly, G.S., Comparison between small capacity mechanical vapor compression (MVC) and reverse osmosis (RO) desalting plants, Desalination, **78**(1990)313-326.

deGunzbourg, J., and Larger, D., Cogeneration applied to very high efficiency thermal seawater desalination plants: A concept, Desalination & Water Reuse, **7**(1998)38-41.

Ettouney, H.M., El-Dessouky, H.T., Faibish, R., and Gowin, P., Economics of thermal and membrane desalination processes, Middle East Energy 2001, Dubai, UAE, Oct. 8-10, 2001.

Hammond, R.P., Eissenbergy, D.M., Emmermann, D.K., Jones, J.E., Sephton, H.H., Standiford, F.C., Scott, R.F., Rider, W.J., Dean, D.W., Seawater desalination plant for southern California, Desalination, **99**(1994)459-481.

Leitner, G., Water desalination what are today's costs, Desalination & Water Reuse, **2**(1992)39-43.

Leitner, G., Editor's corner, Desalination & Water Reuse, **8**(1998)2.

Leitner, G., Developer selected for 25 MGD (94625 m^3/d) Florida west coast seawater desalting plant, Desalination & Water Reuse, **9**(1999)11-16.

Matz, R., and Fisher, U., A comparison of the relative economics of sea water desalination by vapour compression and reverse osmosis for small to medium capacity plants, Desalination, **36**(1981)137-151.

Morin, O.J., Design and operating comparison of MSF and MED systems, Desalination, **93**(1993)69-109.

Morin, O.J., Desalting Plant Cost Update: 2000, Proceedings of the IDA world congress on desalination and water reuse, Sand Diego, USA, Vol., III, pp. 341-359, 1999

Pappas, C., For Cyprus 10 year "BOOT" contract, Desalination & Water Reuse, **7**(1997)28-34

Veza, J.M., Mechanical vapour compression desalination plants: A case study, Desalination **101**(1995)1-10.

Wade, N.M., Technical and economic evaluation of distillation and reverse osmosis desalination processes, Desalination, **93**(1993)343-363.

Zimerman, Development of large capacity high efficiency mechanical vapor compression (MVC) units, Desalination, **96**(1994)51-58.

Problems

1. Determine the unit product cost for 54,000 m^3/d MSF unit using the following economic data:
   - Plant capital = \$145,000,000
   - Electric power cost = \$0.05/m^3.
   - Annual interest rate = 7%.
   - Economic life = 30 years.
   - Specific electric power consumption = 4.2 kWh/m^3.
   - Heating steam cost = \$1.5/MkJ.
   - Latent heat of heating steam = 2200 kJ/kg.
   - Plant factor = 0.9
   - Chemicals cost = \$0.019/m^3.
   - Labor cost = \$0.095/m^3.
   - Plant performance ratio = 8.4
2. If the unit product cost for a 27,000 m^3/d MSF unit must be below \$1.34/m^3, determine the plant capital using the following economic data:
   - Electric power cost = \$0.07/m^3.
   - Annual interest rate = 7%.
   - Economic life = 30 years.
   - Specific electric power consumption = 4.5 kWh/m^3.
   - Heating steam cost = \$1.5/MkJ.
   - Latent heat of heating steam = 2200 kJ/kg.
   - Plant factor = 0.9
   - Chemicals cost = \$0.025/m^3.
   - Labor cost = \$0.1/m^3.
   - Plant performance ratio = 8.6
3. If the plant capital in problem 3 is set at \$75,000,000. Determine the unit product cost for plant factors of 0.5, 0.6, 0.7, and 0.8.
4. Determine the unit product cost for an MEE plant with a production capacity of 4,000 m^3/d:
   - Plant capital = \$12,000,000
   - Electric power cost = \$0.07/m^3.
   - Annual interest rate = 7%.

- Economic life = 30 years.
- Specific electric power consumption = 4.5 kWh/m^3.
- Heating steam cost = \$1.5/MkJ.
- Latent heat of heating steam = 2200 kJ/kg.
- Plant factor = 0.9
- Chemicals cost = \$0.025/m^3.
- Labor cost = \$0.1/m^3.
- Plant performance ratio = 8.

5. Repeat problem 2, if the plant capacity is increased to 12,000 m^3/d with a total capital of \$28,000,000.
6. Determine the unit product cost for an RO plant with a production capacity of 10,000 m^3/d:
   - Plant capital = \$12,000,000
   - Electric power cost = \$0.07/m^3.
   - Annual interest rate = 7%.
   - Economic life = 20 years.
   - Specific electric power consumption = 4.5 kWh/m^3.
   - Plant factor = 0.9
   - Chemicals cost = \$0.035/m^3.
   - Labor cost = \$0.08/m^3.
   - Annual membrane replacement cost = \$1,700,000
7. Determine the unit product cost for an RO plant with a production capacity of 50,000 m^3/d:
   - Plant capital = \$53,000,000
   - Electric power cost = \$0.07/m^3.
   - Annual interest rate = 7%.
   - Economic life = 20 years.
   - Specific electric power consumption = 4.5 kWh/m^3.
   - Plant factor = 0.9
   - Chemicals cost = \$0.035/m^3.
   - Labor cost = \$0.07/m^3.
   - Annual membrane replacement cost = \$5,000,000
8. Repeat problem 7 for plant factors of 0.6, 0.7, and 0.8.

Appendix A

Thermodynamic Properties

A.1. *Seawater Density*

The density correlation for seawater is given by

$$\rho = 10^3 (A_1 F_1 + A_2 F_2 + A_3 F_3 + A_4 F_4) \quad (A.1)$$

where

$B = ((2)(X)/1000-150)/150$

$G_1 = 0.5$

$G_2 = B$

$G_3 = 2 B^2 - 1$

$A_1 = 4.032219 G_1 + 0.115313 G_2 + 3.26x10^{-4} G_3$

$A_2 = - 0.108199 G_1 + 1.571x10^{-3} G_2 - 4.23x10^{-4} G_3$

$A_3 = - 0.012247 G_1 + 1.74x10^{-3} G_2 - 9x10^{-6} G_3$

$A_4 = 6.92x10^{-4} G_1 - 8.7x10^{-5} G_2 - 5.3x10^{-5} G_3$

$A = ((2)(T) - 200)/160$

$F_1 = 0.5,\ F_2 = A,\ F_3 = 2 A^2 - 1,\ F_4 = 4 A^3 - 3 A$

In the above equations ρ is the seawater density in kg/m^3, X is is the seawater salinity in ppm, and T is the seawater temperature in °C. This correlation is valid over the following ranges: $0 \le X \le 160000$ ppm and $10 \le T \le 180$ °C. Variations in the seawater density as a function of temperature and salinity are given in Table A.1 and Fig. A.1.

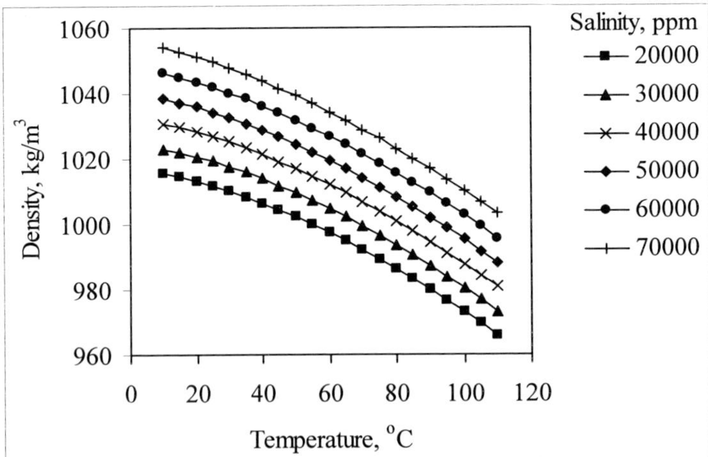

Figure A.1: Variation in density of seawater as a function of temperature and salinity.

Table A.1: Variation in seawater density (kg/m^3) as a function of temperature (°C) and salinity (ppm)

	Salinity ppm						
T (°C)	10000	20000	30000	40000	50000	60000	70000
10	1008	1015	1023	1031	1038	1046	1054
15	1007	1014	1022	1030	1037	1045	1053
20	1006	1013	1021	1028	1036	1044	1051
25	1004	1012	1019	1027	1034	1042	1050
30	1003	1010	1018	1025	1033	1040	1048
35	1001	1008	1016	1023	1031	1038	1046
40	999	1007	1014	1021	1029	1036	1044
45	997	1004	1012	1019	1027	1034	1042
50	995	1002	1010	1017	1024	1032	1039
55	993	999.9	1007	1015	1022	1029	1037
60	990	997.5	1005	1012	1020	1027	1034
65	988	994.9	1002	1010	1017	1024	1032
70	985	992.2	999.5	1007	1014	1022	1029
75	982	989.3	996.6	1004	1011	1019	1026
80	979	986.3	993.7	1001	1008	1016	1023
85	976	983.2	990.6	997.9	1005	1013	1020
90	973	980	987.4	994.7	1002	1010	1017
95	969	976.7	984	991.4	998.8	1006	1014
100	966	973.2	980.6	988	995.4	1003	1010
105	962	969.6	977	984.4	991.9	999.3	1007
110	958	965.9	973.3	980.8	988.3	995.7	1003

A.2. Seawater Specific Heat at Constant Pressure

The seawater specific heat at constant pressure is given by the following correlation

$$C_p = (A + BT + CT^2 + DT^3) \times 10^{-3} \quad \text{(A.2)}$$

The variables A, B, C and D are evaluated as a function of the water salinity as follows:

$$A = 4206.8 - 6.6197\ s + 1.2288 \times 10^{-2}\ s^2$$
$$B = -1.1262 + 5.4178 \times 10^{-2}\ s - 2.2719 \times 10^{-4}\ s^2$$
$$C = 1.2026 \times 10^{-2} - 5.3566 \times 10^{-4}\ s + 1.8906 \times 10^{-6}\ s^2$$
$$D = 6.8777 \times 10^{-7} + 1.517 \times 10^{-6}\ s - 4.4268 \times 10^{-9}\ s^2$$

where Cp in kJ/kg °C, T in °C, and s is the water salinity in gm/kg. The above correlation is valid over salinity and temperature ranges of $20000 \leq X \leq 160000$ ppm and $20 \leq T \leq 180$ °C, respectively. Variations in the seawater specific heat as a function of temperature and salinity are given in Table A.2 and Fig. A.2.

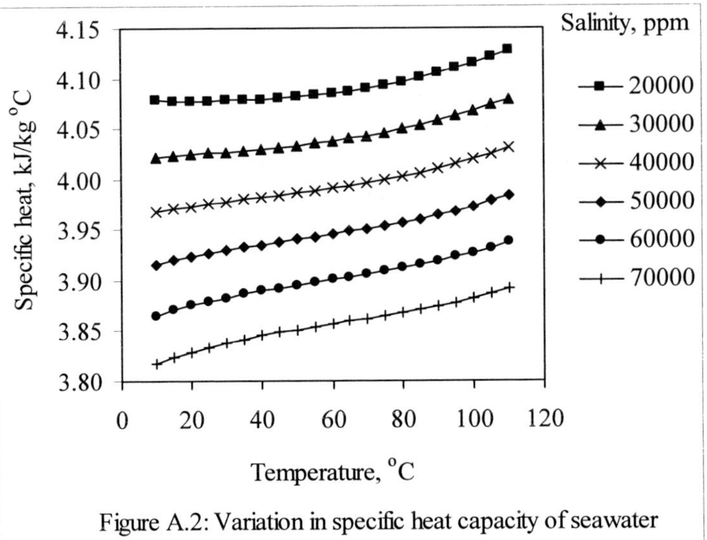

Figure A.2: Variation in specific heat capacity of seawater as a function of temperature and salinity.

Table A.2: Variation in seawater specific heat (kJ/kg °C) as a function of temperature (°C) and salinity (ppm)

	Salinity ppm						
T (°C)	10000	20000	30000	40000	50000	60000	70000
10	4.14	4.078	4.022	3.968	3.916	3.866	3.818
15	4.13	4.078	4.023	3.971	3.92	3.871	3.824
20	4.13	4.078	4.025	3.973	3.923	3.875	3.829
25	4.13	4.078	4.026	3.976	3.927	3.879	3.834
30	4.13	4.078	4.027	3.978	3.93	3.883	3.838
35	4.13	4.078	4.029	3.98	3.933	3.887	3.842
40	4.13	4.079	4.03	3.982	3.935	3.89	3.845
45	4.13	4.08	4.032	3.984	3.938	3.893	3.849
50	4.13	4.082	4.033	3.986	3.94	3.895	3.851
55	4.13	4.083	4.035	3.989	3.943	3.898	3.854
60	4.13	4.085	4.038	3.991	3.945	3.901	3.857
65	4.14	4.087	4.04	3.994	3.948	3.903	3.86
70	4.14	4.09	4.043	3.997	3.951	3.906	3.862
75	4.14	4.093	4.046	4	3.954	3.909	3.865
80	4.15	4.097	4.05	4.003	3.957	3.912	3.868
85	4.15	4.101	4.053	4.007	3.961	3.915	3.871
90	4.15	4.105	4.058	4.011	3.964	3.919	3.874
95	4.16	4.11	4.062	4.015	3.969	3.923	3.878
100	4.17	4.116	4.068	4.02	3.973	3.927	3.882
105	4.17	4.122	4.073	4.025	3.978	3.932	3.887
110	4.18	4.129	4.08	4.031	3.984	3.937	3.892

A.3. Seawater Dynamic Viscosity

The correlation for the dynamic viscosity of seawater is given by

$$\mu = (\mu_W)\,(\mu_R) \times 10^{-3} \tag{A.3}$$

with

$$Ln(\mu_W) = -\,3.79418 + 604.129/(139.18+T)$$

$$\mu_R = 1 + A\,s + B\,s^2$$

$$A = 1.474\text{x}10^{-3} + 1.5\text{x}10^{-5}\,T - 3.927\text{x}10^{-8}\,T^2$$

$$B = 1.0734\text{x}10^{-5} - 8.5\text{x}10^{-8}\,T + 2.23\text{x}10^{-10}\,T^2$$

where μ in kg/m s, T in ^{o}C, and s in gm/kg. The above correlation is valid over the following ranges $0 \leq s \leq 130$ gm/kg and $10 \leq T \leq 180$ ^{o}C. Variations in the seawater viscosity as a function of temperature and salinity are given in Table A.3 and Fig. A.3.

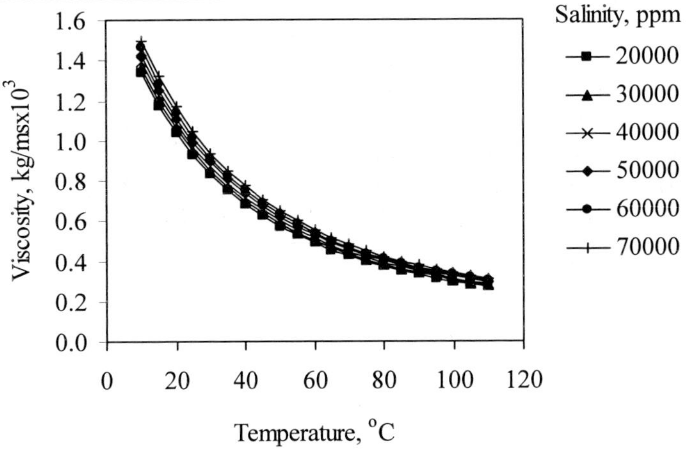

Figure A.3: Variation in viscosity of seawater as a function of temperature and salinity.

Table A.3: Variation in seawater viscosity (kg/m s) as a function of temperature (°C) and salinity (ppm)

	Salinity ppm						
T (°C)	10000	20000	30000	40000	50000	60000	70000
10	1.31	1.338	1.365	1.395	1.428	1.463	1.5
15	1.15	1.175	1.199	1.226	1.255	1.286	1.319
20	1.02	1.04	1.062	1.086	1.112	1.14	1.169
25	0.91	0.928	0.948	0.969	0.993	1.018	1.044
30	0.82	0.833	0.851	0.871	0.892	0.915	0.939
35	0.74	0.753	0.77	0.788	0.807	0.827	0.849
40	0.67	0.684	0.7	0.716	0.734	0.753	0.772
45	0.61	0.625	0.639	0.655	0.671	0.688	0.706
50	0.56	0.573	0.587	0.601	0.616	0.632	0.649
55	0.52	0.529	0.541	0.555	0.569	0.584	0.599
60	0.48	0.489	0.501	0.514	0.527	0.541	0.555
65	0.44	0.455	0.466	0.478	0.49	0.503	0.516
70	0.41	0.424	0.435	0.446	0.457	0.469	0.482
75	0.39	0.397	0.407	0.417	0.428	0.439	0.451
80	0.36	0.372	0.382	0.392	0.402	0.413	0.424
85	0.34	0.35	0.359	0.369	0.379	0.389	0.399
90	0.32	0.33	0.339	0.348	0.357	0.367	0.377
95	0.3	0.313	0.321	0.329	0.338	0.347	0.357
100	0.29	0.296	0.304	0.312	0.321	0.33	0.338
105	0.27	0.282	0.289	0.297	0.305	0.313	0.322
110	0.26	0.268	0.275	0.283	0.291	0.298	0.307

A.4. Seawater Thermal Conductivity

The seawater thermal conductivity is given by

$$\mathrm{Log}_{10}(k) = \mathrm{Log}_{10}(240 + A\ s) + 0.434\left(2.3 - \frac{343.5 + B\ s}{T + 273.15}\right)\left(1 - \frac{T + 273.15}{647.3 + C\ s}\right)^{1/3} \qquad (A.4)$$

where k is the thermal conductivity in W/m °C, s is the salinity in gm/kg, T is the temperature in °C. The constants A, B, and C are equal to 2×10^{-4}, 3.7×10^{-2}, and 3×10^{-2}, respectively. The above correlation valid over the following ranges, $0 \le s \le 160$ gm/kg and $20 \le T \le 180$ °C. Variations in the seawater thermal conductivity as a function of temperature and salinity are given in Table A.4 and Fig. A.4.

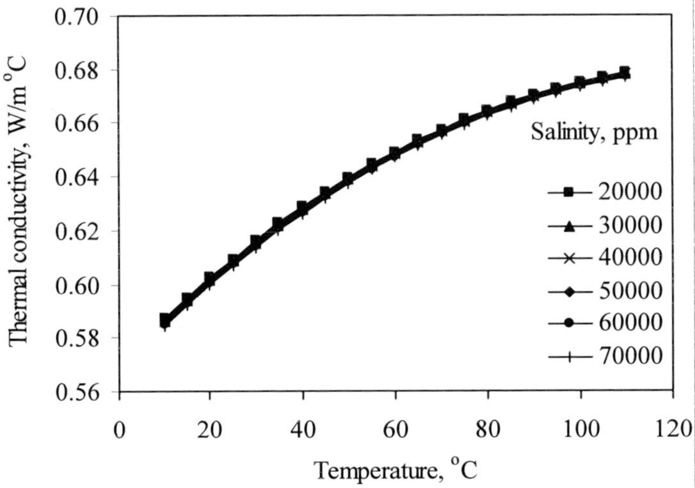

Figure A.4: Variation in thermal conductivity of seawater as a function of temperature and salinity.

Table A.4: Variation in seawater thermal conductivity (kW/m°C) as a function of temperature (°C) and salinity (gm/kg)

	Salinity (gm/kg)						
T (°C)	10	20	30	40	50	60	70
10	0.5877	0.5872	0.5866	0.586	0.5855	0.5849	0.5844
15	0.5955	0.595	0.5944	0.5939	0.5933	0.5928	0.5922
20	0.603	0.6024	0.6019	0.6013	0.6008	0.6003	0.5997
25	0.61	0.6095	0.609	0.6084	0.6079	0.6074	0.6068
30	0.6168	0.6162	0.6157	0.6152	0.6147	0.6141	0.6136
35	0.6231	0.6226	0.6221	0.6216	0.621	0.6205	0.62
40	0.6291	0.6286	0.6281	0.6276	0.6271	0.6266	0.6261
45	0.6347	0.6343	0.6338	0.6333	0.6328	0.6323	0.6318
50	0.6401	0.6396	0.6391	0.6386	0.6381	0.6377	0.6372
55	0.645	0.6446	0.6441	0.6436	0.6432	0.6427	0.6422
60	0.6497	0.6492	0.6488	0.6483	0.6478	0.6474	0.6469
65	0.654	0.6535	0.6531	0.6527	0.6522	0.6518	0.6513
70	0.658	0.6575	0.6571	0.6567	0.6563	0.6558	0.6554
75	0.6616	0.6612	0.6608	0.6604	0.66	0.6596	0.6591
80	0.665	0.6646	0.6642	0.6638	0.6634	0.663	0.6626
85	0.6681	0.6677	0.6673	0.6669	0.6665	0.6661	0.6657
90	0.6708	0.6704	0.6701	0.6697	0.6693	0.6689	0.6686
95	0.6733	0.6729	0.6725	0.6722	0.6718	0.6715	0.6711
100	0.6754	0.6751	0.6747	0.6744	0.674	0.6737	0.6733
105	0.6773	0.677	0.6766	0.6763	0.676	0.6756	0.6753
110	0.6789	0.6786	0.6783	0.6779	0.6776	0.6773	0.677

A.5. Enthalpy of Saturated Liquid Water

The correlation for enthalpy of saturated liquid water is given by

$$H = -0.033635409 + 4.207557011\,T - 6.200339\text{x}10^{-4}\,T^2 + 4.459374\text{x}10^{-6}\,T^3 \quad (A.5)$$

In the above equation, T is the saturation temperature in °C and H is the enthalpy in kJ/kg. Values for the calculated enthalpy over a temperature range of 5-200 °C are given in Table A.5. The table also includes values extracted from the steam tables. The percentage errors for the calculated versus the steam table values are less than 0.04%. Figure A.5 show variations in the calculated and the steam tables values for the liquid water enthalpy as a function of temperature.

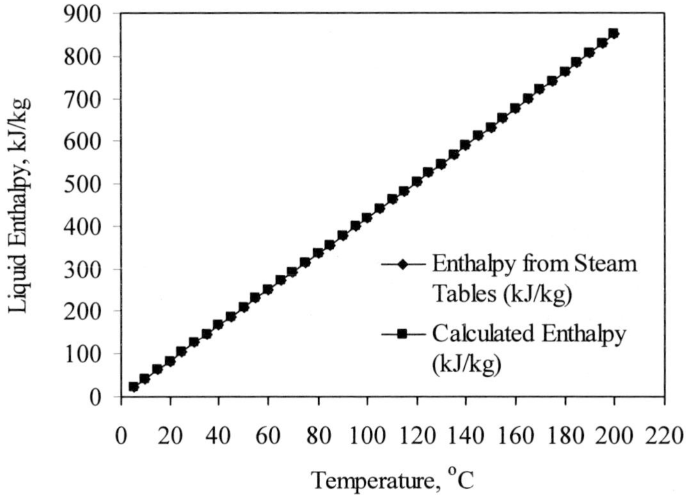

Figure A.5: Variation in enthalpy of liquid water as a function of temperature.

Table A.5: Variation in liquid water enthalpy (kJ/kg) as a function of temperature (°C).

T (°C)	Calculated Enthalpy (kJ/kg)	Enthalpy from Steam Tables (kJ/kg)	Percentage Error
5	20.98921	20.98	0.043881
10	41.98439	41.99	0.013359
15	62.95526	62.98	0.039278
20	83.90517	83.94	0.041498
25	104.8374	104.87	0.031042
30	125.7554	125.77	0.011571
35	146.6625	146.66	0.001714
40	167.562	167.54	0.013126
45	188.4572	188.42	0.019755
50	209.3516	209.31	0.019852
55	230.2483	230.2	0.020993
60	251.1509	251.11	0.016283
65	272.0626	272.03	0.011978
70	292.9868	292.96	0.009133
75	313.9267	313.91	0.005335
80	334.8859	334.88	0.001764
85	355.8676	355.88	0.00349
90	376.8751	376.9	0.006605
95	397.9118	397.94	0.007079
100	418.9811	419.02	0.009283
105	440.0863	440.13	0.009938
110	461.2307	461.27	0.00853
115	482.4176	482.46	0.008783
120	503.6505	503.69	0.007839
125	524.9327	524.96	0.005205
130	546.2674	546.29	0.004128
135	567.6582	567.67	0.002083
140	589.1082	589.11	0.000305
145	610.6209	610.61	0.001782
150	632.1995	632.18	0.003091
155	653.8475	653.82	0.004212
160	675.5682	675.53	0.005657
165	697.3649	697.32	0.006441
170	719.241	719.2	0.005699
175	741.1998	741.16	0.005365
180	763.2446	763.21	0.004534
185	785.3788	785.36	0.002399
190	807.6058	807.61	0.000517
195	829.9289	829.96	0.003747
200	852.3514	852.43	0.00922

A.6. Enthalpy of Saturated Water Vapor

The correlation for the water vapor enthalpy is given by

$$H" = 2501.689845 + 1.806916015\ T + 5.087717x10^{-4}\ T^2 - 1.1221x10^{-5}\ T^3 \qquad (A.6)$$

In the above equation, T is the saturation temperature in °C and H" is the vapor enthalpy in kJ/kg. Values for the calculated enthalpy over a temperature range of 0.01–200 °C are given in Table A.6. The table also includes values extracted from the steam tables. The percentage errors for the calculated versus the steam table values are less than 0.017%. Figure A.6 show variations in the calculated and the steam table values for the enthalpy of water vapor as a function of temperature.

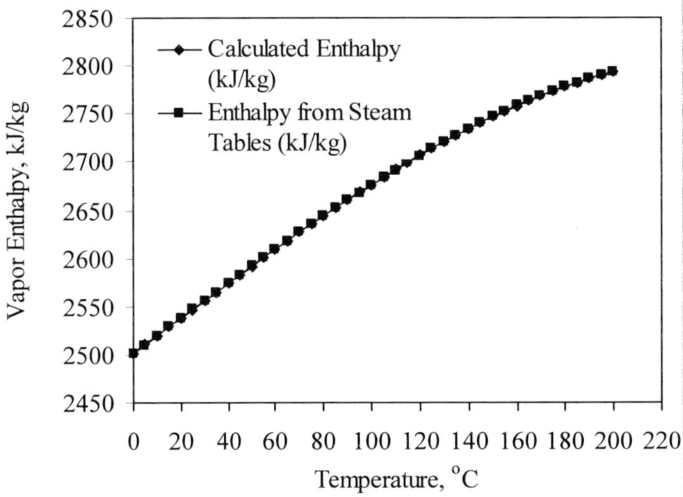

Figure A.6: Variation in enthalpy of water vapor as a function of temperature.

Table A.6: Variation in water vapor enthalpy (kJ/kg) as a function of temperature (°C)

T (°C)	Calculated Enthalpy (kJ/kg)	Enthalpy from Steam Tables (kJ/kg)	Percentage Error
0.01	2501.708	2501.35	0.014309
5	2510.736	2510.54	0.007797
10	2519.799	2519.74	0.002328
15	2528.87	2528.91	0.001574
20	2537.942	2538.06	0.004653
25	2547.005	2547.17	0.006462
30	2556.052	2556.25	0.007736
35	2565.074	2565.28	0.008028
40	2574.062	2574.26	0.007677
45	2583.009	2583.19	0.007014
50	2591.905	2592.06	0.005982
55	2600.742	2600.86	0.004523
60	2609.513	2609.59	0.002964
65	2618.207	2618.24	0.001246
70	2626.818	2626.8	0.000691
75	2635.337	2635.28	0.002146
80	2643.754	2643.66	0.003561
85	2652.063	2651.93	0.004997
90	2660.253	2660.09	0.006137
95	2668.318	2668.13	0.007044
100	2676.248	2676.05	0.007406
105	2684.036	2683.83	0.007659
110	2691.672	2691.47	0.007492
115	2699.148	2698.96	0.006966
120	2706.456	2706.3	0.005773
125	2713.588	2713.46	0.004715
130	2720.535	2720.46	0.002746
135	2727.288	2727.26	0.00103
140	2733.84	2733.87	0.001109
145	2740.181	2740.26	0.002881
150	2746.304	2746.44	0.004958
155	2752.2	2752.39	0.006918
160	2757.86	2758.09	0.008344
165	2763.276	2763.53	0.00918
170	2768.44	2768.7	0.009375
175	2773.344	2773.58	0.008513
180	2777.978	2778.16	0.006543
185	2782.335	2782.43	0.003413
190	2786.406	2786.37	0.001289
195	2790.182	2789.96	0.007972
200	2793.656	2793.18	0.017047

A.7. Latent Heat of Water Evaporation

The correlation for latent heat of water evaporation is given by

$$\lambda = 2501.897149 - 2.407064037\ T + 1.192217\text{x}10^{-3}\ T^2 - 1.5863\text{x}10^{-5}\ T^3 \quad (A.7)$$

In the above equation, T is the saturation temperature in °C and λ is the latent heat in kJ/kg. Values for the calculated enthalpy over a temperature range of 5–200 °C are given in Table A.7. The table also includes values extracted from the steam tables. The percentage errors for the calculated versus the steam table values are less than 0.026%. Figure A.7 show variations in the calculated and the steam table values for the latent heat of water as a function of temperature.

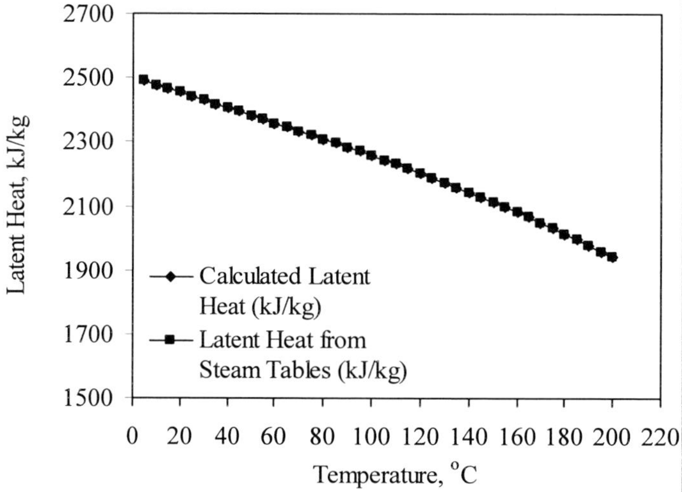

Figure A.7: Variation in latent heat of water evaporation as a function of temperature.

Table A.7: Variation in latent heat of water evaporation in (kJ/kg) as a function of temperature (°C)

T (°C)	Calculated Latent Heat (kJ/kg)	Latent Heat from Steam Tables (kJ/kg)	Percentage Error
5	2489.89	2489.56	0.013241
10	2477.93	2477.75	0.007259
15	2466.006	2465.93	0.003078
20	2454.106	2454.12	0.000577
25	2442.218	2442.3	0.003365
30	2430.33	2430.48	0.006175
35	2418.43	2418.62	0.007845
40	2406.507	2406.72	0.008854
45	2394.548	2394.77	0.009271
50	2382.542	2382.75	0.008746
55	2370.476	2370.66	0.007767
60	2358.339	2358.48	0.005984
65	2346.119	2346.21	0.00389
70	2333.804	2333.84	0.001563
75	2321.381	2321.37	0.000489
80	2308.84	2308.78	0.002614
85	2296.169	2296.05	0.005166
90	2283.354	2283.19	0.007192
95	2270.385	2270.19	0.008602
100	2257.25	2257.03	0.009743
105	2243.936	2243.7	0.010528
110	2230.432	2230.2	0.010415
115	2216.726	2216.5	0.010206
120	2202.806	2202.61	0.008904
125	2188.66	2188.5	0.007316
130	2174.276	2174.17	0.004888
135	2159.643	2159.59	0.002441
140	2144.748	2144.76	0.00058
145	2129.579	2129.65	0.00334
150	2114.125	2114.26	0.006395
155	2098.373	2098.57	0.009369
160	2082.313	2082.56	0.01187
165	2065.931	2066.21	0.013499
170	2049.216	2049.5	0.013838
175	2032.157	2032.42	0.01295
180	2014.74	2014.95	0.010402
185	1996.955	1997.07	0.005742
190	1978.79	1978.76	0.001499
195	1960.232	1960	0.011812
200	1941.269	1940.75	0.026741

A.8. Entropy of Saturated Liquid Water

The correlation for entropy of saturated liquid water is given by

$$S = -0.00057846 + 0.015297489\,T - 2.63129\text{x}10^{-5}\,T^2 + 4.11959\text{x}10^{-8}\,T^3 \qquad (A.8)$$

In the above equation, T is the saturation temperature in °C and S is the entropy of saturated liquid water in kJ/kg °C. Values for the calculated entropy over a temperature range of 5–200 °C are given in Table A.8. The table also includes values extracted from the steam tables. The percentage errors for the calculated versus the steam table values are less than 0.4%. Figure A.8 show variations in the calculated and the steam table values for the saturation entropy of water vapor as a function of temperature.

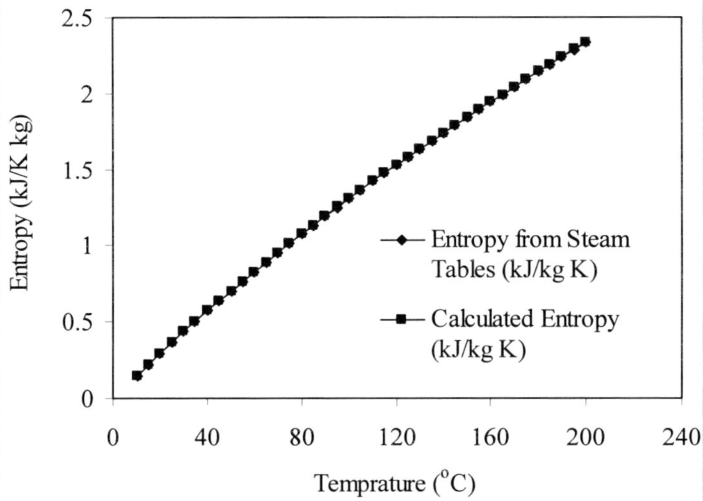

Figure A.8: Variation in entropy of saturated liquid water as a function of temperature

Table A.8: Variation in the entropy of saturated liquid water (kJ/kg °C) as a function of temperature (°C)

T (°C)	Entropy from Steam Table (kJ/kg °C)	Calculated Entropy (kJ/kg °C)	Percentage Error
10	0.149806	0.151	0.790506
15	0.223103	0.2245	0.62249
20	0.295176	0.2966	0.4802
25	0.366057	0.3673	0.338446
30	0.435777	0.4369	0.257064
35	0.504367	0.5052	0.16496
40	0.571857	0.5724	0.094865
45	0.638279	0.6386	0.050283
50	0.703663	0.7037	0.005227
55	0.768041	0.7679	0.018345
60	0.831443	0.8311	0.041239
65	0.8939	0.8934	0.055935
70	0.955443	0.9548	0.067316
75	1.016103	1.0154	0.069199
80	1.07591	1.0752	0.066068
85	1.134897	1.1342	0.061435
90	1.193093	1.1924	0.058103
95	1.250529	1.25	0.042348
100	1.307237	1.3068	0.033462
105	1.363247	1.3629	0.025497
110	1.418591	1.4184	0.013459
115	1.473298	1.4733	0.000109
120	1.527401	1.5275	0.006489
125	1.580929	1.5812	0.017124
130	1.633914	1.6343	0.023596
135	1.686387	1.6869	0.0304
140	1.738379	1.739	0.035735
145	1.789919	1.7906	0.038009
150	1.841041	1.8417	0.035802
155	1.891773	1.8924	0.033127
160	1.942148	1.9426	0.023281
165	1.992195	1.9924	0.010267
170	2.041947	2.0418	0.007203
175	2.091434	2.0909	0.025519
180	2.140686	2.1395	0.055425
185	2.189735	2.1878	0.088431
190	2.238611	2.2358	0.125732
195	2.287346	2.2835	0.168425
200	2.33597	2.3308	0.22182

A.9. Entropy of Saturated Water Vapor

The correlation for entropy of saturated water vapor is given by

$$S = 9.149505306 - 2.581012 \times 10^{-2}\, T + 9.625687 \times 10^{-5}\, T^2 - 1.786615 \times 10^{-7}\, T^3 \qquad (A.9)$$

In the above equation, T is the saturation temperature in °C and S is the entropy of saturated water vapor in kJ/kg K. Values for the calculated entropy over a temperature range of 0.01–200 °C are given in Table A.9. The table also includes values extracted from the steam tables. The percentage errors for the calculated versus the steam table values are less than 0.4%. Figure A.9 show variations in the calculated and the steam table values for the saturation entropy of water vapor as a function of temperature.

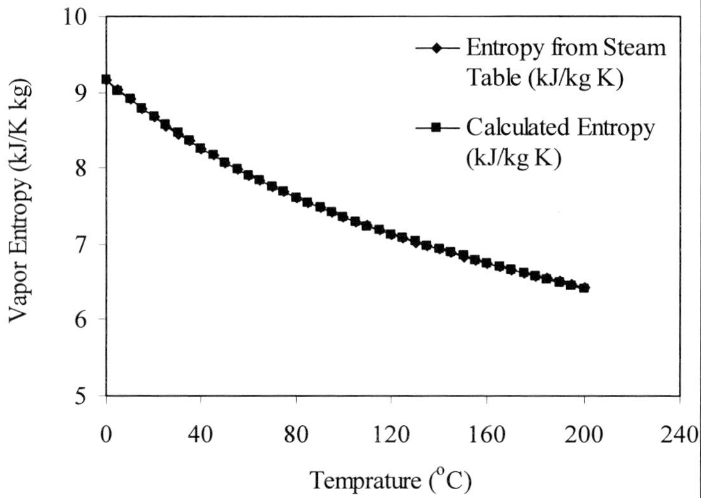

Figure A.9: Variation in entropy of saturated water vapor as a function of temperature

Table A.9: Variation in the entropy of saturated water vapor (kJ/kg K) as a function of temperature (°C)

T (°C)	Entropy from Steam Table (kJ/kg °C)	Calculated Entropy (kJ/kg K)	Percentage Error
0.01	9.1562	9.14925	0.07594
5	9.0257	9.02284	0.03170
10	8.9007	8.90085	0.00170
15	8.7813	8.78341	0.02401
20	8.6671	8.67038	0.03780
25	8.5579	8.56162	0.04348
30	8.4533	8.45701	0.04388
35	8.353	8.35641	0.04077
40	8.2569	8.25968	0.03363
45	8.1647	8.16669	0.02437
50	8.0762	8.07731	0.01373
55	7.9912	7.99140	0.00251
60	7.9095	7.90883	0.00845
65	7.8309	7.82947	0.01829
70	7.7552	7.75317	0.02612
75	7.6824	7.67982	0.03361
80	7.6121	7.60926	0.03725
85	7.5444	7.54138	0.04003
90	7.479	7.47603	0.03970
95	7.4158	7.41308	0.03665
100	7.3548	7.35240	0.03263
105	7.2958	7.29385	0.02671
110	7.2386	7.23730	0.01794
115	7.1832	7.18262	0.00812
120	7.1295	7.12966	0.00228
125	7.0774	7.07831	0.01279
130	7.0269	7.02841	0.02150
135	6.977	6.97985	0.04079
140	6.9298	6.93248	0.03861
145	6.8832	6.88617	0.04309
150	6.8378	6.84078	0.04364
155	6.7934	6.79619	0.04113
160	6.7501	6.75226	0.03206
165	6.7078	6.70886	0.01578
170	6.6663	6.66584	0.00684
175	6.6256	6.62309	0.03794
180	6.5857	6.58045	0.07969
185	6.5464	6.53781	0.13128
190	6.5078	6.49502	0.19645
195	6.4697	6.45195	0.27442
200	6.4322	6.40846	0.36903

A.10. Saturation Pressure of Water Vapor

The correlation for the water vapor saturation pressure is given by

$$\ln(P/P_c) = \left(\frac{T_c}{T+273.15} - 1\right) \sum_{i=1}^{8} f_i \left(0.01(T+273.15-338.15)\right)^{(i-1)} \qquad (A.10)$$

where T_c = 647.286 K and P_c = 22089 kPa and the values of f_i are given in the following tables

f_1	f_2	f_3	f_4
-7.419242	0.29721	-0.1155286	0.008685635
f_5	f_6	f_7	f_8
0.001094098	-0.00439993	0.002520658	-0.000521868

where P is kPa and T is °C. Values for the calculated vapor pressure over a temperature range of 5-200 °C are given in table A.10. The table also includes values extracted from the steam tables. The percentage errors for the calculated versus the steam table values are less than 0.05%. Figure A.10 shows variations in the calculated and the steam table values for the vapor pressure of water as a function of temperature.

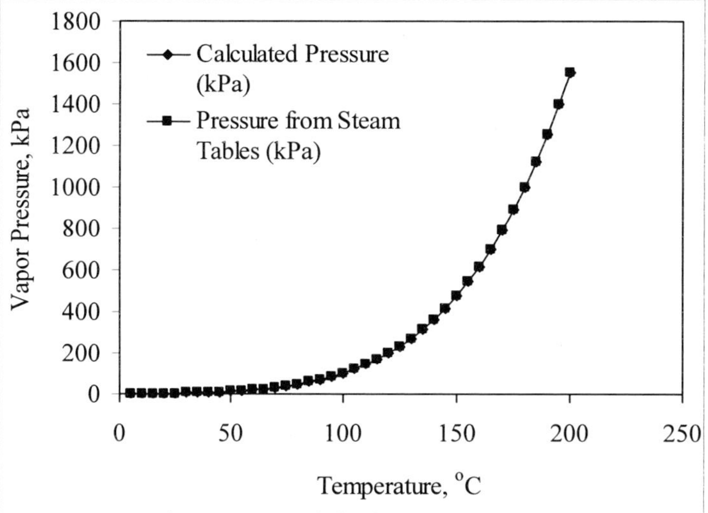

Figure A.10: Variation in saturation pressure of water vapor as a function of temperature.

Table A.10: Variation in saturation pressure of water vapor (kPa) as a function of temperature (°C)

T (°C)	Calculated Pressure (kPa)	Pressure from Steam Tables (kPa)	Percentage Error
5	0.872	0.8721	1.04E-04
10	1.228	1.2276	1.23E-03
15	1.705	1.705	3.76E-03
20	2.339	2.339	2.12E-02
25	3.169	3.169	1.61E-03
30	4.246	4.246	5.81E-04
35	5.628	5.628	8.74E-04
40	7.384	7.384	5.03E-03
45	9.593	9.593	2.26E-03
50	12.349	12.35	5.41E-03
55	15.758	15.758	8.65E-04
60	19.940	19.941	2.93E-03
65	25.033	25.03	1.05E-02
70	31.188	31.19	6.73E-03
75	38.577	38.58	7.77E-03
80	47.389	47.39	1.63E-03
85	57.833	57.83	5.57E-03
90	70.138	70.14	3.22E-03
95	84.552	84.55	2.72E-03
100	101.348	101.3	4.72E-02
105	120.817	120.8	1.40E-02
110	143.275	143.3	1.77E-02
115	169.059	169.1	2.45E-02
120	198.529	198.5	1.46E-02
125	232.069	232.1	1.32E-02
130	270.086	270.1	5.30E-03
135	313.007	313	2.39E-03
140	361.287	361.3	3.59E-03
145	415.399	415.4	1.34E-04
150	475.843	475.9	1.21E-02
155	543.137	543.1	6.79E-03
160	617.825	617.8	4.03E-03
165	700.471	700.5	4.08E-03
170	791.663	791.7	4.68E-03
175	892.008	892	8.53E-04
180	1002.135	1002.2	6.50E-03
185	1122.695	1122.7	4.07E-04
190	1254.361	1254.4	3.13E-03
195	1397.823	1397.8	1.65E-03
200	1553.795	1553.8	2.95E-04

A.11. Saturation Temperature of Water Vapor

The correlation for the saturation temperature of water vapor is given by

$$T = \left(42.6776 - \frac{3892.7}{(\ln(P/1000) - 9.48654)}\right) - 273.15 \qquad (A.11)$$

where P is in kPa and T is in °C. Values for the calculated saturation temperature over a pressure range of 0.8721-1553.8 kPa and a temperature range of 5-200 °C are given in Table A.11. The table also includes values extracted from the steam tables. The percentage errors for the calculated versus the steam table values are less than 0.28%. Figure A.11 shows variations in the calculated and the steam table values for the vapor pressure of water as a function of temperature.

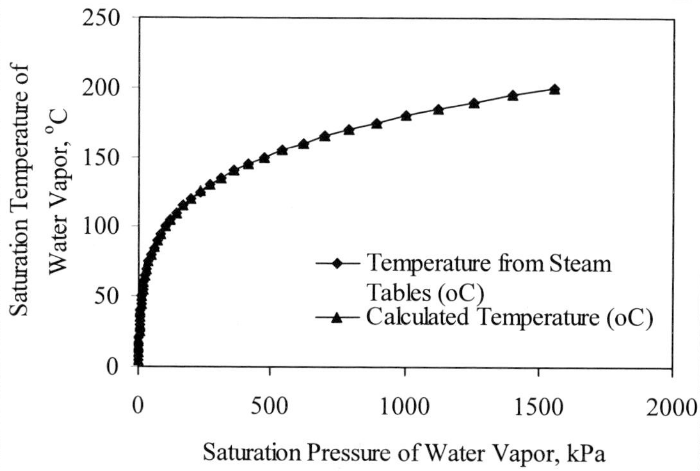

Figure A.11: Variation in saturation temperature of water vapor saturation temperature as a function of saturation pressure.

Table A.11: Variation in saturation temperature of water vapor (°C) as a function of saturation pressure (kPa)

P (kPa)	Calculated Temperature (°C)	Temperature from Steam Tables (°C)	Percentage Error
0.8721	5.004311	5	0.086223
1.2276	9.977515	10	0.224847
1.705	14.95766	15	0.282249
2.339	19.94942	20	0.252887
3.169	24.93937	25	0.242502
4.246	29.93817	30	0.206098
5.628	34.94124	35	0.167899
7.384	39.94826	40	0.129338
9.593	44.95588	45	0.098035
12.35	49.96858	50	0.062849
15.758	54.98005	55	0.036268
19.941	59.99471	60	0.008819
25.03	65.00619	65	0.009523
31.19	70.02479	70	0.035412
38.58	75.03961	75	0.052814
47.39	80.05217	80	0.065212
57.83	85.06355	85	0.074765
70.14	90.07795	90	0.086612
84.55	95.08715	95	0.091739
101.3	100.0839	100	0.083897
120.8	105.1006	105	0.095797
143.3	110.1155	110	0.105
169.1	115.1213	115	0.105474
198.5	120.1106	120	0.092129
232.1	125.1186	125	0.094893
270.1	130.1129	130	0.086868
313	135.1048	135	0.077651
361.3	140.0991	140	0.070757
415.4	145.0877	145	0.060451
475.9	150.0796	150	0.053085
543.1	155.0579	155	0.037346
617.8	160.0422	160	0.026344
700.5	165.0267	165	0.016166
791.7	170.0065	170	0.00382
892	174.9822	175	0.010192
1002.2	179.962	180	0.021134
1122.7	184.9349	185	0.0352
1254.4	189.9109	190	0.046896
1397.8	194.883	195	0.059982
1553.8	199.8582	200	0.070913

A.12. Specific Volume of Saturated Water Vapor

The correlation for the specific volume of saturated water vapor is given by

$$V = V_c\left(\frac{T_c}{T + 273.15} - 1\right) \exp\left(\sum_{i=1}^{6} f_i (T + 273.15)^{(i-1)}\right) \tag{A.12}$$

where T_c = 647.286 K and V_c = 0.003172222 m^3/kg and the values of f_i are given in the following tables

f_1	f_2	f_3	f_4	f_5	f_6
83.63213098	-0.668265339	0.002495964	-5.04185E-06	5.34205E-09	-2.3279E-12

where V is in m^3/kg and T is in °C. Values for the calculated saturation vapor volumes over a temperature range of 5-200 °C are given in Table A.12. The table also includes values extracted from the steam tables. The percentage errors for the calculated versus the steam table values are less than 0.025%. Figure A.12 shows variations in the calculated and the steam table values for the saturation volume of water vapor as a function of temperature.

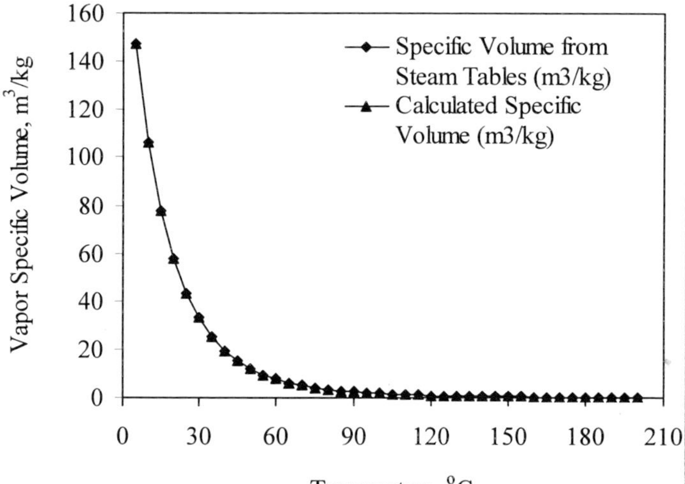

Figure A.12: Variation in water vapor specific volume as a function of temperature.

Table A.12: Variation in water vapor specific volume (m^3/kg) as a function of temperature (°C)

T (°C)	Calculated Specific Volume (m^3/kg)	Specific Volume from Steam Tables (m^3/kg)	Percentage Error
5	147.07980	147.117	0.025285
10	106.37933	106.376	0.003132
15	77.93664	77.925	0.014937
20	57.79982	57.7897	0.017517
25	43.36557	43.3593	0.014456
30	32.89601	32.8932	0.008551
35	25.21619	25.2158	0.001542
40	19.52198	19.5229	0.004735
45	15.25660	15.2581	0.009804
50	12.03025	12.0318	0.012911
55	9.56703	9.56835	0.013787
60	7.66972	7.67071	0.012893
65	6.19591	6.19656	0.010432
70	5.04182	5.04217	0.00686
75	4.13113	4.13123	0.002424
80	3.40722	3.40715	0.001964
85	2.82774	2.82757	0.006085
90	2.36078	2.36056	0.009443
95	1.98209	1.98186	0.011582
100	1.67311	1.6729	0.012267
105	1.41953	1.41936	0.011962
110	1.21027	1.21014	0.010649
115	1.03666	1.03658	0.007642
120	0.89189	0.89186	0.003796
125	0.77059	0.77059	0.000286
130	0.66848	0.6685	0.003484
135	0.58213	0.58217	0.007239
140	0.50880	0.50885	0.010324
145	0.44627	0.44632	0.012189
150	0.39273	0.39278	0.012698
155	0.34672	0.34676	0.0112
160	0.30703	0.30706	0.008363
165	0.27268	0.27269	0.004259
170	0.24283	0.24283	0.001703
175	0.21682	0.2168	0.010016
180	0.19407	0.19405	0.012725
185	0.17412	0.17409	0.016859
190	0.15656	0.15654	0.011531
195	0.14106	0.14105	0.004138
200	0.12733	0.12736	0.023873

A.13. Specific Volume of Saturated Liquid Water

The correlation for the specific volume of saturated liquid water is given by

$$V = V_c\left(\frac{T_c}{T + 273.15} - 1\right) \exp\left(\sum_{i=1}^{6} f_i (T + 273.15)^{(i-1)}\right) \quad \text{(A.13)}$$

where T_c = 647.286 K and V_c = 0.003172222 m^3/kg and the values of f_i are given in the following tables

f_1	f_2	f_3	f_4	f_5	f_6
-2.781015567	0.002543267	9.845047E-06	3.636115E-09	-5.358938E-11	7.019341E-14

In the above equation V is in m^3/kg and T is in °C. Values for the calculated saturation volumes over a temperature range of 5-200 °C are given in Table A.13. The table also includes values extracted from the steam tables. The percentage errors for the calculated versus the steam table values are less than 0.05%. Figure A.13 shows variations in the calculated and the steam table values for the saturation volume of liquid water as a function of temperature.

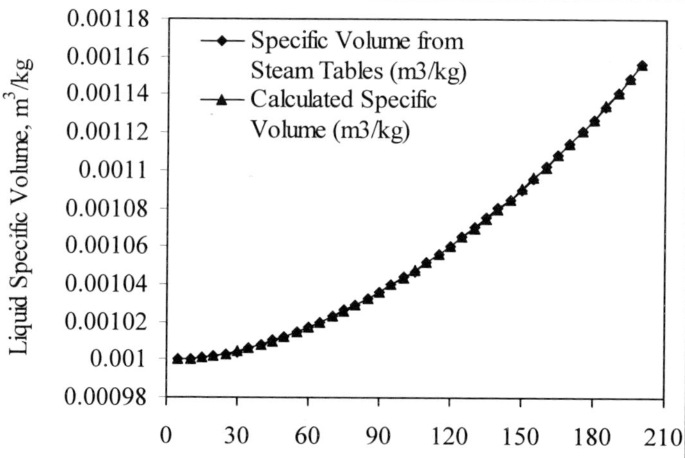

Figure A.13: Variation in liquid water specific volume as a function of temperature.

Table A.13: Variation in liquid water specific volume (m^3/kg) as a function of temperature (°C)

T (°C)	Calculated Specific Volume (m^3/kg)	Specific Volume from Steam Tables (m^3/kg)	Percentage Error
5	0.001000	0.001	0.013868
10	0.001000	0.001	0.024433
15	0.001001	0.001001	0.009436
20	0.001002	0.001002	0.017042
25	0.001003	0.001003	8.72E-05
30	0.001004	0.001004	0.040541
35	0.001006	0.001006	0.003518
40	0.001008	0.001008	0.012434
45	0.001010	0.00101	0.008456
50	0.001012	0.001012	0.014415
55	0.001015	0.001015	0.043332
60	0.001017	0.001017	0.014757
65	0.001020	0.00102	0.008914
70	0.001023	0.001023	0.016466
75	0.001026	0.001026	0.008502
80	0.001029	0.001029	0.014466
85	0.001033	0.001032	0.052013
90	0.001036	0.001036	0.007174
95	0.001040	0.00104	0.023063
100	0.001044	0.001044	0.038921
105	0.001048	0.001047	0.054927
110	0.001052	0.001052	0.028004
115	0.001056	0.001056	0.001294
120	0.001060	0.00106	0.039654
125	0.001065	0.001065	0.00099
130	0.001070	0.00107	0.022361
135	0.001075	0.001075	0.030203
140	0.001080	0.00108	0.022294
145	0.001085	0.001085	0.001649
150	0.001090	0.00109	0.041944
155	0.001096	0.001096	0.007601
160	0.001102	0.001102	0.008834
165	0.001108	0.001108	0.007062
170	0.001114	0.001114	0.013203
175	0.001121	0.001121	0.037027
180	0.001127	0.001127	0.021395
185	0.001134	0.001134	0.01066
190	0.001141	0.001141	0.020023
195	0.001149	0.001149	0.037698
200	0.001156	0.001156	0.011881

A.14. Dynamic Viscosity of Saturated Liquid Water

The correlation for the dynamic viscosity of saturated liquid water is given by

$$\mu = \exp(-3.79418 + 604.129/(139.18+T)) \times 10^{-3} \qquad (A.14)$$

where μ in kg/m s, and T in °C. The above correlation is valid over a temperature range of 10-115 °C. Variations in the dynamic viscosity of saturated water as a function of temperature are given in Table A.14 and Fig. A.14.

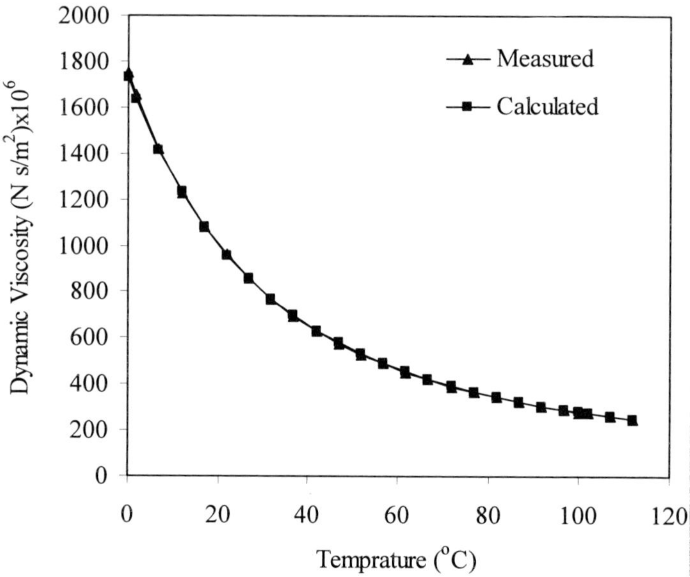

Figure A.14: Variation in dynamic viscosity of saturated liquid water as a function of temperature

Table A.14: Variation in liquid water dynamic viscosity as a function of temperature

T (°C)	Calculated Dynamic Viscosity (N.s/m^2)x10^6	Measured Dynamic Viscosity (N.s/m^2)x10^6	Percentage Error
11.85	1228.605	1225	0.294307
16.85	1080.795	1080	0.073643
21.85	958.3665	959	0.066061
26.85	855.9817	855	0.11482
31.85	769.6022	769	0.078304
36.85	696.1335	695	0.16309
41.85	633.1779	631	0.345152
46.85	578.8578	577	0.321977
51.85	531.6884	528	0.698561
56.85	490.4848	489	0.30365
61.85	454.2936	453	0.285561
66.85	422.3411	420	0.557407
71.85	393.9953	389	1.284127
76.85	368.7358	365	1.023506
81.85	346.1318	343	0.913062
86.85	325.8241	324	0.562993
91.85	307.5116	306	0.493979
96.85	290.9404	289	0.671431
100	281.2965	279	0.823129
101.85	275.8956	274	0.691832
106.85	262.194	260	0.843854
111.85	249.679	248	0.677015

A.15. Dynamic Viscosity of Saturated Water Vapor

The correlation for the dynamic viscosity of saturated water vapor is given by

$$\mu = \exp(-3.609417664 + 275.928958/(-227.0446083 - 0.896081232T - 0.002291383T^2)) \times 10^{-3} \quad (A.15)$$

where μ in kg/m s, and T in °C. The above correlation is valid over the following a temperature range of 10-180 °C. Variations in the saturated water dynamic viscosity as a function of temperature are given in Table A.15 and Fig. A.15.

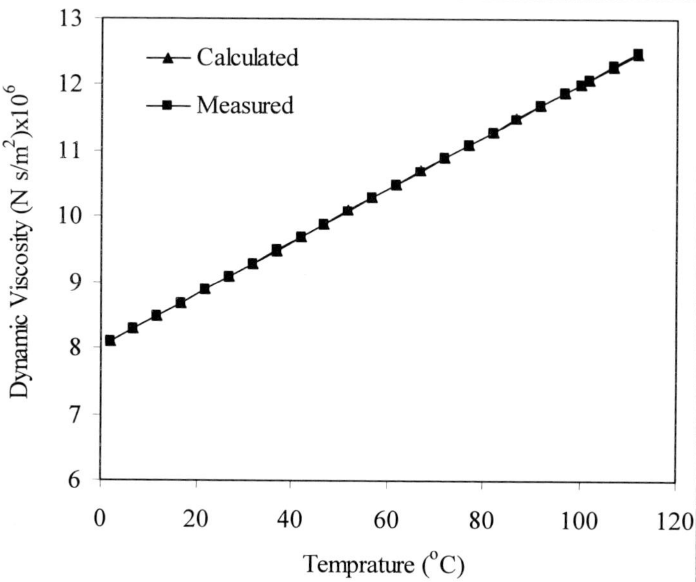

Figure A.15: Variation in dynamic viscosity of saturated water vapor as a function of temperature

Table A.15: Variation in the dynamic viscosity of saturated water vapor as a function of temperature.

T (°C)	Calculated Dynamic Viscosity (N.s/m^2)x10^6	Measured Dynamic Viscosity (N.s/m^2)x10^6	Percentage Error
1.85	8.100136375	8.09	0.125295
6.85	8.294284892	8.29	0.051687
11.85	8.490083562	8.49	0.000984
16.85	8.687322522	8.69	0.030811
21.85	8.885795492	8.89	0.047295
26.85	9.085300391	9.09	0.051701
31.85	9.285639887	9.29	0.046933
36.85	9.486621888	9.49	0.035597
41.85	9.688059977	9.69	0.020021
46.85	9.88977379	9.89	0.002287
51.85	10.09158933	10.09	0.015752
56.85	10.29333925	10.29	0.032451
61.85	10.49486302	10.49	0.046359
66.85	10.69600714	10.69	0.056194
71.85	10.89662523	10.89	0.060838
76.85	11.09657809	11.09	0.059315
81.85	11.29573373	11.29	0.050786
86.85	11.4939674	11.49	0.034529
91.85	11.69116147	11.69	0.009936
96.85	11.88720544	11.89	0.023503
100	12.01007572	12.02	0.082565
101.85	12.08199576	12.09	0.066205
106.85	12.27543577	12.29	0.118505
111.85	12.46743551	12.49	0.18066

A.16. Surface Tension of Saturated Liquid Water

The correlation for the surface tension is given by

$$\sigma = 7.5798x10^{-2} - 1.4691x10^{-4}\ T - 2.2173x10^{-7}\ T^2 \qquad (A.16)$$

where σ in N/m, and T in °C. The above correlation is valid over the following a temperature range of 0-136 °C. Variations in the saturated water surface tension as a function of temperature are given in Table A.16 and Fig. A.16.

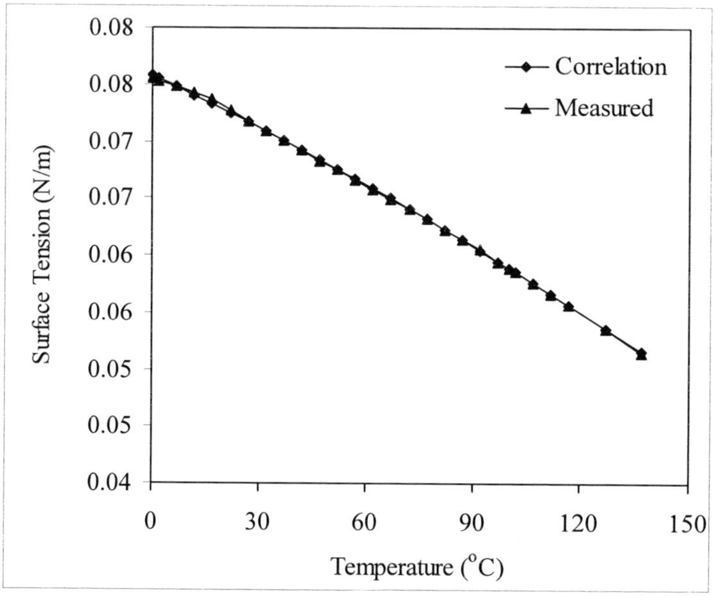

Figure A.16: Variation in surface tension of saturated liquid water as a function of temperature.

Table A.16: Variation in the surface tension of saturated liquid water as a function of temperature.

T (°C)	Measured	Correlation	error%
0	7.55E-02	7.58E-02	3.95E-01
1.85	7.53E-02	7.55E-02	3.00E-01
6.85	7.48E-02	7.48E-02	2.47E-02
11.85	7.43E-02	7.40E-02	3.68E-01
16.85	7.37E-02	7.33E-02	5.97E-01
21.85	7.27E-02	7.25E-02	2.99E-01
26.85	7.17E-02	7.17E-02	8.63E-03
31.85	7.09E-02	7.09E-02	8.22E-03
36.85	7.00E-02	7.01E-02	1.19E-01
41.85	6.92E-02	6.93E-02	8.91E-02
46.85	6.83E-02	6.84E-02	1.88E-01
51.85	6.75E-02	6.76E-02	1.26E-01
56.85	6.66E-02	6.67E-02	1.95E-01
61.85	6.58E-02	6.59E-02	9.65E-02
66.85	6.49E-02	6.50E-02	1.33E-01
71.85	6.41E-02	6.41E-02	3.23E-03
76.85	6.32E-02	6.32E-02	2.36E-03
81.85	6.23E-02	6.23E-02	1.93E-02
86.85	6.14E-02	6.14E-02	5.47E-02
91.85	6.05E-02	6.04E-02	1.10E-01
96.85	5.95E-02	5.95E-02	1.69E-02
100	5.89E-02	5.89E-02	1.75E-02
101.85	5.86E-02	5.85E-02	1.11E-01
106.85	5.76E-02	5.76E-02	5.35E-02
111.85	5.66E-02	5.66E-02	1.39E-02
116.85	5.56E-02	5.56E-02	7.28E-03
126.85	5.36E-02	5.36E-02	1.01E-02
136.85	5.15E-02	5.15E-02	7.91E-02

A.17. Enthalpy of LiBr Water Solution

The enthalpy correlation for saturated LiBr-H_2O solution is given by

$$H(X,T) = X \sum_{i=0}^{4} a_i T^i + (1-X) \sum_{i=0}^{2} b_i T^i + X(1-X) \sum_{i=0}^{4} \sum_{j=0}^{3} c_{ij} (2X-1)^i \, T^j \qquad (A.17)$$

where X is the mass fraction of LiBr and T is the solution temperature. The constants in the above relation are as follows:

$a_0 = 508.668$, $a_1 = 18.6241$, $a_2 = 0.0985946$, $a_3 = -2.500979x10^{-5}$,
$a_4 = 4.15801x10^{-8}$, $b_1 = 1.617155702$, $b_2 = 4.10187485$, $b_3 = 0.000717667$,
$c_{00} = -1021.61$, $c_{10} = -533.08$, $c_{20} = 483.628$, $c_{30} = 1155.13$, $c_{40} = 640.622$,
$c_{01} = 36.8773$, $c_{11} = 40.2847$, $c_{21} = 39.9142$, $c_{31} = 33.3572$, $c_{41} = 13.1032$,
$c_{02} = -0.186051$, $c_{12} = -0.191198$, $c_{22} = 0.199213$, $c_{32} = -0.178258$,
$c_{42} = -0.0775101$, $c_{03} = -7.51277E{-6}$, $c_{13} = 0$, $c_{23} = 0$, $c_{33} = 0$, $c_{43} = 0$

The enthalpy of the saturated LiBr solution are shown in Table A.17 and Fig. A.17 for a temperature range of 10-170 °C and LiBr mass fraction of 0.25-0.75.

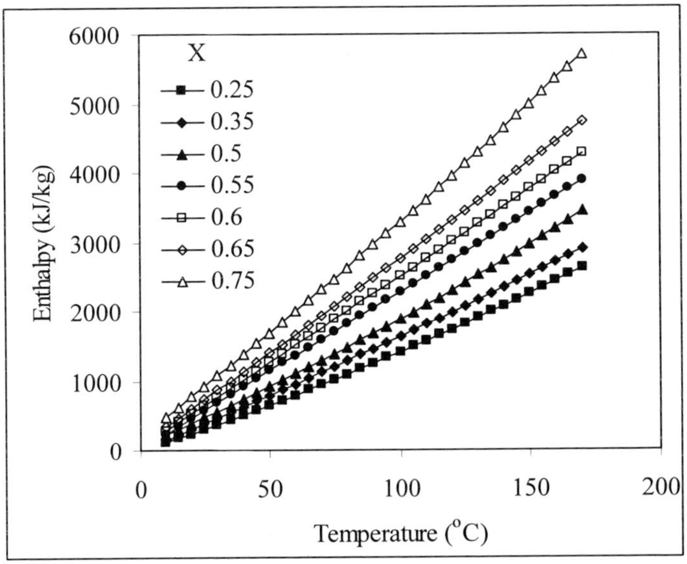

Fig. A.17. Variation in the enthalpy of LiBr solution as a function of temperature and mass fraction

Table A.17: Variation in enthalpy of saturated LiBr solution as a function of temperature (°C) and LiBr mass fraction

	LiBr mass fraction										
T (°C)	0.25	0.3	0.35	0.4	0.45	0.5	0.55	0.6	0.65	0.7	0.75
10	113.11	127.53	142.71	159.68	180.03	205.86	239.56	283.4	339.01	406.79	485.08
15	176.27	198.39	221.34	246.24	274.78	309.13	351.66	404.51	469.08	545.27	630.68
20	240.44	270.01	300.47	333.1	369.71	412.53	463.93	525.95	599.69	684.61	777.52
25	305.64	342.4	380.12	420.26	464.8	516.05	576.37	647.69	730.85	824.79	925.58
30	371.85	415.54	460.26	507.72	560.04	619.67	688.96	769.74	862.54	965.81	1074.9
35	439.07	489.43	540.9	595.45	655.43	723.39	801.71	892.07	994.74	1107.6	1225.3
40	507.3	564.07	622.02	683.46	750.95	827.19	914.58	1014.7	1127.4	1250.3	1377
45	576.53	639.44	703.63	771.74	846.61	931.08	1027.6	1137.6	1260.7	1393.7	1529.8
50	646.75	715.55	785.71	860.28	942.39	1035	1140.7	1260.7	1394.3	1538	1683.8
55	717.97	792.38	868.26	949.07	1038.3	1139.1	1254	1384.1	1528.5	1683	1838.9
60	790.17	869.94	951.27	1038.1	1134.3	1243.2	1367.3	1507.7	1663.1	1828.8	1995.2
65	863.36	948.22	1034.7	1127.4	1230.4	1347.3	1480.7	1631.6	1798.2	1975.3	2152.6
70	937.53	1027.2	1118.7	1216.9	1326.6	1451.5	1594.3	1755.7	1933.8	2122.6	2311.1
75	1012.7	1106.9	1203	1306.7	1422.9	1555.7	1707.9	1880	2069.7	2270.6	2470.8
80	1088.8	1187.3	1287.8	1396.7	1519.2	1659.9	1821.5	2004.5	2206.2	2419.4	2631.5
85	1165.9	1268.4	1373.1	1486.9	1615.7	1764.2	1935.3	2129.2	2343	2568.9	2793.3
90	1243.9	1350.3	1458.8	1577.3	1712.2	1868.5	2049.1	2254.2	2480.3	2719.1	2956.3
95	1323	1432.8	1544.9	1667.9	1808.8	1972.8	2162.9	2379.3	2617.9	2870	3120.3

Table A.17 (continued): Variation in enthalpy of saturated LiBr solution as a function of temperature (°C) and LiBr mass fraction

	LiBr mass fraction										
T (°C)	0.25	0.3	0.35	0.4	0.45	0.5	0.55	0.6	0.65	0.7	0.75
100	1402.9	1516	1631.4	1758.7	1905.4	2077.1	2276.9	2504.6	2756	3021.6	3285.4
105	1483.9	1599.8	1718.4	1849.8	2002.1	2181.4	2390.8	2630.1	2894.5	3174	3451.5
110	1565.8	1684.4	1805.8	1941	2098.9	2285.7	2504.8	2755.7	3033.4	3327	3618.7
115	1648.6	1769.7	1893.6	2032.5	2195.7	2390	2618.9	2881.5	3172.6	3480.7	3787
120	1732.4	1855.6	1981.8	2124.1	2292.6	2494.3	2732.9	3007.5	3312.3	3635.1	3956.4
125	1817.1	1942.2	2070.4	2215.9	2389.5	2598.6	2847	3133.7	3452.3	3790.2	4126.7
130	1902.8	2029.5	2159.5	2307.9	2486.4	2702.9	2961.1	3260	3592.8	3945.9	4298.2
135	1989.5	2117.4	2248.9	2400.1	2583.4	2807.2	3075.3	3386.5	3733.6	4102.4	4470.6
140	2077.1	2206.1	2338.8	2492.5	2680.4	2911.4	3189.4	3513.1	3874.7	4259.5	4644.2
145	2165.6	2295.4	2429	2585	2777.5	3015.6	3303.6	3639.9	4016.3	4417.2	4818.7
150	2255	2385.3	2519.7	2677.8	2874.5	3119.8	3417.8	3766.8	4158.2	4575.7	4994.3
155	2345.4	2476	2610.7	2770.7	2971.7	3224	3532	3893.9	4300.5	4734.8	5170.9
160	2436.8	2567.3	2702.2	2863.8	3068.8	3328.1	3646.2	4021.1	4443.1	4894.6	5348.6
165	2529.1	2659.3	2794	2957	3166	3432.2	3760.5	4148.5	4586.1	5055	5527.3
170	2622.3	2751.9	2886.3	3050.5	3263.2	3536.3	3874.7	4276	4729.5	5216.1	5707

A.18. Boiling Temperature of LiBr Water Solution

The correlation for the boiling temperature of saturated LiBr-H_2O solution is given by

$$T(P,X) = \sum_{i=0}^{10} a_i X^i + T_w \sum_{i=0}^{10} b_i X^i \qquad (A.18)$$

where T_w is the saturation temperature of pure water at pressure P, X is the mass fraction of LiBr in the solution, P is the pressure, and T is the boiling temperature of the LiBr-H_2O solution. The constants in the above relation are as follows:

$a_0 = 0$, $a_1 = 16.634856$, $a_2 = -553.38169$, $a_3 = 11228.338$, $a_4 = -110283.9$,
$a_5 = 621094.64$, $a_6 = -2111256.7$, $a_7 = 4385190.1$, $a_8 = -5409811.5$,
$a_9 = 3626674.2$, $a_{10} = -1015305.9$, $b_0 = 1$, $b_1 = -0.068242821$, $b_2 = 5.873619$
$b_3 = -102.78186$, $b_4 = 930.32374$, $b_5 = -4822.394$, $b_6 = 15189.038$,
$b_7 = -29412.863$, $b_8 = 34100.528$, $b_9 = -21671.48$, $b_{10} = 5799.56$

The boiling temperature of saturated LiBr solution is shown in Table A.18 and Fig. A.18 for a temperature range of 10-170 °C and LiBr mass fraction of 0.25-0.75.

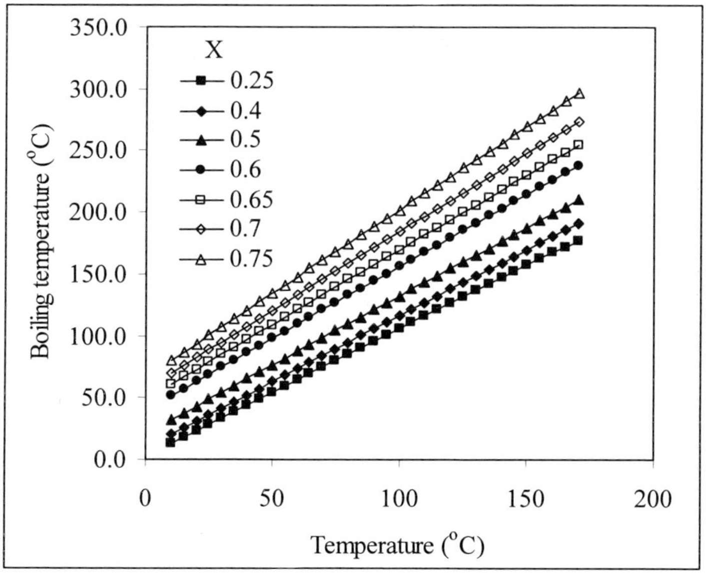

Fig. A.18. Boiling temperature of LiBr solution as a function of temperature and salt concentration

Table A.18: Variation in the boiling temperature of LiBr solution as a function of the salt mass fraction and temperature

	LiBr mass fraction										
T (°C)	0.25	0.3	0.35	0.4	0.45	0.5	0.55	0.6	0.65	0.7	0.75
10	13.5	15.1	17.1	20.0	24.7	31.8	41.0	51.1	60.5	69.4	80.0
15	18.7	20.2	22.3	25.4	30.2	37.4	46.7	56.9	66.5	75.8	86.8
20	23.8	25.4	27.6	30.7	35.6	42.9	52.4	62.7	72.6	82.2	93.5
25	28.9	30.6	32.8	36.1	41.1	48.5	58.1	68.6	78.7	88.6	100.3
30	34.0	35.8	38.1	41.4	46.5	54.1	63.8	74.4	84.7	94.9	107.1
35	39.2	40.9	43.3	46.7	52.0	59.7	69.5	80.3	90.8	101.3	113.9
40	44.3	46.1	48.6	52.1	57.5	65.2	75.2	86.1	96.8	107.7	120.6
45	49.4	51.3	53.8	57.4	62.9	70.8	80.9	91.9	102.9	114.0	127.4
50	54.5	56.5	59.1	62.8	68.4	76.4	86.5	97.8	109.0	120.4	134.2
55	59.7	61.6	64.3	68.1	73.8	81.9	92.2	103.6	115.0	126.8	141.0
60	64.8	66.8	69.6	73.5	79.3	87.5	97.9	109.5	121.1	133.2	147.7
65	69.9	72.0	74.8	78.8	84.7	93.1	103.6	115.3	127.1	139.5	154.5
70	75.0	77.2	80.1	84.2	90.2	98.7	109.3	121.1	133.2	145.9	161.3
75	80.2	82.4	85.3	89.5	95.7	104.2	115.0	127.0	139.3	152.3	168.1
80	85.3	87.5	90.5	94.8	101.1	109.8	120.7	132.8	145.3	158.7	174.8
85	90.4	92.7	95.8	100.2	106.6	115.4	126.4	138.7	151.4	165.0	181.6
90	95.5	97.9	101.0	105.5	112.0	120.9	132.1	144.5	157.4	171.4	188.4

Table A.18 (continued): Variation in the boiling temperature of LiBr solution as a function of the salt mass fraction and temperature

	LiBr mass fraction										
T (°C)	0.25	0.3	0.35	0.4	0.45	0.5	0.55	0.6	0.65	0.7	0.75
95	100.7	103.1	106.3	110.9	117.5	126.5	137.8	150.3	163.5	177.8	195.2
100	105.8	108.2	111.5	116.2	122.9	132.1	143.5	156.2	169.6	184.2	201.9
105	110.9	113.4	116.8	121.6	128.4	137.7	149.2	162.0	175.6	190.5	208.7
110	116.0	118.6	122.0	126.9	133.8	143.2	154.8	167.9	181.7	196.9	215.5
115	121.2	123.8	127.3	132.3	139.3	148.8	160.5	173.7	187.7	203.3	222.3
120	126.3	129.0	132.5	137.6	144.8	154.4	166.2	179.5	193.8	209.7	229.0
125	131.4	134.1	137.8	142.9	150.2	160.0	171.9	185.4	199.9	216.0	235.8
130	136.5	139.3	143.0	148.3	155.7	165.5	177.6	191.2	205.9	222.4	242.6
135	141.7	144.5	148.3	153.6	161.1	171.1	183.3	197.1	212.0	228.8	249.4
140	146.8	149.7	153.5	159.0	166.6	176.7	189.0	202.9	218.1	235.2	256.1
145	151.9	154.8	158.8	164.3	172.0	182.2	194.7	208.8	224.1	241.5	262.9
150	157.0	160.0	164.0	169.7	177.5	187.8	200.4	214.6	230.2	247.9	269.7
155	162.2	165.2	169.3	175.0	183.0	193.4	206.1	220.4	236.2	254.3	276.5
160	167.3	170.4	174.5	180.4	188.4	199.0	211.8	226.3	242.3	260.7	283.2
165	172.4	175.6	179.8	185.7	193.9	204.5	217.5	232.1	248.4	267.0	290.0
170	177.5	180.7	185.0	191.0	199.3	210.1	223.1	238.0	254.4	273.4	296.8

Appendix B

Thermodynamic Losses

B.1 Boiling Point Elevation

The correlation for the boiling point elevation of seawater is

$$BPE = A\,X + B\,X^2 + C\,X^3 \tag{B.1}$$

with

$$A = (8.325\text{x}10^{-2} + 1.883\text{x}10^{-4}\,T + 4.02\text{x}10^{-6}\,T^2)$$
$$B = (-\,7.625\text{x}10^{-4} + 9.02\text{x}10^{-5}\,T - 5.2\text{x}10^{-7}\,T^2)$$
$$C = (1.522\text{x}10^{-4} - 3\text{x}10^{-6}\,T - 3\text{x}10^{-8}\,T^2)$$

where T is the temperature in °C and X is the salt weight percentage. The above equation is valid over the following ranges: $1 \leq X \leq 16\%$, $10 \leq T \leq 180°C$. Variations in the boiling point elevation as a function of the seawater temperature and salinity are given in Table B.1 and Fig. B.1.

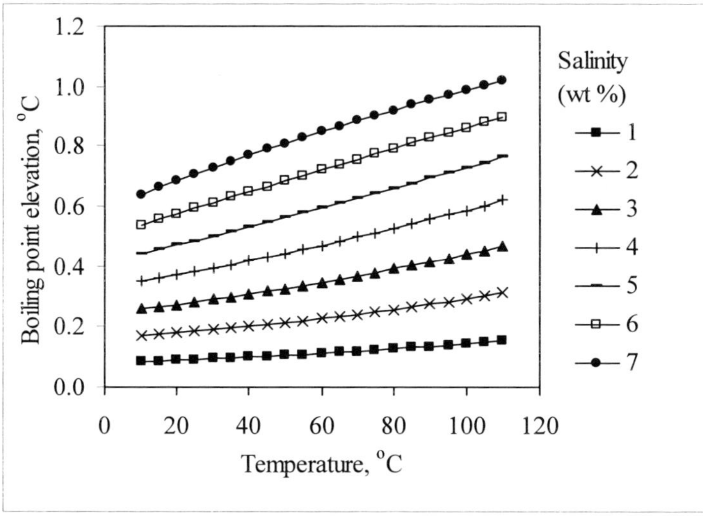

Fig. B.1: Variation in boiling point elevation of seawater as a function of temperature and salinity.

Table B.1: Variation in seawater boiling point elevation (°C) as a function of temperature (°C) and salinity (wt%)

	Salinity (wt%)						
temperature °C	1	2	3	4	5	6	7
10	0.085	0.171	0.258	0.348	0.441	0.538	0.639
15	0.087	0.175	0.266	0.359	0.456	0.556	0.662
20	0.089	0.180	0.273	0.370	0.470	0.575	0.684
25	0.091	0.185	0.281	0.381	0.485	0.593	0.706
30	0.093	0.190	0.290	0.393	0.500	0.612	0.727
35	0.096	0.195	0.298	0.405	0.516	0.630	0.748
40	0.099	0.201	0.307	0.417	0.531	0.648	0.769
45	0.101	0.207	0.316	0.430	0.546	0.666	0.789
50	0.104	0.213	0.326	0.443	0.562	0.684	0.809
55	0.108	0.220	0.336	0.456	0.578	0.703	0.829
60	0.111	0.227	0.346	0.469	0.594	0.721	0.848
65	0.115	0.234	0.357	0.483	0.610	0.739	0.866
70	0.118	0.241	0.368	0.497	0.627	0.756	0.885
75	0.122	0.249	0.379	0.511	0.643	0.774	0.903
80	0.126	0.257	0.391	0.525	0.660	0.792	0.921
85	0.130	0.265	0.402	0.540	0.677	0.810	0.938
90	0.135	0.274	0.415	0.555	0.694	0.828	0.955
95	0.139	0.283	0.427	0.571	0.711	0.845	0.971
100	0.144	0.292	0.440	0.587	0.728	0.863	0.987
105	0.149	0.301	0.453	0.603	0.746	0.880	1.003
110	0.154	0.311	0.467	0.619	0.764	0.898	1.018

B.2 Non-Equilibrium Allowance in MEE

The correlation for the non-equilibrium allowance in the MEE process is developed by Miyatake et al. (1973),

$$(NEA)_j = 33\ (\Delta T_j)^{0.55}/T_{v_j} \tag{B.2}$$

where, $\Delta Tj = T_{j-1} - T_j$, is temperature difference of boiling brine in effects j and j–1, T_{v_j} is the vapor temperature in effect j and is given by $T_{v_j} = T_j - (BPE)_j$. All temperatures in the above correlation are in °C. Results for the non-equilibrium allowance are shown in Table B.2 and Fig. B.2.

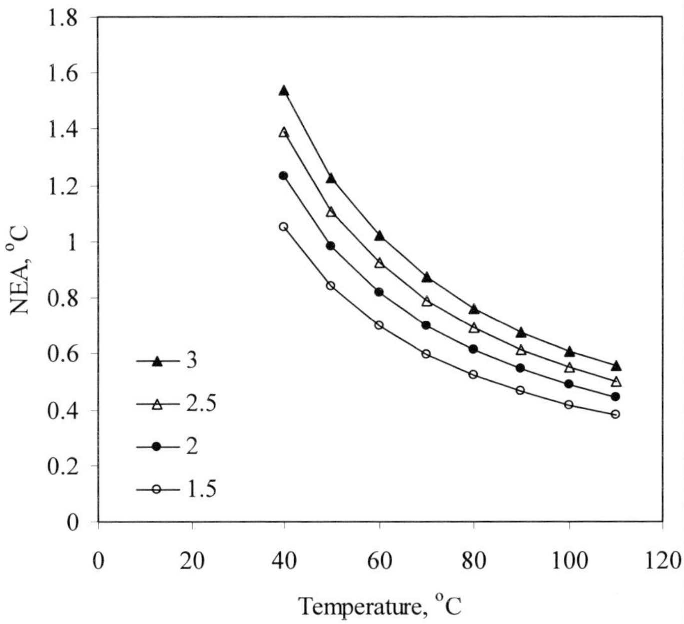

Fig. B.2 Variation in non-equilibrium allowance in MEE as a function of brine temperature and temperature drop per effect for salinity of 40000 ppm

Table B.2: Variation in the non-equilibrium allowance in multiple effect evaporation as a function of effect temperature drop and temperature for a salinity of 40000 ppm.

NEA (°C)	T_j	ΔT	T_{vj}	BPE (°C)
1.042	40	1.5	39.6	0.417
0.832	50	1.5	49.6	0.443
0.693	60	1.5	59.5	0.469
0.593	70	1.5	69.5	0.497
0.519	80	1.5	79.5	0.525
0.461	90	1.5	89.4	0.555
0.415	100	1.5	99.4	0.587
0.377	110	1.5	109.4	0.619
1.221	40	2	39.6	0.417
0.975	50	2	49.6	0.443
0.812	60	2	59.5	0.469
0.695	70	2	69.5	0.497
0.608	80	2	79.5	0.525
0.540	90	2	89.4	0.555
0.486	100	2	99.4	0.587
0.442	110	2	109.4	0.619
1.380	40	2.5	39.6	0.417
1.102	50	2.5	49.6	0.443
0.918	60	2.5	59.5	0.469
0.786	70	2.5	69.5	0.497
0.687	80	2.5	79.5	0.525
0.611	90	2.5	89.4	0.555
0.549	100	2.5	99.4	0.587
0.499	110	2.5	109.4	0.619
1.557	40	3	39.6	0.417
1.232	50	3	49.6	0.443
1.020	60	3	59.5	0.469
0.871	70	3	69.5	0.497
0.761	80	3	79.5	0.525
0.675	90	3	89.4	0.555
0.607	100	3	99.4	0.587
0.551	110	3	109.4	0.619

B.3 Non-Equilibrium Allowance in MSF

Lior (1986) developed correlations for the non-equilibrium allowance for the MSF system. The following two equations give values for NEA as a function of the brine temperature, gate height, the brine flow rate per unit length of the chamber width, and the stage temperature drop;

$$(NEA)_{10} = (0.9784)^{T_i}\,(15.7378)^{H}\,(1.3777)^{V_b \times 10^{-6}} \tag{B.3a}$$

and

$$NEA = (NEA_{10}/(0.5\Delta T + NEA_{10})^{0.3281L}\,(0.5\,\Delta T + NEA_{10}) \tag{B.3b}$$

Eq. B.3a is valid for 10 ft stage length and Eq. B.3b is applicable for stages of any other lengths. In the above equation, T_i is the stage temperature in °C, H is the height of the brine pool in m, V_b is the brine flow rate per unit length of the chamber width in kg/(m s), and ΔT is the stage temperature drop in °C Results for the equations B.3a and B.3b are shown in Fig. B.3 and Table B.3.

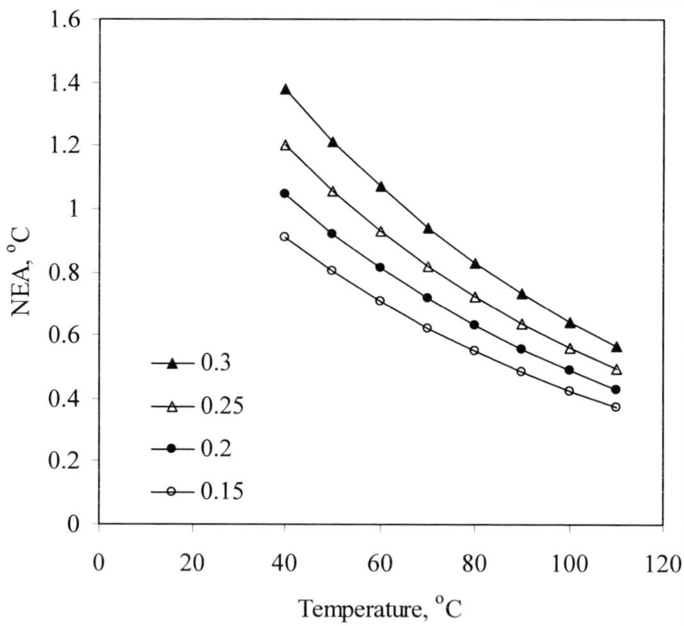

Fig. B.3 Variation in non-equilibrium allowance in MSF as a function of brine temperature and brine height for brine weir load of 180 kg/m/s and stage length = 10 ft.

Table B.3: Variations in non-equilibrium allowance for MSF as a function of flashing temperature and brine height for a weir load of 180 kg/m s and stage length equal to 10 ft.

NEA (°C)	T_j (°C)	H (m)	NEA (°C)	T_j (°C)	H (m)
0.63	40	0.15	0.83	40	0.25
0.51	50	0.15	0.67	50	0.25
0.41	60	0.15	0.54	60	0.25
0.33	70	0.15	0.43	70	0.25
0.26	80	0.15	0.35	80	0.25
0.21	90	0.15	0.28	90	0.25
0.17	100	0.15	0.22	100	0.25
0.14	110	0.15	0.18	110	0.25
0.72	40	0.2	0.95	40	0.3
0.58	50	0.2	0.77	50	0.3
0.47	60	0.2	0.62	60	0.3
0.38	70	0.2	0.50	70	0.3
0.30	80	0.2	0.40	80	0.3
0.24	90	0.2	0.32	90	0.3
0.20	100	0.2	0.26	100	0.3
0.16	110	0.2	0.21	110	0.3

B.4 Demister Pressure Drop

The correlation for pressure drop in the demister, ΔP_p, is developed by El-Dessouky et al. (2000) for industrial type wire pads. The ranges of the experimental variables were V (0.98-7.5 m/s), ρ_p (80.317-208.16 kg/m^3), L (100-200 mm), δ_w (0.2-.32 mm), and d_d (1-5 mm). This correlation is given by

$$\Delta P_p = 3.88178\,(\rho_p)^{0.375798}\,(V)^{0.81317}\,(\delta_w)^{-1.56114147} \tag{B.4}$$

where ΔP_p is the demister pressure drop in Pa/m, δ_w is the wire diameter in mm, d_d is the diameter of entrained droplets in mm, L is the mesh pad thickness in mm, V is the vapor velocity in the demister in m/s, and ρ is the demister density in kg/m^3. In Eq. B.4 the subscript p denotes the demister. Results for the demister pressure drop is shown in Table B.4. Figures illustrating the demister pressure are given in Chapter 9.

Table B.4: Variations in the demister pressure drop as a function of the vapor velocity, packing density, and wire diameter.

Δp (Pa/m)	V (m/s)	ρ_p (kg/m3)	δ_w (m)	Δp (Pa/m)	V (m/s)	ρ_p (kg/m3)	δ_w (m)
190.1495	1.37	80.317	0.28	799.5571	5.16	208.16	0.28
299.9815	2.4	80.317	0.28	832.1666	5.42	208.16	0.28
370.3531	3.11	80.317	0.28	429.5316	1.36	176.35	0.2
419.028	3.62	80.317	0.28	597.2926	2.04	176.35	0.2
456.3063	4.02	80.317	0.28	656.1643	2.29	176.35	0.2
594.874	5.57	80.317	0.28	342.3253	1.46	176.35	0.24
689.6065	6.68	80.317	0.28	434.9566	1.96	176.35	0.24
527.0877	4.8	80.317	0.28	581.3096	2.8	176.35	0.24
631.9565	6	80.317	0.28	672.578	3.35	176.35	0.24
352.5772	2.26	140.6	0.28	769.0008	3.95	176.35	0.24
441.4914	2.98	140.6	0.28	839.5147	4.4	176.35	0.24
520.6471	3.65	140.6	0.28	931.4795	5	176.35	0.24
590.3651	4.26	140.6	0.28	885.7715	4.7	176.35	0.24
673.5824	5.01	140.6	0.28	383.9096	2.26	176.35	0.28
751.298	5.73	140.6	0.28	480.7253	2.98	176.35	0.28
804.1878	6.23	140.6	0.28	566.9154	3.65	176.35	0.28
327.6714	1.86	176.35	0.28	642.8291	4.26	176.35	0.28
434.2842	2.63	176.35	0.28	733.4416	5.01	176.35	0.28
517.1443	3.26	176.35	0.28	818.0635	5.73	176.35	0.28
644.0559	4.27	176.35	0.28	875.6535	6.23	176.35	0.28
720.3196	4.9	176.35	0.28	243.6977	1.67	176.35	0.32
799.4394	5.57	176.35	0.28	301.5393	2.17	176.35	0.32
829.6542	5.83	176.35	0.28	374.2175	2.83	176.35	0.32
333.4171	1.76	208.16	0.28	471.4877	3.76	176.35	0.32
452.1815	2.56	208.16	0.28	546.6402	4.51	176.35	0.32
543.5242	3.21	208.16	0.28	601.2238	5.07	176.35	0.32
676.3193	4.2	208.16	0.28	659.4141	5.68	176.35	0.32
761.546	4.86	208.16	0.28	761.5104	6.78	176.35	0.32
772.7173	2.8	176.35	0.2	721.0738	6.34	176.35	0.32
848.1905	3.14	176.35	0.2	809.5738	7.31	176.35	0.32
935.0569	3.54	176.35	0.2	687.5985	5.98	176.35	0.32
965.0183	3.68	176.35	0.2	790.6104	7.1	176.35	0.32

B.5 Pressure Drop in Connecting Lines

The pressure drop in the lines connecting the vapor space in effect i and the evaporator tubes of the next effect is calculated from the Unwin formula, ORNL (1967),

$$\Delta P = \frac{0.0001306 \ M^2 \ L \, (1 + \frac{3.6}{\delta_i})}{\rho_v \delta_i^5} \tag{B.5}$$

where M is the mass flow rate of the vapor stream (kg/s), L is the tube length (m), δ_i is the tube inner diameter (m), ρ_v is the vapor density (kg/m^3), and ΔP is the pressure drop (Pa/m). Results for the pressure drop in connecting lines are given in Table B.5 and Fig. B.5.

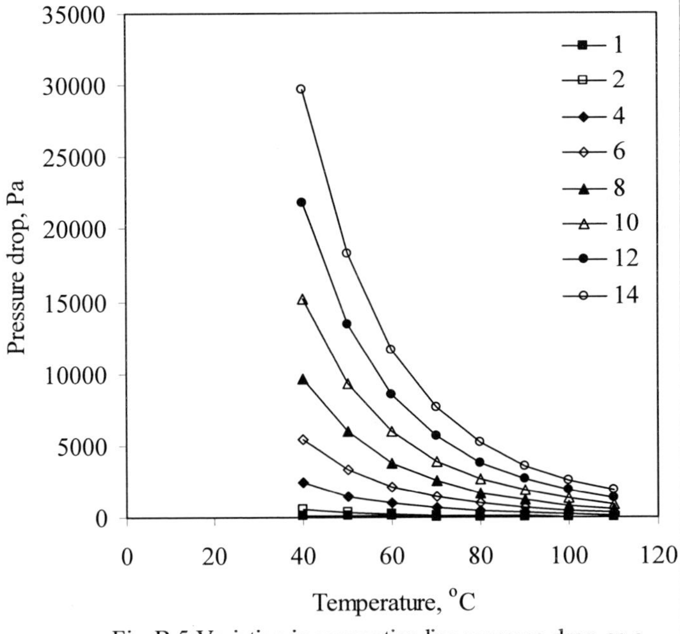

Fig. B.5 Variation in connection line pressure drop as a function of flow rate and and temperature (δ_i =0.2 m, L = 1 m)

Table B.5: Variation in connecting line pressure drop as a function of temperature and flow rate for L = 1 m and δ_i = 0.2 m.

T (°C)	ΔP (Pa/m)	M (kg/s)	ρ_v (kg/m^3)
40	151.3807	1	0.051224324
50	93.28704	1	0.083123816
60	59.47389	1	0.13038284
70	39.09619	1	0.198340913
80	26.42084	1	0.293494667
90	18.3064	1	0.423588291
100	12.97389	1	0.597691043
110	9.384879	1	0.826262685
40	605.5229	2	0.051224324
50	373.1482	2	0.083123816
60	237.8956	2	0.13038284
70	156.3848	2	0.198340913
80	105.6834	2	0.293494667
90	73.22558	2	0.423588291
100	51.89554	2	0.597691043
110	37.53951	2	0.826262685
40	2422.092	4	0.051224324
50	1492.593	4	0.083123816
60	951.5823	4	0.13038284
70	625.5391	4	0.198340913
80	422.7334	4	0.293494667
90	292.9023	4	0.423588291
100	207.5822	4	0.597691043
110	150.1581	4	0.826262685
40	5449.706	6	0.051224324
50	3358.334	6	0.083123816
60	2141.06	6	0.13038284
70	1407.463	6	0.198340913
80	951.1502	6	0.293494667
90	659.0303	6	0.423588291
100	467.0599	6	0.597691043
110	337.8556	6	0.826262685
40	9688.366	8	0.051224324
50	5970.371	8	0.083123816
60	3806.329	8	0.13038284
70	2502.156	8	0.198340913
80	1690.934	8	0.293494667
90	1171.609	8	0.423588291
100	830.3287	8	0.597691043
110	600.6322	8	0.826262685

Table B.5 (Continued): Variation in connecting line pressure drop as a function of temperature and flow rate for L = 1 m and δ_i = 0.2 m.

T (°C)	ΔP (Pa/m)	M (kg/s)	ρ_v (kg/m^3)
40	15138.07	10	0.051224324
50	9328.704	10	0.083123816
60	5947.389	10	0.13038284
70	3909.619	10	0.198340913
80	2642.084	10	0.293494667
90	1830.64	10	0.423588291
100	1297.389	10	0.597691043
110	938.4879	10	0.826262685
40	21798.82	12	0.051224324
50	13433.33	12	0.083123816
60	8564.241	12	0.13038284
70	5629.852	12	0.198340913
80	3804.601	12	0.293494667
90	2636.121	12	0.423588291
100	1868.239	12	0.597691043
110	1351.423	12	0.826262685
40	29670.62	14	0.051224324
50	18284.26	14	0.083123816
60	11656.88	14	0.13038284
70	7662.854	14	0.198340913
80	5178.484	14	0.293494667
90	3588.054	14	0.423588291
100	2542.882	14	0.597691043
110	1839.436	14	0.826262685

B.6 Gravitational Pressure Drop

The gravitational pressure drop during condensation inside the evaporator tubes is given by

$$\Delta P = \left(\rho_v \alpha + (1-\alpha)\rho_\ell\right) g\, L \sin(\theta) \tag{B.6}$$

where ΔP is the pressure drop in Pa, g is the gravitational acceleration (m/s^2), L is length of the evaporator tubes (m), and θ is the inclination angle. The expression for α is given by Zivi (1964), $\alpha = \dfrac{1}{\left(1+\dfrac{1-\chi}{\chi}\left(\dfrac{\rho_v}{\rho_\ell}\right)^{0.5}\right)}$, where χ is the vapor mass fraction, which is greater than zero and less than 1. In Eq. B.6, ρ_v and ρ_ℓ are the density of vapor and liquid streams at saturation conditions and are given in (kg/m^3). Results for the above losses are given in Table B.6 and Fig. B.6.

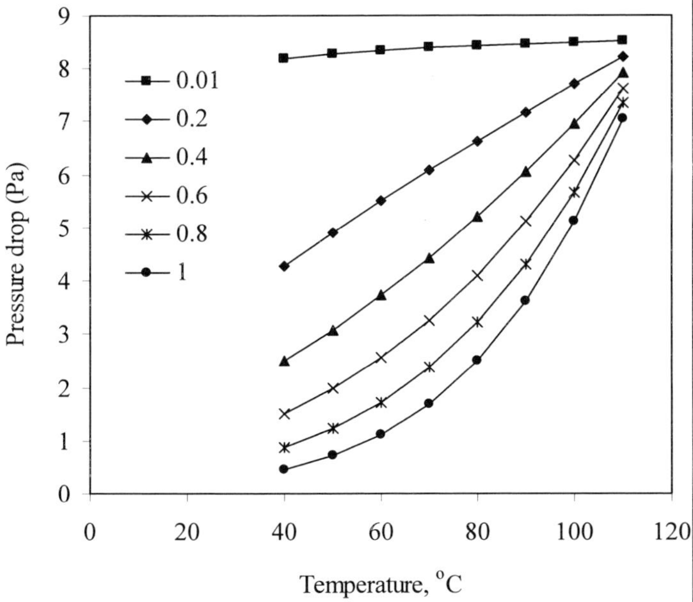

Fig. B.6 Variations in the gravitational pressure drop as a function of the vapor temperature and the vapor mass fraction. ($\theta = 5^{o}$ and L = 10 m)

Table B.6: Variation in gravitational pressure drop as a function of temperature and vapor mass fraction for $\theta = 5^{o}$ and L = 10 m.

ΔP (Pa)	T (^{o}C)	χ	ρ_{ℓ} (kg/m^3)	ρ_v (kg/m^3)
3522.797	40	0.01	992.19	0.051224
4016.468	50	0.01	988.00	0.083124
4473.981	60	0.01	983.14	0.130383
4886.44	70	0.01	977.68	0.198341
5249.686	80	0.01	971.68	0.293495
5563.216	90	0.01	965.18	0.423588
5829.014	100	0.01	958.23	0.597691
6050.516	110	0.01	950.84	0.826263
237.1867	40	0.2	992.19	0.051224
299.3431	50	0.2	988.00	0.083124
370.8429	60	0.2	983.14	0.130383
451.7166	70	0.2	977.68	0.198341
541.7984	80	0.2	971.68	0.293495
640.7363	90	0.2	965.18	0.423588
748.0056	100	0.2	958.23	0.597691
862.9278	110	0.2	950.84	0.826263
90.79624	40	0.4	992.19	0.051224
115.2305	50	0.4	988.00	0.083124
143.686	60	0.4	983.14	0.130383
176.336	70	0.4	977.68	0.198341
213.3048	80	0.4	971.68	0.293495
254.669	90	0.4	965.18	0.423588
300.4589	100	0.4	958.23	0.597691
350.6601	110	0.4	950.84	0.826263

Table B.6 (Continued): Variation in gravitational pressure drop as a function of temperature and vapor mass fraction for $\theta = 5^{o}$ and L = 10 m.

ΔP (Pa)	T (^{o}C)	χ	ρ_{ℓ} (kg/m^3)	ρ_v (kg/m^3)
40.830	40	0.6	992.19	0.051224
51.987	50	0.6	988.00	0.083124
65.076	60	0.6	983.14	0.130383
80.222	70	0.6	977.68	0.198341
97.539	80	0.6	971.68	0.293495
117.127	90	0.6	965.18	0.423588
139.076	100	0.6	958.23	0.597691
163.466	110	0.6	950.84	0.826263
15.630	40	0.8	992.19	0.051224
20.012	50	0.8	988.00	0.083124
25.214	60	0.8	983.14	0.130383
31.316	70	0.8	977.68	0.198341
38.401	80	0.8	971.68	0.293495
46.553	90	0.8	965.18	0.423588
55.864	100	0.8	958.23	0.597691
66.426	110	0.8	950.84	0.826263
1.052	40	0.99	992.19	0.051224
1.492	50	0.99	988.00	0.083124
2.090	60	0.99	983.14	0.130383
2.895	70	0.99	977.68	0.198341
3.963	80	0.99	971.68	0.293495
5.361	90	0.99	965.18	0.423588
7.168	100	0.99	958.23	0.597691
9.473	110	0.99	950.84	0.826263

B.7 Acceleration Pressure Drop

The acceleration pressure drop during condensation inside the evaporator tubes is calculated from the following relation

$$\Delta P = \frac{M^2}{A^2}\left(\frac{\chi_1^2}{\alpha_1 \rho_{v_1}} + \frac{(1-\chi_1)^2}{(1-\alpha_1)\rho_{\ell_1}} - \frac{\chi_2^2}{\alpha_2 \rho_{v_2}} - \frac{(1-\chi_2)^2}{(1-\alpha_2)\rho_{\ell_2}}\right) \tag{B.7}$$

where ΔP is the pressure drop in Pa, M is the mass flow rate in kg/s, A is the cross section area in m^2, ρ_v is the vapor density in kg/m^3, ρ_ℓ is the liquid density in kg/m^3, χ is the vapor phase mass fraction, α is $\alpha = \frac{1}{\left(1 + \frac{1-\chi}{\chi}\left(\frac{\rho_v}{\rho_\ell}\right)^{0.5}\right)}$. The subscripts 1 and 2 refer to the inlet and outlet conditions. Results for the above correlation are shown in Table B.7 and Fig. B.7.

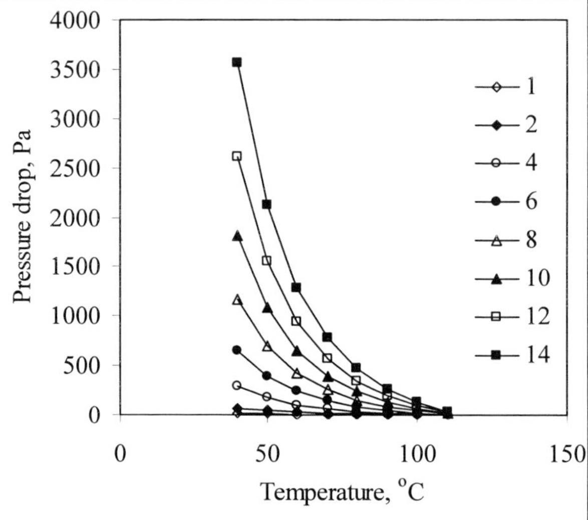

Fig. B.7. Variation in the acceleration pressure drop as a function of temperature and flow rate. ($A = 1\ m^2$, $\chi_1 = 0.99$, $\chi_2 = 0.01$)

Table B.7: Variations in the acceleration pressure drop as a function of the mass flow rate and temperature for a cross section area of 1 m2 and vapor mass fractions of 0.01 and 0.99.

T (°C)	ΔP (Pa)	M (kg/s)	ρ_ℓ (kg/m^3)	ρ_V (kg/m^3)	α_1	α_2
40	19.13	1	992.19	0.05	0.9573	0.0023
50	11.79	1	988.00	0.08	0.9662	0.0029
60	7.52	1	983.14	0.13	0.9728	0.0036
70	4.94	1	977.68	0.20	0.9778	0.0045
80	3.34	1	971.68	0.29	0.9817	0.0054
90	2.31	1	965.18	0.42	0.9847	0.0065
100	1.64	1	958.23	0.60	0.9871	0.0077
110	1.19	1	950.84	0.83	0.9890	0.0091
40	76.52	2	992.19	0.05	0.9573	0.0023
50	47.15	2	988.00	0.08	0.9662	0.0029
60	30.06	2	983.14	0.13	0.9728	0.0036
70	19.76	2	977.68	0.20	0.9778	0.0045
80	13.35	2	971.68	0.29	0.9817	0.0054
90	9.25	2	965.18	0.42	0.9847	0.0065
100	6.55	2	958.23	0.60	0.9871	0.0077
110	4.74	2	950.84	0.83	0.9890	0.0091
40	306.09	4	992.19	0.05	0.9573	0.0023
50	188.62	4	988.00	0.08	0.9662	0.0029
60	120.25	4	983.14	0.13	0.9728	0.0036
70	79.04	4	977.68	0.20	0.9778	0.0045
80	53.41	4	971.68	0.29	0.9817	0.0054
90	37.00	4	965.18	0.42	0.9847	0.0065
100	26.22	4	958.23	0.60	0.9871	0.0077
110	18.96	4	950.84	0.83	0.9890	0.0091
40	688.70	6	992.19	0.05	0.9573	0.0023
50	424.39	6	988.00	0.08	0.9662	0.0029
60	270.55	6	983.14	0.13	0.9728	0.0036
70	177.84	6	977.68	0.20	0.9778	0.0045
80	120.17	6	971.68	0.29	0.9817	0.0054
90	83.25	6	965.18	0.42	0.9847	0.0065
100	58.99	6	958.23	0.60	0.9871	0.0077
110	42.66	6	950.84	0.83	0.9890	0.0091
40	1224.36	8	992.19	0.05	0.9573	0.0023
50	754.47	8	988.00	0.08	0.9662	0.0029
60	480.98	8	983.14	0.13	0.9728	0.0036
70	316.16	8	977.68	0.20	0.9778	0.0045
80	213.64	8	971.68	0.29	0.9817	0.0054
90	148.01	8	965.18	0.42	0.9847	0.0065
100	104.87	8	958.23	0.60	0.9871	0.0077
110	75.85	8	950.84	0.83	0.9890	0.0091

Table B.7 (continued): Variations in the acceleration pressure drop as a function of the mass flow rate and temperature for a cross section area of 1 m^2 and vapor mass fractions of 0.01 and 0.99.

T (°C)	ΔP (Pa)	M (kg/s)	ρ_ℓ (kg/m^3)	ρ_v (kg/m^3)	α_1	α_2
40	1913.06	10	992.19	0.05	0.9573	0.0023
50	1178.87	10	988.00	0.08	0.9662	0.0029
60	751.53	10	983.14	0.13	0.9728	0.0036
70	494.00	10	977.68	0.20	0.9778	0.0045
80	333.81	10	971.68	0.29	0.9817	0.0054
90	231.26	10	965.18	0.42	0.9847	0.0065
100	163.87	10	958.23	0.60	0.9871	0.0077
110	118.51	10	950.84	0.83	0.9890	0.0091
40	2754.80	12	992.19	0.05	0.9573	0.0023
50	1697.57	12	988.00	0.08	0.9662	0.0029
60	1082.21	12	983.14	0.13	0.9728	0.0036
70	711.36	12	977.68	0.20	0.9778	0.0045
80	480.69	12	971.68	0.29	0.9817	0.0054
90	333.01	12	965.18	0.42	0.9847	0.0065
100	235.97	12	958.23	0.60	0.9871	0.0077
110	170.65	12	950.84	0.83	0.9890	0.0091
40	3749.59	14	992.19	0.05	0.9573	0.0023
50	2310.58	14	988.00	0.08	0.9662	0.0029
60	1473.01	14	983.14	0.13	0.9728	0.0036
70	968.24	14	977.68	0.20	0.9778	0.0045
80	654.27	14	971.68	0.29	0.9817	0.0054
90	453.27	14	965.18	0.42	0.9847	0.0065
100	321.18	14	958.23	0.60	0.9871	0.0077
110	232.28	14	950.84	0.83	0.9890	0.0091

References

El-Dessouky, H.T., Alatiqi, I.M., **Ettouney, H.M.**, and Al-Deffeeri, N.S., Performance of wire mesh mist eliminator, Chem. Eng. & Proc., **39**(2000)129-139.

Lior, N., Formulas for calculating the approach to equilibrium in open channel flash evaporators for saline water, Desalination, **60**(1986)223.

Miyatake, O., Murakami, K., Kawata, Y., and Fujii, Fundamental Experiments with Flash Evaporation, Heat Transfer Jpn. Res., **2**(1973)89-100.

Oak Ridge National Laboratory (ORNL), United States Department of the Interior, Research and Development Progress Report No. 315, December 1967.

Zivi, S.M., Estimation of steady-state steam void fraction by means of the principle of minimum entropy production, Trans. ASME, J. of Heat Transfer, **86**(1964), 247-252.

Appendix C

Heat Transfer Coefficients

C.1 Falling Film on the Tube Outside Surface

The heat transfer coefficient of boiling thin film of water flowing over the outside surface of smooth horizontal tubes was developed by Han and Fletcher (1985),

$$h = 0.0004\ (\rho^2\ g\ k^3/\mu^2)^{1/3}\ Re^{0.2}\ Pr^{0.65}\ q^{0.4} \quad \text{(C.1)}$$

The relationship is valid over the following parameter range; $770 \leq Re \leq 7000$, $1.3 \leq Pr \leq 3.6$, $30 \leq q \leq 80$ kW/m^2, and $49 \leq T \leq 127$ °C. In the above equation Re and Pr are Reynolds and Prandtl numbers respectively, q, is the heat flux, μ is the viscosity, ρ is the density and k is the thermal conductivity of the fluid. Table C.1 and Fig. C.1 show variations in the heat transfer coefficient as a function of the system temperature and the heat flux.

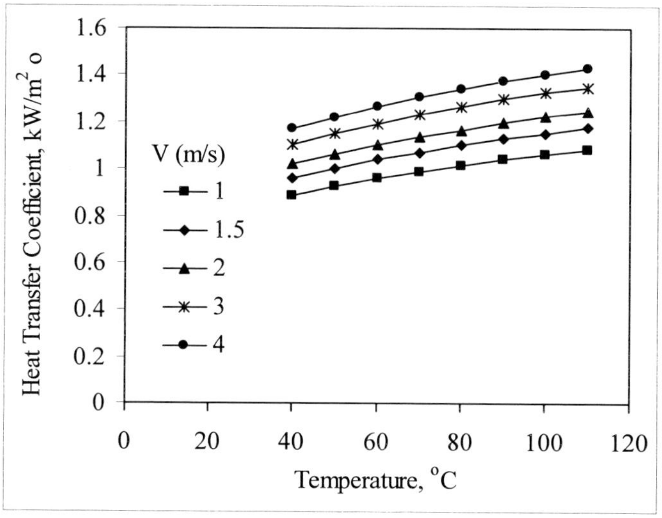

Fig. C.1. Variation in the heat transfer coefficient for boiling falling film

Table C.1: The heat transfer coefficient for boiling falling film for $\delta_o = 0.03$ m, $\delta_i = 0.025$ m, q = 80 kW/m^2, and X = 40000 ppm.

T (°C)	h (kW/m^2 °C)	V (m/s)	Re	Pr	k (kW/m °C)	μ (kg/m s)	Cp (kJ/kg °C)	ρ (kg/m^3)
40	0.89	1	42784.24	4.54	6.28E-04	7.16E-04	3.98	1021.37
50	0.92	1	50750.40	3.75	6.39E-04	6.01E-04	3.99	1017.07
60	0.96	1	59113.56	3.16	6.48E-04	5.14E-04	3.99	1012.24
70	0.99	1	67786.84	2.71	6.57E-04	4.46E-04	4.00	1006.89
80	1.02	1	76688.39	2.36	6.64E-04	3.92E-04	4.00	1001.06
90	1.04	1	85742.46	2.08	6.70E-04	3.48E-04	4.01	994.75
100	1.06	1	94879.95	1.86	6.74E-04	3.12E-04	4.02	987.99
110	1.08	1	104038.65	1.68	6.78E-04	2.83E-04	4.03	980.79
40	0.96	1.5	64176.36	4.54	6.28E-04	7.16E-04	3.98	1021.37
50	1.00	1.5	76125.61	3.75	6.39E-04	6.01E-04	3.99	1017.07
60	1.04	1.5	88670.34	3.16	6.48E-04	5.14E-04	3.99	1012.24
70	1.07	1.5	101680.26	2.71	6.57E-04	4.46E-04	4.00	1006.89
80	1.10	1.5	115032.59	2.36	6.64E-04	3.92E-04	4.00	1001.06
90	1.13	1.5	128613.69	2.08	6.70E-04	3.48E-04	4.01	994.75
100	1.15	1.5	142319.92	1.86	6.74E-04	3.12E-04	4.02	987.99
110	1.18	1.5	156057.97	1.68	6.78E-04	2.83E-04	4.03	980.79
40	1.02	2	85568.48	4.54	6.28E-04	7.16E-04	3.98	1021.37
50	1.06	2	101500.81	3.75	6.39E-04	6.01E-04	3.99	1017.07
60	1.10	2	118227.11	3.16	6.48E-04	5.14E-04	3.99	1012.24
70	1.14	2	135573.68	2.71	6.57E-04	4.46E-04	4.00	1006.89
80	1.17	2	153376.79	2.36	6.64E-04	3.92E-04	4.00	1001.06
90	1.20	2	171484.92	2.08	6.70E-04	3.48E-04	4.01	994.75
100	1.22	2	189759.90	1.86	6.74E-04	3.12E-04	4.02	987.99
110	1.25	2	208077.30	1.68	6.78E-04	2.83E-04	4.03	980.79
40	1.10	3	128352.72	4.54	6.28E-04	7.16E-04	3.98	1021.37
50	1.15	3	152251.21	3.75	6.39E-04	6.01E-04	3.99	1017.07
60	1.19	3	177340.67	3.16	6.48E-04	5.14E-04	3.99	1012.24
70	1.23	3	203360.52	2.71	6.57E-04	4.46E-04	4.00	1006.89
80	1.27	3	230065.18	2.36	6.64E-04	3.92E-04	4.00	1001.06
90	1.30	3	257227.37	2.08	6.70E-04	3.48E-04	4.01	994.75
100	1.33	3	284639.85	1.86	6.74E-04	3.12E-04	4.02	987.99
110	1.35	3	312115.95	1.68	6.78E-04	2.83E-04	4.03	980.79
40	1.17	4	171136.95	4.54	6.28E-04	7.16E-04	3.98	1021.37
50	1.22	4	203001.62	3.75	6.39E-04	6.01E-04	3.99	1017.07
60	1.26	4	236454.23	3.16	6.48E-04	5.14E-04	3.99	1012.24
70	1.30	4	271147.36	2.71	6.57E-04	4.46E-04	4.00	1006.89
80	1.34	4	306753.57	2.36	6.64E-04	3.92E-04	4.00	1001.06
90	1.37	4	342969.83	2.08	6.70E-04	3.48E-04	4.01	994.75
100	1.40	4	379519.80	1.86	6.74E-04	3.12E-04	4.02	987.99
110	1.43	4	416154.60	1.68	6.78E-04	2.83E-04	4.03	980.79

C.2 Vapor Condensation Inside Tubes

The heat transfer coefficient for vapor condensation inside horizontal tubes was developed by Shah (1978).

$$h/h_u = 1 + 3.8 / Z^{0.95} \quad (C.2)$$

where $Z = ((1/\chi) - 1)^{0.8}\ Pr^{0.4}$, $h_u = h_\ell\ (1 - \chi)^{0.8}$, $h_\ell = 0.023\ Re^{0.8}\ Pr^{0.4}\ (k_\ell/\delta_i)$, χ is the vapor phase mass fraction and the subscripts i, ℓ, and u denotes the tube inside, the liquid phase, and the local superficial value. The above correlation is valid over the following ranges: $2.8 \le \delta_i \le 40$ mm, $21 \le T \le 355$ °C, $0 \le \chi \le 1$, $0.158 \le q \le 16000$ kW/m^2, $11 \le G \le 4000$ kg/m^2 s, $0.7 \le P \le 1$ bar, $0.0019 \le Pr \le 0.82$, $350 \le Re \le 100000$. Table C.2 and Fig. C.2 show variations in the heat transfer coefficient as a function of the system temperature and vapor fraction.

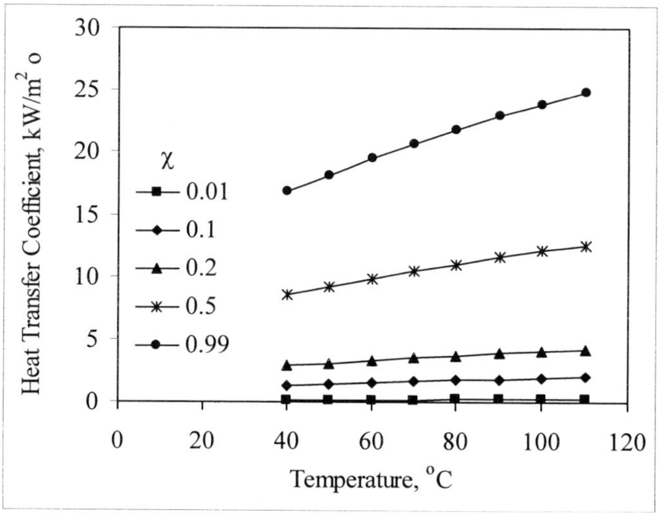

Fig. C.2. Variation in the heat transfer coefficient for vapor condensation inside the tubes

Table C.2: The heat transfer coefficient for vapor condensation inside the tubes, for δ_o = 0.03 m, δ_i = 0.025 m, and X = 40000 ppm.

T (°C)	h (kW/m^2 °C)	χ	Re	Pr	k (kW/m °C)	μ (kg/m s)	Cp (kJ/kg °C)	ρ (kg/m^3)
40	0.15	0.01	427.84	4.54	6.28E-04	7.16E-04	3.98	1021.37
50	0.16	0.01	507.50	3.75	6.39E-04	6.01E-04	3.99	1017.07
60	0.17	0.01	591.14	3.16	6.48E-04	5.14E-04	3.99	1012.24
70	0.18	0.01	677.87	2.71	6.57E-04	4.46E-04	4.00	1006.89
80	0.19	0.01	766.88	2.36	6.64E-04	3.92E-04	4.00	1001.06
90	0.20	0.01	857.42	2.08	6.70E-04	3.48E-04	4.01	994.75
100	0.21	0.01	948.80	1.86	6.74E-04	3.12E-04	4.02	987.99
110	0.22	0.01	1040.39	1.68	6.78E-04	2.83E-04	4.03	980.79
40	1.34	0.1	4278.42	4.54	6.28E-04	7.16E-04	3.98	1021.37
50	1.45	0.1	5075.04	3.75	6.39E-04	6.01E-04	3.99	1017.07
60	1.55	0.1	5911.36	3.16	6.48E-04	5.14E-04	3.99	1012.24
70	1.65	0.1	6778.68	2.71	6.57E-04	4.46E-04	4.00	1006.89
80	1.74	0.1	7668.84	2.36	6.64E-04	3.92E-04	4.00	1001.06
90	1.83	0.1	8574.25	2.08	6.70E-04	3.48E-04	4.01	994.75
100	1.91	0.1	9487.99	1.86	6.74E-04	3.12E-04	4.02	987.99
110	1.98	0.1	10403.86	1.68	6.78E-04	2.83E-04	4.03	980.79
40	2.88	0.2	8556.85	4.54	6.28E-04	7.16E-04	3.98	1021.37
50	3.11	0.2	10150.08	3.75	6.39E-04	6.01E-04	3.99	1017.07
60	3.33	0.2	11822.71	3.16	6.48E-04	5.14E-04	3.99	1012.24
70	3.54	0.2	13557.37	2.71	6.57E-04	4.46E-04	4.00	1006.89
80	3.74	0.2	15337.68	2.36	6.64E-04	3.92E-04	4.00	1001.06
90	3.92	0.2	17148.49	2.08	6.70E-04	3.48E-04	4.01	994.75
100	4.09	0.2	18975.99	1.86	6.74E-04	3.12E-04	4.02	987.99
110	4.25	0.2	20807.73	1.68	6.78E-04	2.83E-04	4.03	980.79
40	8.49	0.5	21392.12	4.54	6.28E-04	7.16E-04	3.98	1021.37
50	9.18	0.5	25375.20	3.75	6.39E-04	6.01E-04	3.99	1017.07
60	9.83	0.5	29556.78	3.16	6.48E-04	5.14E-04	3.99	1012.24
70	10.45	0.5	33893.42	2.71	6.57E-04	4.46E-04	4.00	1006.89
80	11.03	0.5	38344.20	2.36	6.64E-04	3.92E-04	4.00	1001.06
90	11.57	0.5	42871.23	2.08	6.70E-04	3.48E-04	4.01	994.75
100	12.08	0.5	47439.97	1.86	6.74E-04	3.12E-04	4.02	987.99
110	12.55	0.5	52019.32	1.68	6.78E-04	2.83E-04	4.03	980.79
40	16.82	0.99	42356.40	4.54	6.28E-04	7.16E-04	3.98	1021.37
50	18.18	0.99	50242.90	3.75	6.39E-04	6.01E-04	3.99	1017.07
60	19.47	0.99	58522.42	3.16	6.48E-04	5.14E-04	3.99	1012.24
70	20.69	0.99	67108.97	2.71	6.57E-04	4.46E-04	4.00	1006.89
80	21.84	0.99	75921.51	2.36	6.64E-04	3.92E-04	4.00	1001.06
90	22.92	0.99	84885.03	2.08	6.70E-04	3.48E-04	4.01	994.75
100	23.93	0.99	93931.15	1.86	6.74E-04	3.12E-04	4.02	987.99
110	24.86	0.99	102998.26	1.68	6.78E-04	2.83E-04	4.03	980.79

C.3 *Seawater Flowing Inside Tubes*

The heat transfer coefficient for seawater inside the tubes was developed for desalination plants by Wangnick (1995).

$$h = (3293.5+T(84.24-0.1714\ T) - x\ (8.471+0.1161\ x + 0.2716\ T))/((\delta_i/0.017272)^{0.2}) ((0.656\ V)^{0.8})\ (\delta_i/\delta_o) \quad (C.3)$$

where x is the salt concentration in weight percent, T is the temperature, and δ_i and δ_o are the inside and the outside tube diameter respectively. Table C.3 and Fig. C.3 show variations in the heat transfer coefficient as a function of the system temperature and the velocity. Table C.3 includes also values of the heat transfer coefficient as predicted by the Dittus-Bolter equation. It should be noted that values for Reynolds number, Prandtl number, and other physical properties are the same as those given in Table C.1.

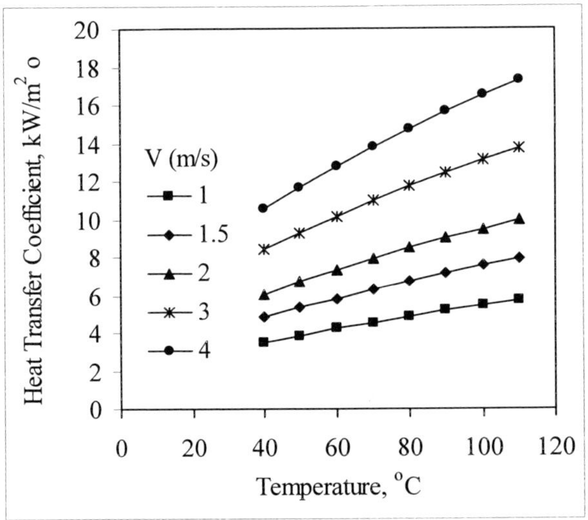

Fig. C.3. Variation in the heat transfer coefficient for seawater flowing inside the tubes

Table C.3: The heat transfer coefficient for seawater flowing inside the tubes, for $\delta_o = 0.03$ m, $\delta_i = 0.025$ m, and X = 40000 ppm.

T (°C)	h (Eq. C.3) (kW/m^2 °C)	h (Dittus-Bolter) (kW/m^2 °C)	V (m/s)
40	3.49	4.19	1
50	3.86	4.58	1
60	4.21	4.97	1
70	4.55	5.33	1
80	4.87	5.68	1
90	5.17	6.01	1
100	5.45	6.32	1
110	5.71	6.62	1
40	4.82	5.79	1.5
50	5.34	6.34	1.5
60	5.83	6.87	1.5
70	6.29	7.38	1.5
80	6.73	7.86	1.5
90	7.15	8.32	1.5
100	7.53	8.75	1.5
110	7.89	9.15	1.5
40	6.07	7.29	2
50	6.72	7.98	2
60	7.34	8.65	2
70	7.92	9.29	2
80	8.48	9.89	2
90	9.00	10.47	2
100	9.48	11.01	2
110	9.94	11.52	2
40	8.39	10.09	3
50	9.29	11.04	3
60	10.15	11.96	3
70	10.96	12.84	3
80	11.72	13.68	3
90	12.44	14.48	3
100	13.11	15.23	3
110	13.74	15.93	3
40	10.57	12.70	4
50	11.70	13.90	4
60	12.78	15.06	4
70	13.80	16.17	4
80	14.76	17.23	4
90	15.66	18.23	4
100	16.51	19.17	4
110	17.30	20.05	4

C.4 Vapor Condensation on the Outside Surface of Tubes

The correlation for the heat transfer coefficient during vapor condensation outside the preheater/condenser tubes was developed by Henning and Wangnick (1995),

$$h = 0.725\,(k_\ell^3 \rho_\ell (\rho_\ell - \rho_v) g \lambda_v / \delta_o \mu \Delta T)^{0.25}\, C_1 C_2 \qquad (C.4)$$

with $C_1 = 1.23795 + 0.353808 N_l - 0.0017035 N_l^2$,

$C_2 = 1 - 34.313 X_{nc} + 1226.8 X_{nc}^2 - 14923 X_{nc}^3$,

$N_l = 0.564\sqrt{N_t}$, and $N_t = \dfrac{4 M_f}{\pi \delta_i^2 \rho_f V_f}$

Variations in the heat transfer coefficient are shown in Table C.4 and Fig. C.4.

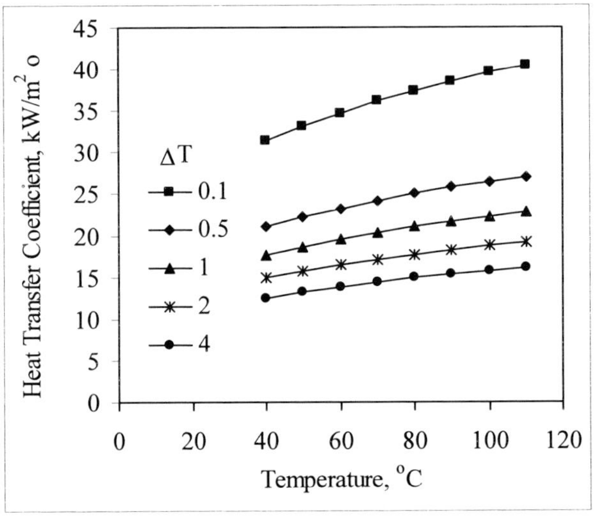

Fig. C.4. Variation in the heat transfer coefficient on the outside surface of tubes

Table C.4: Variation in the heat transfer coefficient during condensation on the outside surface of tubes. Parameters include X_{nc} = 0.015, δ_o = 0.03 m, M_f = 1 kg/s, V_f = 1.5 m/s, δ_i = 0.025 m. Calculated values include N_t = 1.37, N_1 = 0.66, C_1 = 1.47, C_2 = 0.71.

T (°C)	h (kW/m^2 °C)	ΔT (°C)	ρ_v (kg/m^3)	ρ_ℓ (kg/m^3)	k_ℓ (kW/m °C)	λ_v (kJ/kg)	μ_ℓ (kg/m s)
40	31.41	0.1	0.051	991.861	6.30E-04	2406.50	6.55E-04
50	33.12	0.1	0.083	987.683	6.41E-04	2382.52	5.48E-04
60	34.68	0.1	0.130	982.924	6.50E-04	2358.31	4.67E-04
70	36.11	0.1	0.198	977.602	6.58E-04	2333.76	4.04E-04
80	37.41	0.1	0.293	971.734	6.65E-04	2308.77	3.54E-04
90	38.56	0.1	0.424	965.339	6.71E-04	2283.25	3.14E-04
100	39.58	0.1	0.598	958.434	6.76E-04	2257.11	2.81E-04
110	40.47	0.1	0.826	951.037	6.79E-04	2230.25	2.54E-04
40	21.01	0.5	0.051	991.861	6.30E-04	2406.50	6.55E-04
50	22.15	0.5	0.083	987.683	6.41E-04	2382.52	5.48E-04
60	23.19	0.5	0.130	982.924	6.50E-04	2358.31	4.67E-04
70	24.15	0.5	0.198	977.602	6.58E-04	2333.76	4.04E-04
80	25.02	0.5	0.293	971.734	6.65E-04	2308.77	3.54E-04
90	25.79	0.5	0.424	965.339	6.71E-04	2283.25	3.14E-04
100	26.47	0.5	0.598	958.434	6.76E-04	2257.11	2.81E-04
110	27.07	0.5	0.826	951.037	6.79E-04	2230.25	2.54E-04
40	17.66	1	0.051	991.861	6.30E-04	2406.50	6.55E-04
50	18.62	1	0.083	987.683	6.41E-04	2382.52	5.48E-04
60	19.50	1	0.130	982.924	6.50E-04	2358.31	4.67E-04
70	20.31	1	0.198	977.602	6.58E-04	2333.76	4.04E-04
80	21.04	1	0.293	971.734	6.65E-04	2308.77	3.54E-04
90	21.69	1	0.424	965.339	6.71E-04	2283.25	3.14E-04
100	22.26	1	0.598	958.434	6.76E-04	2257.11	2.81E-04
110	22.76	1	0.826	951.037	6.79E-04	2230.25	2.54E-04
40	14.85	2	0.051	991.861	6.30E-04	2406.50	6.55E-04
50	15.66	2	0.083	987.683	6.41E-04	2382.52	5.48E-04
60	16.40	2	0.130	982.924	6.50E-04	2358.31	4.67E-04
70	17.08	2	0.198	977.602	6.58E-04	2333.76	4.04E-04
80	17.69	2	0.293	971.734	6.65E-04	2308.77	3.54E-04
90	18.24	2	0.424	965.339	6.71E-04	2283.25	3.14E-04
100	18.72	2	0.598	958.434	6.76E-04	2257.11	2.81E-04
110	19.14	2	0.826	951.037	6.79E-04	2230.25	2.54E-04
40	12.49	4	0.051	991.861	6.30E-04	2406.50	6.55E-04
50	13.17	4	0.083	987.683	6.41E-04	2382.52	5.48E-04
60	13.79	4	0.130	982.924	6.50E-04	2358.31	4.67E-04
70	14.36	4	0.198	977.602	6.58E-04	2333.76	4.04E-04
80	14.87	4	0.293	971.734	6.65E-04	2308.77	3.54E-04
90	15.33	4	0.424	965.339	6.71E-04	2283.25	3.14E-04
100	15.74	4	0.598	958.434	6.76E-04	2257.11	2.81E-04
110	16.09	4	0.826	951.037	6.79E-04	2230.25	2.54E-04

C.5 Water Flow in Plate Heat Exchanger

The heat transfer coefficient in plate heat exchangers is developed by Buonopane et al. (1974) and is given in terms of variations in the Nusselt number as a function of the Reynolds and Prandtl numbers of the fluid; this is

$$h = 0.2536 \, Re^{0.65} \, Pr^{0.4} \, (k_w/De) \tag{C.5}$$

Where Re is the Reynolds number, which is defined in terms of the effective diameter ($Re = De \, \rho \, v/\mu$), Pr is the Prandtl number, and De is the equivalent diameter, which is defined by $De = 4 \, (wd)/(2(w+d))$. In the above equations, ρ is density, μ is viscosity, k is thermal conductivity, Cp is heat capacity, v is velocity, w is plate width, and d is plate spacing. Variations in the heat transfer coefficient are shown in Table C.5 and Fig. C.5 as a function of temperature and velocity. Values for physical properties, which includes μ, k, Cp, and ρ are given Table C.1.

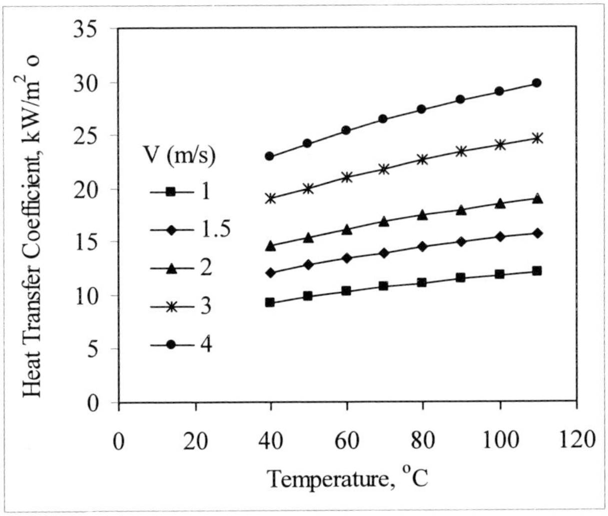

Fig. C.5. Variation in the heat transfer coefficient for seawater flowing in plate heat exchanger

Table C.5: The heat transfer coefficient in plate heat exchanger for w = 0.2 m, d = 0.02 m, De = 0.036 m, X = 40,000 ppm.

T (°C)	h (kW/m^2 °C)	V (m/s)	Re	Pr
40	9.31	1	51859.68	4.54
50	9.80	1	61515.64	3.75
60	10.26	1	71652.80	3.16
70	10.68	1	82165.87	2.71
80	11.07	1	92955.63	2.36
90	11.42	1	103930.25	2.08
100	11.75	1	115006.00	1.86
110	12.04	1	126107.45	1.68
40	12.11	1.5	77789.52	4.54
50	12.76	1.5	92273.46	3.75
60	13.35	1.5	107479.19	3.16
70	13.90	1.5	123248.80	2.71
80	14.41	1.5	139433.44	2.36
90	14.87	1.5	155895.38	2.08
100	15.29	1.5	172509.00	1.86
110	15.66	1.5	189161.18	1.68
40	14.60	2	103719.37	4.54
50	15.38	2	123031.28	3.75
60	16.10	2	143305.59	3.16
70	16.76	2	164331.73	2.71
80	17.37	2	185911.26	2.36
90	17.93	2	207860.50	2.08
100	18.43	2	230012.00	1.86
110	18.89	2	252214.91	1.68
40	19.01	3	155579.05	4.54
50	20.02	3	184546.93	3.75
60	20.96	3	214958.39	3.16
70	21.82	3	246497.60	2.71
80	22.61	3	278866.88	2.36
90	23.33	3	311790.76	2.08
100	23.99	3	345018.00	1.86
110	24.58	3	378322.36	1.68
40	22.91	4	207438.73	4.54
50	24.14	4	246062.57	3.75
60	25.26	4	286611.19	3.16
70	26.30	4	328663.46	2.71
80	27.26	4	371822.51	2.36
90	28.13	4	415721.01	2.08
100	28.92	4	460023.99	1.86
110	29.63	4	504429.82	1.68

C.6 Condensers and Evaporators Overall Heat Transfer Coefficient

Several correlations are available for the overall heat transfer coefficient. Predicted values by theses correlations vary between lows of 2 kW/m^2 ^oC up to highs of 4 kW/m^2 ^oC. Variations depend on the fouling resistance and the surface conditions. Results for these correlations are shown in Fig. C.6. The following is a list of these correlations:

- Fouled condenser, El-Dessouky et al. (1998)

$$U_c = 1x10^{-3}\,(1617.5 + 0.1537\,T + 0.1825\,T^2 - 0.00008026\,T^3) \qquad (C.6)$$

- Fouled evaporator, El-Dessouky et al. (1998)

$$U_e = 1x10^{-3}\,(1939.4 + 1.40562\,T - 0.0207525\,T^2 + 0.0023186\,T^3) \qquad (C.7)$$

- Fouled condenser, Takada et al. (1983)

$$U_c = 0.8\,(3 + 0.05\,(T - 60)) \qquad (C.8)$$

- Clean drop wise condenser, Bromly et al. (1970)

$$U_c = 1x10^{-3}\,(5186 - 90.82\,T + 0.5566\,T^2 - 0.0009159\,T^3)/0.17612 \qquad (C.9)$$

- Clean film wise condenser, Bromly et al. (1970)

$$U_c = 1x10^{-3}\,(-316.2 + 6.62\,T)/0.17612 \qquad (C.10)$$

- Oxidized film wise condenser, Bromly et al. (1970)

$$U_c = 1x10^{-3}\,(-64.37 + 4.625\,T)/0.17612 \qquad (C.11)$$

In the above equations, U_c is the condenser overall heat transfer coefficient (with vapor condensing on the outside surface and seawater flow on the tube side), U_e is the evaporator overall heat transfer coefficient (with water forming a falling film on a horizontal tube bundle and vapor is condensing inside the tubes), and T is the evaporation/condensation temperature. As is shown in Fig. C.6 the units of (U) and (T) are kW/m^2 ^oC and ^oC, respectively. It should be noted that Fig. C.6 includes additional data points by other investigator; however, these investigators did not provide a correlation. In addition, the clean overall heat transfer coefficient by El-Dessouky et al. (1998) is the same as Eqs. (C.6) and C.7), where it uses a fouling resistance of 0.08 m^2 °C/kW.

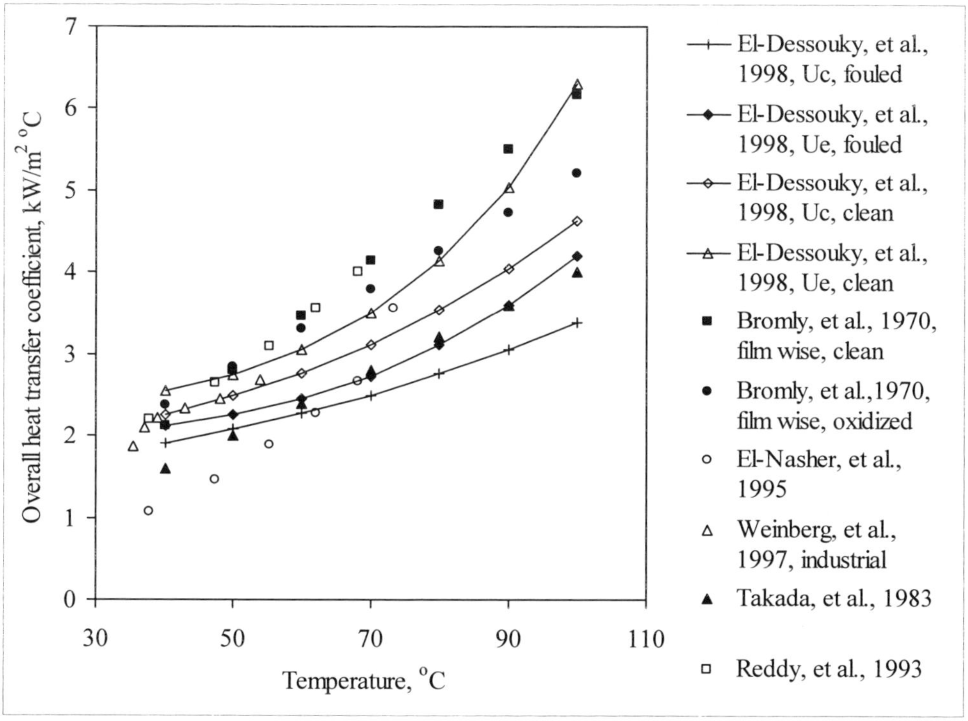

Figure 4: Variations in the overall heat transfer coefficient predicted by various correlations and as a function of temperature.

References

Bromley, L.A., and Read, S.M., Multiple effect flash (MEF) evaporator, Desalination, **70**(1970)3413-391.

Buonopane, R.A., Troupe, R.A., and Morgan, J.C., Heat transfer design method for plate heat exchangers, Chem. Eng. Progress, **59**(7)(1963)57-61.

El-Dessouky, H., Alatiqi, I., Bingulac, S., and Ettouney, H., Steady-state analysis of the multiple effect evaporation desalination process, Chem. Eng. Technol., **21**(1998)15-29.

El-Nashar, A.M., and Qamhiyeh, A.A., Simulation of the steady-stage operation of a mulit-effect stack seawater distillation plant, Desalination, **101**(1995)231-243.

Han, J. and Fletcher, L., Falling film evaporation and boiling in circumferential and axial grooves on horizontal tubes, Ind. Eng. Chem. Process Des. Dev., **24**(1985)570-597.

Henning, S, and Wangnick, K., Comparison of different equations for the calculation of heat transfer coefficients in MSF multi-stage flash evaporators, Proceedings of the IDA World Congress on Desalination and Water Sciences, Abu Dhabi, November, 1995, Vol. III, pp. 515-524.

Reddy, G.P., Husain, A., and Al-Gobaisi, D.M.K., Modelling and optimization of muliple effect horizontal tube falling film evaporators, Proceeding of the IDA World Congress on Desalination and Water Sciences, Madrid, Spain, October, 1997, Vol. I, pp 131-149.

Shah, M.M., Heat transfer, pressure drop, visual observations, test data for ammonia evaporating inside tubes, ASHRAE Trans., Vol. 84, Part 2, 1978.

Takada, M., and Drake, J.C., Application of improved high performance evaporator, Desalination, **45**(1983)3-12.

Wangnick, K., How incorrectly determined physical and constructional properties in the seawater and brine regimes influence the design and size of an MSF desalination plant - stimulus for further thoughts, Proceedings of the IDA World Congress on Desalination and Water Science, Abu Dhabi, November, 1995, Vol. II, pp. 201-218.

Weinberg, J., and Ophir, A., Ashdod experience and other dual purpose desalination plants based on multi effect desalination with aluminum tubes, Symposium on Desalination of Seawater with Nuclear Energy, Taejon, Republic of Korea, May, 1997.

Appendix D

Computer Package

D.1. Main Window

A schematic of main window is displayed in Figs. D.1-D.5. The file menu of the window includes commands for program exit, process selection, calculations of the physical properties, and help. The physical property command includes options for calculations of the seawater and the fresh water properties.

Process selection includes the following:

- Single effect evaporation (SEE)
- Single effect evaporation with mechanical vapor compression (MVC).
- Single effect evaporation with thermal vapor compression (TVC).
- Single effect evaporation with absorption vapor compression (ABVC).
- Single effect evaporation with adsorption vapor compression (ADVC).
- Parallel feed multiple effect evaporation (MEE-P).
- Parallel feed multiple effect evaporation with mechanical vapor compression (MEE-P-MVC).
- Parallel feed multiple effect evaporation with thermal vapor compression (MEE-P-TVC).
- Forward feed multiple effect evaporation.
- Multistage flash desalination with brine circulation (MSF).
- Multistage flash desalination once through process (MSF-OT).

Before using the package the user should keep in mind that the model equations for each system are only valid over a specific range. The package has some built-in features that warn the user if the design or solution parameters are outside practical ranges. However, in some instances and because of the equations non-linearity, the package may return an error and abort calculations. In this case, it is highly advised to start over with the default options and proceed with small variations in the desired parameter. Also, in the event of persistent error the user is encouraged to contact the authors.

The package has windows for system design and flow chart calculations. The design window allows for variations of the feed conditions as well as the design features of the condenser, evaporator, and preheaters. The flow chart window includes only the option to adjust the feed conditions and it uses default values for design of the condenser, evaporator, and preheaters. The flow chart window is intended for educational and training purposes, since it provide the user various elements of the system.

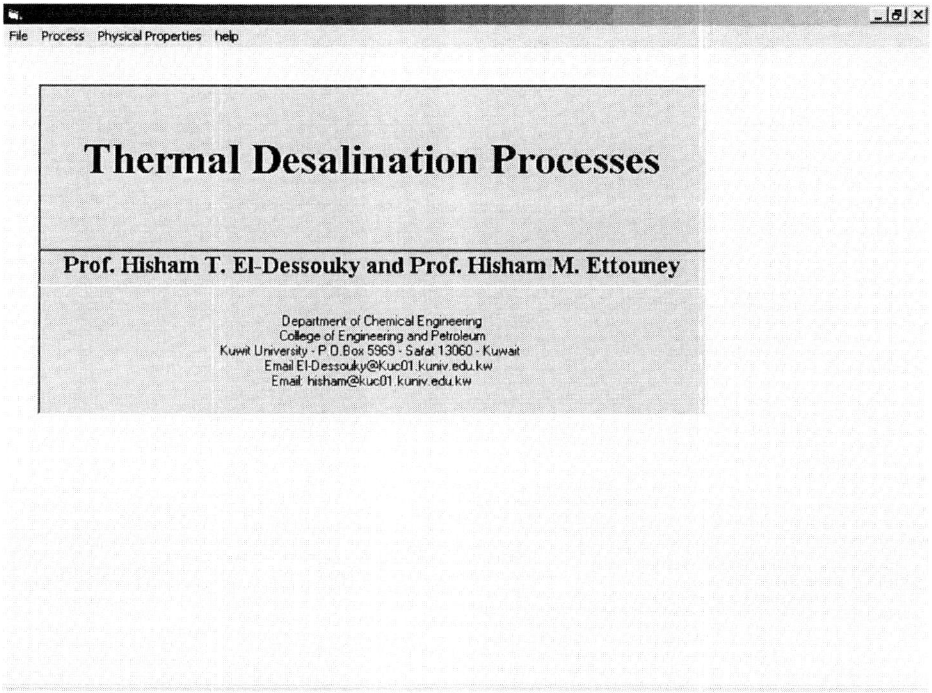

Fig. D.1 Startup window

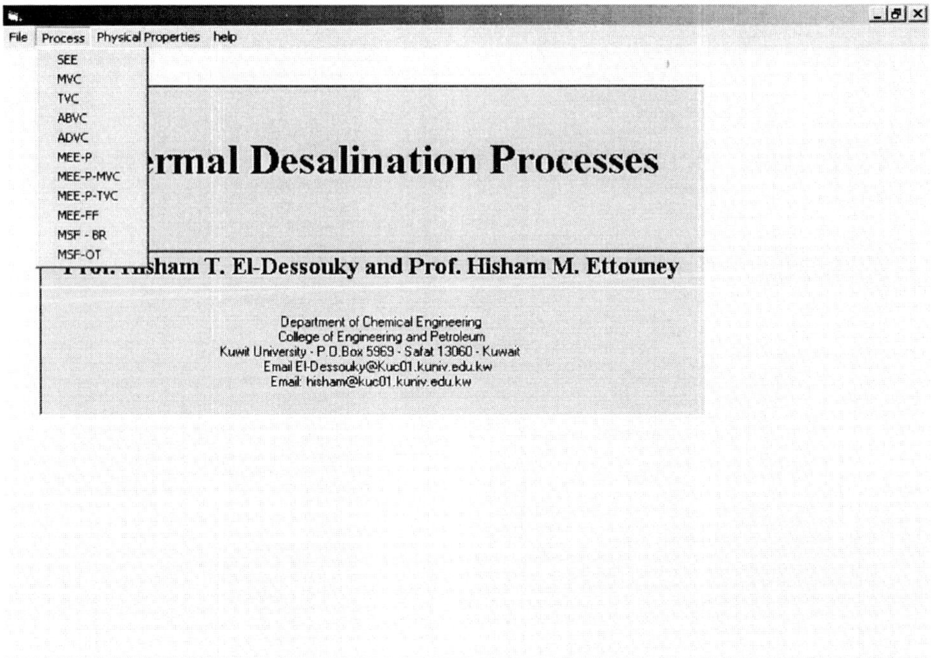

Fig. D.2 Process selection menu

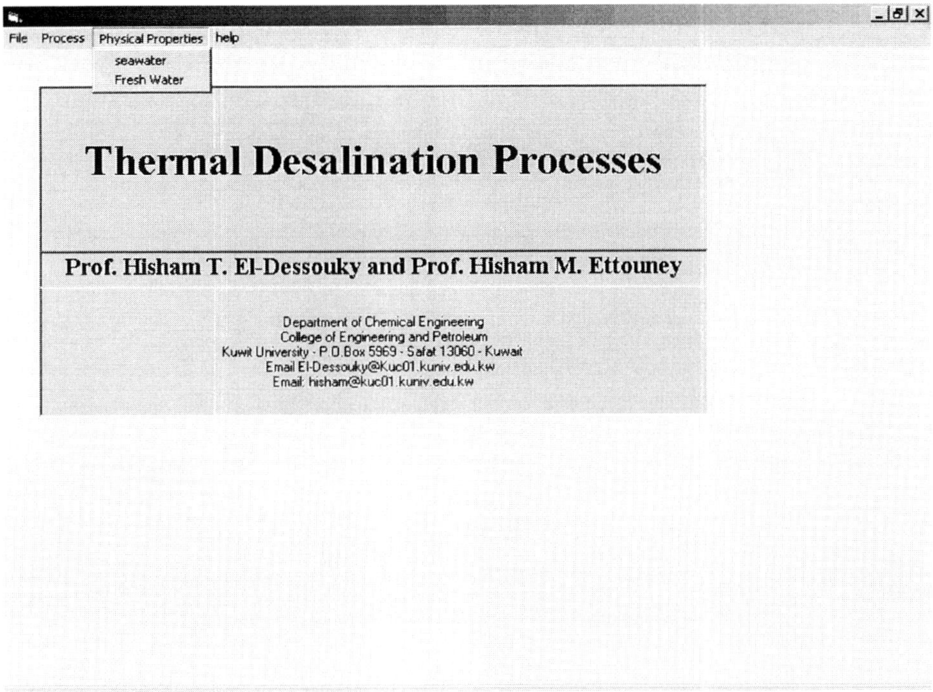

Fig. D.3 Physical properties selection menu

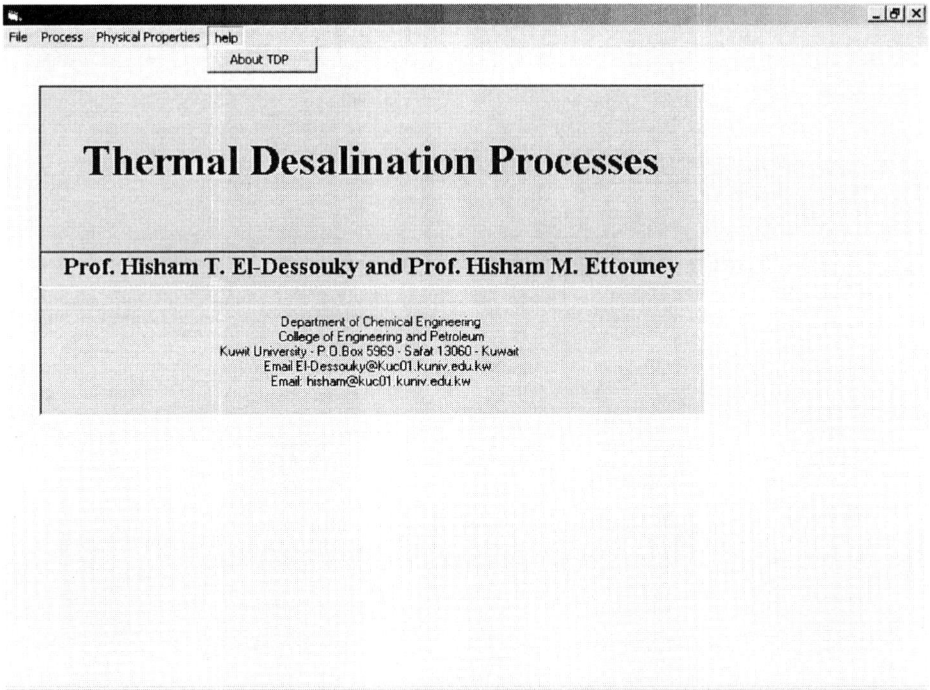

Fig. D.4 Program information in startup window

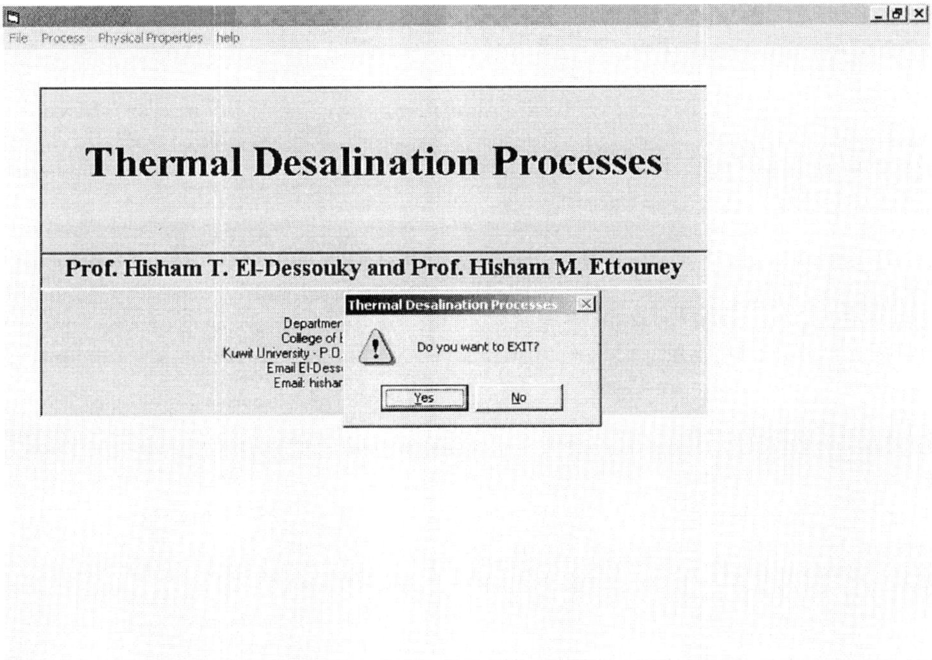

Fig. D.5 Exit command in startup window

D.2. Physical Properties

The physical properties include calculations for seawater and fresh water. The seawater window, Fig. D.6, includes properties of density, specific heat, dynamic viscosity, and thermal conductivity. Calculations are performed as a function of temperature and salinity.

The fresh water window, Fig. D.7, includes properties for liquid water and water vapor. The liquid water properties include enthalpy, entropy, specific volume, viscosity, and surface tension. The water vapor properties include enthalpy, entropy, specific volume, and viscosity. The calculations are performed at saturation conditions and assume zero salinity.

The fresh water window also includes the saturation pressure and latent heat that corresponds to the system temperature. It should be stressed that limits on correlations for fresh water and seawater are given in appendix A.

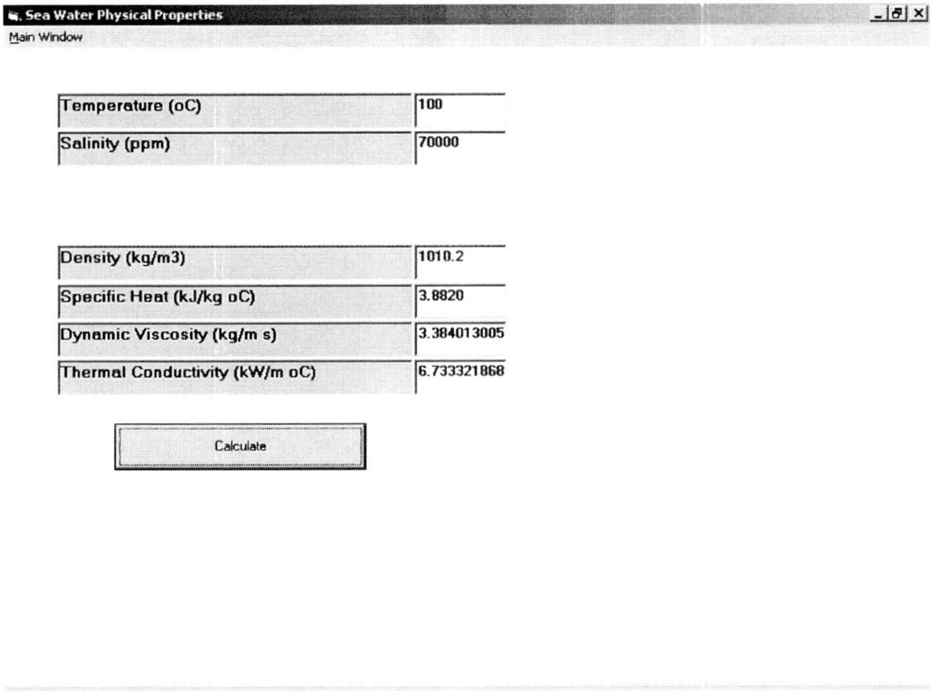

Fig. D.6 Physical properties of seawater

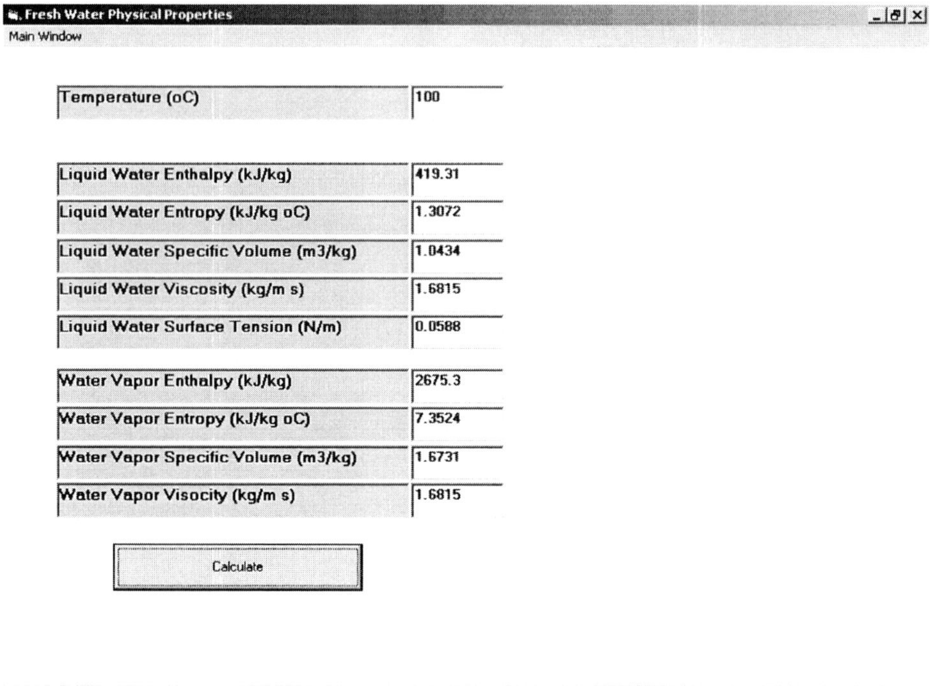

Fig. D.7 Physical properties of fresh water

D.3. *Single Effect Evaporation (SEE)*

The startup window of the SEE process has the process title and includes options to view the process diagram, perform design calculations, perform flow chart calculations, or return to the main window, Figs. D.8-D.9.

The SEE design window include entries for input and design data. The input data includes temperatures, flow rates, and salinity of feed streams. The design data include specifications for the condenser and evaporator tube as well as the demister. The window includes options to view the default design data, results, flow chart calculations, or return to the main window. Also, the window includes an option to print the design form, Fig. D.10. Calculations results include heat transfer area, flow rates, salinity, and temperature of outlet streams, Fig. D.11.

The displays for the flow chart window are shown in Figs. D.12-D.16. Figure D.12 illustrates various elements of the single effect flow chart, which includes the evaporator and condenser as well as various inlet/outlet streams. Figure D.13 shows return message if the user attempts to view the contents of the block diagram. Solution of the system equations require definition of the data shown in Fig. D.14. Solution of the system equations allow to view the content of the block elements of diagram, Figs. D.15-D.16. In Fig. D.15 the contents of the evaporator/condenser unit are displayed and it includes the heat transfer area, the overall heat transfer coefficient, the thermal load, and the performance ratio. In Fig. D.16, the contents of the down condenser are illustrated.

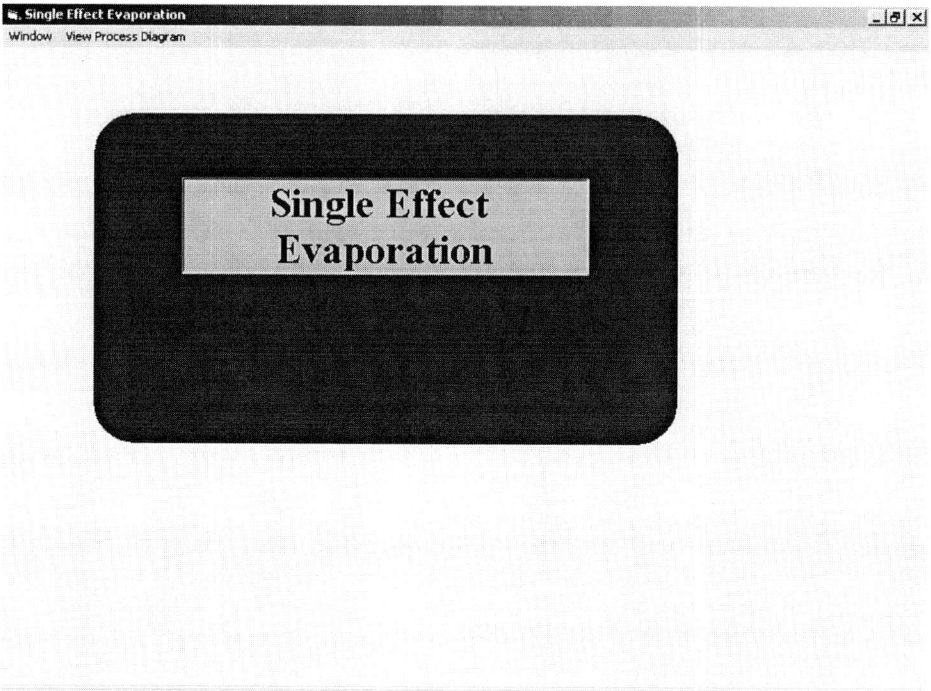

Fig. D.8 Startup window for single effect evaporation

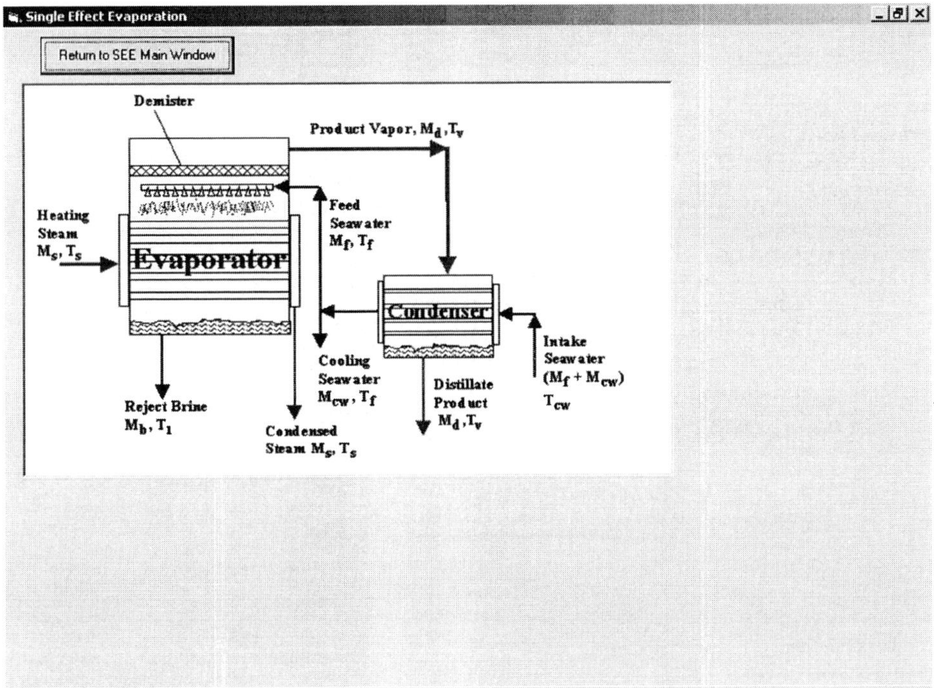

Fig. D.9 Schematic of single effect evaporation

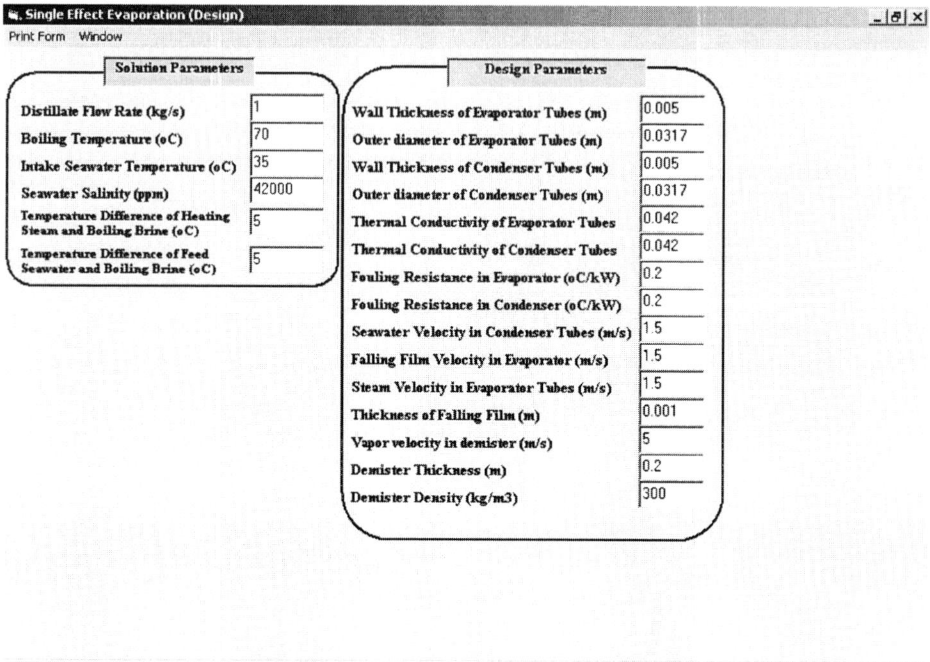

Fig. D.10 Design window for the single effect evaporation

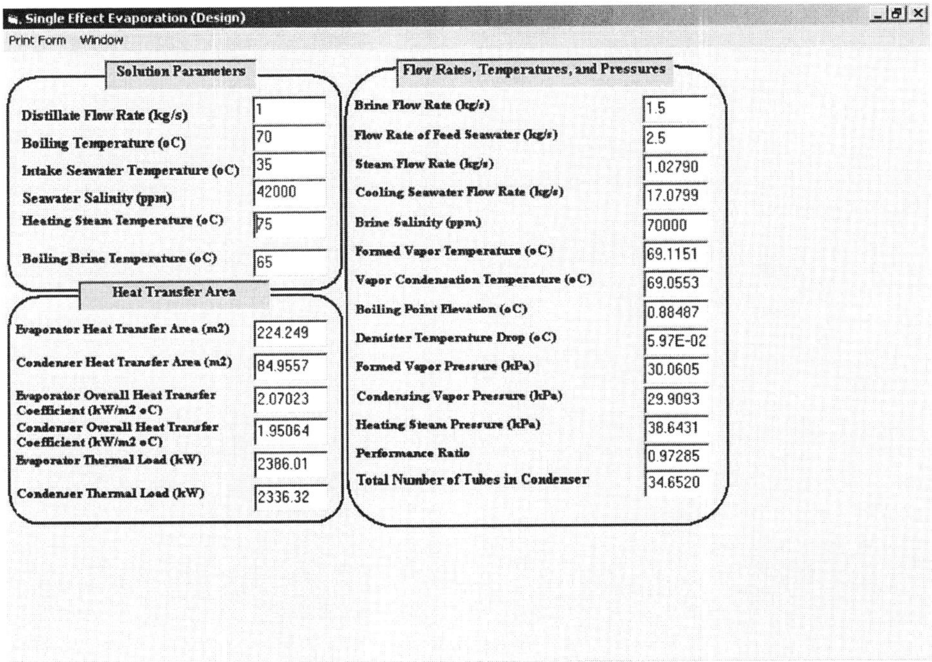

Fig. D.11 Results window for design of the single effect evaporation.

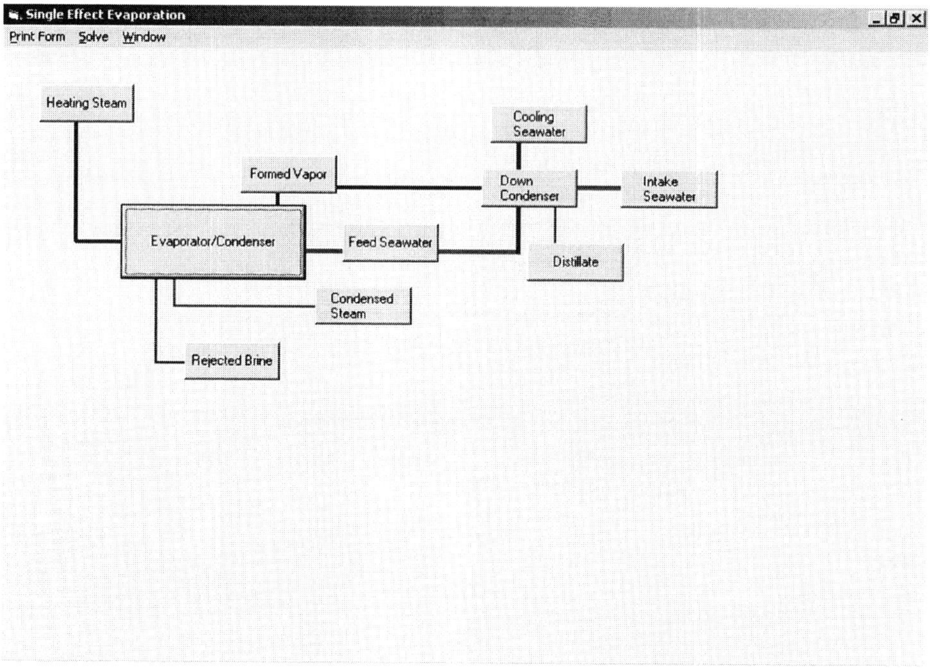

Fig. D.12 Flow chart window for single effect evaporation.

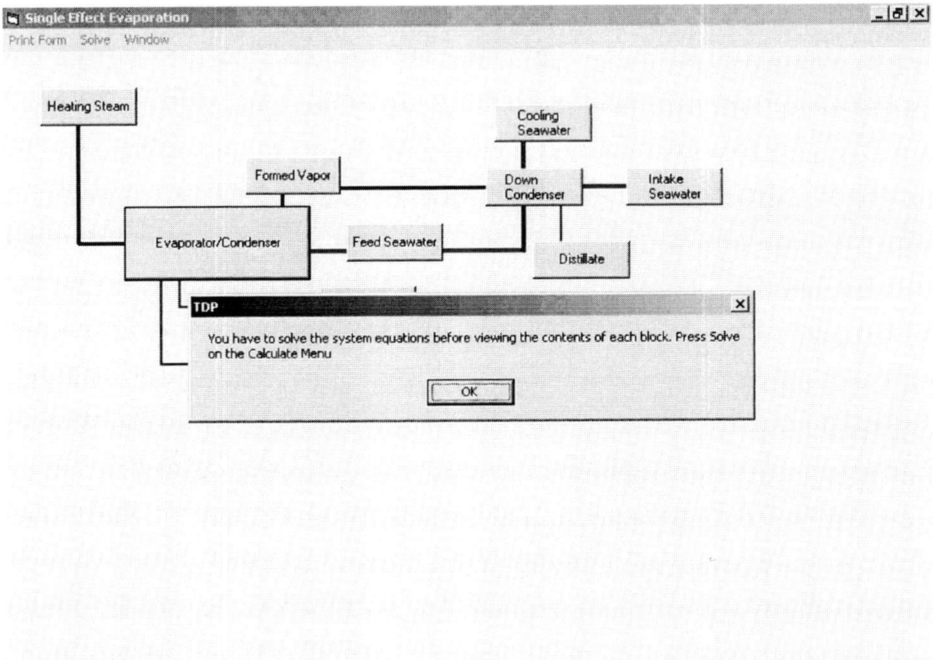

Fig. D.13 Warning message requesting solution of the model equations for the flow chart of the single effect evaporation.

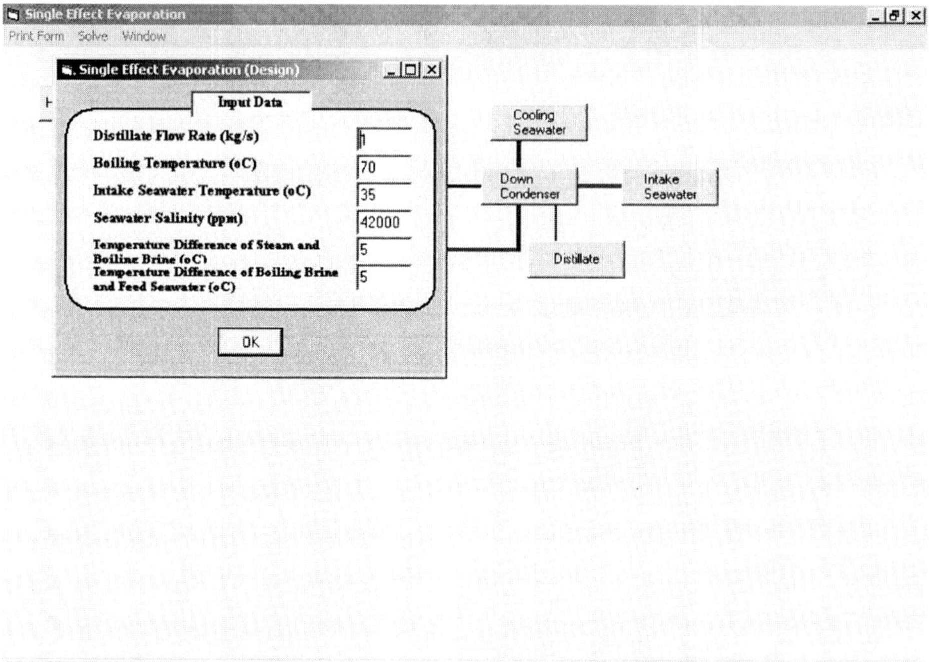

Fig. D.14 Solution parameters for the flow chart of the single effect evaporation process.

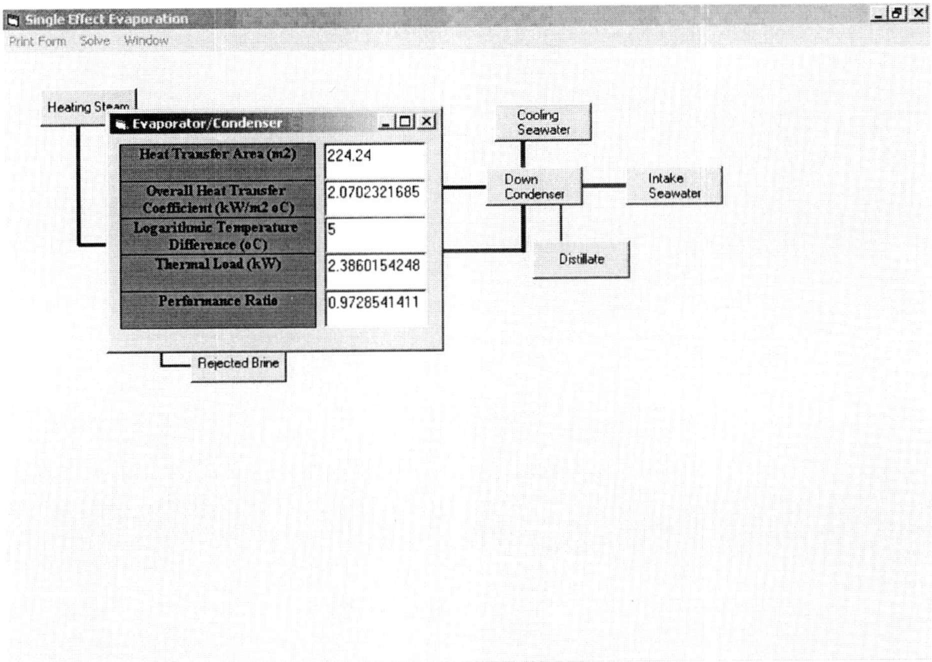

Fig. D.15 Results for the evaporator/condenser unit.

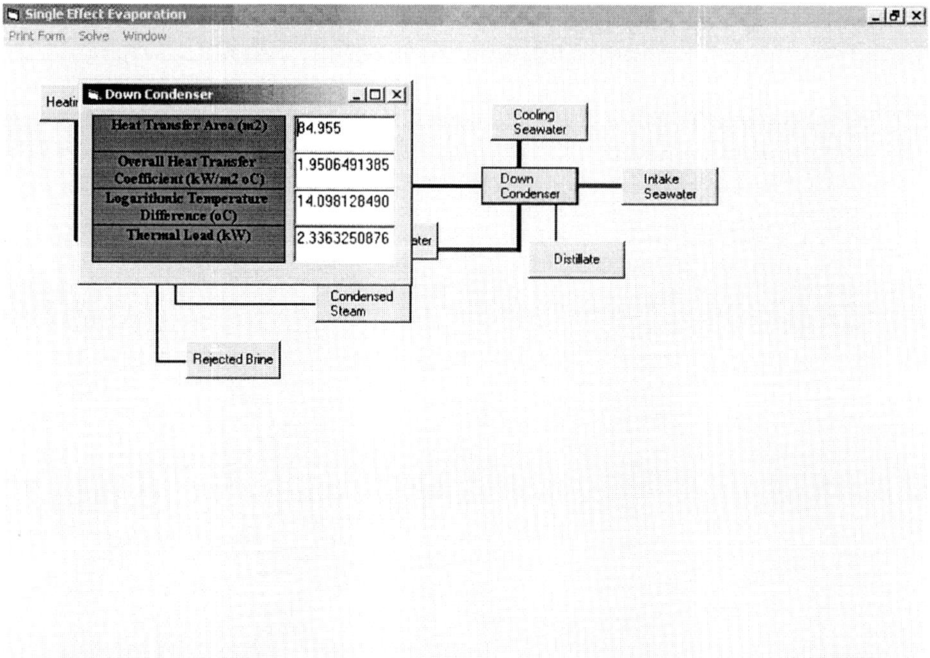

Fig. D.16 Solution results of the down condenser.

D.4. Single Effect Evaporation with Vapor Compression

The package elements for the single effect evaporation with vapor compression are similar to those of the single effect evaporation unit given in the previous section Figs. D.17-D.37. It should be noted that the flow chart calculations are only available for the single effect systems with mechanical or thermal vapor compression systems. As for the absorption or adsorption vapor compression units, the calculations are limited to the design diagram.

As is shown in Figs D.17-D.37 each system has its own specific feature, i.e., the steam jet ejector in the thermal vapor compression or the compressor in the mechanical vapor compression system. In addition, results for each system include specific feature, i.e., the specific power consumption of the compressor, the flow rate of the motive steam in the ejector of the thermal vapor compression unit, or the heat transfer area of the feed preheaters in the mechanical vapor compression unit.

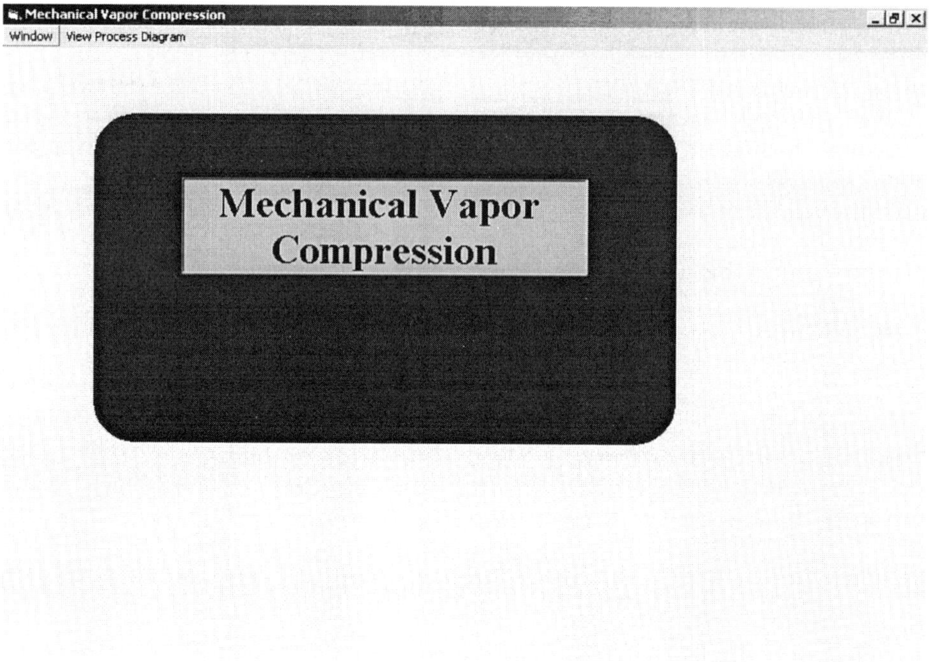

Fig. D.17 Startup window for the single effect mechanical vapor compression.

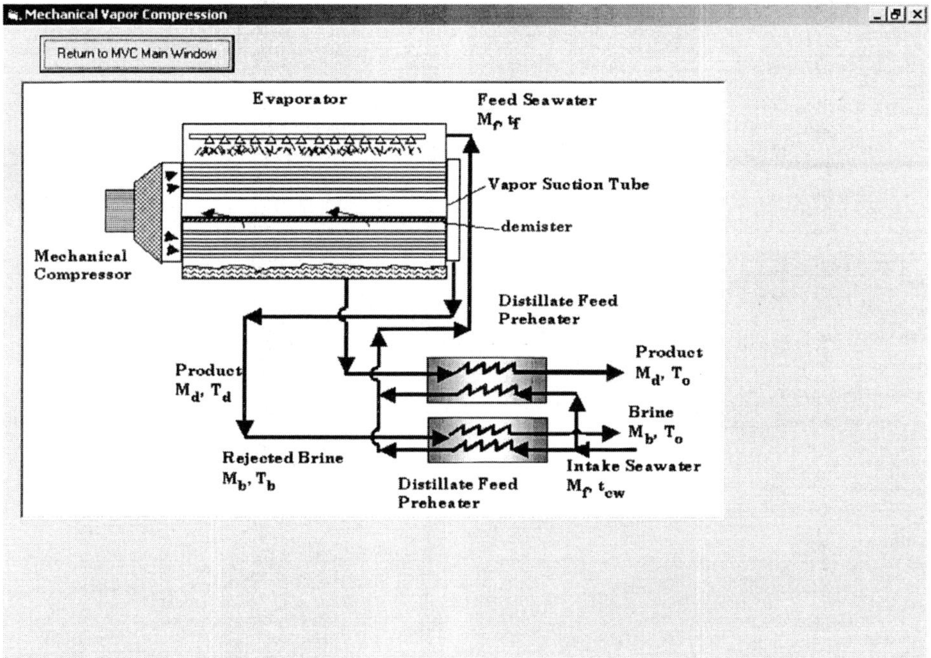

Fig. D.18 Schematic for the single effect mechanical vapor compression.

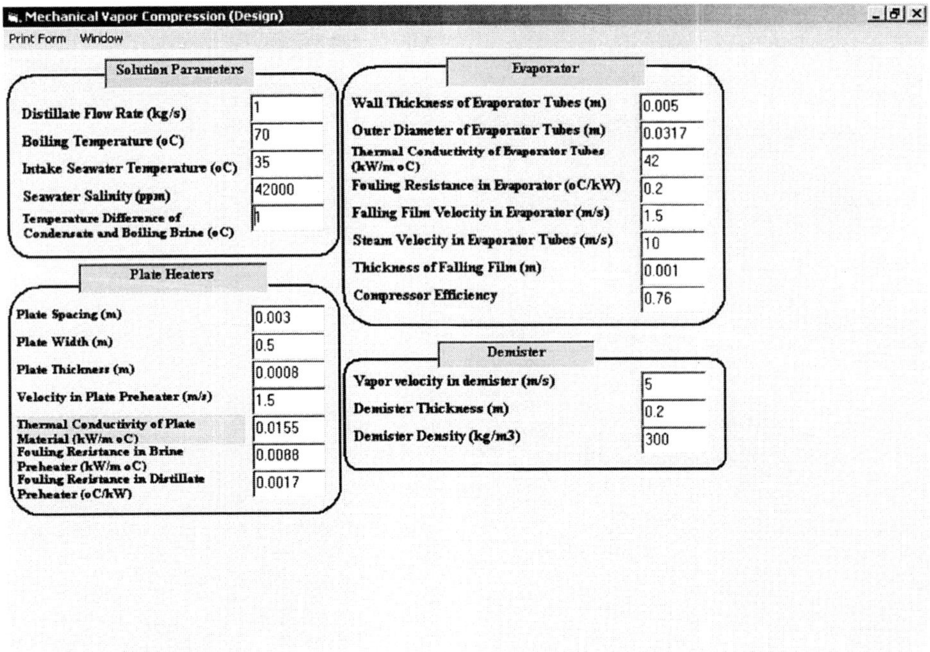

Fig. D.19 Input data for design of the single effect mechanical vapor compression.

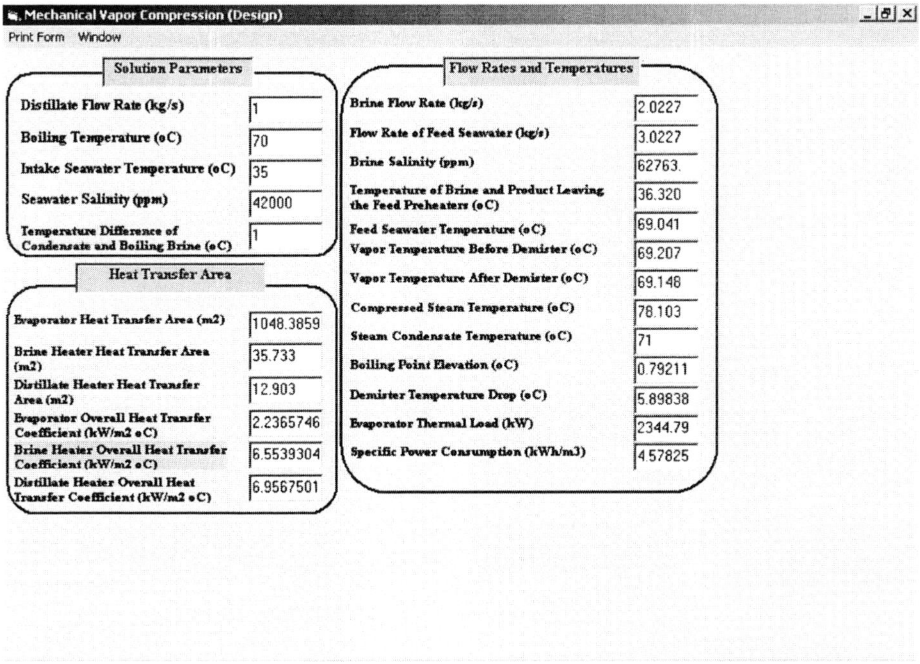

Fig. D.20 Results for design of the single effect mechanical vapor compression.

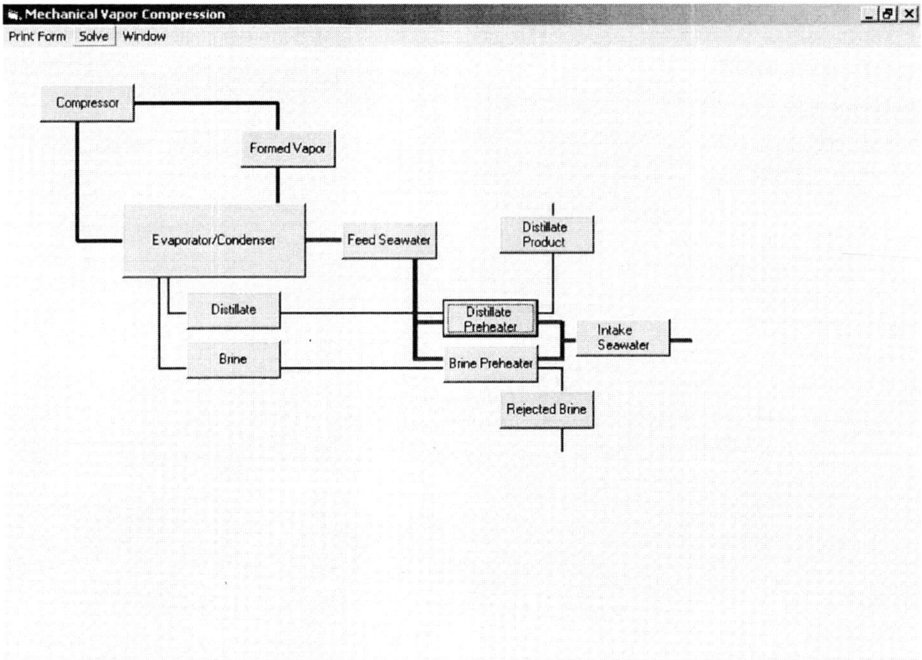

Fig. D.21 Flow chart window for the single effect mechanical vapor compression.

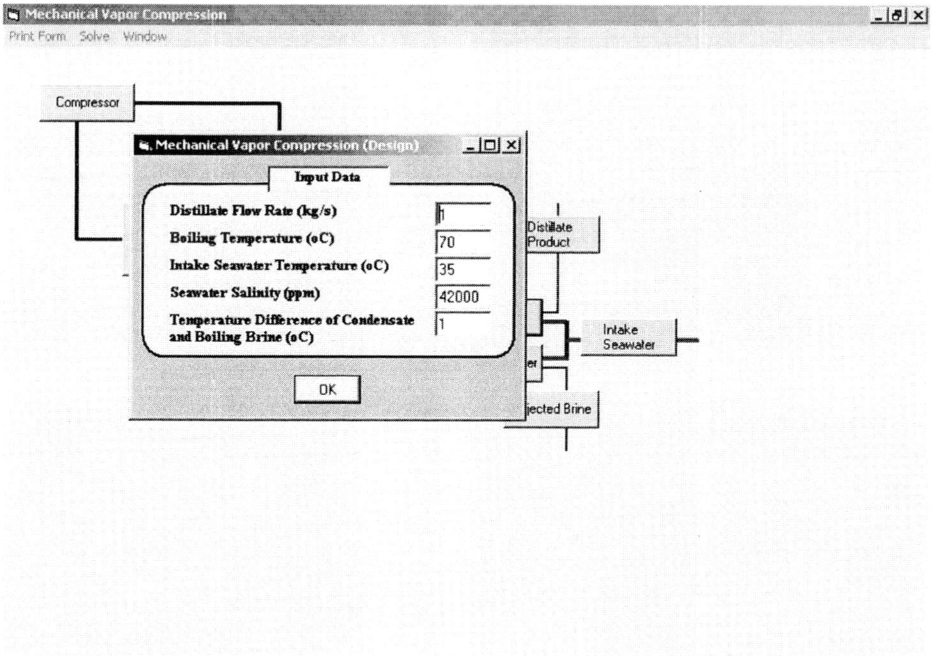

Fig. D.22 Input data for the flow chart calculations of the single effect mechanical vapor compression.

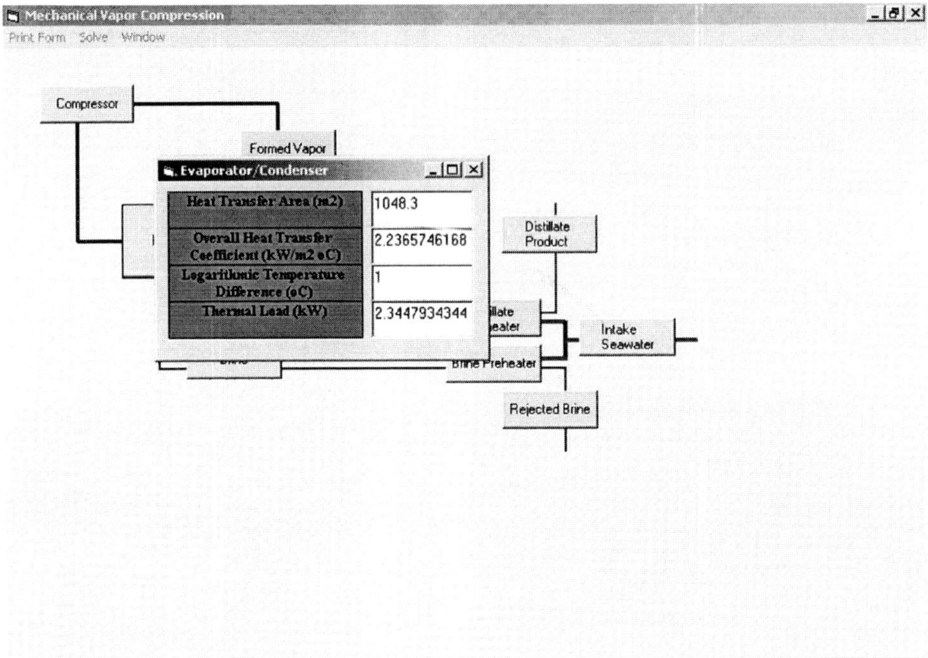

Fig. D.23 Design results of the evaporator/condenser unit for the single effect mechanical vapor compression.

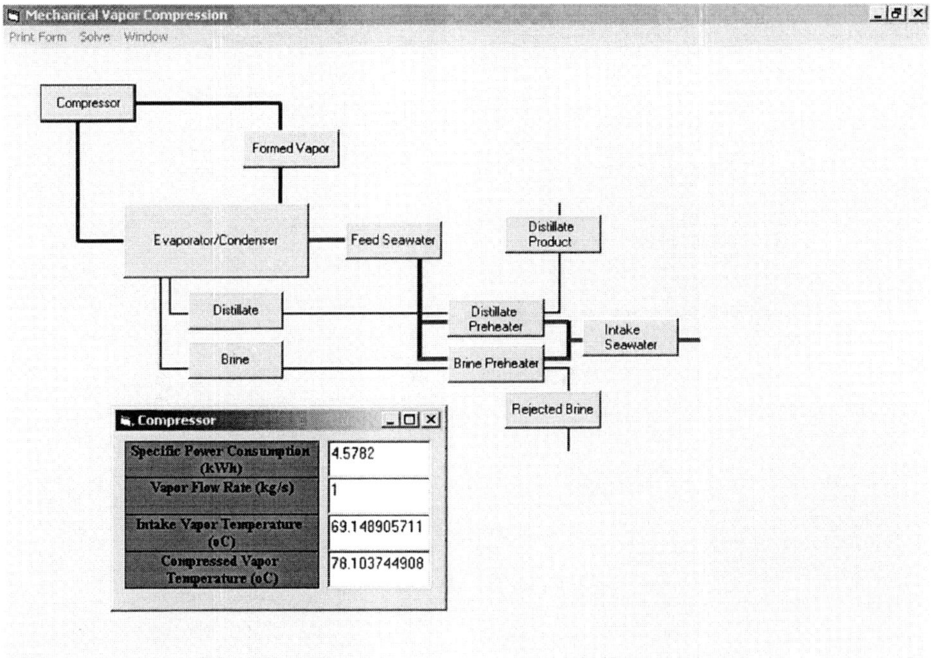

Fig. D.24 Design results for the compressor of the single effect mechanical vapor compression.

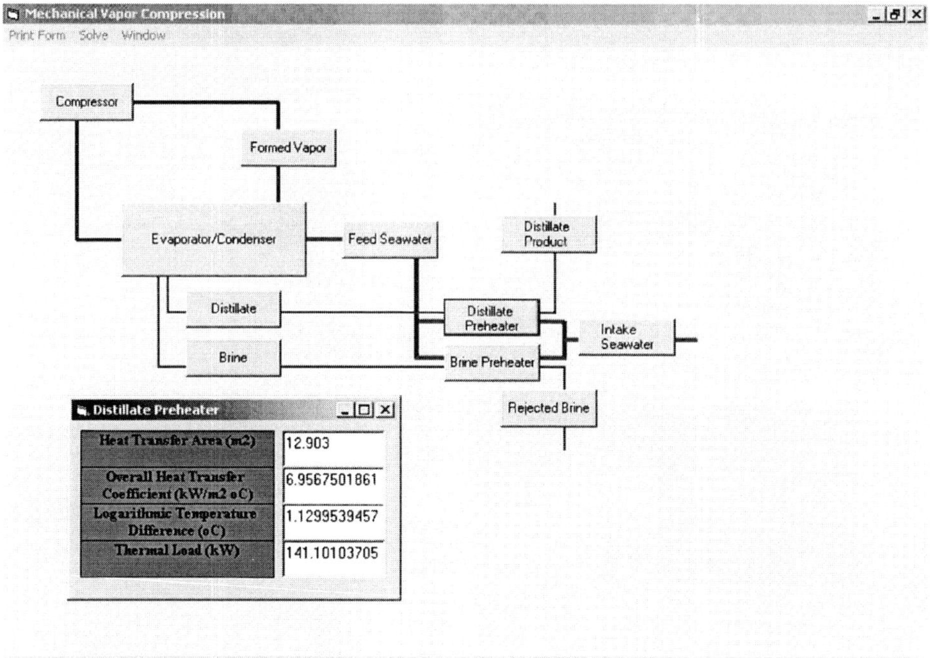

Fig. D.25 Design results for the distillate preheater in the single effect mechanical vapor compression.

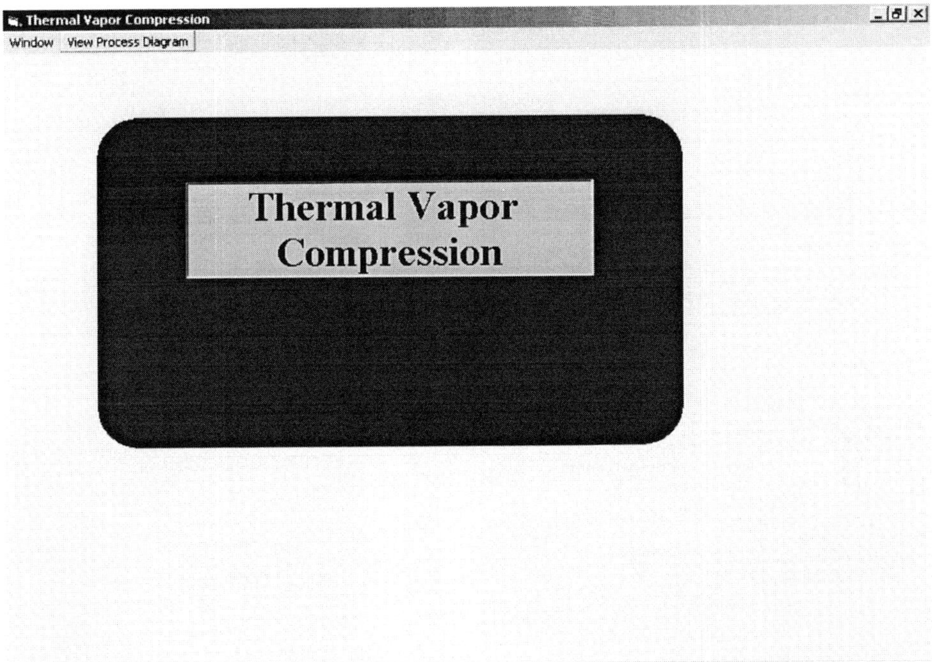

Fig. D.26 Startup window for the single effect thermal vapor compression system.

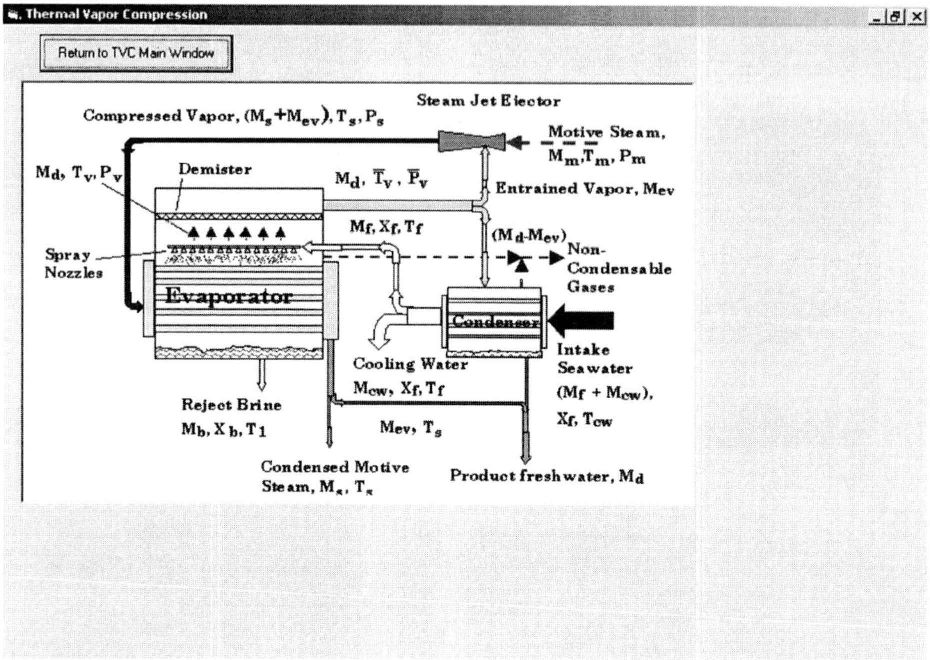

Fig. D.27 Process schematic of the single effect thermal vapor compression system.

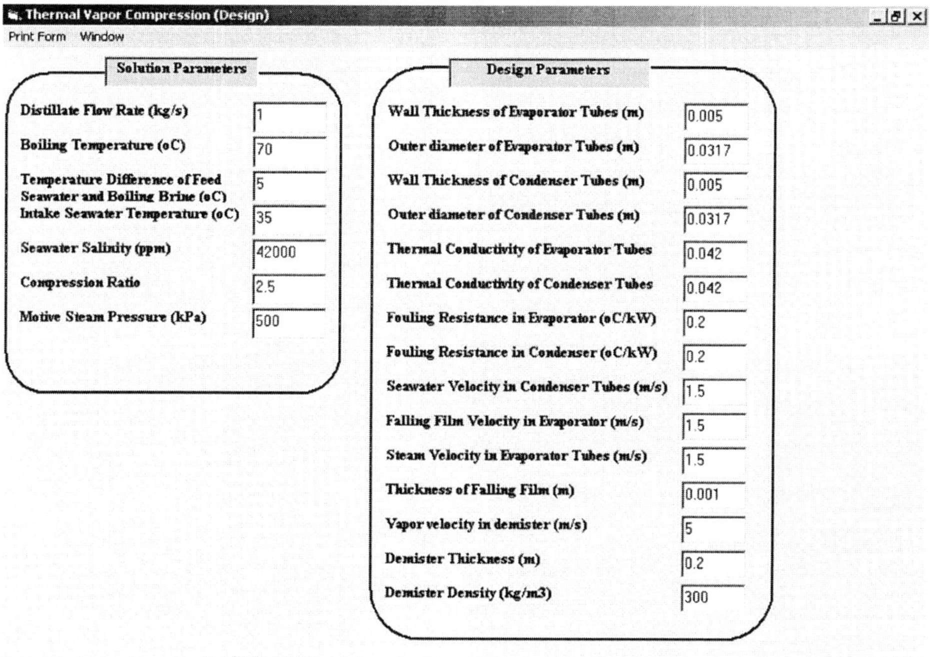

Fig. D.28 Input design data for the single effect thermal vapor compression system.

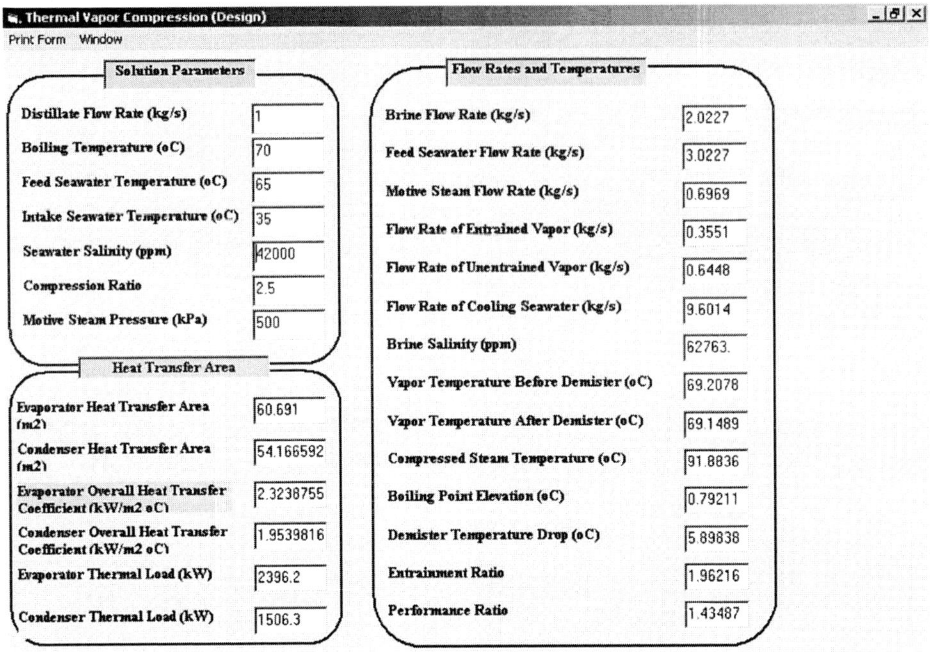

Fig. D.29 Design results for the single effect thermal vapor compression system.

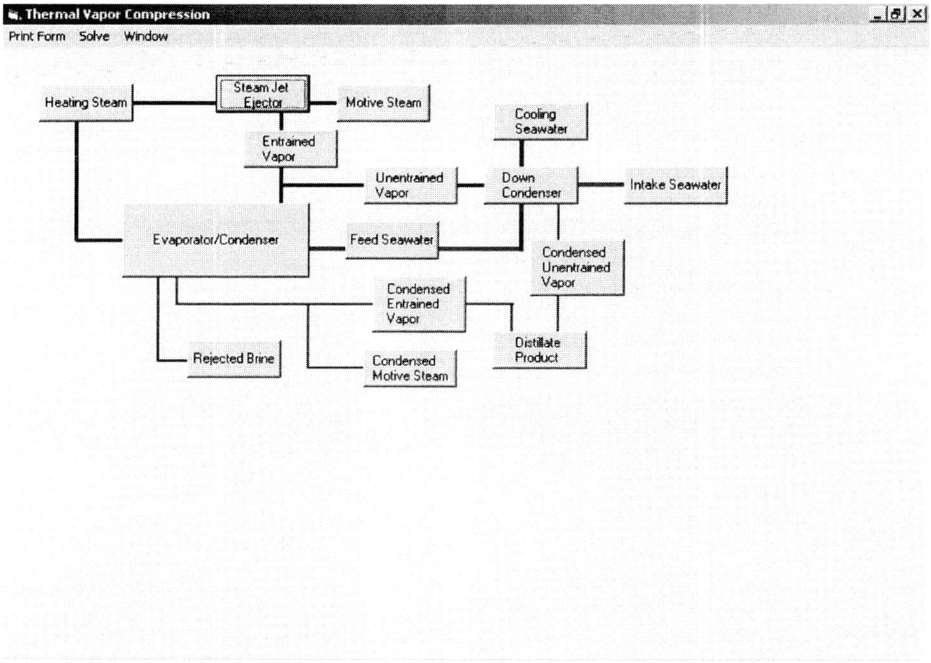

Fig. D.30 Flow chart for the single effect thermal vapor compression system.

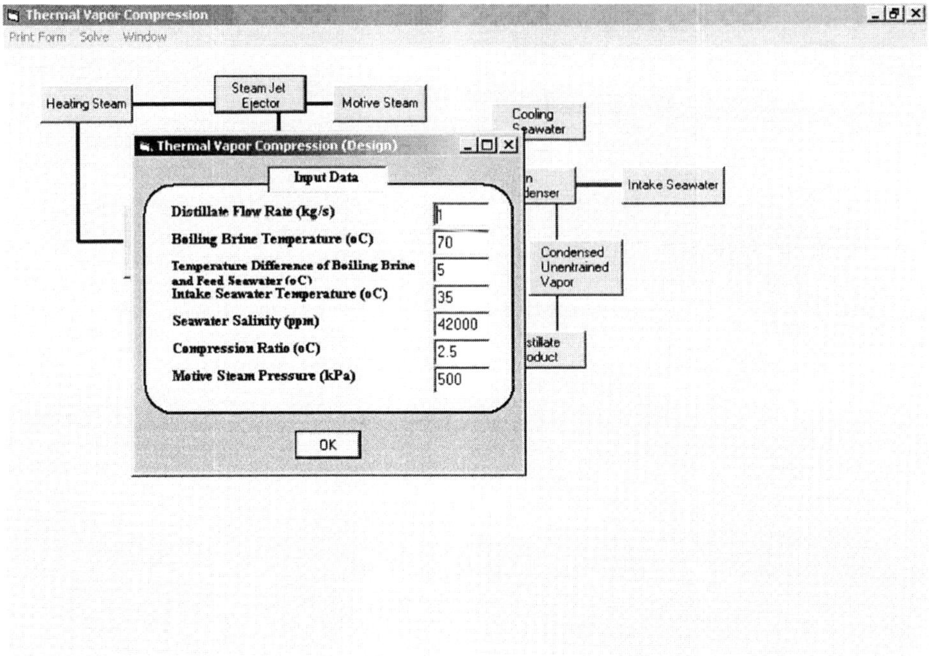

Fig. D.31 Input data for design calculations of the flow chart of the single effect thermal vapor compression system.

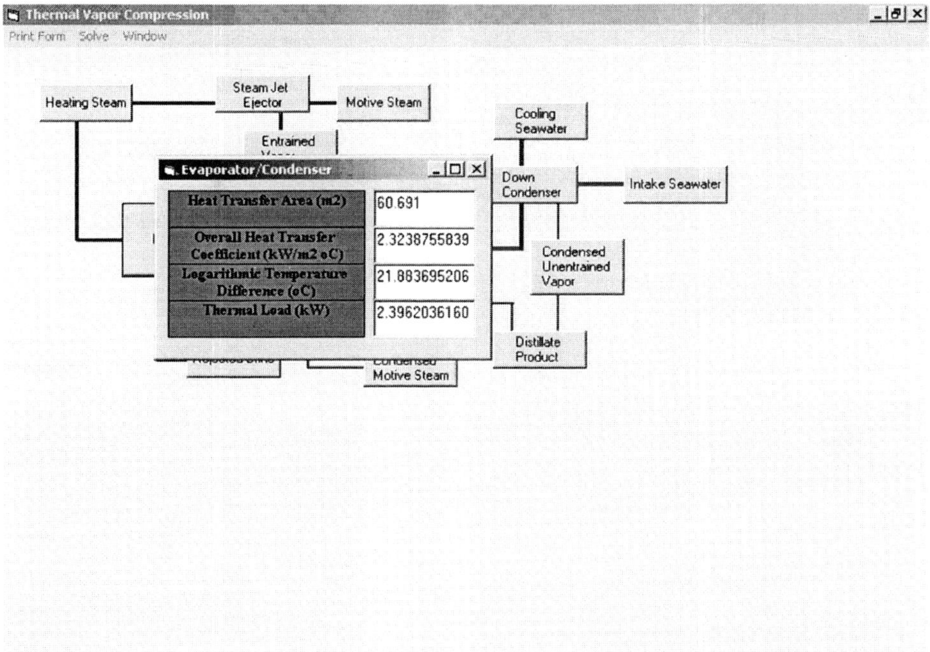

Fig. D.32 Design results for the evaporator/condenser in the flow chart of the single effect thermal vapor compression system.

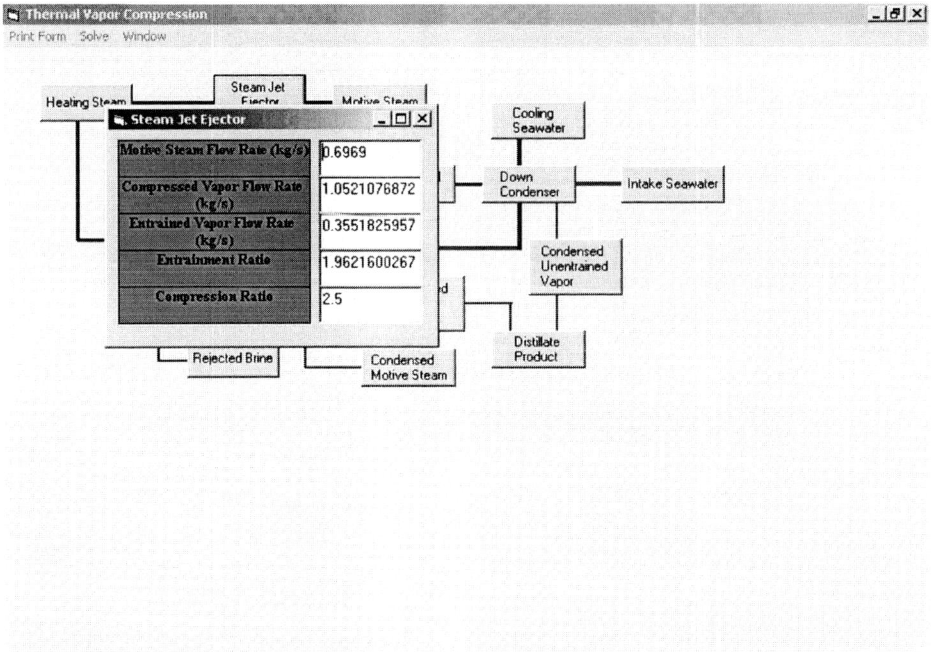

Fig. D.33 Design results for the steam jet ejector in the flow chart of the single effect thermal vapor compression system.

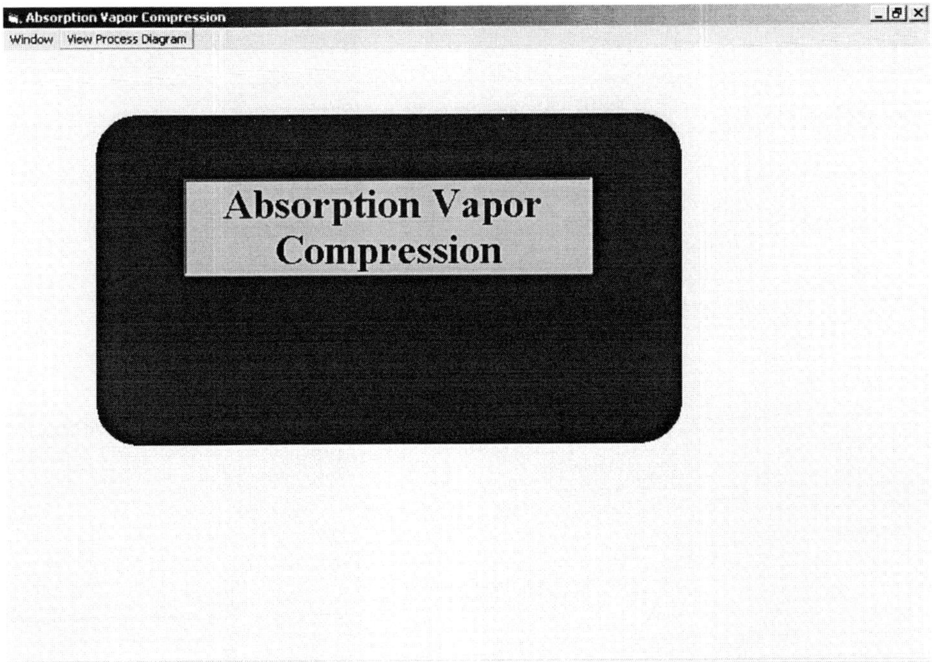

Fig. D.34 Startup window for the single effect absorption vapor compression system.

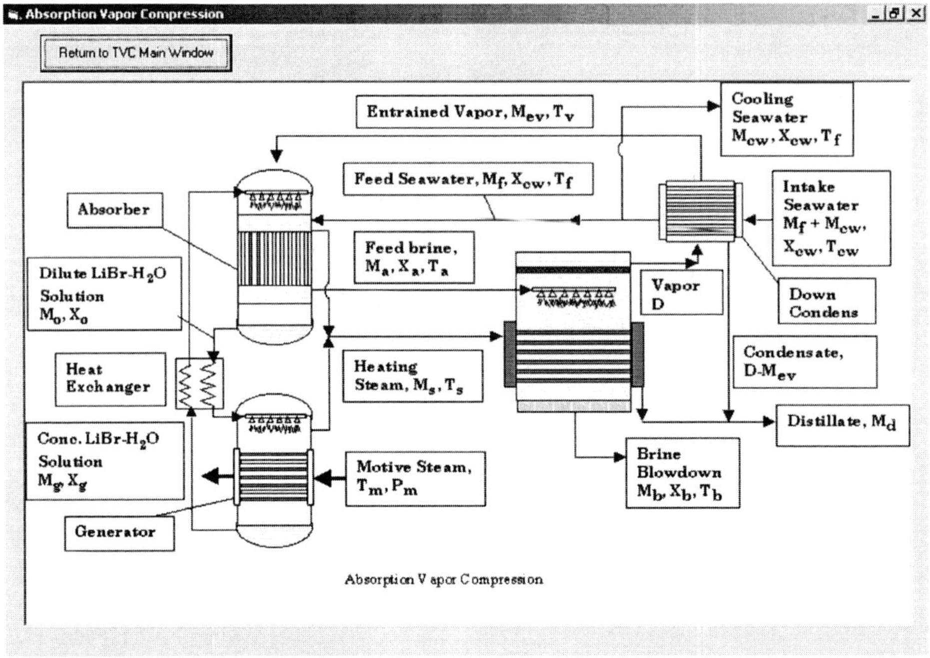

Fig. D.35 Process schematic for the single effect absorption vapor compression system.

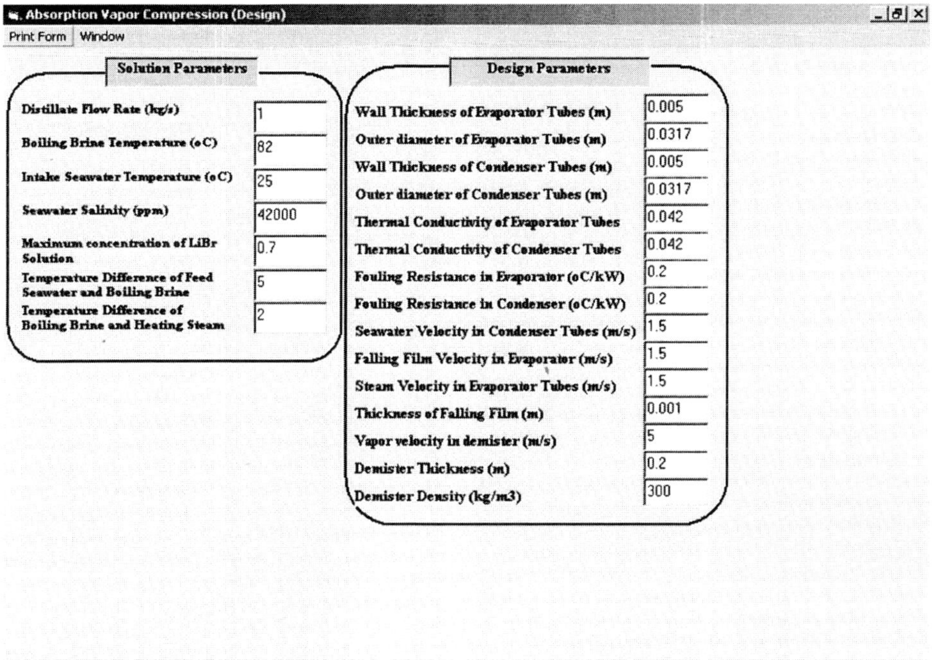

Fig. D.36 Input design data for the single effect evaporation with adsorption vapor compression.

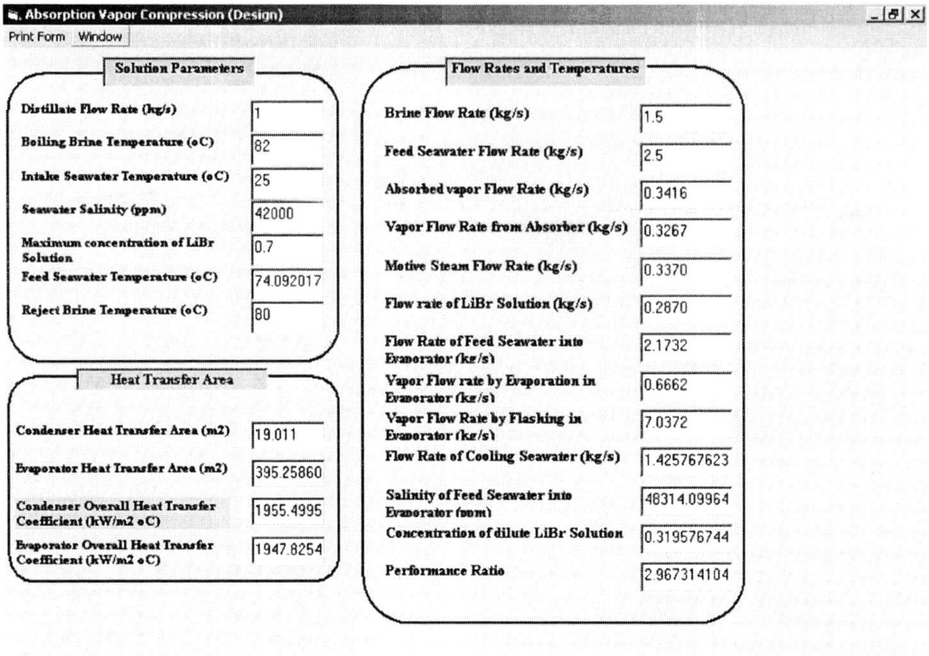

Fig. D.37 Design results for the single effect evaporation with adsorption vapor compression.

D.5. Parallel Feed Multiple Effect Evaporation

Illustrations for the parallel feed multiple effect evaporation system are shown in Figs. D.38-D.40. Figure D.38 shows the main window for the process, which can be used to activate the design calculations, inspect the process diagram, or return to the main window.

The design window includes entries for input data that includes number of effects, which may be varied between 4 and 12. Also, values for temperature, flow rates, salinity, and specifications of the condenser/evaporator tubes are included. The calculations results include the profiles of the brine and distillate streams. Also, the data includes the performance ratio, specific heat transfer area, and the specific flow rate of cooling water.

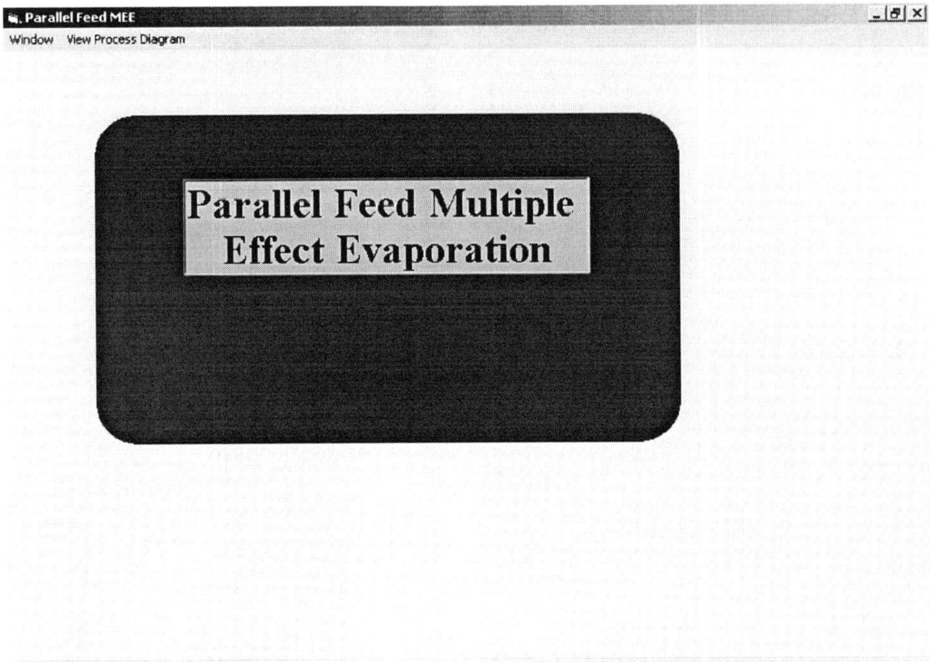

Fig. D.38 Startup window for the parallel feed multiple effect evaporation.

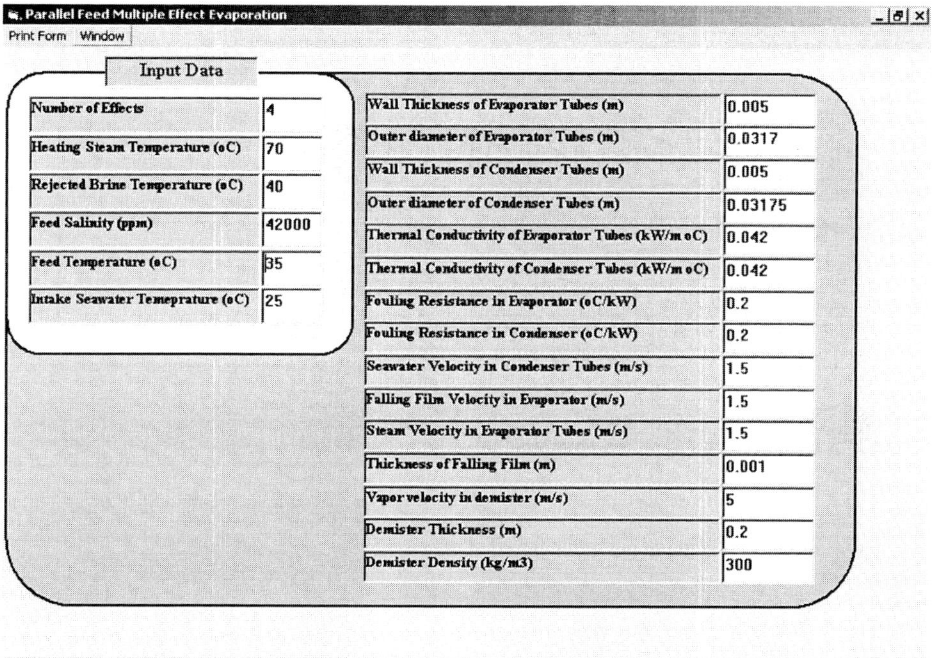

Fig. D.39 Input design data for the parallel feed multiple effect evaporation.

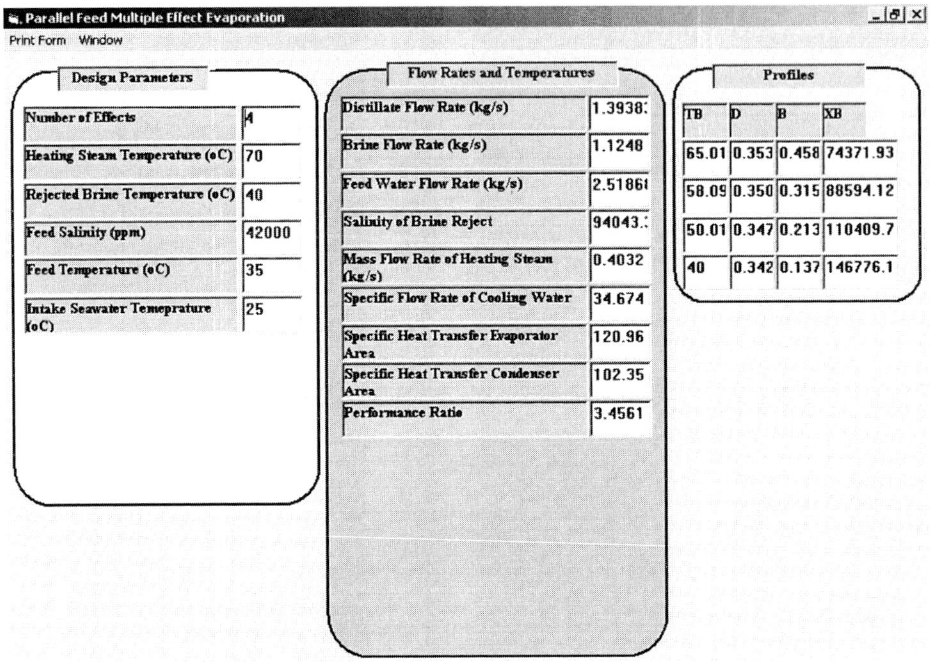

Fig. D.40 Design results for the parallel feed multiple effect evaporation.

D.6. Parallel Feed Multiple Effect Evaporation – Vapor Compression

Illustrations for the vapor compression parallel feed multiple effect evaporation system are shown in Figs. D.41-D.46. As is shown, only the thermal and mechanical vapor compression configurations are included. Also, the windows include the design option; however, in this stage the flow chart options are not available. The design option and results have entries similar to those of the stand-alone system in addition to specific features of the vapor compression unit.

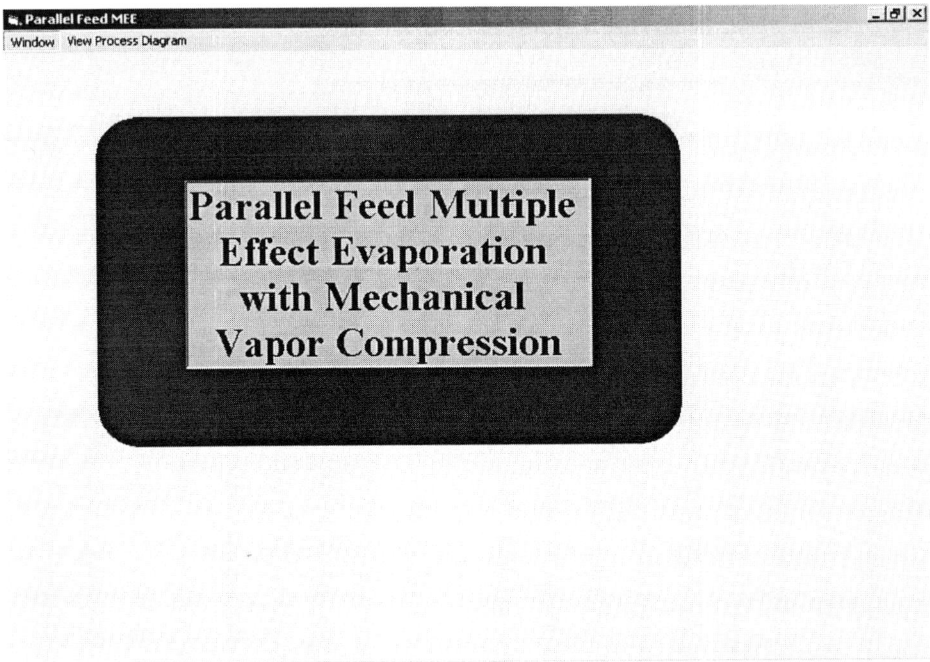

Fig. D.41 Startup window for the parallel feed multiple effect evaporation with mechanical vapor compression

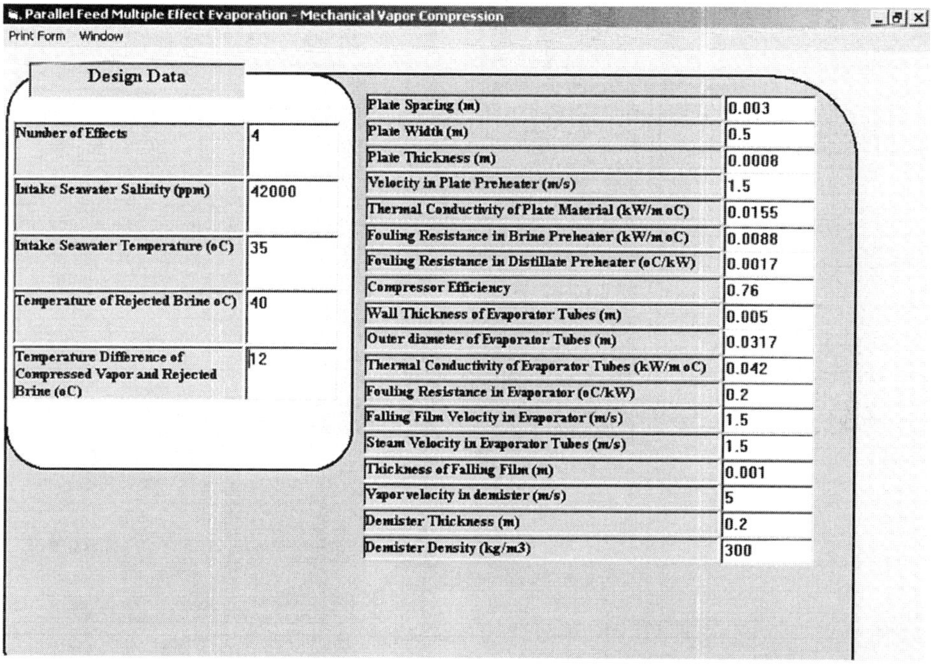

Fig. D.42 Input design data for the parallel feed multiple effect evaporation with mechanical vapor compression

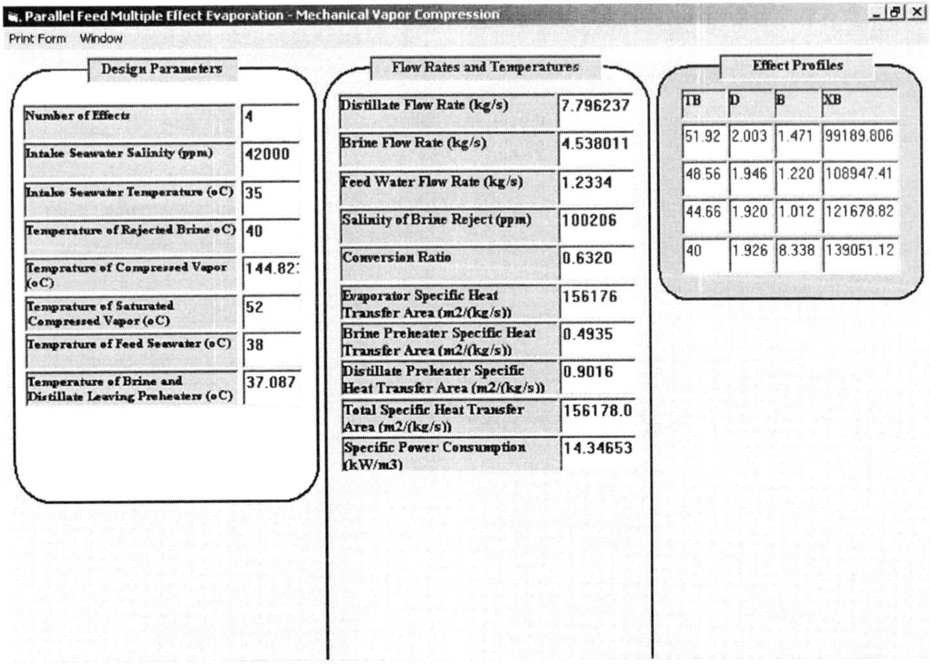

Fig. D.43 Design results for the parallel feed multiple effect evaporation with mechanical vapor compression

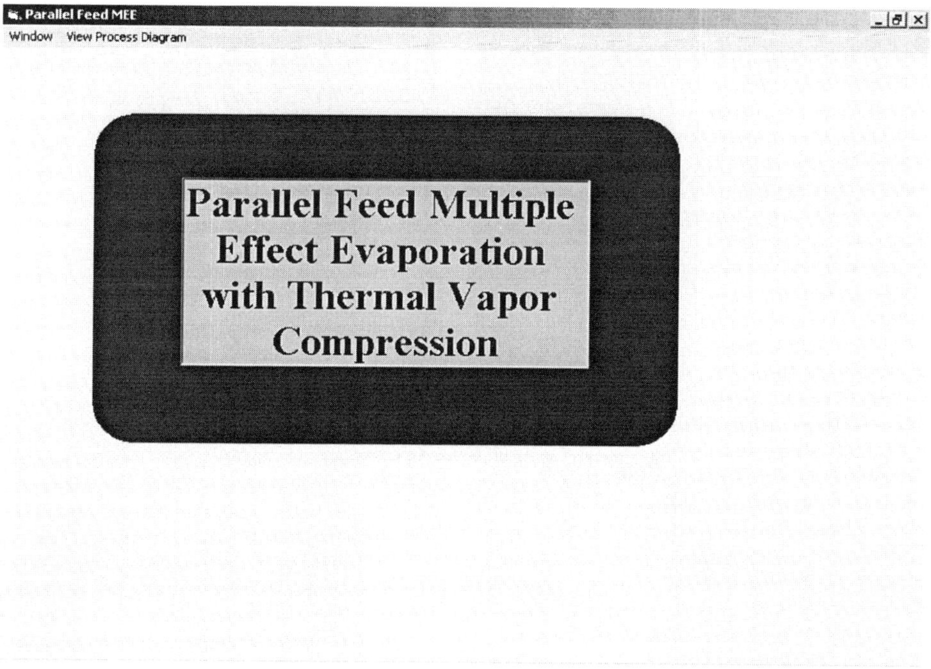

Fig. D.44 Startup window for the parallel feed multiple effect evaporation with thermal vapor compression.

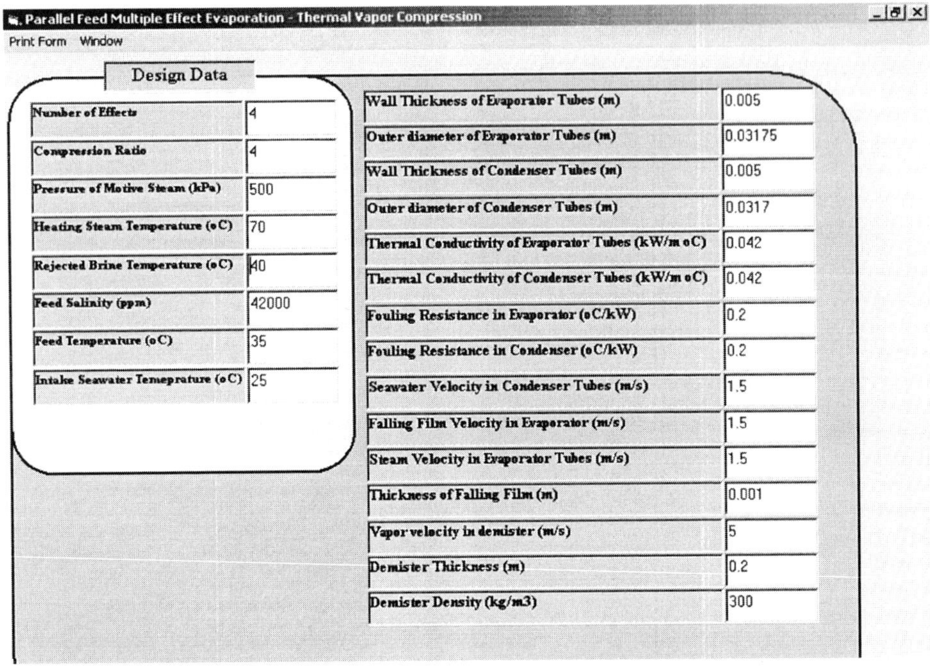

Fig. D.45 Input design data for the parallel feed multiple effect evaporation with thermal vapor compression.

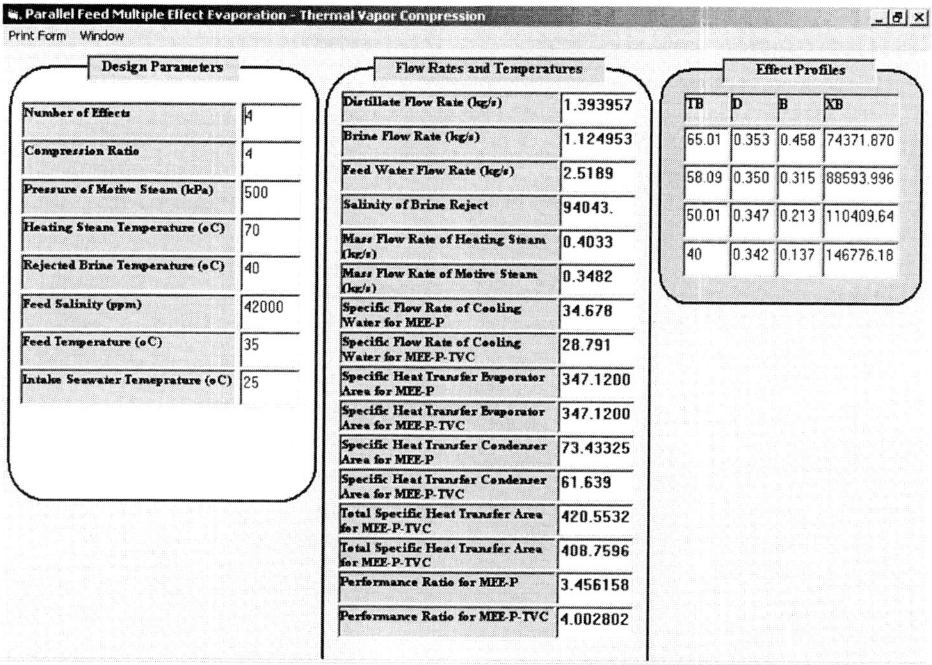

Fig. D.46 Design results for the parallel feed multiple effect evaporation with thermal vapor compression.

D.7. Forward Feed Multiple Effect Evaporation

The forward feed multiple effect evaporation system includes options for design and flow chart calculations. The system main window is shown in Fig. D.47. The design windows, which includes the input data and results, are illustrated in Figs. D.48-D.49. As is shown, contents of these windows are similar to those of the parallel feed unit.

Displays for the flow chart calculations are shown in Figs. D.50-D.53. As is shown the startup window includes the default of 12 effects together with the definitions of various streams. To solve the system equations, it is necessary to define a number of parameters, i.e., number of effects, feed salinity, etc. The solution of the system equations generate a block diagram with the proper number of effects that correspond to those defined previously in the solution window, Fig. D52. As mentioned before, each block in the system diagram includes the specific data of that block. Figure D.53 shows the data of effect number 4, which includes temperatures, salinity, flow rates, and area.

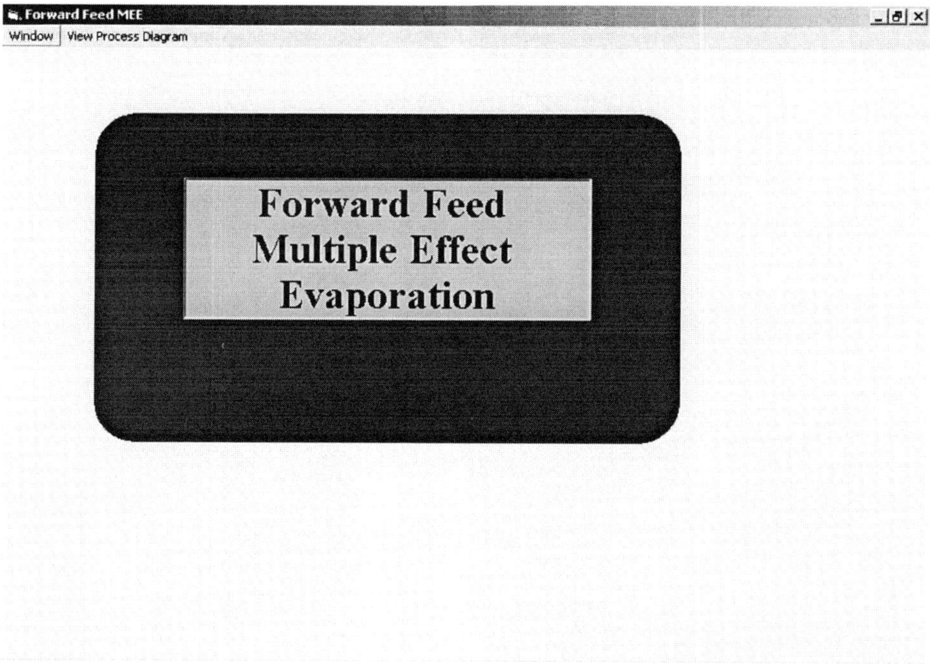

Fig. D.47 Startup window for the forward feed multiple effect evaporation.

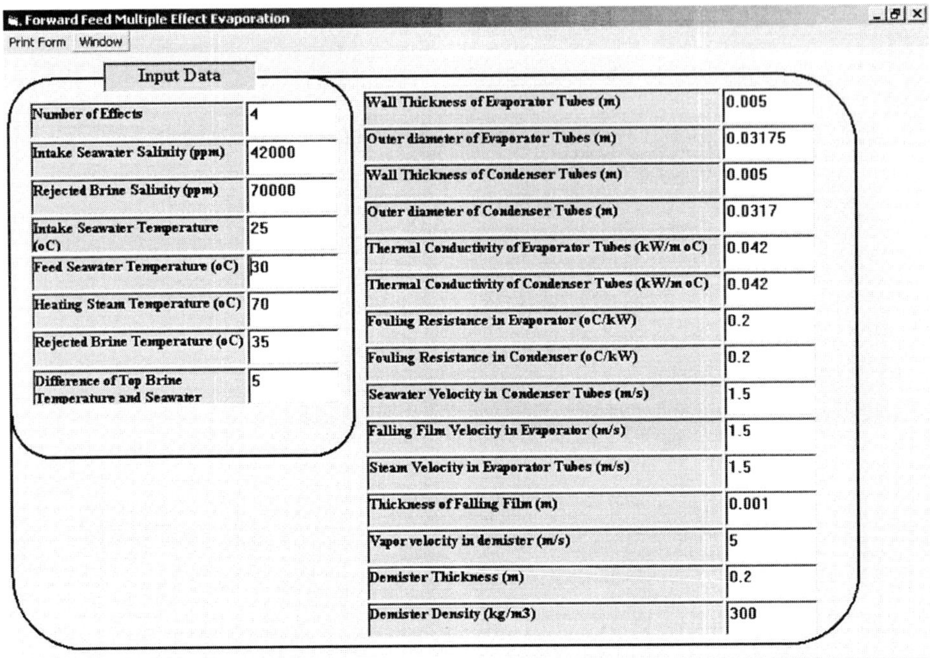

Fig. D.48 Input design data for the forward feed multiple effect evaporation.

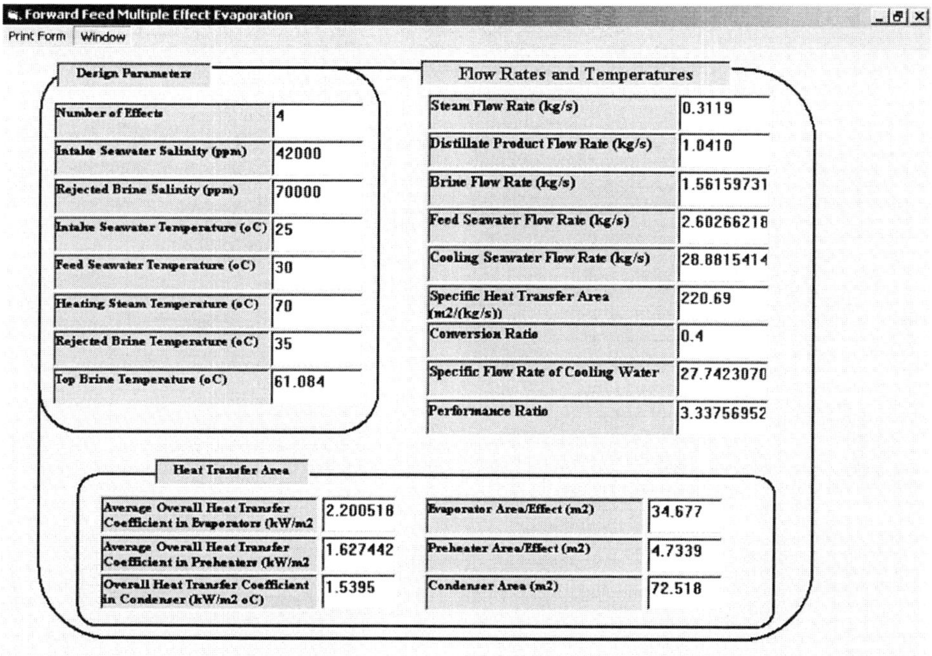

Fig. D.49 Design results for the forward feed multiple effect evaporation.

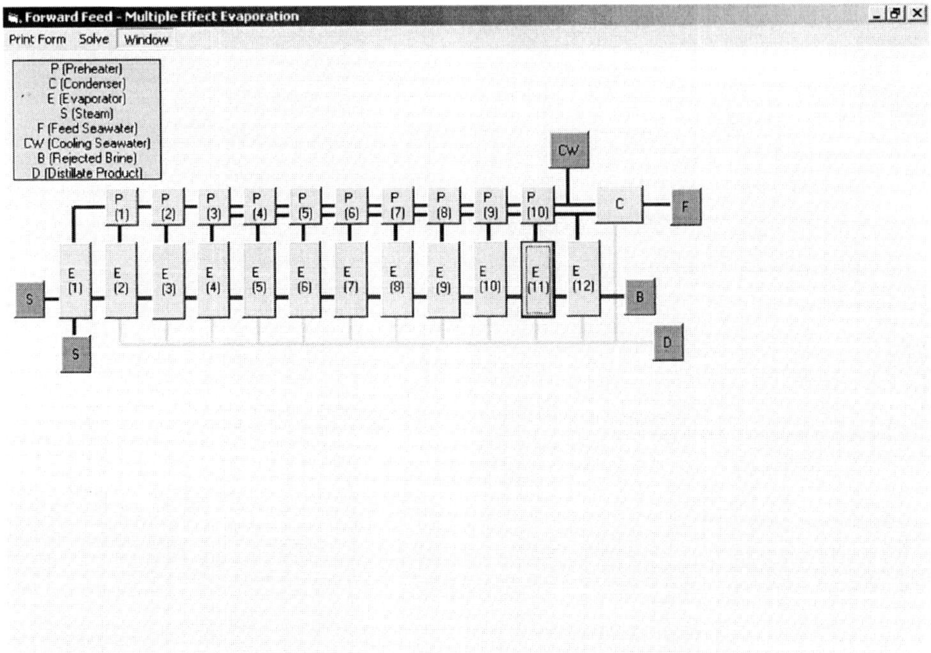

Fig. D.50 Startup window for the flow chart.

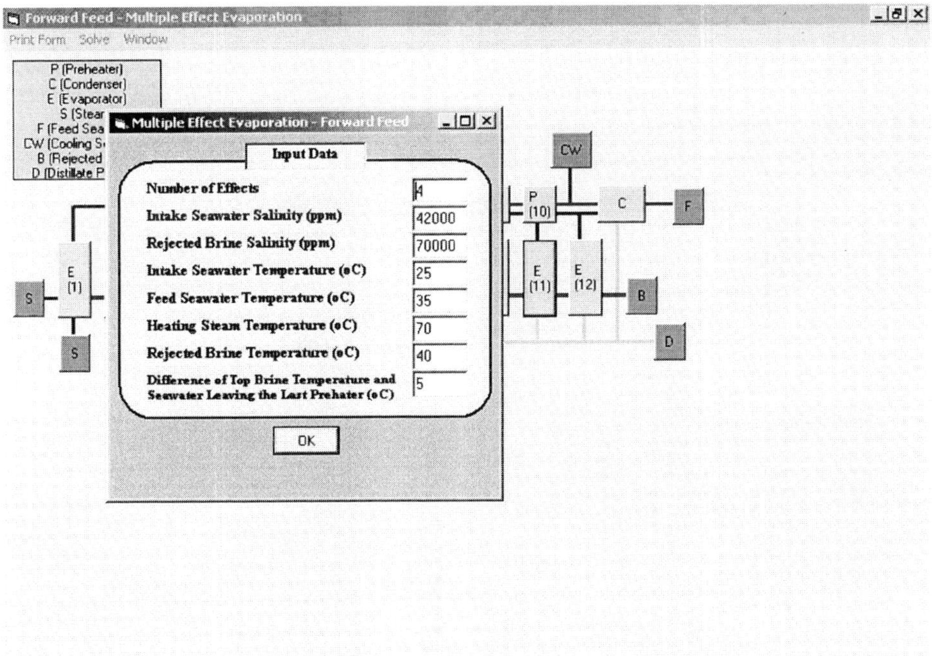

Fig. D.51 Input design data for the flow chart.

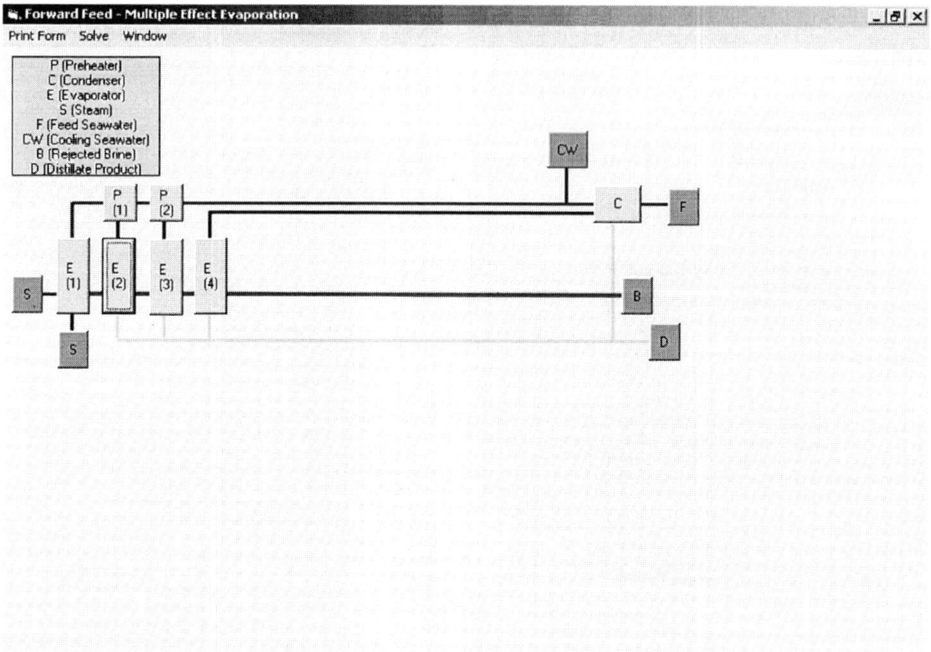

Fig. D.52 Flow chart upon completion of calculations.

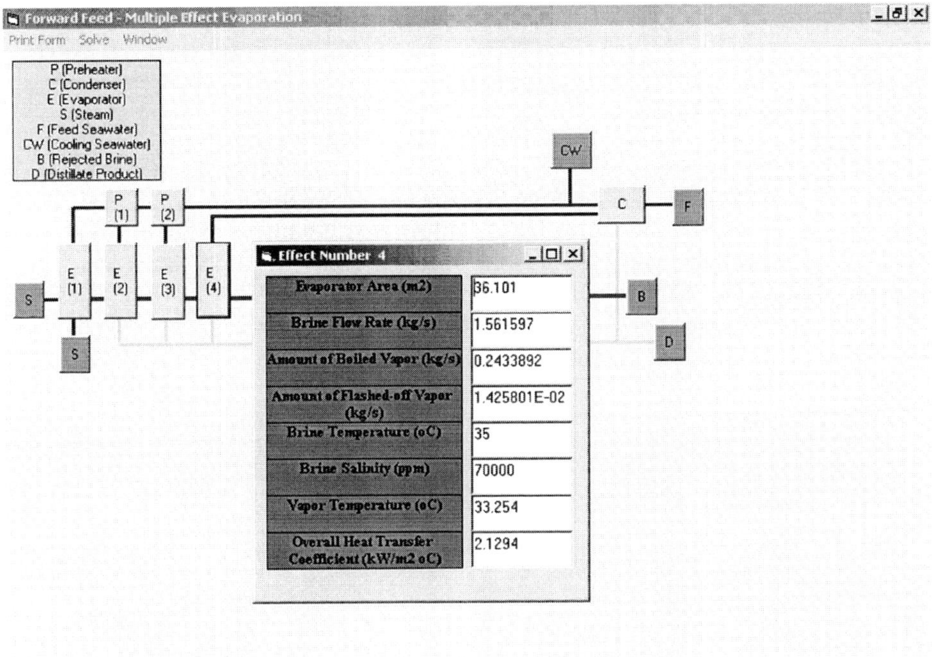

Fig. D.53 Design results for effect number 4.

D.8. Multistage Flash Desalination

The windows for the MSF systems are shown in Figs. D.54-D.60. This system includes only design options for the brine circulation system (or conventional MSF) and the once through system. The windows for the brine circulation system are shown in Figs. D.54-D.57. Figure D.54 shows the main window for MSF with brine circulation. Figure D.56 shows the design input window. A progress bar becomes visible once the calculations start and it is completed, Fig. D.56. Figure D.57 shows the final results together with the system profiles. Similar layout is illustrated for the once through system, Figs. D.58-D.60.

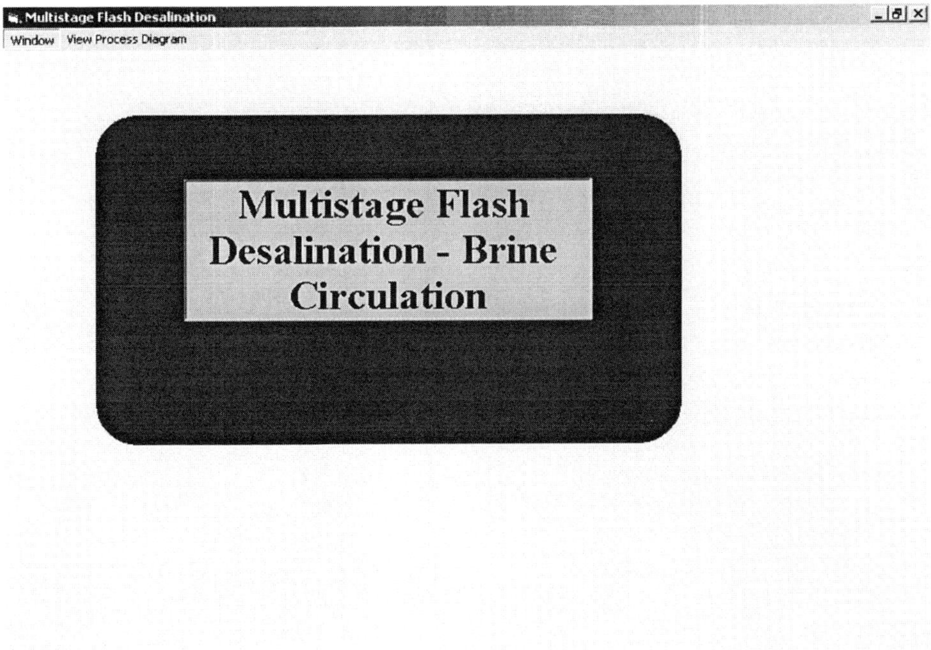

Fig. D.54 Startup window for the MSF system.

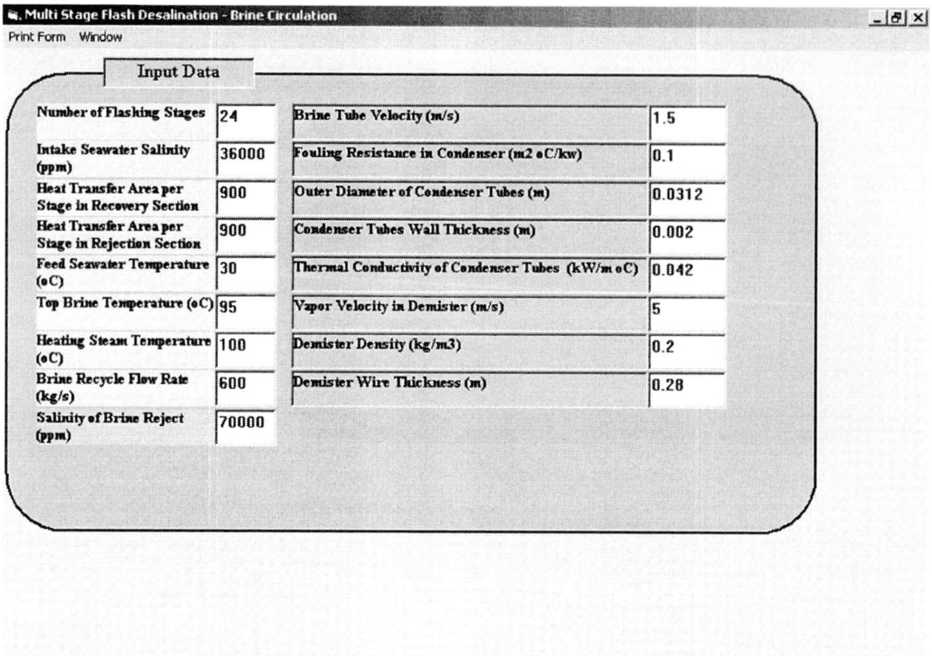

Fig. D.55 Input design data for the MSF system.

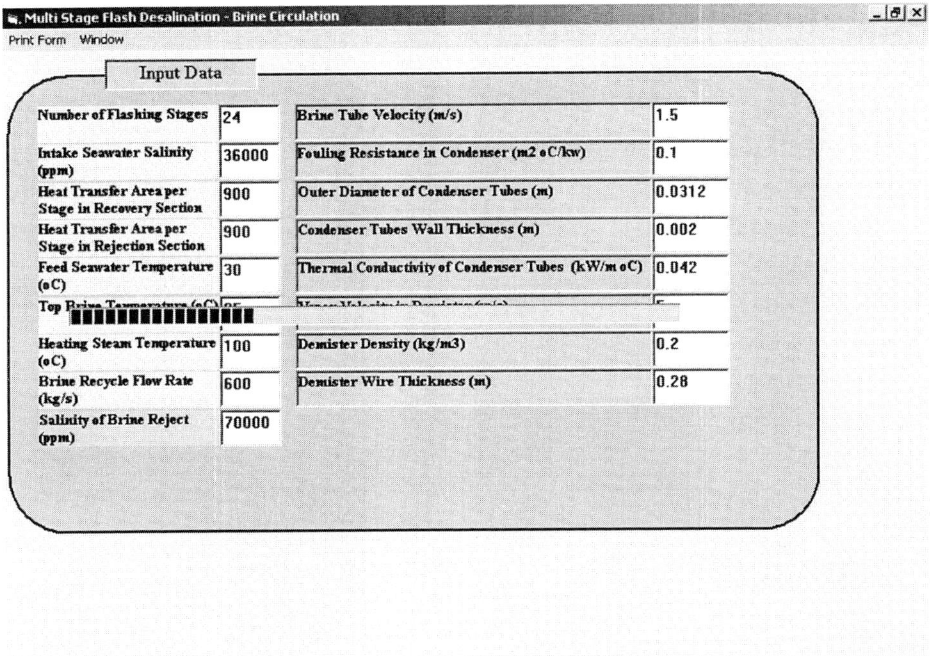

Fig. D.56 Calculation progress for the MSF system.

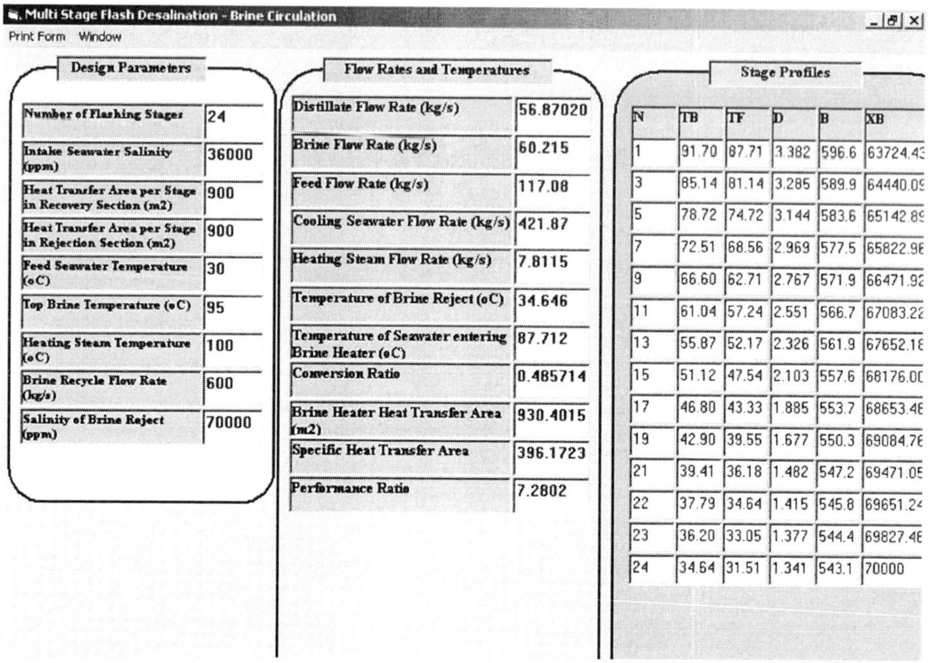

Fig. D.57 Design results and profiles for the MSF system.

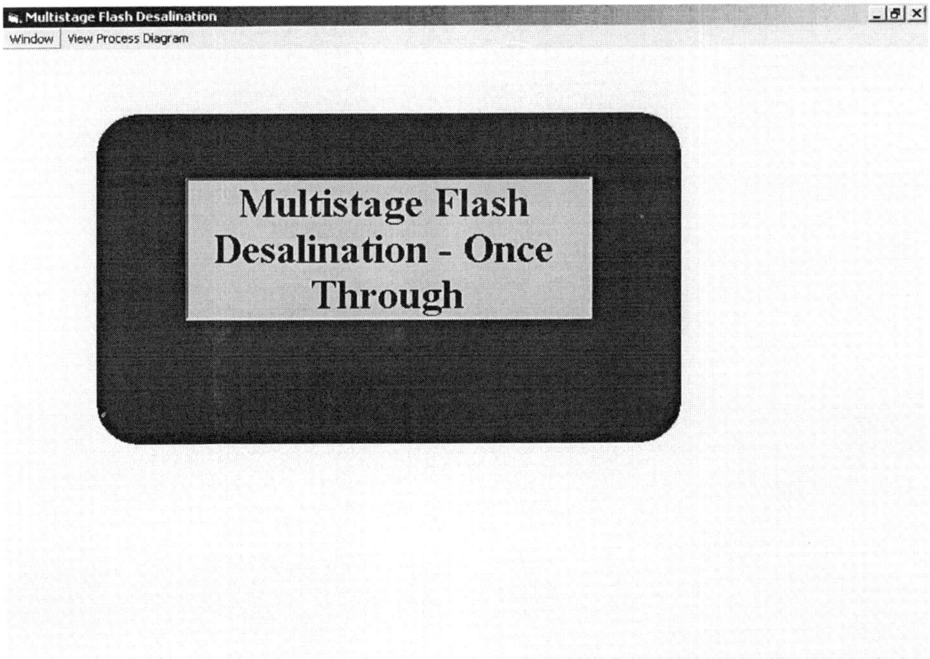

Fig. D.58 Startup window for MSF-OT

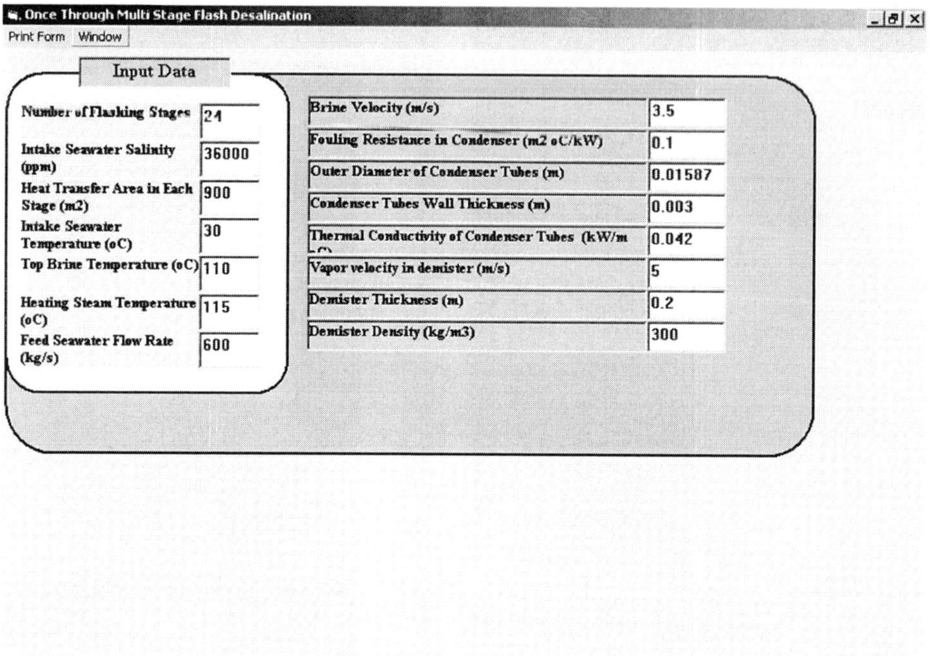

Fig. D.59 Design input data for MSF-OT

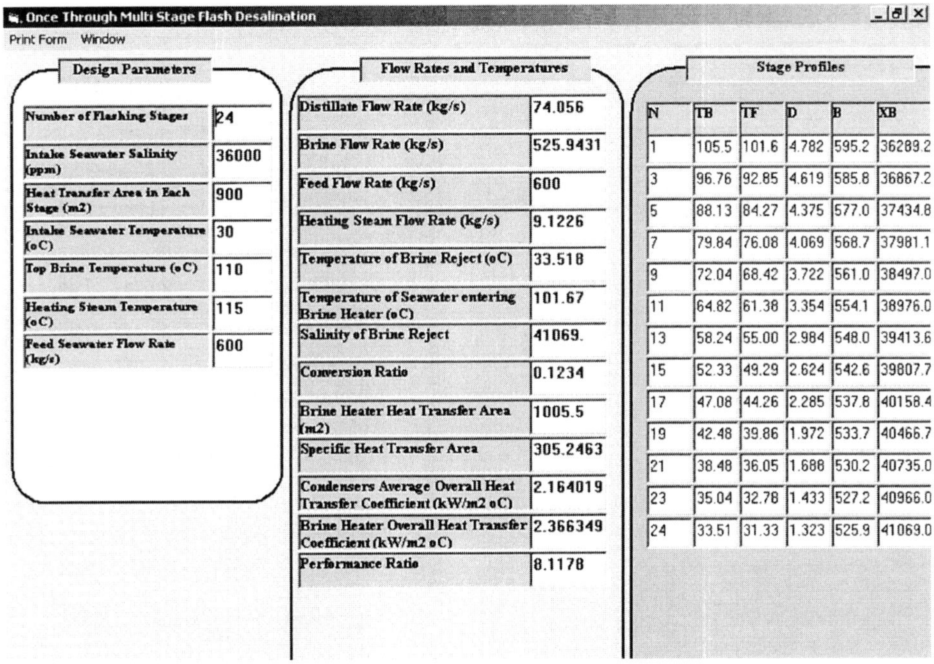

Fig. D.60 Design results for MSF-OT

Index